Year 6A

A Guide to Teaching for Mastery

Series Editor: Tony Staneff

Contents

Introduction

Foreword by the series editor and author, Tony Staneff

For far too long in the UK, maths has been feared by learners – and by many teachers, too. As a result, most learners consistently underachieve. More crucially, negative beliefs about ability, aptitude and the nature of maths are entrenched in children's thinking from an early age.

Yet, as someone who has loved maths all my life, I've always believed that every child has the capacity to succeed in maths. I've also had the great pleasure of leading teams and departments who share that belief and passion. Teaching for mastery, as practised in China and other South-East Asian jurisdictions since the 1980s, has confirmed my conviction that maths really is for everyone and not just those who have a special talent. In recent years, my team and I at Trinity Academy, Halifax, have had the privilege of researching with and working alongside some of the finest mastery practitioners from the UK and beyond, whose impact on learners' confidence, achievement and attitude is an inspiration.

The mastery approach recognises the value of developing the power to think rather than just do. It also recognises the value of making a coherent journey in which whole-class groups tackle concepts in very small steps, one by one. You cannot build securely on loose foundations – and it is just the same with maths: by creating a solid foundation of deep understanding, our children's skills and confidence will be strong and secure. What's more, the mindset of learner and teacher alike is fundamental: everyone can do maths … EVERYONE CAN!

I am proud to have been part of the extensive team responsible for turning the best of the world's practice, research, insights, and shared experiences into *Power Maths*, a unique teaching and learning resource developed especially for UK classrooms. *Power Maths* embodies our vision to help and support primary maths teachers to transform every child's mathematical and personal development. 'Everyone can!' has become our mantra and our passion, and we hope it will be yours, too.

Now, explore and enjoy all the resources you need to teach for mastery, and please get back to us with your *Power Maths* experiences and stories!

What is *Power Maths*?

Created especially for UK primary schools, and aligned with the new National Curriculum, *Power Maths* is a whole-class, textbook-based mastery resource that empowers every child to understand and succeed. *Power Maths* rejects the notion that some people simply 'can't do' maths. Instead, it develops growth mindsets and encourages hard work, practice and a willingness to see mistakes as learning tools.

Best practice consistently shows that mastery of small, cumulative steps builds a solid foundation of deep mathematical understanding. *Power Maths* combines interactive teaching tools, high-quality textbooks and continuing professional development (CPD) to help you equip children with a deep and long lasting understanding. Based on extensive evidence, and developed in partnership with practising teachers, *Power Maths* ensures that it meets the needs of children in the UK.

Power Maths and Mastery

Power Maths makes mastery practical and achievable by providing the structures, pathways, content, tools and support you need to make it happen in your classroom.

To develop mastery in maths children need to be enabled to acquire a deep understanding of maths concepts, structures and procedures, step by step. Complex mathematical concepts are built on simpler conceptual components and when children understand every step in the learning sequence, maths becomes transparent and makes logical sense. Interactive lessons establish deep understanding in small steps, as well as effortless fluency in key facts such as tables and number bonds. The whole class works on the same content and no child is left behind.

Power Maths

- Builds every concept in small, progressive steps.
- Is built with interactive, whole-class teaching in mind.
- Provides the tools you need to develop growth mindsets.
- Helps you check understanding and ensure that every child is keeping up.
- Establishes core elements such as intelligent practice and reflection.

The *Power Maths* approach

Everyone can!

Founded on the conviction that every child can achieve, *Power Maths* enables children to build number fluency, confidence and understanding, step by step.

Child-centred learning

Children master concepts one step at a time in lessons that embrace a Concrete-Pictorial-Abstract (C-P-A) approach, avoid overload, build on prior learning and help them see patterns and connections. Same-day intervention ensures sustained progress.

Continuing professional development

Embedded teacher support and development offer every teacher the opportunity to continually improve their subject knowledge and manage whole-class teaching for mastery.

Whole-class teaching

An interactive, whole-class teaching model encourages thinking and precise mathematical language and allows children to deepen their understanding as far as they can.

Introduction to the author team

Power Maths arises from the work of maths mastery experts who are committed to proving that, given the right mastery mindset and approach, **everyone can do maths**. Based on robust research and best practice from around the world, *Power Maths* was developed in partnership with a group of UK teachers to make sure that it not only meets our children's wide-ranging needs but also aligns with the National Curriculum in England.

Tony Staneff, Series Editor and author

Vice Principal at Trinity Academy, Halifax, Tony also leads a team of mastery experts who help schools across the UK to develop teaching for mastery via nationally recognised CPD courses, problem-solving and reasoning resources, schemes of work, assessment materials and other tools.

A team of experienced authors, including:

- **Josh Lury** – a specialist maths teacher, author and maths consultant with a passion for innovative and effective maths education

- **Trinity Academy, Halifax** (Michael Gosling CEO, Tony Staneff, Emily Fox, Kate Henshall, Rebecca Holland, Stephanie Kirk, Stephen Monaghan, Beth Smith and Rachel Webster)

- **David Board**, **Belle Cottingham**, **Jonathan East**, **Tim Handley**, **Derek Huby**, **Neil Jarrett**, **Timothy Weal**, **Paul Wrangles** – skilled maths teachers and mastery experts

- **Cherri Moseley** – a maths author, former teacher and professional development provider

Professors Liu Jian and Zhang Dan, Series Consultants and authors, and their team of mastery expert authors:

- **Wei Huinv, Huang Lihua, Zhu Dejiang, Zhu Yuhong, Hou Huiying, Yin Lili, Zhang Jing, Zhou Da and Liu Qimeng**

Used by over 20 million children, Professor Liu Jian's textbook programme is one of the most popular in China. He and his author team are highly experienced in intelligent practice and in embedding key maths concepts using a C-P-A approach.

A group of 15 teachers and maths co-ordinators

We have consulted our teacher group throughout the development of *Power Maths* to ensure we are meeting their real needs in the classroom.

Your *Power Maths* resources

To help you teach for mastery, *Power Maths* comprises a variety of high-quality resources.

Pupil Textbooks

Discover, Share, and Think together sections promote discussion and introduce mathematical ideas logically, so that children understand more easily.

Using a Concrete-Pictorial-Abstract approach, clear mathematical models help children to make connections and grasp concepts.

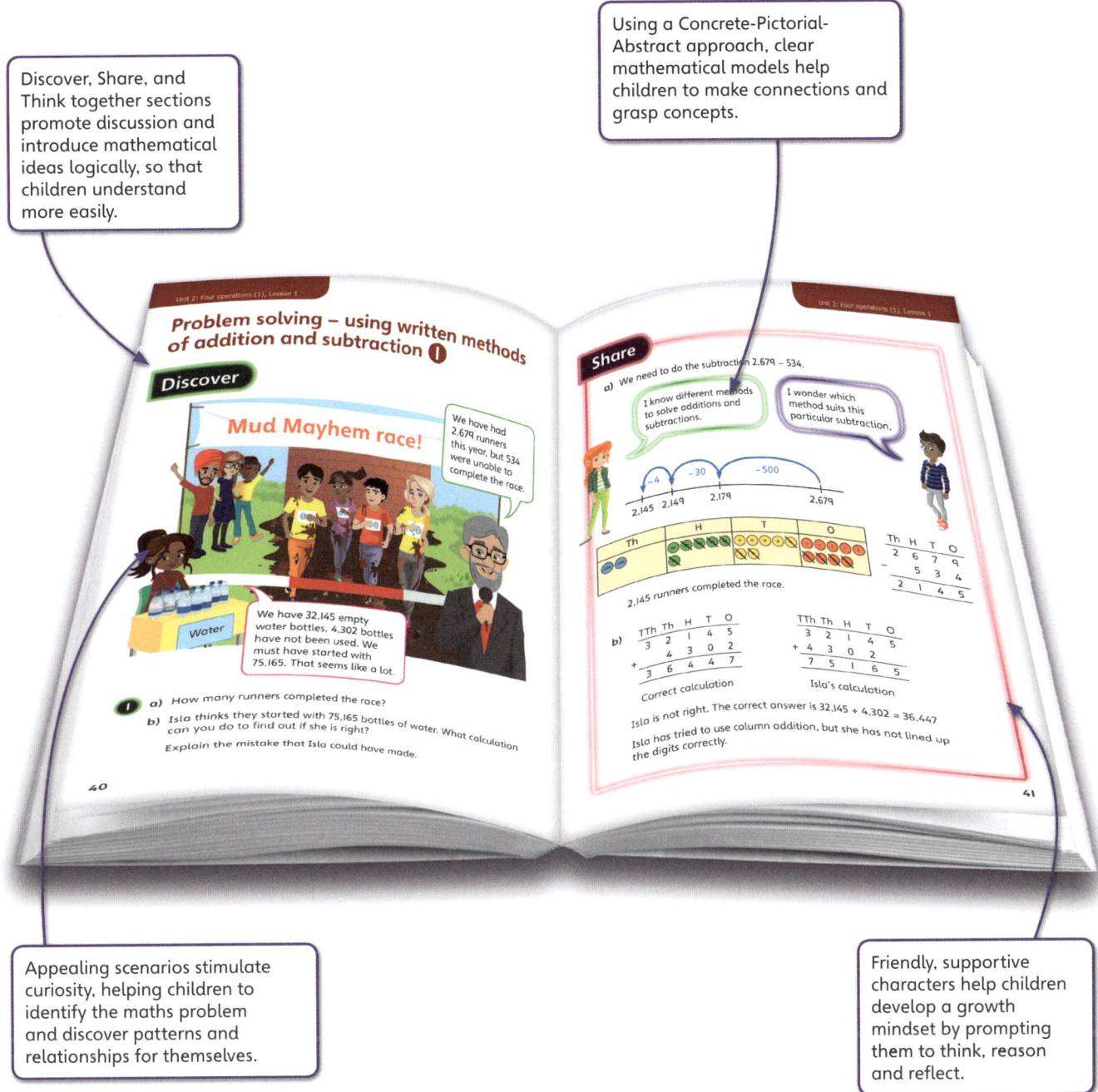

Appealing scenarios stimulate curiosity, helping children to identify the maths problem and discover patterns and relationships for themselves.

Friendly, supportive characters help children develop a growth mindset by prompting them to think, reason and reflect.

The coherent *Power Maths* lesson structure carries through into the vibrant, high-quality textbooks. Setting out the core learning objectives for each class, the lesson structure follows a carefully mapped journey through the curriculum and supports children on their journey to deeper understanding.

Pupil Practice Books

The Practice Books offer just the right amount of intelligent practice for children to complete independently in the final section of each lesson.

The practice questions are for everyone – each question varies one small element to move children on in their thinking. Look at the different parts in question ❶!

Calculations are connected so that children think about the underlying concept. In question ❸, children have to write out the calculation to find the answer. Concepts are presented differently again in question ❹ to challenge children.

Practice questions are finely tuned to move children forward in their thinking and to reveal misconceptions.

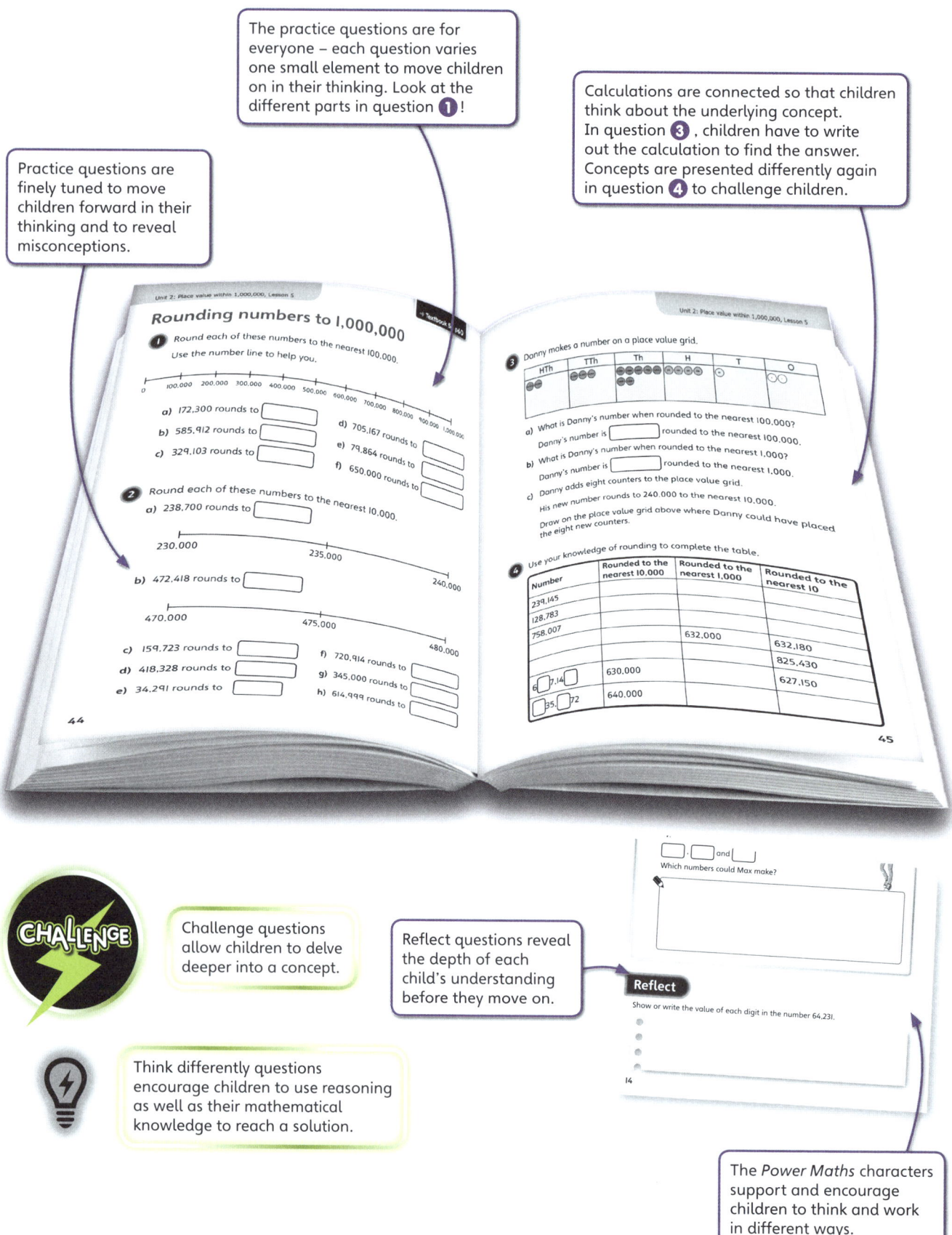

Unit 2: Place value within 1,000,000, Lesson 5

Rounding numbers to 1,000,000

❶ Round each of these numbers to the nearest 100,000.
Use the number line to help you.

0   100,000   200,000   300,000   400,000   500,000   600,000   700,000   800,000   900,000   1,000,000

a) 172,300 rounds to
b) 585,912 rounds to
c) 329,103 rounds to
d) 705,167 rounds to
e) 79,864 rounds to
f) 650,000 rounds to

❷ Round each of these numbers to the nearest 10,000.
a) 238,700 rounds to

230,000   235,000   240,000

b) 472,418 rounds to

470,000   475,000   480,000

c) 159,723 rounds to
d) 418,328 rounds to
e) 34,291 rounds to
f) 720,914 rounds to
g) 345,000 rounds to
h) 614,999 rounds to

44

Unit 2: Place value within 1,000,000, Lesson 5

❸ Danny makes a number on a place value grid.

HTh	TTh	Th	H	T	O

a) What is Danny's number when rounded to the nearest 100,000?
Danny's number is [   ] rounded to the nearest 100,000.
b) What is Danny's number when rounded to the nearest 1,000?
Danny's number is [   ] rounded to the nearest 1,000.
c) Danny adds eight counters to the place value grid.
His new number rounds to 240,000 to the nearest 10,000.
Draw on the place value grid above where Danny could have placed the eight new counters.

❹ Use your knowledge of rounding to complete the table.

Number	Rounded to the nearest 10,000	Rounded to the nearest 1,000	Rounded to the nearest 10
239,145			
128,783			
758,007			
		632,000	
			632,180
			825,430
6☐7,14☐	630,000		
			627,150
☐35,☐72	640,000		

45

CHALLENGE

Challenge questions allow children to delve deeper into a concept.

Reflect questions reveal the depth of each child's understanding before they move on.

[   ] . [   ] and [   ]
Which numbers could Max make?

Reflect

Show or write the value of each digit in the number 64,231.

14

Think differently questions encourage children to use reasoning as well as their mathematical knowledge to reach a solution.

The *Power Maths* characters support and encourage children to think and work in different ways.

Online subscriptions

The online subscription will give you access to additional resources.

eTextbooks

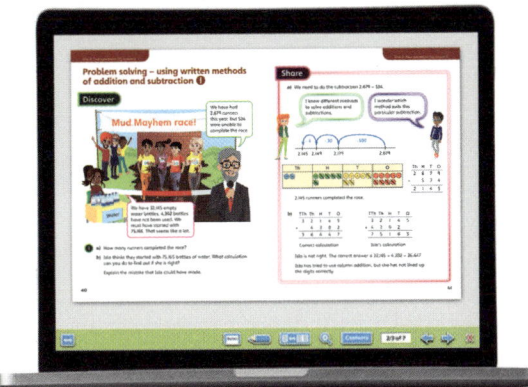

Digital versions of *Power Maths* Textbooks allow class groups to share and discuss questions, solutions and strategies. They allow you to project key structures and representations at the front of the class, to ensure all children are focusing on the same concept.

Teaching tools

Here you will find interactive versions of key *Power Maths* structures and representations.

Power Ups

Use this series of daily activities to promote and check number fluency.

Online versions of Teacher Guide pages

PDF pages give support at both unit and lesson levels. You will also find help with key strategies and templates for tracking progress.

Unit videos

Watch the professional development videos at the start of each unit to help you teach with confidence. The videos explore common misconceptions in the unit, and include intervention suggestions as well as suggestions on what to look out for when assessing mastery in your children.

End of unit Strengthen and Deepen materials

Each Strengthen activity at the end of every unit addresses a key misconception and can be used to support children who need it. The Deepen activities are designed to be 'Low Threshold High Ceiling' and will challenge those children who can understand more deeply. These resources will help you ensure that every child understands and will help you keep the class moving forward together. These printable activities provide an optional resource bank for use after the assessment stage.

Underpinning all of these resources, *Power Maths* is infused throughout with continual professional development, supporting you at every step.

The *Power Maths* teaching model

At the heart of *Power Maths* is a clearly structured teaching and learning process that helps you make certain that every child masters each maths concept securely and deeply. For each year group, the curriculum is broken down into core concepts, taught in units. A unit divides into smaller learning steps – lessons. Step by step, strong foundations of cumulative knowledge and understanding are built.

Quick check on prerequisite skills and a warm-up for children.

Rich assessments show mastery of key skills combined with a child self-assessment and reflection opportunity.

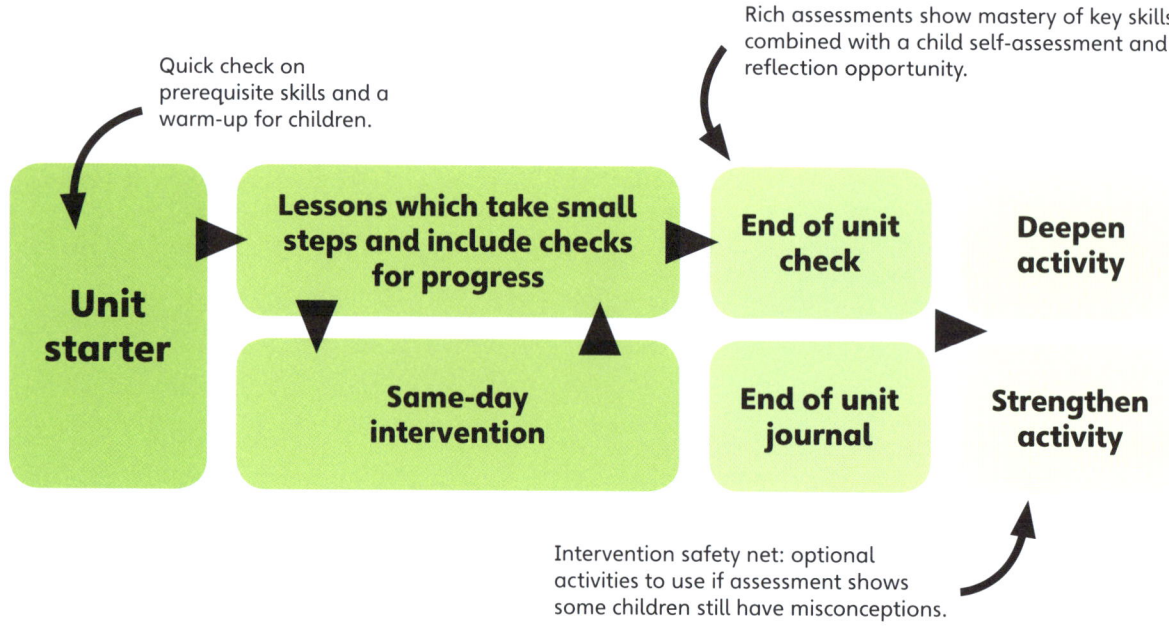

Intervention safety net: optional activities to use if assessment shows some children still have misconceptions.

Unit starter

Each unit begins with a unit starter, which introduces the learning context along with key mathematical vocabulary, structures and representations.

- The Textbooks include a check on readiness and a warm-up task for children to complete.

- Your Teacher Guide gives support right from the start on important structures and representations, mathematical language, common misconceptions and intervention strategies.

- Unit-specific videos develop your subject knowledge and insights so you feel confident and fully equipped to teach each new unit. These are available via the online subscription.

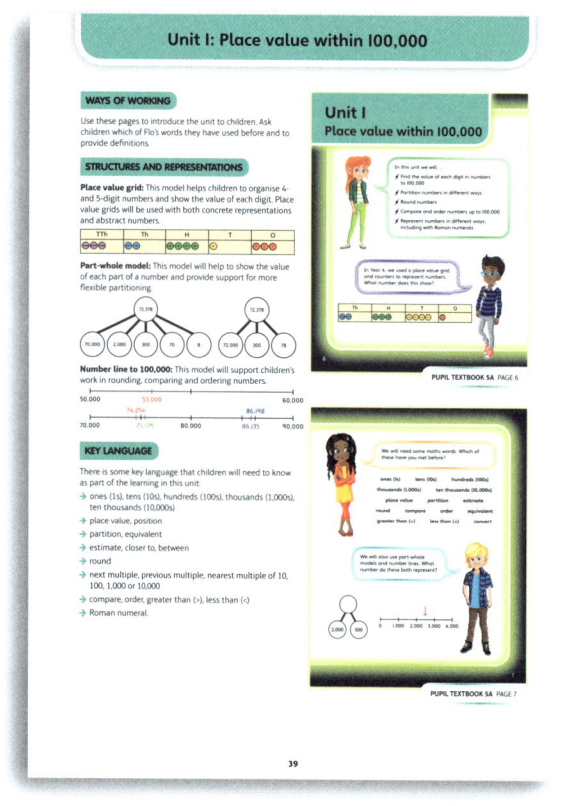

Lesson

Once a unit has been introduced, it is time to start teaching the series of lessons.

- Each lesson is scaffolded with Textbook and Practice Book activities and always begins with a Power Up activity (available via online subscription).
- *Power Maths* identifies lesson by lesson what concepts are to be taught.
- Your Teacher Guide offers lots of support for you to get the most from every child in every lesson. As well as highlighting key points, tricky areas and how to handle them, you will also find question prompts to check on understanding and clarification on why particular activities and questions are used.

Same-day intervention

Same-day interventions are vital in order to keep the class progressing together. Therefore, *Power Maths* provides plenty of support throughout the journey.

- Intervention is focused on keeping up now, not catching up later, so interventions should happen as soon as they are needed.
- Practice questions are designed to bring misconceptions to the surface, allowing you to identify these easily as you circulate during independent practice time.
- Child-friendly assessment questions in the Teacher Guide help you identify easily which children need to strengthen their understanding.

End of unit check and journal

At the end of a unit, summative assessment tasks reveal essential information on each child's understanding. An End of unit check in the Pupil Textbook lets you see which children have mastered the key concepts, which children have not and where their misconceptions lie. The Practice Book includes an End of unit journal in which children can reflect on what they have learnt. Each unit also offers Strengthen and Deepen activities, available via the online subscription.

The Teacher Guide offers support with handling misconceptions.

The End of unit check presents six to nine multiple-choice questions. These questions are designed to reveal misconceptions and help you target areas that need strengthening.

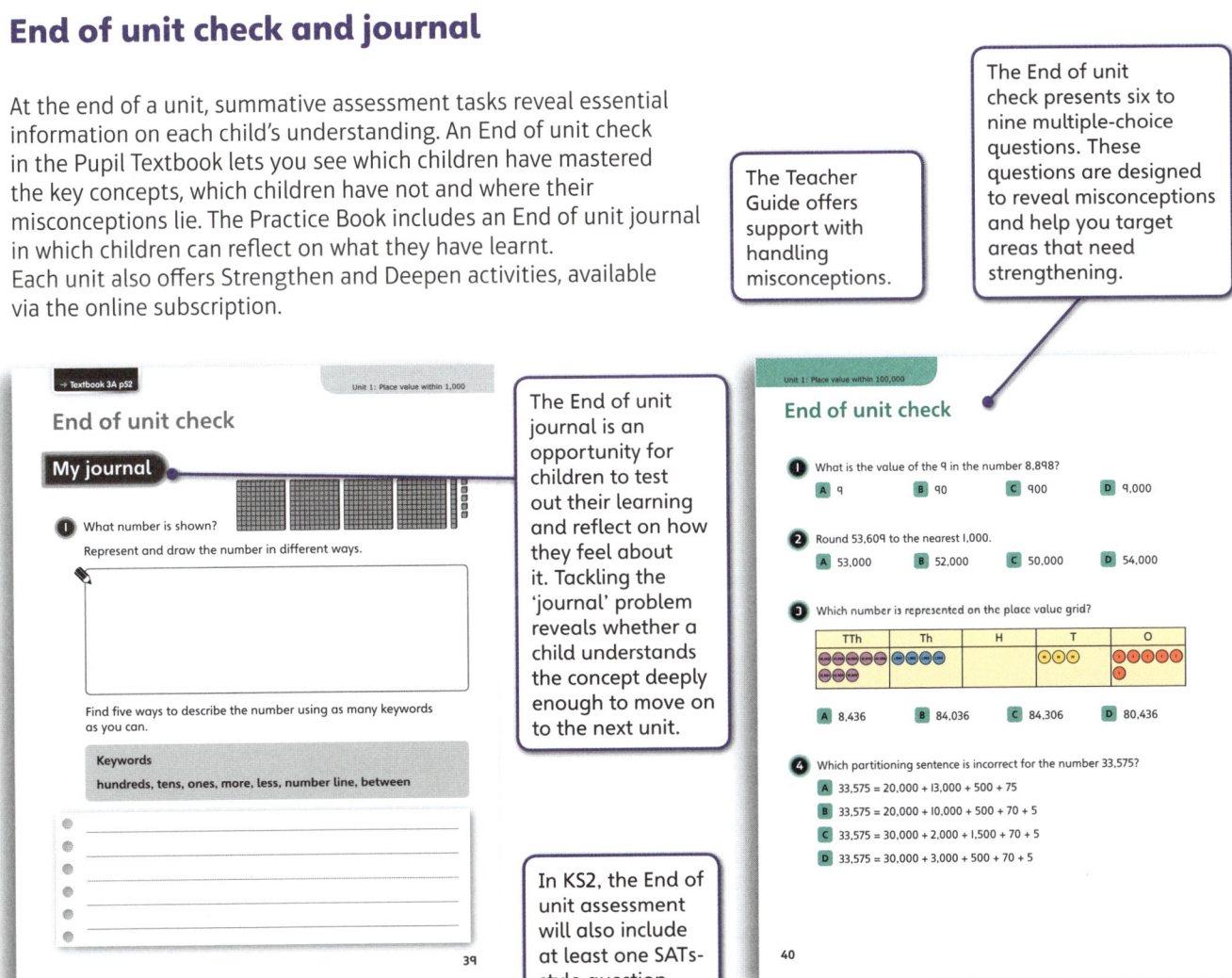

The End of unit journal is an opportunity for children to test out their learning and reflect on how they feel about it. Tackling the 'journal' problem reveals whether a child understands the concept deeply enough to move on to the next unit.

In KS2, the End of unit assessment will also include at least one SATs-style question.

The *Power Maths* lesson sequence

At the heart of *Power Maths* is a unique lesson sequence designed to empower children to understand core concepts and grow in confidence. Embracing the National Centre for Excellence in the Teaching of Mathematics' (NCETM's) definition of mastery, the sequence guides and shapes every *Power Maths* lesson you teach.

Flexibility is built into the *Power Maths* programme so there is no one-to-one mapping of lessons and concepts meaning you can pace your teaching according to your class. While some children will need to spend longer on a particular concept (through interventions or additional lessons), others will reach deeper levels of understanding. However, it is important that the class moves forward together through the termly schedules.

Power Up ⏱ 5 minutes

Each lesson begins with a Power Up activity (available via the online subscription) which supports fluency in key number facts.

The whole-class approach depends on fluency, so the Power Up is a powerful and essential activity.

TOP TIP

If the class is struggling with the task, revisit it later and check understanding.

Power Ups reinforce key skills such as times-tables, number bonds and working with place value.

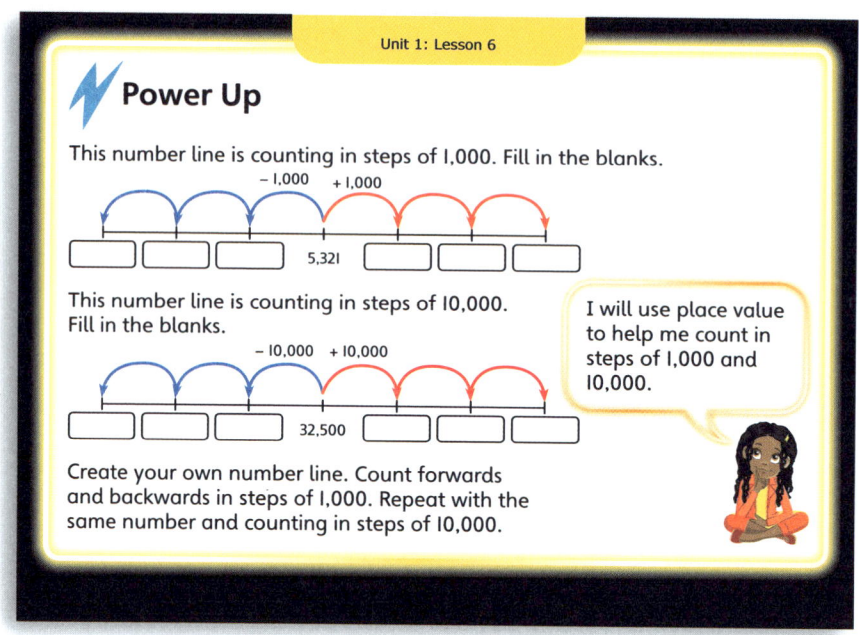

Discover ⏱ 10 minutes

A practical, real-life problem arouses curiosity. Children find the maths through story-telling.

A real-life scenario is provided for the Discover section but feel free to build upon these with your own examples that are more relevant to your class.

TOP TIP

Discover works best when run at tables, in pairs with concrete objects.

Question ❶ a) tackles the key concept and question ❶ b) digs a little deeper. Children have time to explore, play and discuss possible strategies.

Share 🕐 10 minutes

Teacher-led, this interactive section follows the Discover activity and highlights the variety of methods that can be used to solve a single problem.

TOP TIP
Ask children to discuss their methods. Pairs sharing a textbook is a great format for this!

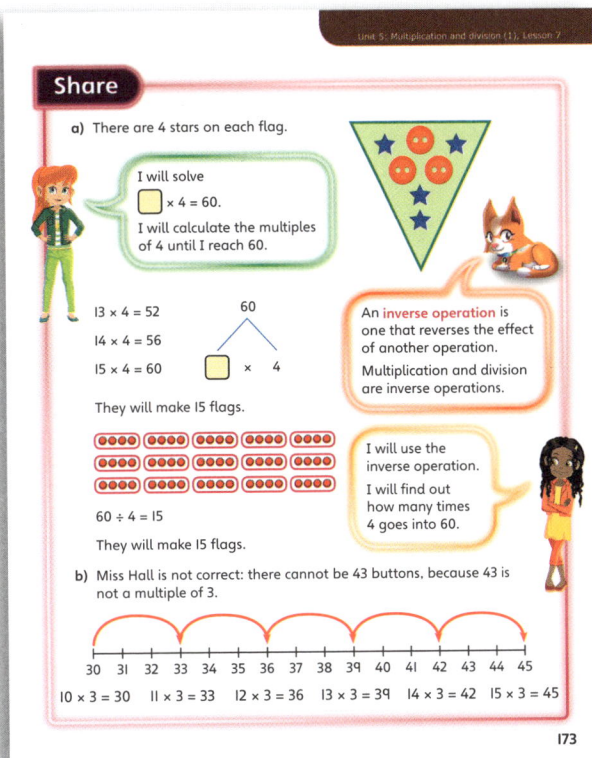

Your Teacher Guide gives target questions for children. The online toolkit provides interactive structures and representations to link concrete and pictorial to abstract concepts.

TOP TIP
Bring children to the front to share and celebrate their solutions and strategies.

Think together

🕐 10 minutes

Children work in groups on the carpet or at tables, using their textbooks or eBooks.

TOP TIP
Make sure children have mini whiteboards or pads to write on if they are not at their tables.

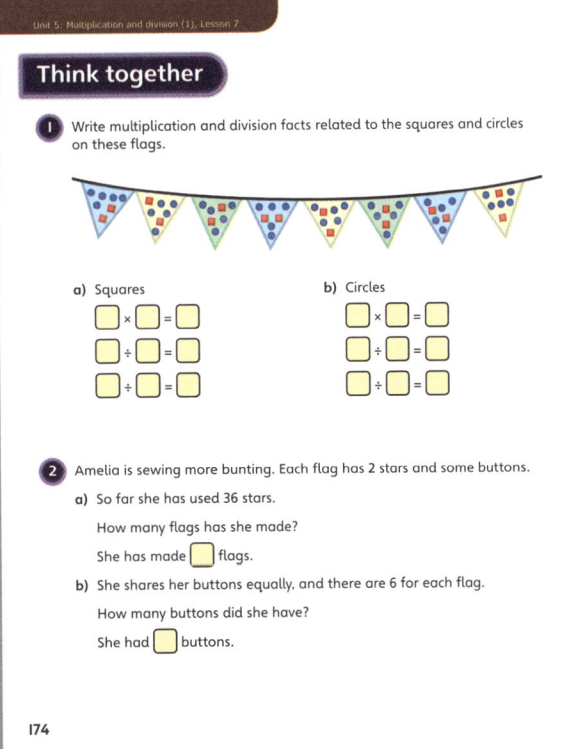

Using the Teacher Guide, model question ❶ for your class.

Question ❷ is less structured. Children will need to think together in their groups, then discuss their methods and solutions as a class.

In questions ❸ and ❹ children try working out the answer independently. The openness of the challenge question helps to check depth of understanding.

13

Practice ⏱ 15 minutes

Using their Practice Books, children work independently while you circulate and check on progress.

Questions follow small steps of progression to deepen learning.

TOP TIP
Some children could work separately with a teacher or assistant.

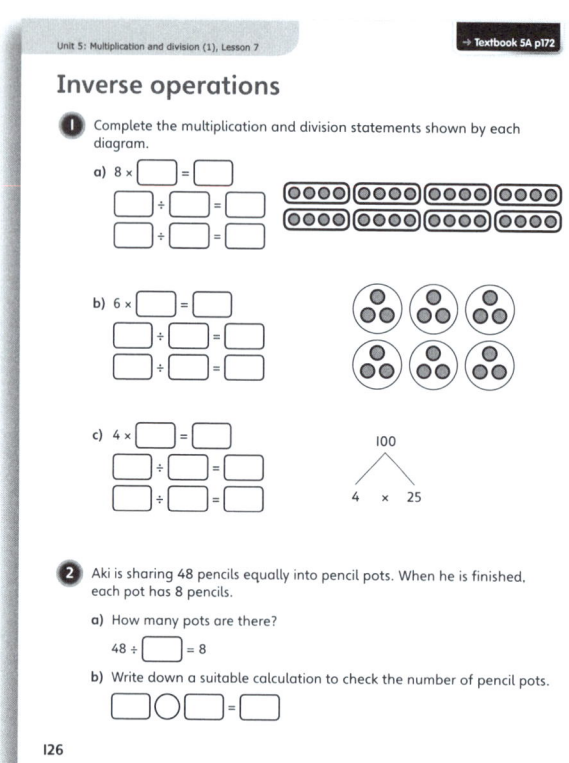

→ Textbook 5A p172

Unit 5: Multiplication and division (1), Lesson 7

Inverse operations

1   Complete the multiplication and division statements shown by each diagram.

a) 8 × ☐ = ☐
    ☐ ÷ ☐ = ☐
    ☐ ÷ ☐ = ☐

b) 6 × ☐ = ☐
    ☐ ÷ ☐ = ☐
    ☐ ÷ ☐ = ☐

c) 4 × ☐ = ☐
    ☐ ÷ ☐ = ☐
    ☐ ÷ ☐ = ☐

        100
       /   \
      4   ×   25

2   Aki is sharing 48 pencils equally into pencil pots. When he is finished, each pot has 8 pencils.

a) How many pots are there?

    48 ÷ ☐ = 8

b) Write down a suitable calculation to check the number of pencil pots.

    ☐ ◯ ☐ = ☐

126

Are some children struggling? If so, work with them as a group, using mathematical structures and representations to support understanding as necessary.

There are no set routines: for real understanding, children need to think about the problem in different ways.

Reflect ⏱ 5 minutes

'Spot the mistake' questions are great for checking misconceptions.

The Reflect section is your opportunity to check how deeply children understand the target concept.

Unit 5: Multiplication and division (1), Lesson 7

6   a) What number did Reena start with?

Reena started with ☐ .

Reena   **CHALLENGE**

I divide my number by 8 and get the answer 2 remainder 7.

b) What number did Andy divide by?

Andy divided by ☐ .

Andy

I start with 69. I divide by a number and get the answer 9 remainder 6.

c) What numbers could Jamie have started with?

Jamie could have started with ☐ or ☐ .

Jamie

I am thinking of a prime number between 50 and 100. I divide by 6 and get a remainder of 1.

Reflect

Show the different strategies required to find the missing values of the following calculations.

18 ÷ ☐ = 3   ☐ ÷ 3 = 18

128

The Practice Books use various approaches to check that children have fully understood each concept.

Looking like they understand is not enough! It is essential that children can show they have grasped the concept.

Using the *Power Maths* Teacher Guide

Think of your Teacher Guides as *Power Maths* handbooks that will guide, support and inspire your day-to-day teaching. Clear and concise, and illustrated with helpful examples, your Teacher Guides will help you make the best possible use of every individual lesson. They also provide wrap-around professional development, enhancing your own subject knowledge and helping you to grow in confidence about moving your children forward together.

There is a Teacher Guide per year group for every term with unit and lesson level guidance and support.

Tips and advice on key elements such as C-P-A approaches, misconceptions, language, modelling growth mindsets and same-day intervention.

Annotations for every Pupil Textbook and Practice Book page, providing prompts for key questions to ask to expose understanding and explanations as to why key questions have been chosen.

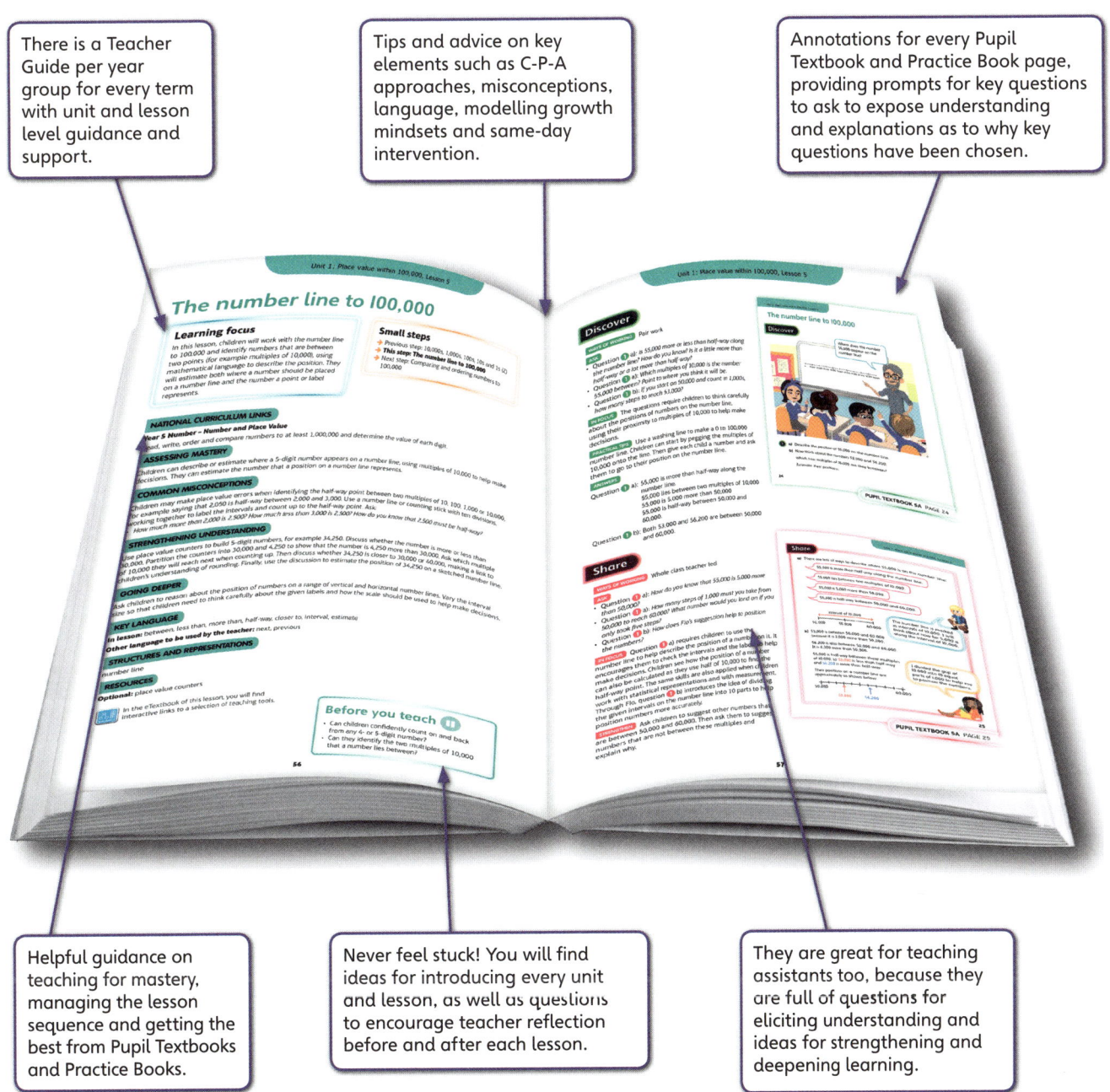

Helpful guidance on teaching for mastery, managing the lesson sequence and getting the best from Pupil Textbooks and Practice Books.

Never feel stuck! You will find ideas for introducing every unit and lesson, as well as questions to encourage teacher reflection before and after each lesson.

They are great for teaching assistants too, because they are full of questions for eliciting understanding and ideas for strengthening and deepening learning.

At the end of each unit, your Teacher Guide helps you identify who has fully grasped the concept, who has not and how to move every child forward. This is covered later in the Assessment strategies section.

Power Maths Year 6, yearly overview

Textbook	Strand	Unit		Number of Lessons
Textbook A / Practice Book A (Term 1)	Number – number and place value	1	Place value within 10,000,000	7
	Number – addition, subtraction, multiplication and division	2	Four operations (1)	10
	Number – addition, subtraction, multiplication and division	2	Four operations (2)	9
	Number – fractions	4	Fractions (1)	11
	Number – fractions	5	Fractions (2)	9
	Geometry – position and direction	6	Geometry – position and direction	4
Textbook B / Practice Book B (Term 2)	Number – fractions (including decimals and percentages)	7	Decimals	9
	Number – fractions (including decimals and percentages)	8	Percentages	9
	Algebra	9	Algebra	11
	Measurement	10	Measure – imperial and metric measures	5
	Measurement	11	Measure – perimeter, area and volume	11
	Ratio and proportion	12	Ratio and proportion	9
Textbook C / Practice Book C (Term 3)	Geometry – properties of shapes	13	Geometry – properties of shapes	12
	Number – number and place value	14	Problem solving	14
	Statistics	15	Statistics	10

Power Maths Year 6, Textbook 6A (Term I) Overview

Strand 1	Strand 2	Unit		Lesson number	Lesson title	NC Objective 1	NC Objective 2	NC Objective 3
Number – number and place value		Unit 1	Place value within 10,000,000	1	Numbers to 1,000,000	Read, write, order and compare numbers up to 10,000,000 and determine the value of each digit		
Number – number and place value		Unit 1	Place value within 10,000,000	2	Numbers to 10,000,000 (1)	Read, write, order and compare numbers up to 10,000,000 and determine the value of each digit		
Number – number and place value		Unit 1	Place value within 10,000,000	3	Numbers to 10,000,000 (2)	Solve number and practical problems that involve all of the above		
Number – number and place value		Unit 1	Place value within 10,000,000	4	Number line to 10,000,000	Read, write, order and compare numbers up to 10,000,000 and determine the value of each digit		
Number – number and place value		Unit 1	Place value within 10,000,000	5	Comparing and ordering numbers to 10,000,000	Read, write, order and compare numbers up to 10,000,000 and determine the value of each digit		
Number – number and place value		Unit 1	Place value within 10,000,000	6	Rounding numbers	Round any whole number to a required degree of accuracy		

Strand 1	Strand 2	Unit		Lesson number	Lesson title	NC Objective 1	NC Objective 2	NC Objective 3
Number – number and place value		Unit 1	Place value within 10,000,000	7	Negative numbers	Use negative numbers in context, and calculate intervals across zero		
Number – addition, subtraction, multiplication and division		Unit 2	Four operations (1)	1	Problem solving – using written methods of addition and subtraction (1)	Solve addition and subtraction multi-step problems in contexts, deciding which operations and methods to use and why		
Number – addition, subtraction, multiplication and division		Unit 2	Four operations (1)	2	Problem solving – using written methods of addition and subtraction (2)	Solve addition and subtraction multi-step problems in contexts, deciding which operations and methods to use and why		
Number – addition, subtraction, multiplication and division		Unit 2	Four operations (1)	3	Multiplying numbers up to 4 digits by a 1-digit number	Multiply multi-digit numbers up to 4 digits by a two-digit whole number using the formal written method of long multiplication		
Number – addition, subtraction, multiplication and division		Unit 2	Four operations (1)	4	Multiplying numbers up to 4 digits by a 2-digit number	Multiply multi-digit numbers up to 4 digits by a two-digit whole number using the formal written method of long multiplication		
Number – addition, subtraction, multiplication and division		Unit 2	Four operations (1)	5	Dividing numbers up to 4 digits by a 2-digit number (1)	Divide numbers up to 4 digits by a two-digit number using the formal written method of short division where appropriate, interpreting remainders according to the context		
Number – addition, subtraction, multiplication and division		Unit 2	Four operations (1)	6	Dividing numbers up to 4 digits by a 2-digit number (2)	Divide numbers up to 4 digits by a two-digit number using the formal written method of short division where appropriate, interpreting remainders according to the context		
Number – addition, subtraction, multiplication and division		Unit 2	Four operations (1)	7	Dividing numbers up to 4 digits by a 2-digit number (3)	Divide numbers up to 4 digits by a two-digit whole number using the formal written method of long division, and interpret remainders as whole number remainders, fractions, or by rounding, as appropriate for the context		
Number – addition, subtraction, multiplication and division		Unit 2	Four operations (1)	8	Dividing numbers up to 4 digits by a 2-digit number (4)	Divide numbers up to 4 digits by a two-digit whole number using the formal written method of long division, and interpret remainders as whole number remainders, fractions, or by rounding, as appropriate for the context		
Number – addition, subtraction, multiplication and division		Unit 2	Four operations (1)	9	Dividing numbers up to 4 digits by a 2-digit number (5)	Divide numbers up to 4 digits by a two-digit whole number using the formal written method of long division, and interpret remainders as whole number remainders, fractions, or by rounding, as appropriate for the context		
Number – addition, subtraction, multiplication and division		Unit 2	Four operations (1)	10	Dividing numbers up to 4 digits by a 2-digit number (6)	Divide numbers up to 4 digits by a two-digit whole number using the formal written method of long division, and interpret remainders as whole number remainders, fractions, or by rounding, as appropriate for the context		
Number – addition, subtraction, multiplication and division		Unit 3	Four operations (2)	1	Common factors	Identify common factors, common multiples and prime numbers		

Strand 1	Strand 2	Unit		Lesson number	Lesson title	NC Objective 1	NC Objective 2	NC Objective 3
Number – addition, subtraction, multiplication and division		Unit 3	Four operations (2)	2	Common multiples	Identify common factors, common multiples and prime numbers		
Number – addition, subtraction, multiplication and division		Unit 3	Four operations (2)	3	Recognising prime numbers up to 100	Identify common factors, common multiples and prime numbers		
Number – multiplication and division		Unit 3	Four operations (2)	4	Squares and cubes	Recognise and use square numbers and cube numbers, and the notation for squared (2) and cubed (3) **(Year 5)**		
Number – addition, subtraction, multiplication and division		Unit 3	Four operations (2)	5	Order of operations	Use their knowledge of the order of operations to carry out calculations involving the four operations		
Number – addition, subtraction, multiplication and division		Unit 3	Four operations (2)	6	Brackets	Use their knowledge of the order of operations to carry out calculations involving the four operations		
Number – addition, subtraction, multiplication and division		Unit 3	Four operations (2)	7	Mental calculations (1)	Perform mental calculations, including with mixed operations and large numbers		
Number – addition, subtraction, multiplication and division		Unit 3	Four operations (2)	8	Mental calculations (2)	Perform mental calculations, including with mixed operations and large numbers		
Number – addition, subtraction, multiplication and division		Unit 3	Four operations (2)	9	Reasoning from known facts	Use their knowledge of the order of operations to carry out calculations involving the four operations	Solve problems involving addition, subtraction, multiplication and division	
Number – fractions		Unit 4	Fractions (1)	1	Simplifying fractions (1)	Use common factors to simplify fractions; use common multiples to express fractions in the same denomination		
Number – fractions		Unit 4	Fractions (1)	2	Simplifying fractions (2)	Use common factors to simplify fractions; use common multiples to express fractions in the same denomination	Compare and order fractions, including fractions > 1	
Number – fractions		Unit 4	Fractions (1)	3	Fractions on a number line	Compare and order fractions, including fractions > 1		
Number – fractions		Unit 4	Fractions (1)	4	Comparing and ordering fractions (1)	Compare and order fractions, including fractions > 1	Use common factors to simplify fractions; use common multiples to express fractions in the same denomination	
Number – fractions		Unit 4	Fractions (1)	5	Comparing and ordering fractions (2)	Compare and order fractions, including fractions > 1	Use common factors to simplify fractions; use common multiples to express fractions in the same denomination	
Number – fractions		Unit 4	Fractions (1)	6	Adding and subtracting fractions (1)	Add and subtract fractions with different denominators and mixed numbers, using the concept of equivalent fractions		
Number – fractions		Unit 4	Fractions (1)	7	Adding and subtracting fractions (2)	Add and subtract fractions with different denominators and mixed numbers, using the concept of equivalent fractions		

Strand 1	Strand 2	Unit		Lesson number	Lesson title	NC Objective 1	NC Objective 2	NC Objective 3
Number – fractions		Unit 4	Fractions (1)	8	Adding fractions	Add and subtract fractions with different denominators and mixed numbers, using the concept of equivalent fractions		
Number – fractions		Unit 4	Fractions (1)	9	Subtracting fractions	Add and subtract fractions with different denominators and mixed numbers, using the concept of equivalent fractions		
Number – fractions		Unit 4	Fractions (1)	10	Problem solving – adding and subtracting fractions (1)	Add and subtract fractions with different denominators and mixed numbers, using the concept of equivalent fractions		
Number – fractions		Unit 4	Fractions (1)	11	Problem solving – adding and subtracting fractions (2)	Add and subtract fractions with different denominators and mixed numbers, using the concept of equivalent fractions		
Year 5 – Number – fractions		Unit 5	Fractions (2)	1	Multiplying a fraction by a whole number	Multiply proper fractions and mixed numbers by whole numbers, supported by materials and diagrams		
Number – fractions		Unit 5	Fractions (2)	2	Multiplying a fraction by a fraction (1)	Multiply simple pairs of proper fractions, writing the answer in its simplest form (for example, $\frac{1}{4} \times \frac{1}{2} = \frac{1}{8}$)		
Number – fractions		Unit 5	Fractions (2)	3	Multiplying a fraction by a fraction (2)	Multiply simple pairs of proper fractions, writing the answer in its simplest form (for example, $\frac{1}{4} \times \frac{1}{2} = \frac{1}{8}$)		
Number – fractions		Unit 5	Fractions (2)	4	Dividing a fraction by a whole number (1)	Divide proper fractions by whole numbers (for example, $\frac{1}{3} \div 2 = \frac{1}{6}$)		
Number – fractions		Unit 5	Fractions (2)	5	Dividing a fraction by a whole number (2)	Divide proper fractions by whole numbers (for example, $\frac{1}{3} \div 2 = \frac{1}{6}$)		
Number – fractions		Unit 5	Fractions (2)	6	Dividing a fraction by a whole number (3)	Divide proper fractions by whole numbers (for example, $\frac{1}{3} \div 2 = \frac{1}{6}$)		
Number – fractions	Number – addition, subtraction, multiplication and division	Unit 5	Fractions (2)	7	Four rules with fractions	Add and subtract fractions with different denominators and mixed numbers, using the concept of equivalent fractions	Multiply simple pairs of proper fractions, writing the answer in its simplest form (for example, $\frac{1}{4} \times \frac{1}{2} = \frac{1}{8}$)	Use their knowledge of the order of operations to carry out calculations involving the four operations
Number – fractions		Unit 5	Fractions (2)	8	Calculating fractions of amounts	Use written division methods in cases where the answer has up to two decimal places		
Number – fractions		Unit 5	Fractions (2)	9	Problem solving – fractions of amounts	Use written division methods in cases where the answer has up to two decimal places		
Geometry – position and direction		Unit 6	Geometry – position and direction	1	Plotting coordinates in the first quadrant	Describe positions on the full coordinate grid (all four quadrants)		
Geometry – position and direction		Unit 6	Geometry – position and direction	2	Plotting coordinates	Describe positions on the full coordinate grid (all four quadrants)		
Geometry – position and direction		Unit 6	Geometry – position and direction	3	Plotting translations and reflections	Draw and translate simple shapes on the coordinate plane, and reflect them in the axes		
Geometry – position and direction		Unit 6	Geometry – position and direction	4	Reasoning about shapes with coordinates	Draw and translate simple shapes on the coordinate plane, and reflect them in the axes		

Mindset: an introduction

Global research and best practice deliver the same message: learning is greatly affected by what learners perceive they can or cannot do. What is more, it is also shaped by what their parents, carers and teachers perceive they can do. Mindset – the thinking that determines our beliefs and behaviours – therefore has a fundamental impact on teaching and learning.

Everyone can!

Power Maths and mastery methods focus on the distinction between 'fixed' and 'growth' mindsets (Dweck, 2007).[1] Those with a fixed mindset believe that their basic qualities (for example, intelligence, talent and ability to learn) are pre-wired or fixed: 'If you have a talent for maths, you will succeed at it. If not, too bad!' By contrast, those with a growth mindset believe that hard work, effort and commitment drive success and that 'smart' is not something you are or are not, but something you become. In short, everyone can do maths!

Key mindset strategies

A growth mindset needs to be actively nurtured and developed. *Power Maths* offers some key strategies for fostering healthy growth mindsets in your classroom.

It is okay to get it wrong

Mistakes are valuable opportunities to re-think and understand more deeply. Learning is richer when children and teachers alike focus on spotting and sharing mistakes as well as solutions.

Praise hard work

Praise is a great motivator, and by focusing on praising effort and learning rather than success, children will be more willing to try harder, take risks and persist for longer.

Mind your language!

The language we use around learners has a profound effect on their mindsets. Make a habit of using growth phrases, such as, 'Everyone can!', 'Mistakes can help you learn' and 'Just try for a little longer'. The king of them all is one little word, 'yet …
I cannot solve this … yet!'
Encourage parents and carers to use the right language too.

Build in opportunities for success

The step-by-small-step approach enables children to enjoy the experience of success. In addition, avoid ability grouping and encourage every child to answer questions and explain or demonstrate their methods to others.

[1]Dweck, C (2007) *The New Psychology of Success*, Ballantine Books: New York

The *Power Maths* characters

The *Power Maths* characters model the traits of growth mindset learners and encourage resilience by prompting and questioning children as they work. Appearing frequently in the Textbooks and Practice Books, they are your allies in teaching and discussion, helping to model methods, alternatives and misconceptions, and to pose questions. They encourage and support your children, too: they are all hardworking, enthusiastic and unafraid of making and talking about mistakes.

Meet the team!

Flexible Flo is open-minded and sometimes indecisive. She likes to think differently and come up with a variety of methods or ideas.

Determined Dexter is resolute, resilient and systematic. He concentrates hard, always tries his best and he'll never give up – even though he doesn't always choose the most efficient methods!

'Let's try again.'

'Mistakes are cool!'

'Have I found all of the solutions?'

'Let's try it this way ...'

'Can we do it differently?'

'I've got another way of doing this!'

'I'm going to try this!'

'I know how to do that!'

'Want to share my ideas?'

Curious Ash is eager, interested and inquisitive, and he loves solving puzzles and problems. Ash asks lots of questions but sometimes gets distracted.

'What if we tried this ...?'

'I wonder ...'

'Is there a pattern here?'

Miaow!   Sparks the Cat

Brave Astrid is confident, willing to take risks and unafraid of failure. She is never scared to jump straight into a problem or question, and although she often makes simple mistakes she is happy to talk them through with others.

Mathematical language

Traditionally, we in the UK have tended to try simplifying mathematical language to make it easier for young children to understand. By contrast, evidence and experience show that by diluting the correct language, we actually mask concepts and meanings for children. We then wonder why they are confused by new and different terminology later down the line! *Power Maths* is not afraid of 'hard' words and avoids placing any barriers between children and their understanding of mathematical concepts. As a result, we need to be planned, precise and thorough in building every child's understanding of the language of maths. Throughout the Teacher Guides you will find support and guidance on how to deliver this, as well as individual explanations throughout the Pupil Textbooks.

Use the following key strategies to build children's mathematical vocabulary, understanding and confidence.

Precise and consistent

Everyone in the classroom should use the correct mathematical terms in full, every time. For example, refer to 'equal parts', not 'parts'. Used consistently, precise maths language will be a familiar and non-threatening part of children's everyday experience.

Full sentences

Teachers and children alike need to use full sentences to explain or respond. When children use complete sentences, it both reveals their understanding and embeds their knowledge.

Stem sentences

These important sentences help children express mathematical concepts accurately, and are used throughout the *Power Maths* books. Encourage children to repeat them frequently, whether working independently or with others. Examples of stem sentences are:

'4 is a part, 5 is a part, 9 is the whole.'

'There are ... groups. There are ... in each group.'

Key vocabulary

The unit starters highlight essential vocabulary for every lesson. In the Pupil Textbooks, characters flag new terminology and the Teacher Guide lists important mathematical language for every unit and lesson. New terms are never introduced without a clear explanation.

Mathematical signs

Mathematical signs are used early on so that children quickly become familiar with them and their meaning. Often, the *Power Maths* characters will highlight the connection between language and particular signs.

The role of talk and discussion

When children learn to talk purposefully together about maths, barriers of fear and anxiety are broken down and they grow in confidence, skills and understanding. Building a healthy culture of 'maths talk' empowers their learning from day one.

Explanation and discussion are integral to the *Power Maths* structure, so by simply following the books your lessons will stimulate structured talk. The following key 'maths talk' strategies will help you strengthen that culture and ensure that every child is included.

Sentences, not words

Encourage children to use full sentences when reasoning, explaining or discussing maths. This helps both speaker and listeners to clarify their own understanding. It also reveals whether or not the speaker truly understands, enabling you to address misconceptions as they arise.

Working together

Working with others in pairs, groups or as a whole class is a great way to support maths talk and discussion. Use different group structures to add variety and challenge. For example, children could take timed turns for talking, work independently alongside a 'discussion buddy', or perhaps play different *Power Maths* character roles within their group.

Think first – then talk

Provide clear opportunities within each lesson for children to think and reflect, so that their talk is purposeful, relevant and focused.

Give every child a voice

Where the 'hands up' model allows only the more confident child to shine, *Power Maths* involves everyone. Make sure that no child dominates and that even the shyest child is encouraged to contribute – and is praised when they do.

Assessment strategies

Teaching for mastery demands that you are confident about what each child knows and where their misconceptions lie: therefore, practical and effective assessment is vitally important.

Formative assessment within lessons

The Think together section will often reveal any confusions or insecurities: try ironing these out by doing the first Think together question as a class. For children who continue to struggle, you or your teaching assistant should provide support and enable them to move on.

→

Performance in Practice can be very revealing: check Practice Books and listen out both during and after practice to identify misconceptions.

→

The Reflect section is designed to check on the all-important depth of understanding. Be sure to review how children performed in this final stage before you teach the next lesson.

End of unit check – Textbook

Each unit concludes with a summative check to help you assess quickly and clearly each child's understanding, fluency, reasoning and problem-solving skills. In KS2 this check also contains a SATs-style question to help children become familiar with answering this type of question.

In KS2 we would suggest the End of unit check is completed independently in children's exercise books, but you can adapt this to suit the needs of your class.

End of unit check – Practice Book

The Practice Book contains further opportunities for assessment, and can be completed by children independently whilst you are carrying out diagnostic assessment with small groups. Your Teacher Guide will advise you on what to do if children struggle to articulate an explanation – or perhaps encourage you to write down something they have explained well. It will also offer insights into children's answers and their implications for the next learning steps. It is split into three main sections, outlined below.

My journal

My journal is designed to allow children to show their depth of understanding of the unit. It can also serve as a way of checking that children have grasped key mathematical vocabulary. Children should have some time to think about how they want to answer the question, and you could ask them to talk to a partner about their ideas. Then children should write their answer in their Practice Book.

Power check

The Power check allows children to self-assess their level of confidence on the topic by colouring in different smiley faces. You may want to introduce the faces as follows:

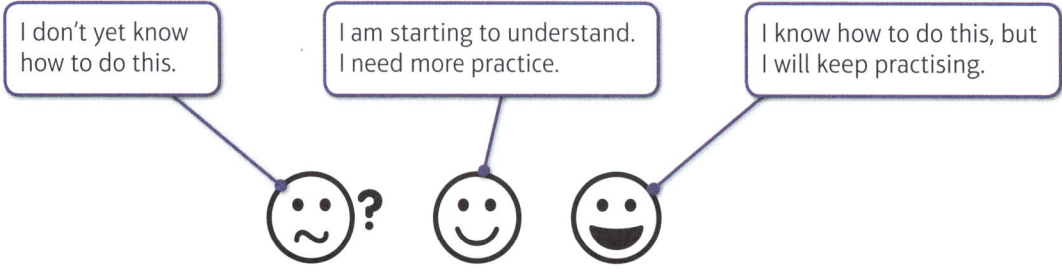

I don't yet know how to do this.

I am starting to understand. I need more practice.

I know how to do this, but I will keep practising.

Power play or Power puzzle

Each unit ends with either a Power play or a Power puzzle. This is an activity, puzzle or game that allows children to use their new knowledge in a fun, informal way. In Key Stage 2 we have also included a deeper level to each game to help challenge those children who have grasped a concept quickly.

How to use diagnostic questions

The diagnostic questions provided in *Power Maths* Textbooks are carefully structured to identify both understanding and misconceptions (if children answer in a particular way, you will know why). The simple procedure below may be helpful:

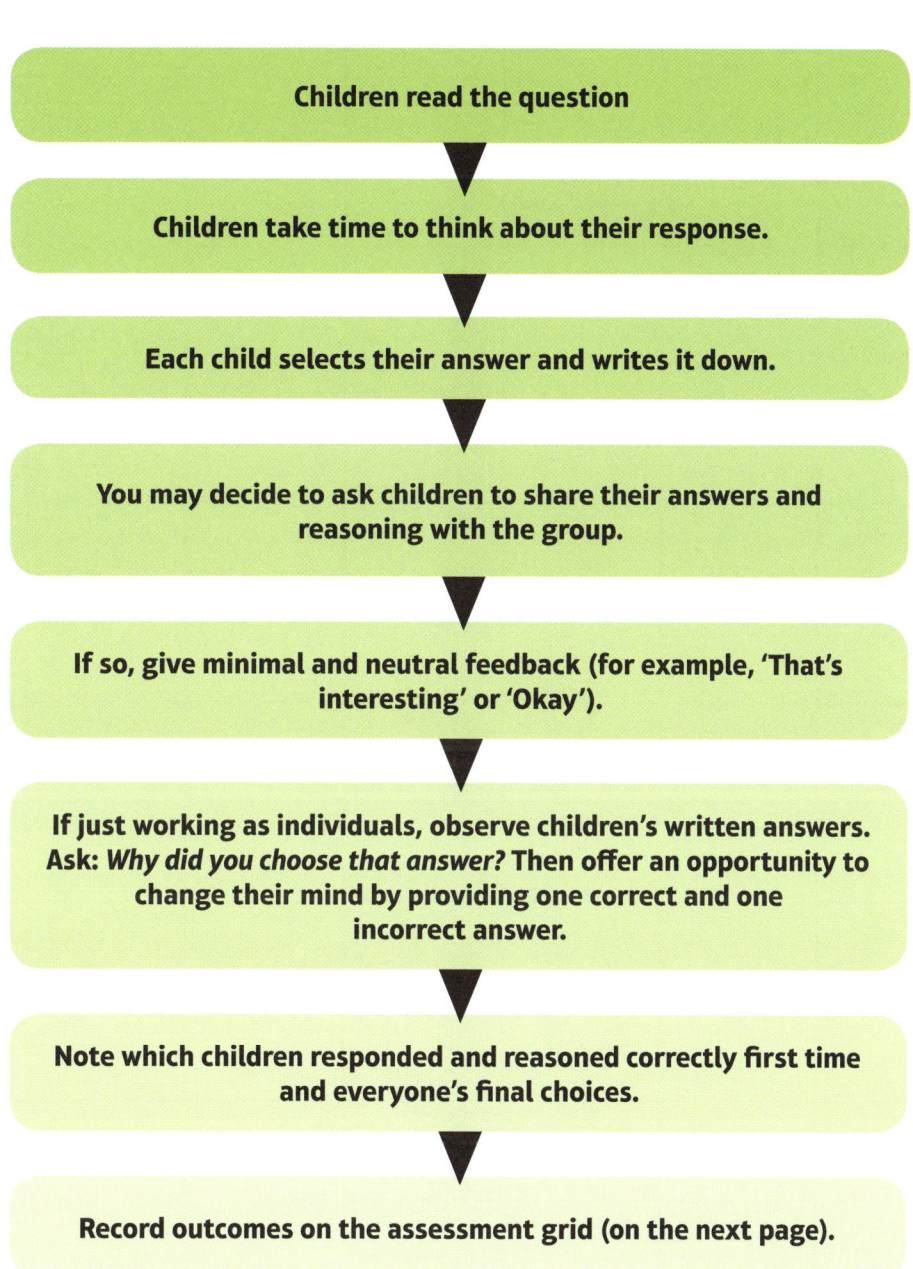

Children read the question

Children take time to think about their response.

Each child selects their answer and writes it down.

You may decide to ask children to share their answers and reasoning with the group.

If so, give minimal and neutral feedback (for example, 'That's interesting' or 'Okay').

If just working as individuals, observe children's written answers. Ask: *Why did you choose that answer?* Then offer an opportunity to change their mind by providing one correct and one incorrect answer.

Note which children responded and reasoned correctly first time and everyone's final choices.

Record outcomes on the assessment grid (on the next page).

Power Maths unit assessment grid

Year ___   **Unit** ___ _____

Record only as much information as you judge appropriate for your assessment of each child's mastery of the unit and any steps needed for intervention.

Name	Diagnostic questions	SATs-style question	My journal	Power check	Power play/puzzle	Mastery	Intervention/ Strengthen

Keeping the class together

Traditionally, children who learn quickly have been accelerated through the curriculum. As a consequence, their learning may be superficial and will lack the many benefits of enabling children to learn with and from each other.

By contrast, *Power Maths'* mastery approach values real understanding and richer, deeper learning above speed. It sees all children learning the same concept in small, cumulative steps, each finding and mastering challenge at their own level. Remember that when you teach for mastery, EVERYONE can do maths! Those who grasp a concept easily have time to explore and understand that concept at a deeper level. The whole class therefore moves through the curriculum at broadly the same pace via individual learning journeys.

For some teachers, the idea that a whole class can move forward together is revolutionary and challenging. However, the evidence of global good practice clearly shows that this approach drives engagement, confidence, motivation and success for all learners, and not just the high flyers. The strategies below will help you keep your class together on their maths journey.

Mix it up

Do not stick to set groups at each table. Every child should be working on the same concept, and mixing up the groupings widens children's opportunities for exploring, discussing and sharing their understanding with others.

Recycling questions

Reuse the Pupil Textbook and Practice Book questions with concrete materials to allow children to explore concepts and relationships and deepen their understanding. This strategy is especially useful for reinforcing learning in same-day interventions.

Strengthen at every opportunity

The next lesson in a *Power Maths* sequence always revises and builds on the previous step to help embed learning. These activities provide golden opportunities for individual children to strengthen their learning with the support of teaching assistants.

Prepare to be surprised!

Children may grasp a concept quickly or more slowly. The 'fast graspers' won't always be the same individuals, nor does the speed at which a child understands a concept predict their success in maths. Are they struggling or just working more slowly?

Depth and breadth

Just as prescribed in the National Curriculum, the goal of *Power Maths* is never to accelerate through a topic but rather to gain a clear, deep and broad understanding.

"Pupils who grasp concepts rapidly should be challenged through being offered rich and sophisticated problems before any acceleration through new content. Those who are not sufficiently fluent with earlier material should consolidate their understanding, including through additional practice, before moving on."

National Curriculum: Mathematics programmes of study: KS1 & 2, 2013

The lesson sequence offers many opportunities for you to deepen and broaden children's learning, some of which are suggested below.

Discover

As well as using the questions in the Teacher Guide, check that children are really delving into why something is true. It is not enough to simply recite facts, such as '6 + 3 = 9'. They need to be able to see why, explain it, and to demonstrate the solution in several ways.

Share

Make sure that every child is given chances to offer answers and expand their knowledge and not just those with the greatest confidence.

Think together

Encourage children to think about how they found the solution and explain it to their partner. Be sure to make concrete materials available on group tables throughout the lesson to support and reinforce learning.

Practice

Avoid any temptation to select questions according to your assessment of ability: practice questions are presented in a logical sequence and it is important that each child works through every question.

Reflect

Open-ended questions allow children to deepen their understanding as far as they can by discovering new ways of finding answers. For example, *Give me another way of working out how high the wall is … And another way?*

Online materials

For each unit you will find additional strengthening activities to support those children who need it and to deepen the understanding of those who need the additional challenge.

Same-day intervention

Since maths competence depends on mastering concepts one-by-one in a logical progression, it is important that no gaps in understanding are ever left unfilled. Same-day interventions – either within or after a lesson – are a crucial safety net for any child who has not fully made the small step covered that day. In other words, intervention is always about keeping up, not catching up, so that every child has the skills and understanding they need to tackle the next lesson. That means presenting the same problems used in the lesson, with a variety of concrete materials to help children model their solutions.

We offer two intervention strategies below, but you should feel free to choose others if they work better for your class.

Within-lesson intervention

The Think together activity will reveal those who are struggling, so when it is time for Practice, bring these children together to work with you on the first Practice questions. Observe these children carefully, ask questions, encourage them to use concrete models and check that they reach and can demonstrate their understanding.

After-lesson intervention

You might like to use Think together before an assembly, giving you or teaching assistants time to recap and expand with slow graspers during assembly time. Teaching assistants could also work with strugglers at other convenient points in the school day.

The role of practice

Practice plays a pivotal role in the *Power Maths* approach. It takes place in class groups, smaller groups, pairs and independently, so that children always have the opportunities for thinking as well as the models and support they need to practise meaningfully and with understanding.

Intelligent practice

In *Power Maths*, practice never equates to the simple repetition of a process. Instead we embrace the concept of intelligent practice, in which all children become fluent in maths through varied, frequent and thoughtful practice that deepens and embeds conceptual understanding in a logical, planned sequence. To see the difference, take a look at the following examples.

Traditional practice

- Repetition can be rote – no need for a child to think hard about what they are doing.

- Praise may be misplaced.

- Does this prove understanding?

Intelligent practice

- Varied methods – concrete, pictorial and abstract.

- Calculations expressed in different ways, requiring thought and understanding.

- Constructive feedback.

All practice questions are designed to move children on and reveal misconceptions.

Simple, logical steps build onto earlier learning.

C-P-A runs throughout – different ways of modelling and understanding the same concept.

Conceptual variation – children work on different representations of the same maths concept.

Friendly characters offer support and encourage children to try different approaches.

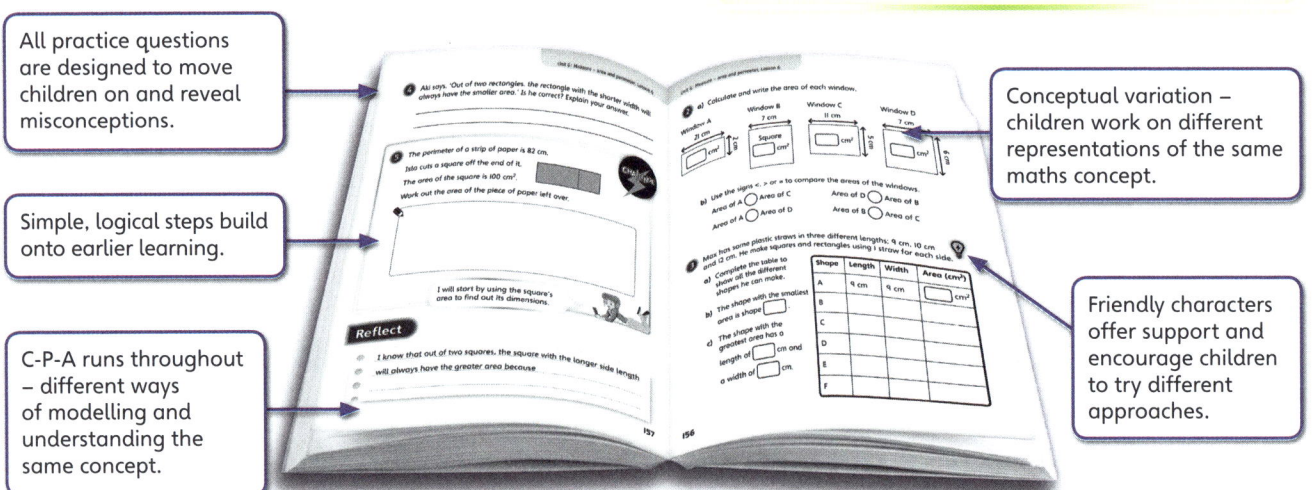

A carefully designed progression

The Practice Books provide just the right amount of intelligent practice for children to complete independently in the final sections of each lesson. It is really important that all children are exposed to the Practice questions, and that children are not directed to complete different sections. That is because each question is different and has been designed to challenge children to think about the maths they are doing. The questions become more challenging so children grasping concepts more quickly will start to slow down as they progress. Meanwhile, you have the chance to circulate and spot any misconceptions before they become barriers to further learning.

Homework and the role of carers

While *Power Maths* does not prescribe any particular homework structure, we acknowledge the potential value of practice at home. For example, practising fluency in key facts, such as number bonds and times-tables, is an ideal homework task, and carers could work through uncompleted Practice Book questions with children at either primary stage.

However, it is important to recognise that many parents and carers may themselves lack confidence in maths, and few, if any, will be familiar with mastery methods. A Parents' and Carers' Evening that helps them understand the basics of mindsets, mastery and mathematical language is a great way to ensure that children benefit from their homework. It could be a fun opportunity for children to teach their families that everyone can do maths!

Structures and representations

Unlike most other subjects, maths comprises a wide array of abstract concepts – and that is why children and adults so often find it difficult. By taking a Concrete-Pictorial-Abstract (C-P-A) approach, *Power Maths* allows children to tackle concepts in a tangible and more comfortable way.

Non-linear stages

Concrete

Replacing the traditional approach of a teacher working through a problem in front of the class, the concrete stage introduces real objects that children can use to 'do' the maths – any familiar object that a child can manipulate and move to help bring the maths to life. It is important to appreciate, however, that children must always understand the link between models and the objects they represent. For example, children need to first understand that three cakes could be represented by three pretend cakes, and then by three counters or bricks. Frequent practice helps consolidate this essential insight. Although they can be used at any time, good concrete models are an essential first step in understanding.

Pictorial

This stage uses pictorial representations of objects to let children 'see' what particular maths problems look like. It helps them make connections between the concrete and pictorial representations and the abstract maths concept. Children can also create or view a pictorial representation together, enabling discussion and comparisons. The *Power Maths* teaching tools are fantastic for this learning stage, and bar modelling is invaluable for problem solving throughout the primary curriculum.

Abstract

Our ultimate goal is for children to understand abstract mathematical concepts, signs and notation and, of course, some children will reach this stage far more quickly than others. To work with abstract concepts, a child needs to be comfortable with the meaning of, and relationships between, concrete, pictorial and abstract models and representations. The C-P-A approach is not linear, and children may need different types of models at different times. However, when a child demonstrates with concrete models and pictorial representations that they have grasped a concept, we can be confident that they are ready to explore or model it with abstract signs such as numbers and notation.

Use at any time and with any age to support understanding.

Practical aspects of *Power Maths*

One of the key underlying elements of *Power Maths* is its practical approach, allowing you to make maths real and relevant to your children, no matter their age.

Manipulatives are essential resources for both key stages and *Power Maths* encourages teachers to use these at every opportunity, and to continue the Concrete-Pictorial-Abstract approach right through to Year 6.

The Textbooks and Teacher Guides include lots of opportunities for teaching in a practical way to show children what maths means in real life.

Discover and Share

The Discover and Share sections of the Textbook give you scope to turn a real-life scenario into a practical and hands-on section of the lesson. Use these sections as inspiration to get active in the classroom. Where appropriate, use the Discover contexts as a springboard for your own examples that have particular resonance for your children – and allow them to get their hands dirty trying out the mathematics for themselves.

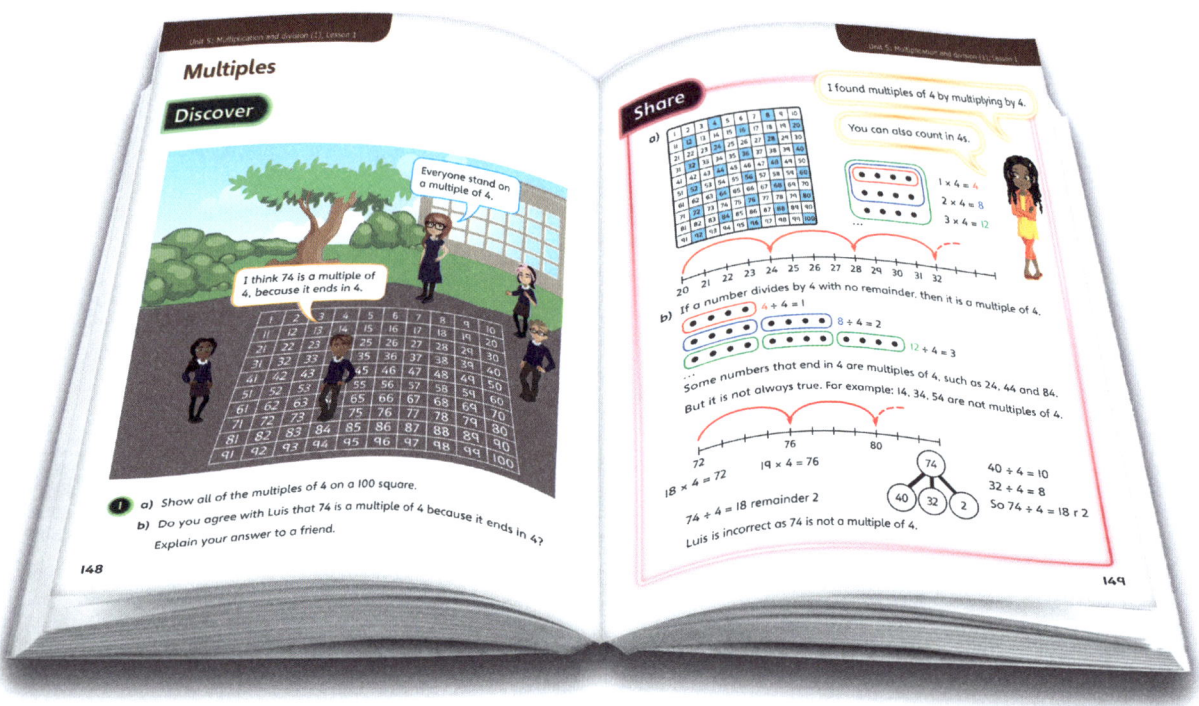

Unit videos

Every unit has a video which incorporates real-life classroom sequences.

These videos show you how the reasoning behind mathematics can be carried out in a practical manner by showing real children using various concrete and pictorial methods to come to the solution. You can see how using these practical models, such as part-whole and bar models, helps them to find and articulate their answer.

Mastery tips

Mastery Experts give anecdotal advice on where they have used hands-on and real-life elements to inspire their children.

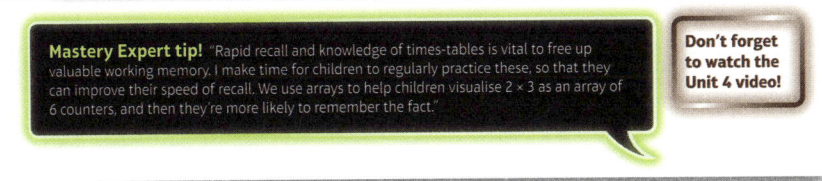

Mastery Expert tip! "Rapid recall and knowledge of times-tables is vital to free up valuable working memory. I make time for children to regularly practice these, so that they can improve their speed of recall. We use arrays to help children visualise 2 × 3 as an array of 6 counters, and then they're more likely to remember the fact."

Don't forget to watch the Unit 4 video!

Concrete-Pictorial-Abstract (C-P-A) approach

Each Share section uses various methods to explain an answer, helping children to access abstract concepts by using concrete tools, such as counters. Remember this isn't a linear process, so even children who appear confident using the more abstract method can deepen their knowledge by exploring the concrete representations. Encourage children to use all three methods to really solidify their understanding of a concept.

Pictorial representation – drawing the problem in a logical way that helps children visualise the maths

Concrete representation – using manipulatives to represent the problem. Encourage children to physically use resources to explore the maths.

Abstract representation – using words and calculations to represent the problem.

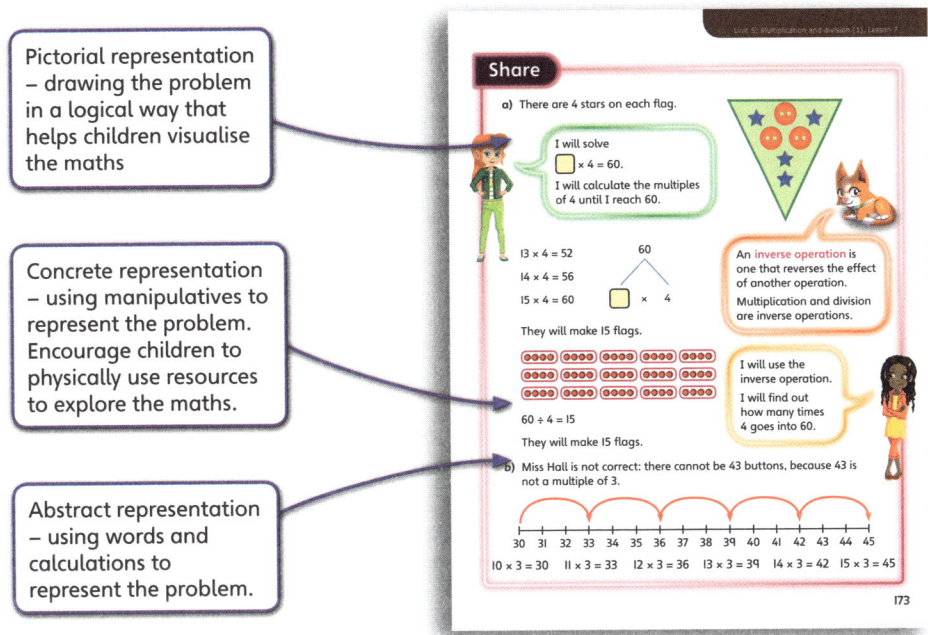

Practical tips

Every lesson suggests how to draw out the practical side of the Discover context.

You'll find these in the Discover section of the Teacher Guide for each lesson.

PRACTICAL TIPS You could use balls, counters or cubes under plastic cups to re-enact the artwork and help children get a feel for this activity.

Resources

Every lesson lists the practical resources you will need or might want to use. There is also a summary of all of the resources used throughout the term on page 34 to help you be prepared.

RESOURCES

Mandatory: cubes, counters, number lines
Optional: balls, plastic cups

List of practical resources

Year 6A Mandatory resources

Resource	Lesson
Coordinate grids with the first quadrant	**Unit 6** lesson 1
Coordinate grids with all four quadrants	**Unit 6** lessons 2, 3, 4
Counters	**Unit 3** lesson 4
Paper plates	**Unit 5** lesson 4
Paper strips	**Unit 5** lesson 4
Place value grids and place value counters	**Unit 1** lessons 2, 6

Year 6A Optional resources

Resource	Lesson
100 square	**Unit 2** lesson 6 **Unit 3** lessons 4, 9
Apples	**Unit 5** lesson 8
Base 10 equipment	**Unit 1** lessons 1, 3 **Unit 2** lessons 1, 2 **Unit 3** lessons 6, 7
Baskets	**Unit 5** lesson 8
Bead strings	**Unit 1** lesson 4 **Unit 3** lessons 5, 6
Bean bags	**Unit 3** lesson 2
Bingo game	**Unit 3** lesson 8
Card	**Unit 3** lesson 6
Concrete materials to represent length of bamboo (e.g. sticks, wool, ribbon, card, cubes)	**Unit 5** lesson 6
Cooking cups	**Unit 4** lesson 6
Counters	**Unit 1** lessons 3, 5, 6 **Unit 2** lessons 1, 2, 3, 4, 5, 6 **Unit 3** lessons 1, 3, 5, 6 **Unit 5** lessons 3, 5, 6, 8, 9 **Unit 6** lessons 1, 4
Digit cards	**Unit 1** lessons 1, 4, 5 **Unit 3** lesson 9
Flash cards	**Unit 1** lesson 1
'Follow Me' cards	**Unit 3** lessons 1, 8
Fraction circles	**Unit 5** lesson 1
Fraction strips	**Unit 4** lesson 6 **Unit 5** lessons 1, 6
Fraction walls	**Unit 5** lesson 3
Hoops	**Unit 3** lesson 2
Large paper	**Unit 3** lesson 6
Matchsticks	**Unit 6** lessons 1, 2, 4
Mini whiteboards	**Unit 2** lessons 8, 9, 10
Mirror	**Unit 6** lesson 3

Resource	Lesson
Multilink cubes	**Unit 3** lessons 4, 5, 6 **Unit 5** lesson 9
Multiplication grids	**Unit 3** lessons 1, 2, 4, 9
Multiplication square	**Unit 2** lesson 6
Number lines	**Unit 1** lessons 2, 7 **Unit 3** lesson 8 **Unit 4** lessons 5, 6, 9
Number lines with fractional divisions	**Unit 4** lesson 3
Paper circles	**Unit 5** lessons 4, 5, 6
Paper squares	**Unit 5** lesson 4, 6
Paper strips	**Unit 5** lessons 5, 6, 8
Paper plates	**Unit 5** lesson 2
Pictures of pandas	**Unit 5** lesson 6
Place value cards	**Unit 1** lessons 1, 3, 4
Place value counters	**Unit 1** lesson 3 **Unit 2** lessons 1, 2, 3, 4, 5, 6
Place value grids	**Unit 1** lessons 3, 5, 6 **Unit 2** lessons 1, 2
Play coins	**Unit 1** lessons 3 **Unit 5** lesson 9
Real house sale advertisements	**Unit 3** lesson 8
Square templates	**Unit 5** lessons 2, 3
Tens frames and counters	**Unit 3** lesson 5
Whiteboard pens	**Unit 5** lessons 5, 6

Variation helps visualisation

Children find it much easier to visualise and grasp concepts if they see them presented in a number of ways, so be prepared to offer and encourage many different representations.

For example, the number six could be represented in various ways:

Getting started with *Power Maths*

As you prepare to put *Power Maths* into action, you might find the tips and advice below helpful.

STEP 1: Train up!

A practical, up-front, full-day professional development course will give you and your team a brilliant head-start as you begin your *Power Maths* journey. You will learn more about the ethos, how it works and why.

STEP 2: Check out the progression

Take a look at the yearly and termly overviews. Next take a look at the unit overview for the unit you are about to teach in your Teacher Guide, remembering that you can match your lessons and pacing to your class.

STEP 3: Explore the context

Take a little time to look at the context for this unit: what are the implications for the unit ahead? (Think about key language, common misunderstandings and intervention strategies, for example.) If you have the online subscription, don't forget to watch the corresponding unit video.

STEP 4: Prepare for your first lesson

Familiarise yourself with the objectives, essential questions to ask and the resources you will need. The Teacher Guide offers tips, ideas and guidance on individual lessons to help you anticipate children's misconceptions and challenge those who are ready to think more deeply.

STEP 5: Teach and reflect

Deliver your lesson – and enjoy!

Afterwards, reflect on how it went … Did you cover all five stages?
Does the lesson need more time? How could you improve it?
What percentage of your class do you think mastered the concept?
How can you help those that didn't?

Unit I
Place value within 10,000,000

Mastery Expert tip! "I made sure children each had their own place value grid they could access throughout this unit. It was really useful to ensure children could independently secure their understanding of any new numbers they met."

Don't forget to watch the Unit 1 video!

WHY THIS UNIT IS IMPORTANT

This unit develops children's understanding of place value and properties of numbers up to 10,000,000. It is an important unit as the number sense they develop now will support their learning in future units.

Children will recap their understanding of place value and properties of numbers to 1,000,000 before investigating the same properties of numbers up to 10,000,000. They will investigate the partitioning of larger numbers and will use them in different contexts. Following this, children will develop their understanding and use of number lines up to 10,000,000 and will plot numbers on to partially completed number lines.

Having developed their understanding of the properties of numbers up to 10,000,000, children will use this to compare and order numbers and to round them to any degree of accuracy up to the nearest 1,000,000. Finally, children will investigate negative numbers, how they compare to positive numbers and their use in contexts.

WHERE THIS UNIT FITS

→ **Unit 1: Place value within 10,000,000**

→ Unit 2: Four operations (1)

In this unit, children extend their knowledge of numbers from within 1,000,000 to within 10,000,000, before they go on to work with the four operations in the next two units. This includes looking at place value, ordering and comparing numbers and rounding. They will also look at number lines and negative numbers. Before they start this unit, it is expected that children understand the place value of numbers within 1,000,000, can use number lines, including counting in 10s, 100s, 1,000s and 10,000s and can round numbers within 1,000,000.

ASSESSING MASTERY

Children will demonstrate fluent understanding of place value within numbers up to 10,000,000. They will be able to read, write and partition numbers accurately and use this, and their understanding of place value, to compare and order numbers up to 10,000,000. They will demonstrate confident fluency using number lines and will be able to complete number lines using unlabelled intervals to calculate what each interval is worth, as well as calculating intervals across 0. Children will be able to use their understanding of place value to round to powers of 10, up to 1,000,000, and will be able to recognise and use negative numbers in real-life contexts.

COMMON MISCONCEPTIONS	STRENGTHENING UNDERSTANDING	GOING DEEPER
Children may assume that the only way to partition a number is into its place value headings (for example, only partitioning 136 as 100, 30 and 6).	Give children opportunities to create numbers using different representations, such as base 10 equipment, which they can then partition in different ways.	Children could be given, or create their own, missing number equations. For example: $1,456 = 500 + 500 + \_\_\_ + 156$
Children may assume that negative numbers work in a similar way to positive numbers. For example, thinking that ‾5 must be greater than ‾1 because 5 is greater than 1.	Provide a number line from ‾10 up to 10. Place the correct number of objects above the positive numbers. Ask: • *What do you notice about the objects as you move down to 0? What will happen to the numbers below 0?*	Encourage children to count down from a given positive number. Ask: • *What do you notice about the positive numbers as you count down? What will happen with the negative numbers?*

Unit I: Place value within 10,000,000

PUPIL TEXTBOOK 6A PAGE 6

WAYS OF WORKING

Use these pages to introduce the focus to children. You can use the characters to explore different ways of working too.

STRUCTURES AND REPRESENTATIONS

Place value grid: Place value grids are used with both counters and numbers in this unit to help children read numbers and recognise the value of each digit in numbers up to 10,000,000.

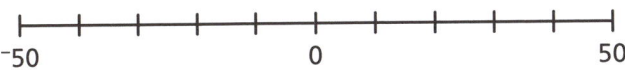

M	HTh	TTh	Th	H	T	O
4	5	9	0	I	2	4

Part-whole model: Part-whole models are used to help children partition numbers.

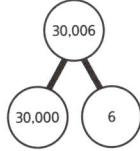

Number line: Number lines are used to help children plot numbers from 0 to 10,000,000, work out differences and round numbers. Later in the unit, they are used to show negatives and work out intervals across 0.

```
+---+---+---+---+---+---+---+---+---+---+
-50              0                    50
```

Bar chart: A bar chart is used in this unit to give context to children's learning about number.

Bar model: A bar model is used to help children understand how to work out the value of unlabelled intervals on a number line.

KEY LANGUAGE

There is some key language that children will need to know as a part of the learning in this unit.

→ ones (1s), tens (10s), hundreds (100s), thousands (1,000s), ten thousands (10,000s), hundred thousands (100,000s), millions (1,000,000s), ten million (10,000,000)

→ place value

→ partition/partitioned/partitioning

→ interval

→ estimate

→ compare/comparison/comparing

→ order/ordering

→ less than (<), greater than (>), equal to (=)

→ rounding/rounded/round up/round down/rounds

→ negative, positive

→ odd, even

→ accurate/accurately, exactly, approximately

Unit I
Place value within 10,000,000

In this unit we will ...
- ⚡ Learn to read and write numbers to 10,000,000
- ⚡ Partition, compare and order numbers up to 10,000,000
- ⚡ Round numbers
- ⚡ Work with negative numbers

Do you remember what this is called? We will use it to help identify the place value of digits in a number.

M	HTh	TTh	Th	H	T	O
I	0	0	0	0	0	0

6

PUPIL TEXTBOOK 6A PAGE 6

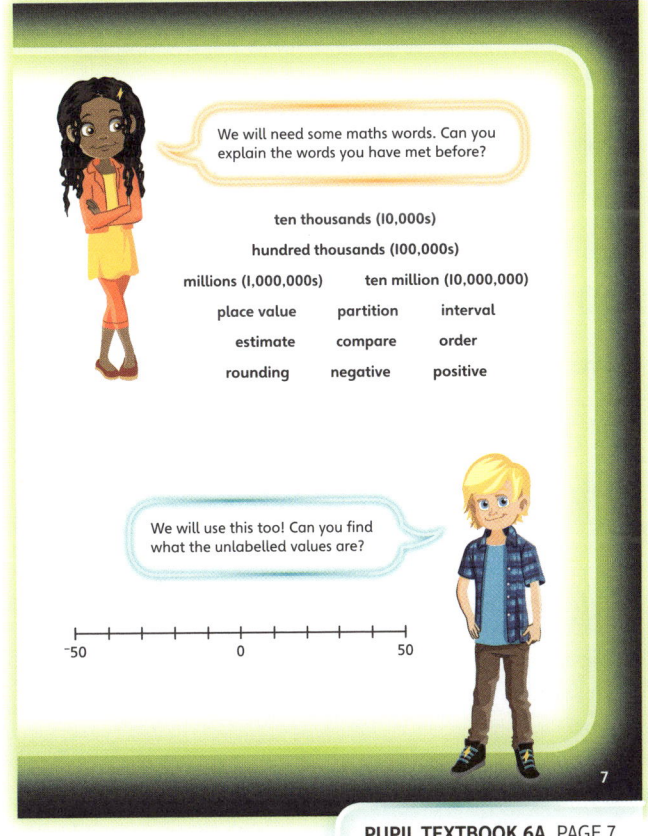

We will need some maths words. Can you explain the words you have met before?

ten thousands (10,000s)

hundred thousands (100,000s)

millions (1,000,000s)   ten million (10,000,000)

place value   partition   interval

estimate   compare   order

rounding   negative   positive

We will use this too! Can you find what the unlabelled values are?

```
+---+---+---+---+---+---+---+
-50          0            50
```

7

PUPIL TEXTBOOK 6A PAGE 7

Numbers to 1,000,000

Learning focus

In this lesson, children will learn about the place value of numbers to 1,000,000. They will learn to read and write these numbers fluently and identify their place value.

Small steps

→ **This step: Numbers to 1,000,000**
→ Next step: Numbers to 10,000,000 (1)

NATIONAL CURRICULUM LINKS

Year 6 Number – Number and Place Value
- Read, write, order and compare numbers up to 10,000,000 and determine the value of each digit.
- Solve number and practical problems that involve all of the above.

ASSESSING MASTERY

Children can recognise, read and write numbers up to 1,000,000. They can use what they know about place value to describe the value of each digit in a number and can solve problems using this understanding.

COMMON MISCONCEPTIONS

Children may neglect to include 0s as place holders in numbers when writing them. For example, they may write 132,045 as 13,245. Show children both numbers and ask:
- *Can you build both of these numbers? What is the same and what is different about them?*
- *What is the value of each digit in each number? Why is the 0 interesting?*

STRENGTHENING UNDERSTANDING

As children may misspell the mathematical vocabulary when writing the names of numbers, it would be beneficial to have the vocabulary displayed on the classroom wall. For children who may struggle with numbers up to 1,000,000, give them opportunities to investigate the place value of numbers up to 100,000 before the lesson using base 10 equipment and place value cards.

GOING DEEPER

Children can be encouraged to design their own hidden number challenges. Ask: *Can you create a set of clues where there are many possible answers? Can you create a set of clues where only one solution is possible? Can you write each of your clues so they require a different mathematical skill to solve?*

KEY LANGUAGE

In lesson: ones (1s), tens (10s), hundreds (100s), thousands (1,000s), ten thousands (10,000s), hundred thousands (100,000s), odd, place value, multiple

Other language to be used by the teacher: million (1,000,000), even

STRUCTURES AND REPRESENTATIONS

place value grid, part-whole model, number line, table

RESOURCES

Optional: base 10 equipment, place value cards, digit cards, flash cards

 In the eTextbook of this lesson, you will find interactive links to a selection of teaching tools.

Before you teach

- Are children confident with the place value of numbers up to 100,000?
- Are children more confident reading numbers than writing them? How will you support children who find writing and spelling challenging?

Discover

WAYS OF WORKING Pair work

ASK

• Question ❶ a): *What do you know about the number Richard has made?*
• Question ❶ a): *Is there more than one solution? How can you prove it?*
• Question ❶ b): *What can you say is true and not true about Lexi's number?*

IN FOCUS Question ❶ a) allows children the opportunity to reason with numbers. They should be able to recognise that there is more than one possible solution to the clues given. In contrast, question ❶ b) narrows the focus, as it has only one solution. Ensure children are able to use their knowledge of number, and the clues given, to find the mystery number.

PRACTICAL TIPS The game in the picture can be repeated easily in the classroom using digit cards. Encourage children to set similar challenges for each other.

ANSWERS

Question ❶ a): There are six possible solutions:
627,489; 628,479; 726,489; 728,469; 629,487; 628,497.

Question ❶ b): Lexi has made the number 96,287.

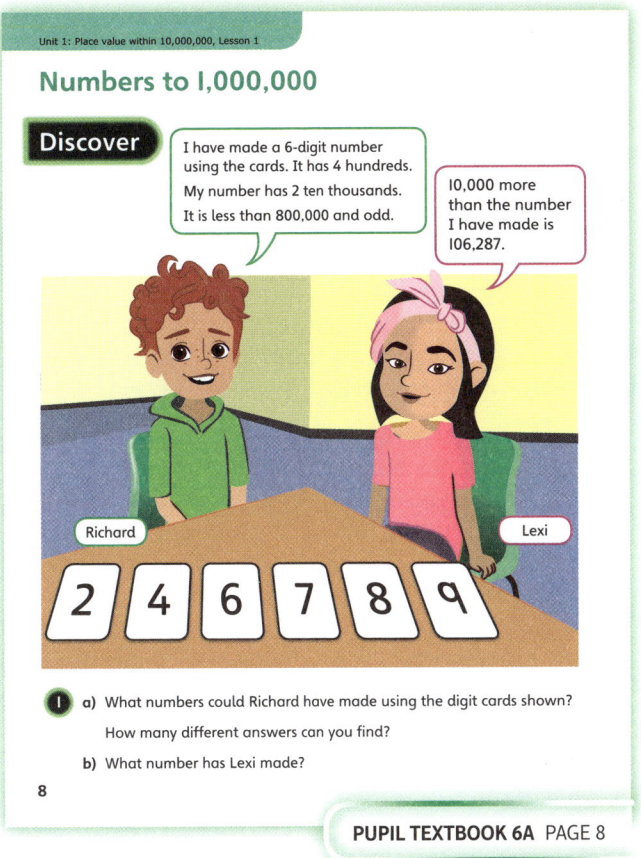

PUPIL TEXTBOOK 6A PAGE 8

Share

WAYS OF WORKING Whole class teacher led

ASK

• Questions ❶ a) and ❶ b): *How does the place value grid help you to solve this problem?*
• Question ❶ a): *How many solutions did you find?*
• Question ❶ b): *What is each digit worth in the solution?*

IN FOCUS Question ❶ a) will help children to recap the place values they have met before. It will also develop their reasoning, showing how to make progress through the problem given in the **Discover** section. You can use Flo's comment to spark discussion about the different methods children used when solving the problem.

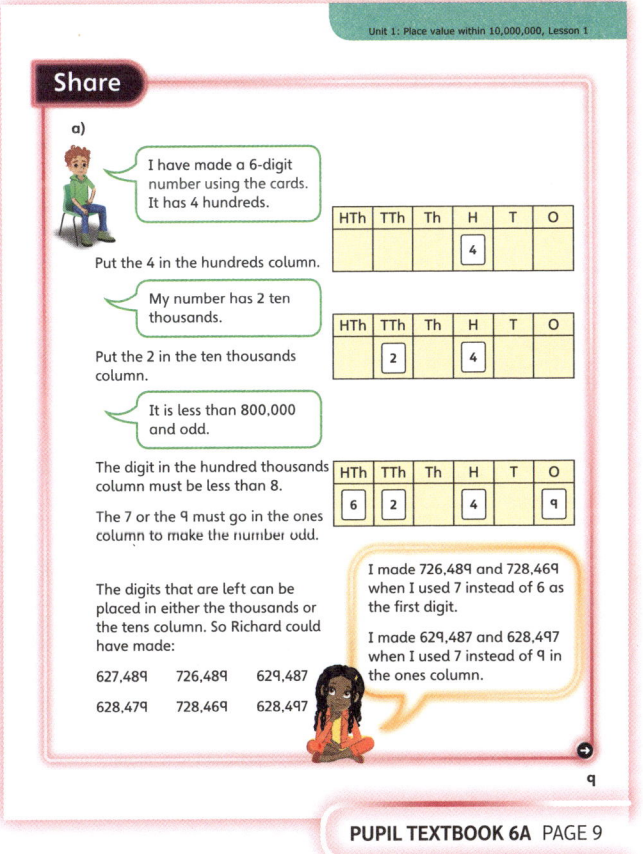

PUPIL TEXTBOOK 6A PAGE 9

Think together

Whole class teacher led (I do, We do, You do)

ASK

- Question **1**: *How do you know how to read each number?*
- Question **1**: *How can you tell what to write?*
- Question **2**: *Do the place value columns help?*
- Question **3**: *How can you begin to work this out?*
- Question **3**: *Using the digit cards, what other numbers can you make that will appear on the number line? Prove it.*

IN FOCUS In question **1** children practise saying and writing different numbers up to 1,000,000. Watch for their responses when reading and writing numbers that contain place-holding 0s to ensure they do not omit those digits or confuse the place value of neighbouring digits. In question **2**, children practise their recognition of the place value of specific digits in given numbers. Question **3** encourages children to begin ordering and comparing numbers up to 1,000,000. Observe carefully the numbers children create and which they choose as their solution, as this will give a good idea of their understanding of place value.

STRENGTHEN In question **1**, if children are struggling, it may help to provide flash cards showing pictorial representations of the different place values.

DEEPEN While solving question **2**, deepen children's reasoning about the place value of the digits by asking:
- *How can you prove that your answers are correct?*
- *What representations could you use to prove your answers?*

ASSESSMENT CHECKPOINT Question **1** assesses children's ability to read and write numbers up to 1,000,000, while questions **2** and **3** assess their understanding of place value. Question **3** also assesses whether children are able to compare and order numbers on a number line. In question **3**, look for children using reasoning to make sensible estimates. For example, children should be able to recognise that the number will be between 180,000 and 190,000. They may then identify that, as it is lower than half-way, the first three digits will be 182 or 184.

ANSWERS

Question **1**: Children should say the numbers aloud.
32,567 – thirty-two thousand, five hundred and sixty-seven
491,062 – four hundred and ninety-one thousand and sixty-two
295,000 – two hundred and ninety-five thousand
30,006 – thirty thousand and six

Question **2** a): 50,000 or 5 ten thousands

Question **2** b): 5 or 5 ones

Question **2** c): 500 or 5 hundreds and 5 or 5 ones

Question **2** d): 500,000 or 5 hundred thousands

Question **3**: Lexi has made a 6-digit number starting with 182, for example 182,904, 182,094 or 182,409. Also accept 184,901 and 184,910.

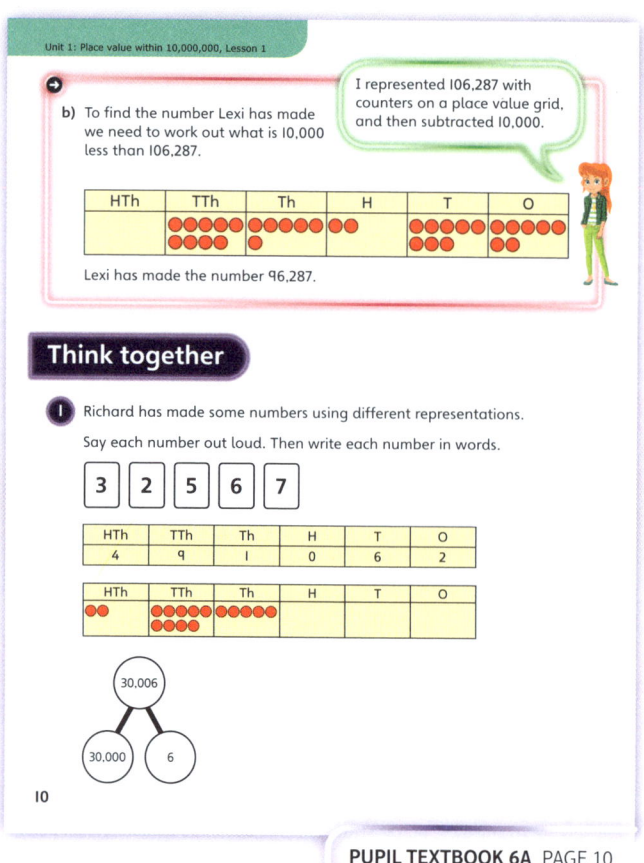

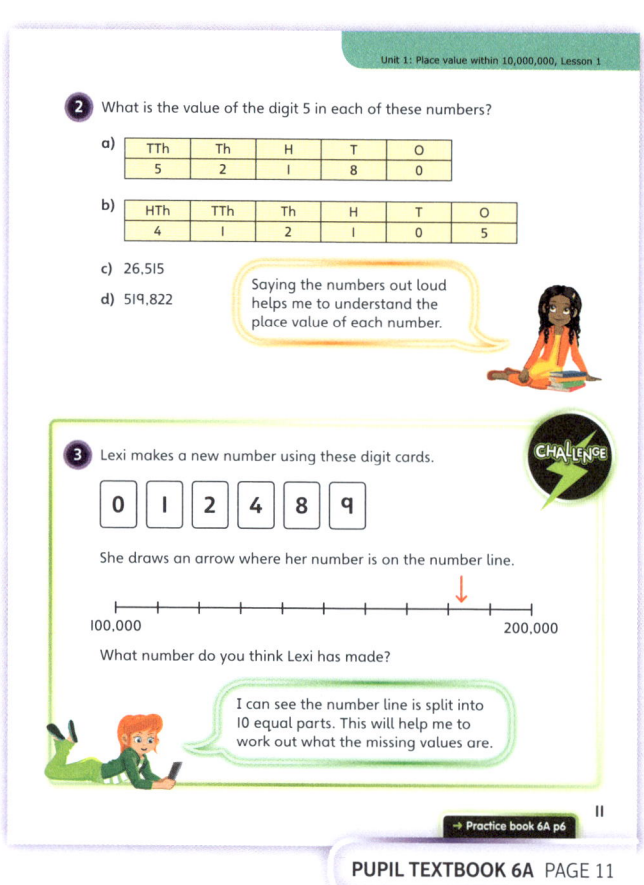

Practice

WAYS OF WORKING Independent thinking

IN FOCUS Question ① scaffolds children's independent understanding of, and ability to recognise, representations of the place value of different numbers up to 1,000,000. Question ② is important, as it helps to build the link between two abstract ways of writing numbers. This allows you to check that children are not only able to read the numbers written as words, but also understand them in order to rewrite them in digits. In questions ④ and ⑤, children then begin using their knowledge to reason and solve problems and puzzles about numbers.

STRENGTHEN In question ⑥, if children are struggling to count on or back, ask:
- *What resources could you use to represent the numbers?*
- *How could you use them to find x more or x less than the given number?*

DEEPEN Question ⑦ deepens children's understanding of the properties of numbers by including clues that draw on other areas of mathematics. Children should be encouraged to give justifications for their solutions, potentially using resources or pictures as supporting evidence.

ASSESSMENT CHECKPOINT Children should understand the place value of different digits and be able to show this in multiple ways. In question ③, children can be encouraged to justify how they know their answers are correct, in order to ensure their full understanding. Question ⑤ assesses children's ability to compare and order numbers up to 1,000,000. Look for their ability to recognise patterns in the numbers along a number line and use this to complete the unlabelled intervals. They should then be able to use their understanding of place value and number to sensibly estimate where the given number would fall between two intervals.

ANSWERS Answers for the **Practice** part of the lesson appear in the separate **Practice and Reflect answer guide**.

Reflect

WAYS OF WORKING Independent thinking and pair work

IN FOCUS This question requires children to demonstrate their fluency with place value and properties of numbers up to 1,000,000. By comparing their facts about the number and discussing any differences in their ideas, children should strengthen their confidence and identify any remaining uncertainties.

ASSESSMENT CHECKPOINT Children should be able to identify the place value of each digit in the number. They may also be able to offer suggestions such as '10,000 more is …', '100 less is …' and so on.

ANSWERS Answers for the **Reflect** part of the lesson appear in the separate **Practice and Reflect answer guide**.

After the lesson

- How confident and fluent were children at reading and writing numbers up to 1,000,000?
- Was this lesson as practical and hands on as it could have been? If not, how will you develop this next time you teach it?

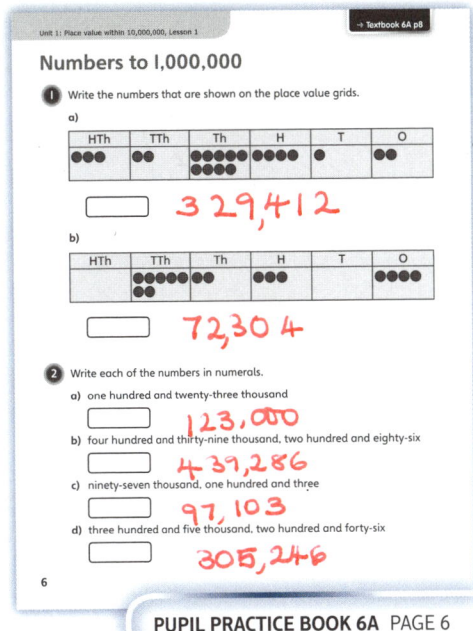

PUPIL PRACTICE BOOK 6A PAGE 6

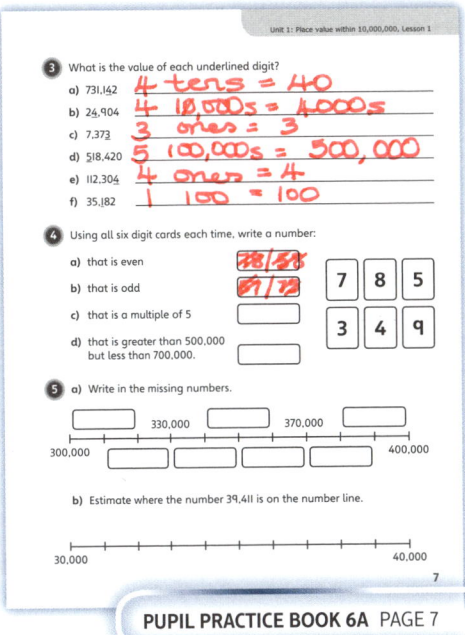

PUPIL PRACTICE BOOK 6A PAGE 7

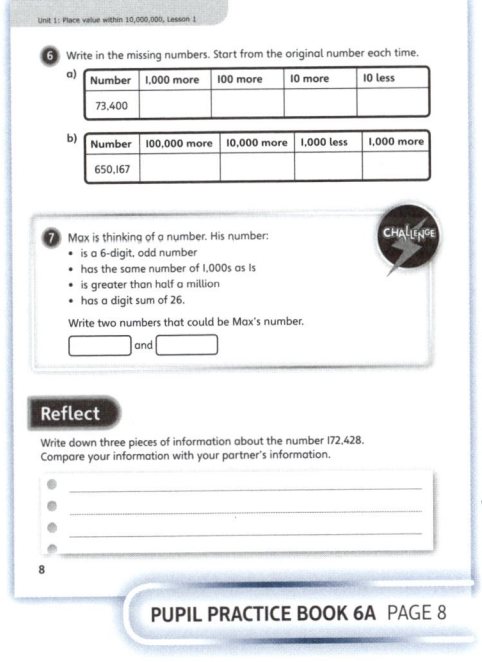

PUPIL PRACTICE BOOK 6A PAGE 8

Numbers to 10,000,000 ①

Learning focus

In this lesson, children will learn about the place value of numbers to 10,000,000. They will learn to read and write these numbers fluently and identify their place value.

Small steps

→ Previous step: Numbers to 1,000,000
→ **This step: Numbers to 10,000,000 (1)**
→ Next step: Numbers to 10,000,000 (2)

NATIONAL CURRICULUM LINKS

Year 6 Number – Number and Place Value
• Read, write, order and compare numbers up to 10,000,000 and determine the value of each digit.
• Solve number and practical problems that involve all of the above.

ASSESSING MASTERY

Children can recognise, describe, read and write numbers up to 10,000,000. They can use their knowledge of place value to describe the value of each digit in a number and can solve problems using this understanding.

COMMON MISCONCEPTIONS

As more digits appear in the numbers children are studying, it is increasingly likely that they may get muddled when they read the numbers, especially when there are place-holding 0s. Ask:
• *How might a place value grid help you to read this number accurately? Can you show me?*

STRENGTHENING UNDERSTANDING

Encourage children to count in steps of 100,000 up to 1,000,000 to help secure their fluency with the pattern of the numbers before they start looking at numbers up to 10,000,000 in this lesson. Also encourage them to write the name of each number in the count in words and numerals. This counting should start at different given numbers, not only 0. A wall display of vocabulary, as in the previous lesson, may also be useful.

GOING DEEPER

Children can use their knowledge of the numbers they have been studying to create 'guess my number' challenges with numbers up to 10,000,000, giving a partner a list of word clues, for example:
• *I have seven millions, thirty-eight thousands, three tens and no ones. In numerals, write what number I am.*

KEY LANGUAGE

In lesson: hundreds (100s), thousands (1,000s), millions (1,000,000s), hundred thousands (100,000s)

Other language to be used by the teacher: ones (1s), tens (10s), ten thousands (10,000s), ten million (10,000,000), multiple, place value, odd, even

STRUCTURES AND REPRESENTATIONS

number line, place value grid

RESOURCES

Mandatory: place value grids, place value counters

Optional: number lines

 In the eTextbook of this lesson, you will find interactive links to a selection of teaching tools.

Before you teach

• Are children fluent at reading and writing numbers up to 1,000,000?

44

Discover

WAYS OF WORKING Pair work

ASK

- Question **1** a): *How can you prove how many 100,000s are in 1,000,000?*
- Question **1** b): *How many 100,000s are in 4,590,124? How would you show this in a place value grid?*

IN FOCUS Question **1** a) builds on the previous lesson by asking children to count up to 1,000,000 in 100,000s. They may find it helpful to use a number line. If children demonstrate they can do this confidently, encourage them to count beyond 1,000,000 by asking them to look at the price of the painting. Focus on the first two digits and ask them to continue counting in 100,000s until they reach £4,500,000. At this point, discuss what values the 4 and 5 have, before children move on to question **1** b).

PRACTICAL TIPS A fun way to begin the lesson would be to host your own mock auction. Give children a set amount of money they can spend. To avoid them going for broke on the first item, you could make it a challenge that they must buy as many items as they can. Once they have finished bidding, discuss the sale prices and compare the numbers.

ANSWERS

Question **1** a): There are ten 100,000s in one million.

Question **1** b): Four million, five hundred and ninety thousand, one hundred and twenty-four; two hundred and thirty-four thousand, five hundred.

M	HTh	TTh	Th	H	T	O
●●●●	●●●●●	●●●●● ●●●●	●	●●	●●	●●●●
4	5	9	0	1	2	4

M	HTh	TTh	Th	H	T	O
	●●	●●●	●●●●	●●●●●		
	2	3	4	5	0	0

Share

WAYS OF WORKING Whole class teacher led

ASK

- Question **1** a): *How did you prove how many 100,000s there are in 1,000,000?*
- Question **1** b): *How do the place value grids help you to read the numbers? What do the grids tell you?*

IN FOCUS Question **1** a) recaps children's understanding of place value and its link to multiples of 100,000. Continue this pattern of thought when children look at the place value grids in question **1** b) by asking them which digit is a multiple of 10, 100, 1,000 and so on. It is worth spending time exploring the place value grids in this way, as it encourages children to not only begin reading and writing numbers beyond 1,000,000 up to 10,000,000, but to also think about what each digit represents.

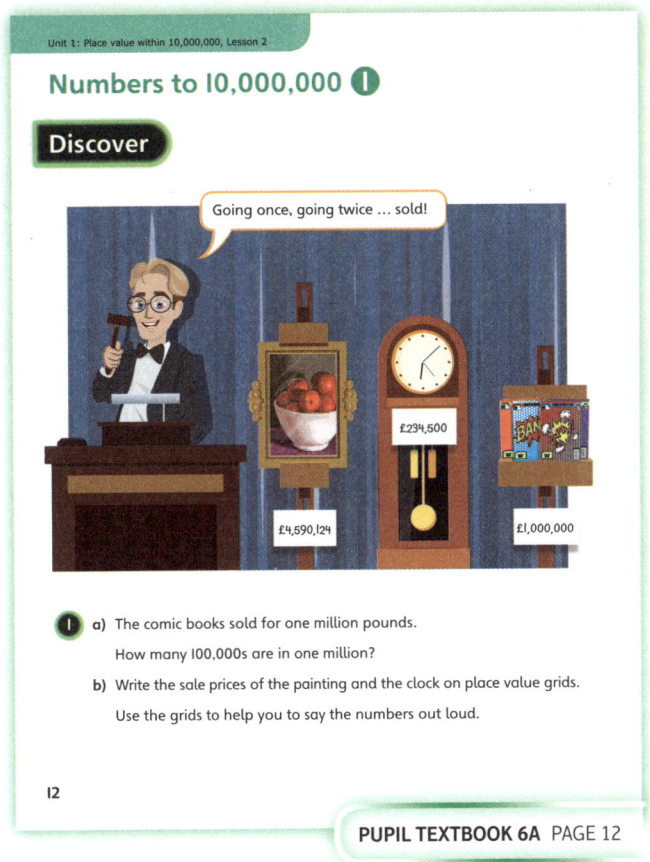

PUPIL TEXTBOOK 6A PAGE 12

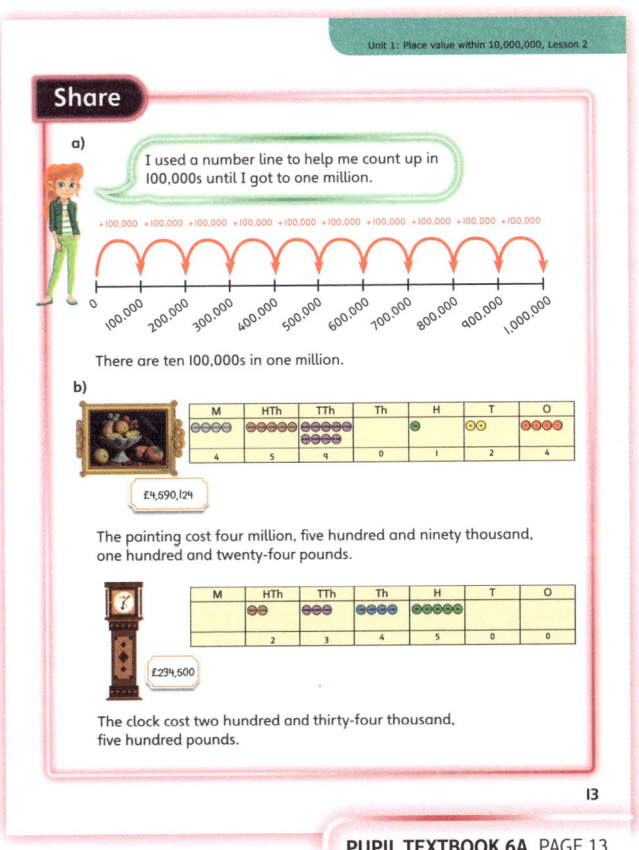

PUPIL TEXTBOOK 6A PAGE 13

Think together

WAYS OF WORKING Whole class teacher led (I do, We do, You do)

ASK

- Question **1**: *How can you prove how many 100,000s there are in 1,200,000?*
- Question **1**: *Is there a link between 100,000 and 1,200,000?*
- Question **2**: *What do you need to be careful with when writing a number in words?*
- Question **3**: *Is there a way of knowing which numbers can be made without needing to test it?*

IN FOCUS Question **1** develops children's understanding of place value and multiples of 100,000. Look for children beginning to make the link between these concepts and recognising the pattern in the numbers. Question **2** offers children the opportunity to practise writing numbers beyond 1,000,000 up to 10,000,000 in numerals and in words, while considering the meaning of place value. For question **3**, give children place value grids and place value counters so they can investigate the numbers using concrete resources. Use Flo's comment to prompt them to notice that the number of counters they need is the same as the sum of the digits.

STRENGTHEN While investigating the numbers in question **3** using place value grids and counters, ask:
- *How will you place the counters?*
- *What number would you make if you placed all your counters in the millions column? Explain.*
- *How can Ash and Flo's comments help you?*

DEEPEN Once children have solved question **3** b), deepen their reasoning by asking:
- *Has Lee listed all the possible numbers he can make or can you find any more?*
- *What patterns, similarities or differences can you find between the numbers?*

ASSESSMENT CHECKPOINT Question **1** assesses whether children can recognise place values in numbers up to 10,000,000 and recognise the link between multiples of 100,000 and 1,000,000s. In question **2**, look for children demonstrating confidence when writing numbers in numerals and in words. Pay particular attention to whether they remember to include place-holding 0s or if these lead to errors. Question **3** assesses children's fluency with place value grids and their ability to reason with numbers up to 10,000,000. It will highlight whether they are equally comfortable working with small numbers and large numbers.

ANSWERS

Question **1**: 12

Question **2** a): 462,305, four hundred and sixty-two thousand, three hundred and five

Question **2** b): 5,104,309, five million, one hundred and four thousand, three hundred and nine

Question **3** a): 2,411,301

Question **3** b): Lee can make 5,502 or 1,304,202 or 1,304,220.

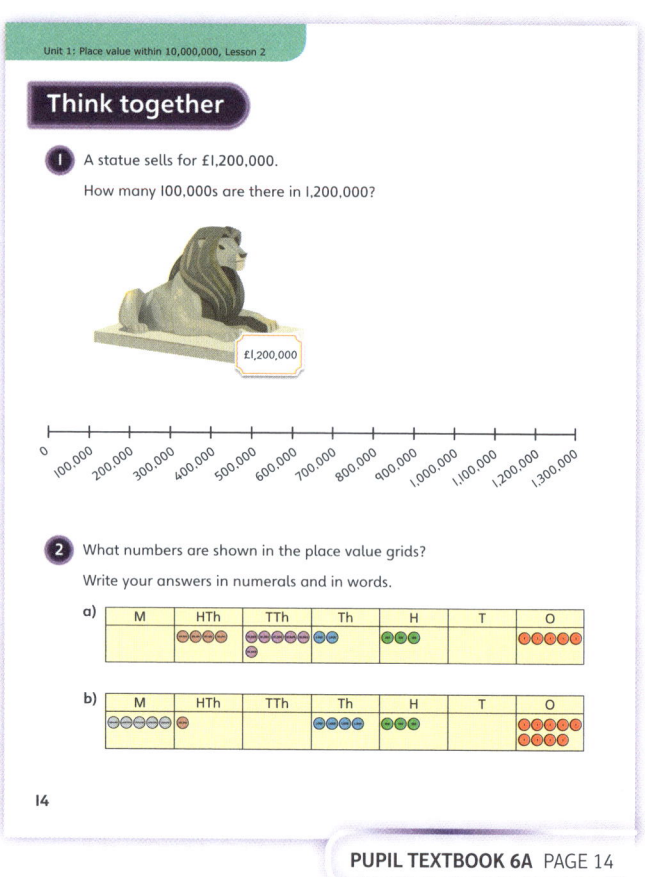

PUPIL TEXTBOOK 6A PAGE 14

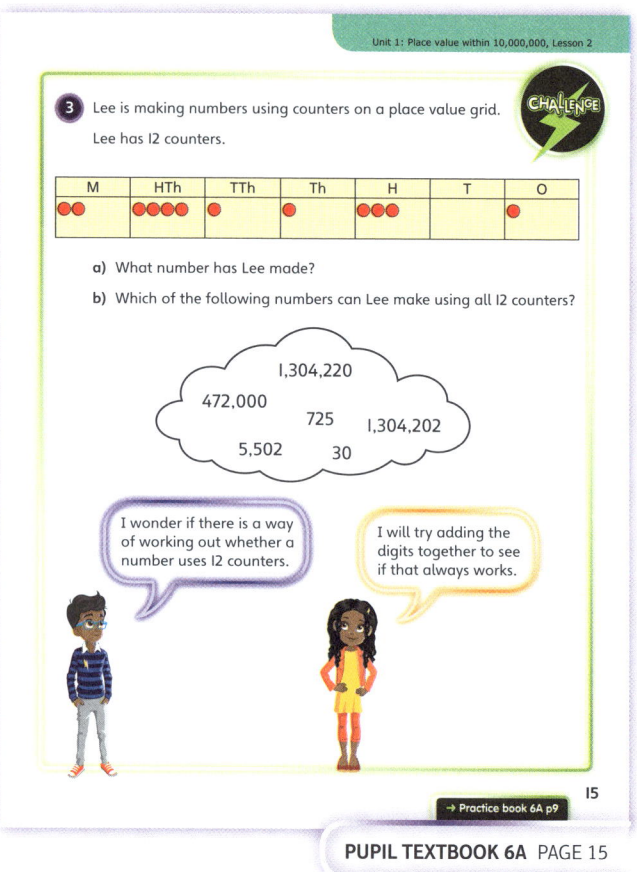

PUPIL TEXTBOOK 6A PAGE 15

Practice

IN FOCUS Question **1** offers the opportunity for children to independently count in 100,000s to 1,000,000 and above. This will help to secure their understanding of the relationship these numbers share regarding place value. Question **2** allows children to develop their ability to write numbers in both numerals and words using their understanding of pictorial representations. This is reversed in question **3**, where children have to draw pictorial representations, using the scaffolding of a place value grid to help secure their understanding of the individual digits in numbers. This scaffolding is removed in question **4**, when children work only in the abstract, writing numbers in numerals using only the written names. Question **5** gives children facts about digits, which they have to arrange to create a number. Look for children remembering to include place-holding 0s for the 100,000s and 10s.

STRENGTHEN In question **3**, if children struggle to draw the counters on the place value grids, it may be beneficial to give them real counters to manipulate without fear of making multiple mistakes in their drawings.

DEEPEN Question **6** gives children the opportunity to generalise using their knowledge of number and number patterns. Push their reasoning further by asking:
• *Can any digit be odd for it to be an odd number?*
• *Why must it be the 1s you look at to determine whether a number is odd or even?*
• *Can you prove your ideas with a picture?*

ASSESSMENT CHECKPOINT Question **1** assesses children's understanding of the values of multiples of 100,000 and how these link to 1,000,000s. It is important to observe children's understanding in question **2** to ensure they have secure knowledge of what each part of the place value grid is worth. At this point in the lesson, children should be confident and fluent when reading and writing numbers up to 10,000,000; this is further assessed in questions **3** and **4**.

ANSWERS Answers for the **Practice** part of the lesson appear in the separate **Practice and Reflect answer guide**.

Reflect

IN FOCUS The question requires children to demonstrate their knowledge and understanding of place value from ones to millions. Further draw out their understanding of the given number by asking them to draw it, say it, make it and write it.

ASSESSMENT CHECKPOINT Children should be able to correctly determine the place values of all digits and not be tripped up by the place-holding 0.

ANSWERS Answers for the **Reflect** part of the lesson appear in the separate **Practice and Reflect answer guide**.

After the lesson ⏸

• How confident were children with numbers in the millions?
• If some children still lack confidence, what other manipulatives could you use to help cement their conceptual understanding of numbers in the millions?

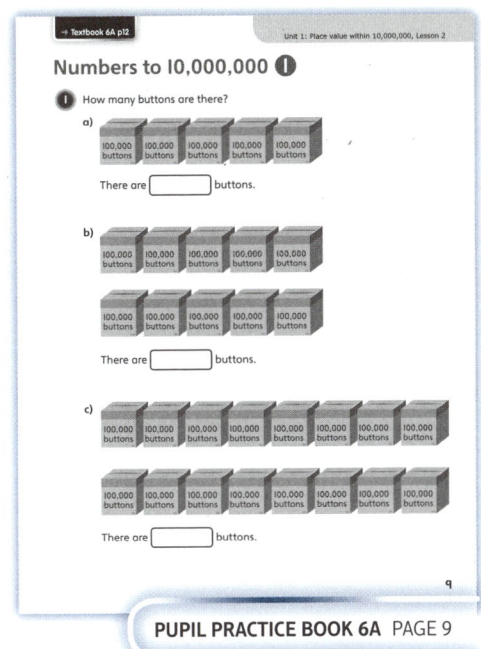

PUPIL PRACTICE BOOK 6A PAGE 9

PUPIL PRACTICE BOOK 6A PAGE 10

PUPIL PRACTICE BOOK 6A PAGE 11

Numbers to 10,000,000 ②

Learning focus

In this lesson, children will use their understanding of place value and numbers up to 10,000,000 to partition numbers and solve problems in real-life contexts.

Small steps

→ Previous step: Numbers to 10,000,000 (1)
→ **This step: Numbers to 10,000,000 (2)**
→ Next step: Number line to 10,000,000

NATIONAL CURRICULUM LINKS

Year 6 Number – Number and Place Value
- Read, write, order and compare numbers up to 10,000,000 and determine the value of each digit.
- Solve number and practical problems that involve all of the above.

ASSESSING MASTERY

Children can use their fluent understanding of the place value of numbers up to 10,000,000 to read, understand and solve problems with real-life contexts. They are able to explain the reasoning behind their solutions using the vocabulary and representations they have learnt about in the previous two lessons.

COMMON MISCONCEPTIONS

Children may assume that the only way to partition a number is into its place value headings (for example, 136 partitioned as 100, 30 and 6), which could lead to confusion in the Challenge questions in this lesson. Base 10 equipment or place value counters may help children to become more flexible in partitioning. Ask:
- *Is that the only way you can break this number into parts?*
- *Can you show me another way using a resource?*

STRENGTHENING UNDERSTANDING

Before starting the lesson, children can strengthen their fluency with the new numbers they have been learning about, and their place value, by counting up and down in different steps from a given number. Ask:
- *Can you count up in 10,000s from 360?*
- *Can you count down in 1,000,000s from 9,780,002?*

GOING DEEPER

Children can be encouraged to create their own missing number equalities to challenge their partner with. For example, 300,000 + _ + 5,000 + 34 = 376,034. Ask: *Can you write a question like this for your partner? How can you make your question easier or more challenging?*

KEY LANGUAGE

In lesson: ones (1s), tens (10s), hundreds (100s), thousands (1,000s), ten thousands (10,000s), hundred thousands (100,000s), millions (1,000,000s), partition/partitioned, place value

Other language to be used by the teacher: ten million (10,000,000)

STRUCTURES AND REPRESENTATIONS

place value grid, part-whole model

RESOURCES

Optional: base 10 equipment, place value grids, place value counters, place value cards, play coins

 In the eTextbook of this lesson, you will find interactive links to a selection of teaching tools.

Before you teach

- Were children able to reason confidently about place value within numbers up to 10,000,000 when working with them out of context?

Discover

ASK

- Questions **1** a) and **1** b): *What mathematical representation may be helpful when writing how much money each player has?*
- Questions **1** a) and **1** b): *Can you show the amounts with counters in a place value grid?*
- Questions **1** a) and **1** b): *Are there other ways to partition the numbers?*
- Question **1** b): *What do you notice about the number of 1,000s?*

IN FOCUS Both questions give children the opportunity to recognise the place value of numbers up to 10,000,000 when used in a real-life context and how these numbers can be partitioned. It is important to discuss how the mathematics changes and stays the same when numbers are used in context. Question **1** b) requires children to recognise where the denominations that Aki has will bridge into the next place value column. In this example, the eleven £1,000 notes bridge into the 10,000s.

PRACTICAL TIPS Children could be given toy money to investigate the numbers with. How many different ways can they make Jamie and Aki's amounts using different notes?

ANSWERS

Question **1** a): Jamie has £4,520,123.

Question **1** b): Aki has £2,071,000.

Numbers to 10,000,000 ❷

Discover

1 a) In the game, how much money does Jamie have?

   b) How much money does Aki have?

16

PUPIL TEXTBOOK 6A PAGE 16

Share

ASK

- Questions **1** a) and **1** b): *Did you choose to use place value grids? Explain why or why not.*
- Questions **1** a) and **1** b): *How do the place value grids help to make writing the numbers easier?*
- Questions **1** a) and **1** b): *What is similar or different about the two numbers in the place value grids?*

IN FOCUS Use this opportunity to make sure children are ready to apply their mathematical knowledge and understanding to contextual scenarios by discussing what mathematical information has been used from the **Discover** picture and how it has been manipulated as non-contextual numbers before being turned back into pounds.

In question **1** b), spend time exploring the bridging of 10 thousands into the ten thousands column, to ensure all children understand this before moving on.

DEEPEN If children are confident with the place value grids, challenge them to write the amounts of money using part-whole models. Ask:

- *Is the part-whole model clearer than the place value grid? Explain your ideas.*

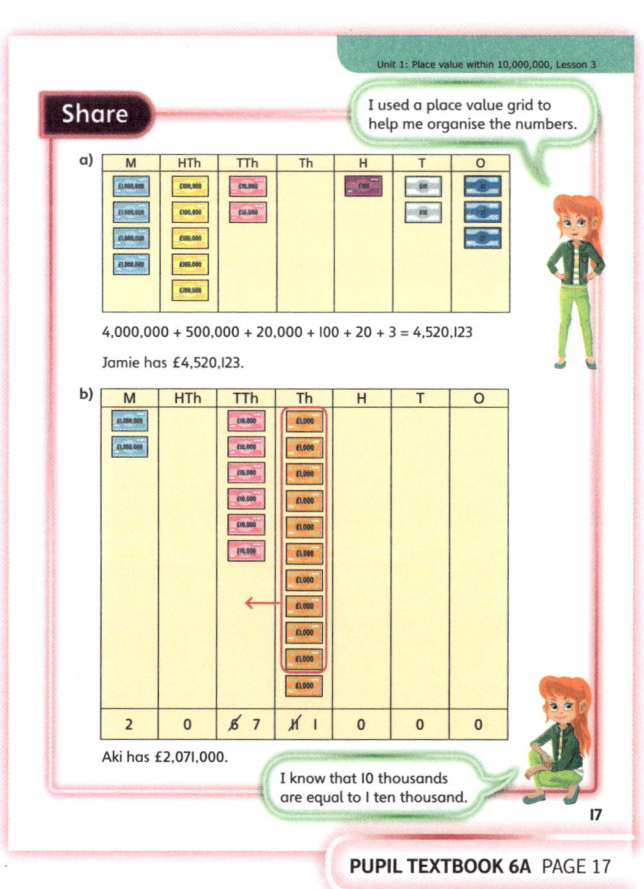

PUPIL TEXTBOOK 6A PAGE 17

Think together

WAYS OF WORKING Whole class teacher led (I do, We do, You do)

ASK

- Question **1**: *How can you use the representations to work out how much money there is?*
- Question **2**: *How can you prove your part-whole models are completed accurately?*
- Question **3**: *Is it always necessary to use a place value grid? Explain your ideas.*

IN FOCUS Questions **1** and **2** give children the opportunity to practise partitioning numbers up to 10,000,000, both in a contextual scenario and using different representations, with the introduction of part-whole models in question **2**. Question **3** ensures children's fluency with partitioned numbers by removing pictorial representations, although children may find it helpful to draw their own. It also recaps their understanding of numbers when written as words.

STRENGTHEN In question **3**, if children are struggling to write the numbers that have been partitioned, it may help to provide place value cards to enable them to build each number. Ask:

- *Can you find the right cards to help build this number?*
- *What does the number look like once you have built it? Can you read it to me?*

DEEPEN When solving question **3**, children could be asked to demonstrate their deep understanding of all the concepts covered in this lesson and the previous one by asking:

- *Choose one of the numbers in the list. How many different ways can you represent that number and the place values of its digits?*
- *How many different concrete, pictorial and abstract ways do you know to represent the number?*

ASSESSMENT CHECKPOINT Questions **1** and **2** assess children's understanding of partitioning and the place values of digits in numbers up to 10,000,000. In question **3**, assess children's fluency with place value and different abstract ways of representing numbers up to 10,000,000 by seeing if they are confident working entirely in the abstract, if they use pictorial representations or if they require concrete resources for additional support.

ANSWERS

Question **1**: £276,302

Question **2** a): Part-whole model showing
3,150,260 = 3,000,000 + 100,000 + 50,000 + 200 + 60

Question **2** b): Part-whole model showing
706,053 = 700,000 + 6,000 + 50 + 3

Question **3**: 7,691,712; 570,209; 3,047,039; 4,038,200; 759,421; 4,300,916; 399,710

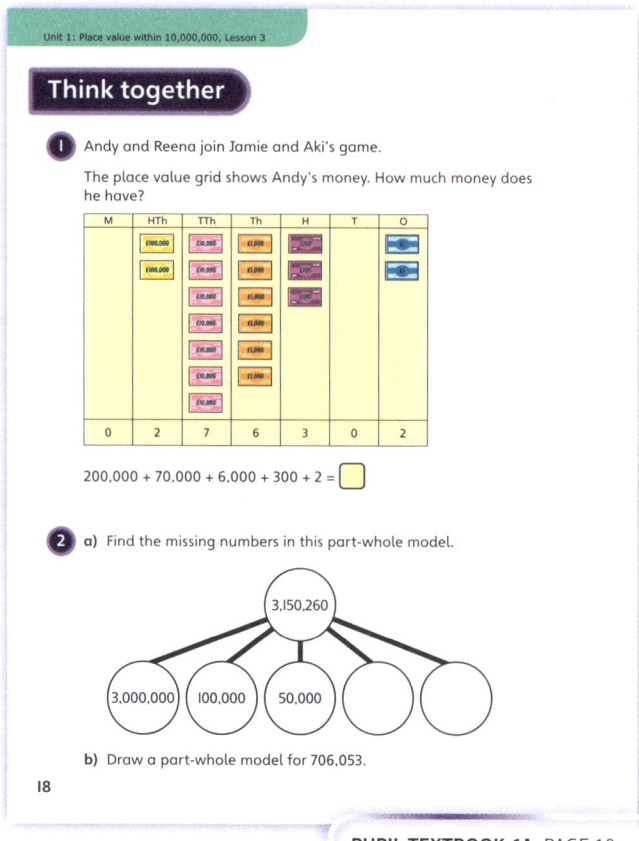

Think together

1 Andy and Reena join Jamie and Aki's game.

The place value grid shows Andy's money. How much money does he have?

M	HTh	TTh	Th	H	T	O
	£100,000	£10,000	£1,000	£100		💷
	£100,000	£10,000	£1,000	£100		💷
		£10,000	£1,000	£100		
		£10,000	£1,000			
		£10,000	£1,000			
		£10,000	£1,000			
		£10,000				
0	2	7	6	3	0	2

200,000 + 70,000 + 6,000 + 300 + 2 = ☐

2 a) Find the missing numbers in this part-whole model.

3,150,260
3,000,000   100,000   50,000   ◯   ◯

b) Draw a part-whole model for 706,053.

18

PUPIL TEXTBOOK 6A PAGE 18

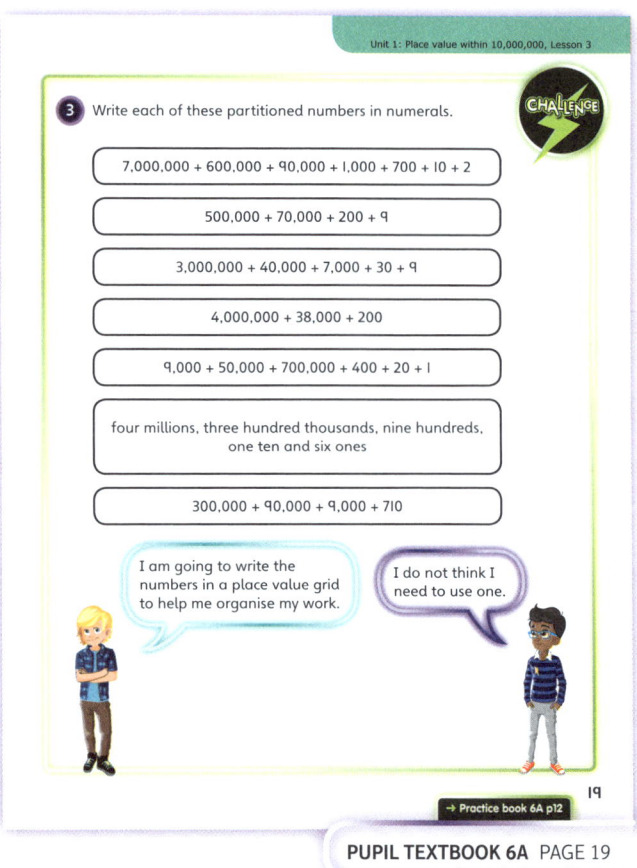

3 Write each of these partitioned numbers in numerals.

7,000,000 + 600,000 + 90,000 + 1,000 + 700 + 10 + 2

500,000 + 70,000 + 200 + 9

3,000,000 + 40,000 + 7,000 + 30 + 9

4,000,000 + 38,000 + 200

9,000 + 50,000 + 700,000 + 400 + 20 + 1

four millions, three hundred thousands, nine hundreds, one ten and six ones

300,000 + 90,000 + 9,000 + 710

I am going to write the numbers in a place value grid to help me organise my work.

I do not think I need to use one.

→ Practice book 6A p12

19

PUPIL TEXTBOOK 6A PAGE 19

Practice

WAYS OF WORKING Independent thinking, pair work

IN FOCUS Question ❶ gives children further opportunity to develop their understanding of place value in numbers up 10,000,000 in a real-life context. As they work through the question, scaffolding for partitioning is gradually reduced. It may be valuable for children to work through question ❸ in pairs, as pictorial representations are removed entirely and they are required to both partition numbers and find numbers using given partitions. Question ❸ g) in particular will test their fluency with the numbers they have been using. As the numbers are not in order, children will need to ensure they carefully check when finding the solution.

STRENGTHEN In question ❺, if children are struggling to recognise the value of the underlined digits, ask:
- *What mathematical representations have you used to help you recognise the place value of digits before?*
- *How could a place value grid help you now?*

DEEPEN Question ❹ requires children to demonstrate their understanding of how digits in a number change when they are adding or subtracting. Deepen their understanding by asking them to generalise based on what they observe. Ask:
- *What do you notice changes and stays the same? Does this happen every time? Why?*
- *Can you predict how a number will change before you add to, or subtract from, it?*

ASSESSMENT CHECKPOINT Questions ❶, ❷ and ❸ assess children's recognition of numbers when partitioned and whether they are able to use the partitions to create totals and vice versa. By the time they reach question ❺, look for children being able to identify the value of each digit in a number without partitioning. Question ❻ allows you to assess children's fluency with different ways of partitioning numbers, not only into their separate place value headings.

ANSWERS Answers for the **Practice** part of the lesson appear in the separate **Practice and Reflect answer guide**.

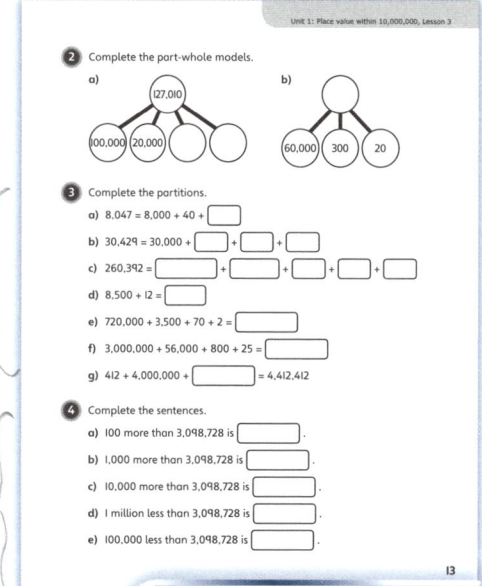

PUPIL PRACTICE BOOK 6A PAGE 12

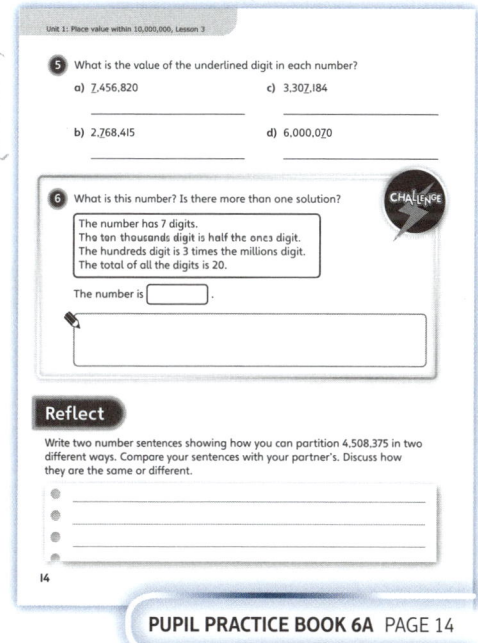

PUPIL PRACTICE BOOK 6A PAGE 13

Reflect

WAYS OF WORKING Independent thinking and pair work

IN FOCUS This question requires children to show flexibility with large numbers. They should now recognise that there are many ways to partition numbers, not only into their place value headings.

ASSESSMENT CHECKPOINT Look for children confidently discussing how the number can be made in different ways and making accurate references to place value.

ANSWERS Answers for the **Reflect** part of the lesson appear in the separate **Practice and Reflect answer guide**.

After the lesson

- How fluent are children with partitioning numbers up to 10,000,000?
- What learning aids could you display in your classroom to help secure children's use of the written vocabulary they have learnt?

PUPIL PRACTICE BOOK 6A PAGE 14

51

Number line to 10,000,000

Learning focus

In this lesson, children will accurately identify and estimate where numbers up to 10,000,000 lie on a number line. They will use their understanding of place value to help them achieve this.

Small steps

→ Previous step: Numbers to 10,000,000 (2)
→ **This step: Number line to 10,000,000**
→ Next step: Comparing and ordering numbers to 10,000,000

NATIONAL CURRICULUM LINKS

Year 6 Number – Number and Place Value
- Read, write, order and compare numbers up to 10,000,000 and determine the value of each digit.
- Solve number and practical problems that involve all of the above.

ASSESSING MASTERY

Children can use their understanding of place value to help them accurately identify, or estimate, where a number up to 10,000,000 lies on a number line.

COMMON MISCONCEPTIONS

Instead of using their knowledge of place value to help them estimate as accurately as possible, children may put numbers into three groups – big, small and medium – placing all low numbers very close to 0, all large numbers very close to 10,000,000 and any others near the centre. Ask:
- *Is that number really close to 0?*
- *Can you first estimate where 10 is? 100? 1,000?*

Children may miscalculate, and so misinterpret, the unlabelled intervals on a number line. Ask:
- *If you count in those steps, does your number line work? Show me.*

STRENGTHENING UNDERSTANDING

Children can be given cards with different numbers up to 10,000,000. Encourage them to place the cards along a piece of string in order of size, smallest first. Ask: *How do you know which number comes first?*

Discuss whether the gaps between numbers should be the same. If not, why not? To show this practically, use four or five numbers up to 50 along a bead string and discuss why the gaps between them are different.

GOING DEEPER

Children can be given more number lines where the intervals do not represent a power of 10. For example, can they place numbers along a number line from 0 to 10,000 that has been divided into 8 equal intervals?

KEY LANGUAGE

In lesson: ten million (10,000,000), accurate/accurately, interval, half-way, approximately, exactly, scale, estimate, divide/dividing

Other language to be used by the teacher: accuracy, division, increase

STRUCTURES AND REPRESENTATIONS

number line, bar chart

RESOURCES

Optional: place value cards, bead strings

 In the eTextbook of this lesson, you will find interactive links to a selection of teaching tools.

Before you teach

- How will you ensure children retain their confidence with number lines when working with very large numbers?
- How will you support abstract understanding when numbers are conceptually harder to envisage?
- How will you build awareness of what approximations are sensible when reading a number line marked in millions?

Discover

WAYS OF WORKING Pair work

ASK

• Question ❶ a): *What would be an unrealistic estimate for this point on the number line? What makes it an unrealistic estimate?*
• Question ❶ a): *What clues can you use to know how far the probe has travelled?*
• Question ❶ b): *How will you work out how far the probe has travelled?*
• Question ❶ b): *What do you need to know about the number line to answer this question?*

IN FOCUS Question ❶ a) encourages children to estimate. It is important to discuss what makes an 'accurate' estimate and what makes an unrealistic estimate. Discuss with children what clues they can use to help them make a more accurate estimate (for example, the end values and the number of intervals). Question ❶ b) gives children the opportunity to begin calculating using a number line.

PRACTICAL TIPS On the playground, draw a long number line with a child standing at either end, holding number cards 0 and 10,000,000. Ask the other children to stand at points on the line where given values (for example, 4,000,000) would be.

ANSWERS

Question ❶ a): The probe is approximately 8,500,000 miles from Earth. Because the scale is quite small, the answer can only be an estimate.

Question ❶ b): The probe travelled 3,000,000 miles between 8 June and 11 August.

Share

WAYS OF WORKING Whole class teacher led

ASK

• Question ❶ a): *Why is it important to know what each interval on the number line represents?*
• Question ❶ a): *How close were you with your estimate? Explain why you were close or far off.*
• Question ❶ b): *How did you use the number line to solve this?*
• Question ❶ b): *How did knowing the value of the intervals help?*

IN FOCUS When working through these questions, it is important that children recognise why it is essential to know the value of each interval on a number line. To illustrate this, count along the number line in different steps, such as 100,000s or 2,000,000s to show that you will not reach 10,000,000 at the end.

DEEPEN To help children's later fluency and reasoning, it may be beneficial to show the same number line but divided into different numbers of intervals, for example, 0 to 10,000,000 divided into four equal intervals.

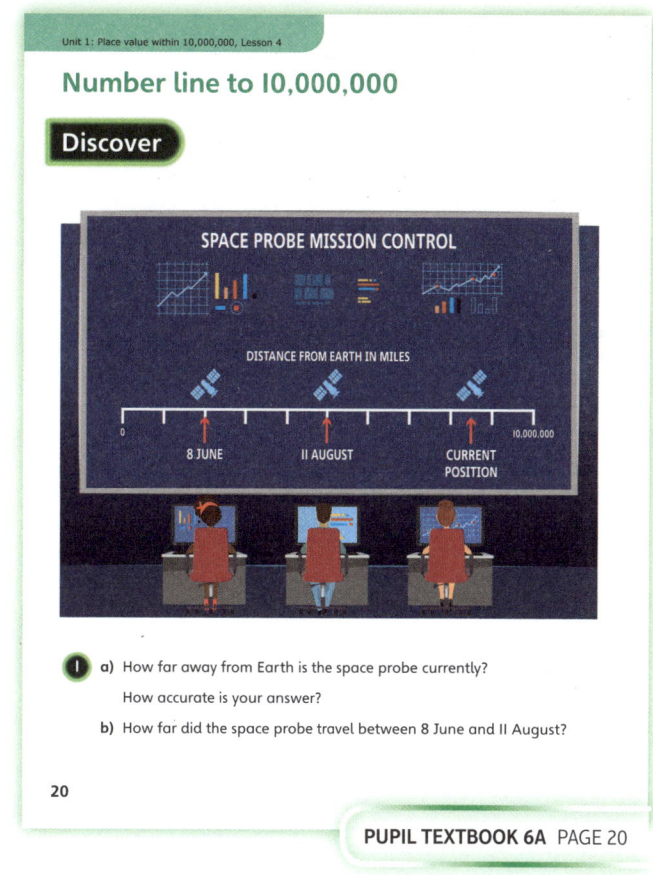

PUPIL TEXTBOOK 6A PAGE 20

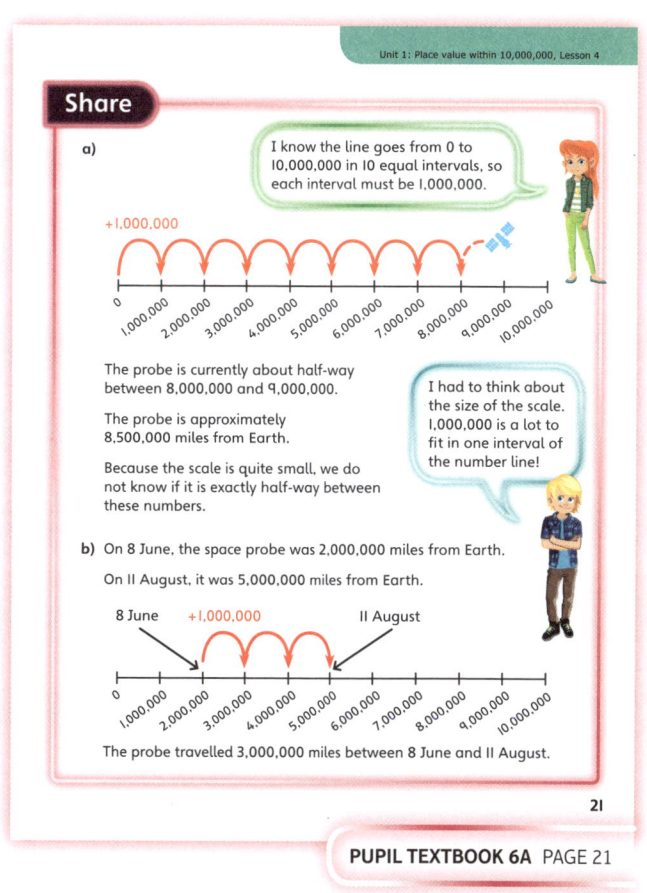

PUPIL TEXTBOOK 6A PAGE 21

Think together

WAYS OF WORKING Whole class teacher led (I do, We do, You do)

ASK

• Questions ❶ and ❷: *What do you need to know first?*
• Questions ❶ and ❷: *Will the answers be exact? Why?*
• Question ❸ a): *What is Kate's mistake?*

IN FOCUS Questions ❶ and ❷ allow children to begin interpreting different number lines and finding unlabelled numbers along them. Discuss the fact that the scales are not the same. Question ❸ a) helps to tackle the potential problem of children misinterpreting the unlabelled intervals on a number line.

STRENGTHEN In question ❷ b), if children are struggling to interpret the value of the intervals, ask:
• *What is the difference between 700,000 and 800,000? How many intervals are there? How can these facts help you to work out what each interval represents?*

DEEPEN In question ❸ b), as only the start and end numbers are labelled, children need to show a deeper understanding of how number lines work as they decide how best to go about solving the problems. Use Dexter and Ash's comments to help scaffold their reasoning. Ask:
• *How many intervals would be useful?*
• *What will each interval represent?*

ASSESSMENT CHECKPOINT Questions ❶, ❷ and ❸ b) assess children's ability to accurately place or estimate numbers on a number line. Look for children understanding why the large numbers on the scale mean readings can only be estimates of the precise numbers. Use question ❸ a) to draw out whether children are accurately calculating the value of intervals.

ANSWERS

Question ❶: Allow reasonable estimates, as these answers can only be approximate due to the size of the numbers.
On 10 July, the probe is 1,200,000 miles from Earth.
On 18 July the probe is 1,650,000 miles from Earth.
On 25 July the probe is 1,880,000 miles from Earth.

Question ❷ a): A = 12,500 miles; B= 18,000 miles; C = 19,100 miles.

Question ❷ b): D = 705,000 miles, E = 740,000 miles; F = 792,000 miles.

Question ❸ a): Kate has made a mistake with place value. She thinks each interval on the line represents 10,000 instead of 1,000. The arrow is pointing at 260,500.

Question ❸ b): Look for children making a reasonable estimate for A = 300,000; B = 340,000; C = 360,000.

Question ❸ c): They should estimate the position of 370,211 as just over half-way along the interval to the right of C.

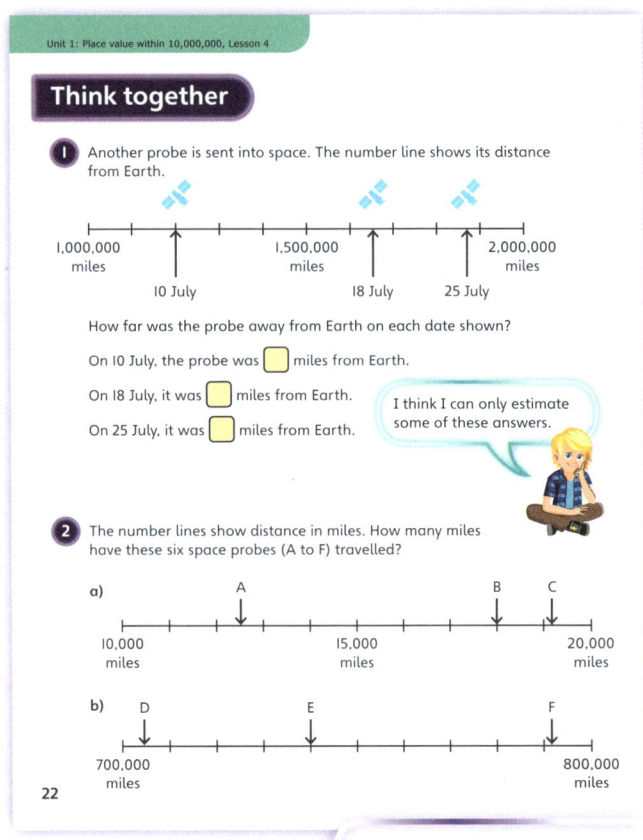

PUPIL TEXTBOOK 6A PAGE 22

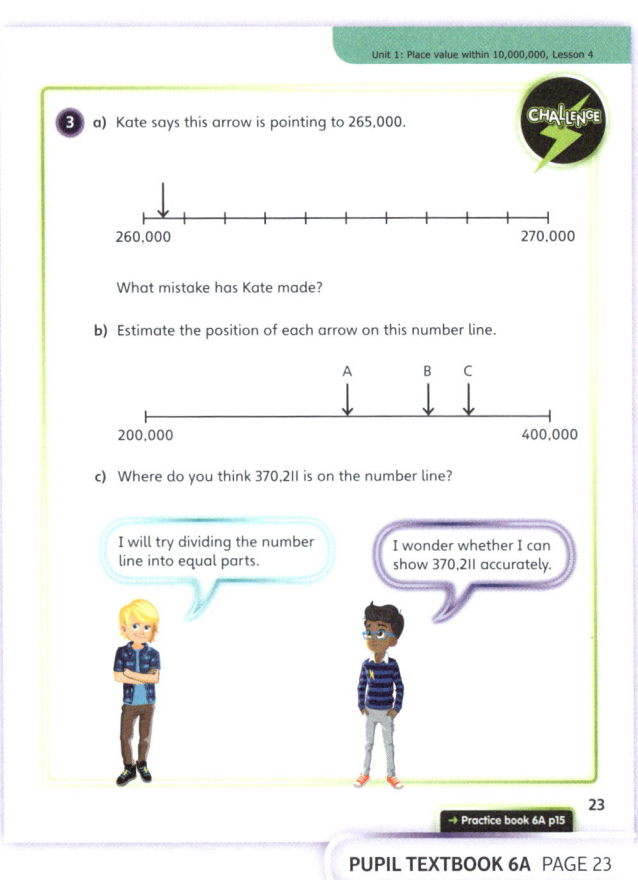

PUPIL TEXTBOOK 6A PAGE 23

Practice

WAYS OF WORKING Independent thinking

IN FOCUS In question **1**, ensure children pay attention to the start and end values, as well as the number of intervals, so they can work out what value each interval represents rather than making an assumption that it is the same for every number line. This will prepare them for questions **2** and **3**, where the number lines again count in different amounts.

STRENGTHEN In question **5**, ask:

• *What do you need to do if the number does not fall on a division?*
• *How did you estimate as accurately as possible previously?*

DEEPEN Question **6** deepens children's reasoning by presenting what they have learnt in a real-life context. They should recognise that the scale on the bar chart is a number line, but with a real-life interpretation. Ask:

• *What other facts can you tell me about the populations of the countries listed?*
• *How can you prove the facts you have noted?*

THINK DIFFERENTLY In question **3**, number lines are omitted. Children should continue with the same thought process as for question **2**, counting from number to number and recognising that the differences are consistent in each step. Question **3** c) may be challenging for some children, as the numbers decrease, but they should realise that the process of reasoning is the same as for increasing number lines.

ASSESSMENT CHECKPOINT Throughout, children should demonstrate independent ability to complete number lines. Look for children counting along the lines and arriving at the correct end numbers to demonstrate that they have correctly worked out what each interval represents. Children may assume that number lines always increase in a power of 10, so pay attention to questions **4** c) and **5** b) as these do not. Questions **4** and **5** assess children's ability to place numbers on number lines and make accurate estimates for numbers that fall between divisions.

ANSWERS Answers for the **Practice** part of the lesson appear in the separate **Practice and Reflect answer guide**.

Reflect

WAYS OF WORKING Independent thinking

IN FOCUS This question requires children to apply what they have learnt to a number line with no divisions. They should work out an appropriate scale, before making a reasoned estimate.

ASSESSMENT CHECKPOINT Methods should show a clear attempt to divide the number line and make an accurate estimate. For example:

• *I would find half-way first.*
• *I would split the number line into 10 equal parts of 10,000.*
• *After that, I would count up to 240,000.*

ANSWERS Answers for the **Reflect** part of the lesson appear in the separate **Practice and Reflect answer guide**.

After the lesson

• What percentage of the class was able to use number lines that were not fully labelled?
• How will you support children who still rely on the intervals being labelled for them?
• What other real-life contexts could you bring into the lesson next time you teach this topic?

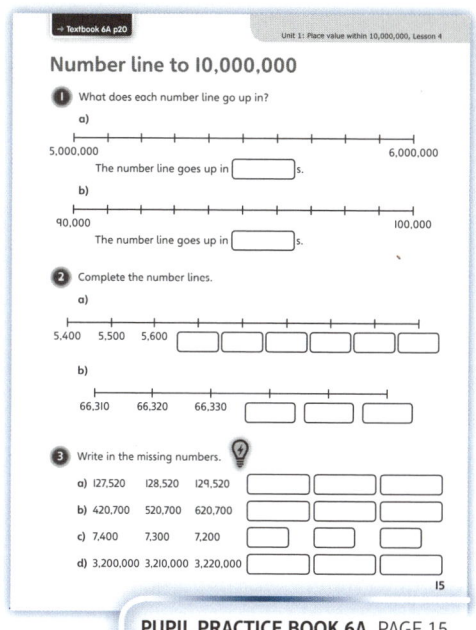

PUPIL PRACTICE BOOK 6A PAGE 15

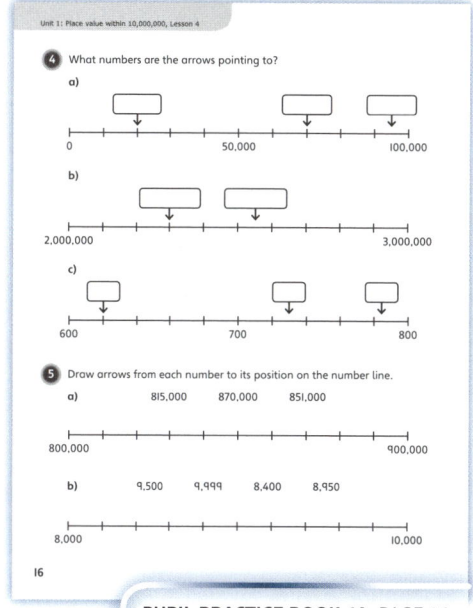

PUPIL PRACTICE BOOK 6A PAGE 16

PUPIL PRACTICE BOOK 6A PAGE 17

Comparing and ordering numbers to 10,000,000

Learning focus

In this lesson, children will use their understanding of place value and numbers up to 10,000,000 to compare and order numbers.

Small steps

→ Previous step: Number line to 10,000,000
→ **This step: Comparing and ordering numbers to 10,000,000**
→ Next step: Rounding numbers

NATIONAL CURRICULUM LINKS

Year 6 Number – Number and Place Value
• Read, write, order and compare numbers up to 10,000,000 and determine the value of each digit.
• Solve number and practical problems that involve all of the above.

ASSESSING MASTERY

Children can confidently use their understanding of place value and numbers up to 10,000,000 to compare the value of numbers. They can explain how they know a number is greater or less than another number and can use this to make accurate comparisons and order numbers correctly.

COMMON MISCONCEPTIONS

Children may compare digits that do not have the same place value; for example, saying that 46,000 is greater than 360,000 because 4 is greater than 3. Children may order numbers in the wrong direction, such as in ascending order instead of descending. Ask:
• *How will you prove that you have put the numbers in the order the question has asked for?*

STRENGTHENING UNDERSTANDING

Give children opportunities to compare and order smaller numbers up to 1,000,000 (met in Year 5) to remind themselves of the process involved, before moving on to bigger numbers with more place value headings.

GOING DEEPER

Set children a game to play in groups of two or three. Provide enough digit cards for each child to make a number in the millions up to 10,000,000. In turn, each child picks a digit card and places it in any position on a place value grid, with the aim of making the largest (or smallest) number compared with the rest of their group.

KEY LANGUAGE

In lesson: sort/sorted, most, least, compare/comparison/comparing, thousands (1,000s), ten thousands (10,000s), hundred thousands (100,000s), millions (1,000,000s), order/ordering, less than (<), greater than (>)

Other language to be used by the teacher: ones (1s), tens (10s), hundreds (100s), greatest, biggest, largest, smallest, lightest, equal to (=), decrease, place value, partition/partitioning, ten million, (10,000,000), increase, reduce, mass, heaviest, ascending, descending

STRUCTURES AND REPRESENTATIONS

place value grid, number line, table

RESOURCES

Optional: digit cards, place value grids, counters

 In the eTextbook of this lesson, you will find interactive links to a selection of teaching tools.

Before you teach

• How fluent are children with place value?

Discover

Pair work

ASK

- Question **1** a): *How will you know which is the most expensive?*
- Question **1** a): *What mathematical structure can you use to compare the numbers? How will it help you?*
- Question **1** b): *How will you decide where to place the boats on the price bar?*

IN FOCUS Question **1** a) introduces the main concept of this lesson: comparing and ordering numbers up to 10,000,000. It is important to focus on the usefulness of place value grids and how they can help to show the comparative values of different numbers. Question **1** b) then draws on children's work from the previous lesson by requiring them to pictorially represent the order on a number line.

PRACTICAL TIPS Children could research real sales listings for items with similar high prices (for example, house advertisements). Children could then draw their own number line and place the values they find on to it.

ANSWERS

Question **1** a): B, A, D, E and C

Question **1** b):

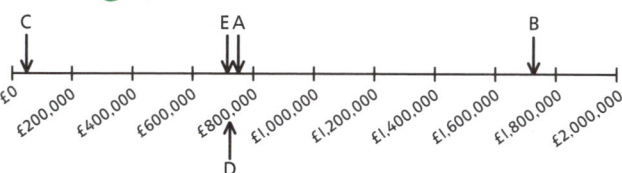

Share

Whole class teacher led

ASK

- Question **1** a): *What did you use to compare the numbers? Explain why.*
- Question **1** a): *Why is a place value grid a useful tool to help compare the numbers?*
- Question **1** a): *Did your order match the correct answer? Explain.*
- Question **1** b): *Are the boats labelled correctly on your number line?*
- Question **1** b): *How did you make sure that you completed your number line as accurately as possible?*

IN FOCUS In question **1** a), talk through the place value grid to ensure children understand how it has been completed. Discuss how a systematic process – working from left to right to compare the digits in each place value column – has been used to work out the correct order. Use the pictorial representation of the number line in question **1** b) to reinforce children's understanding of ordering.

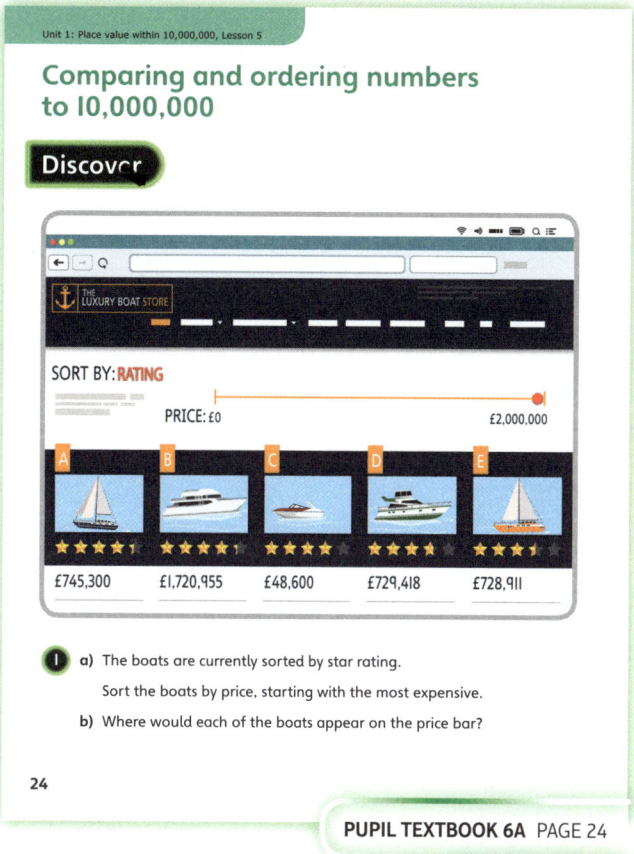

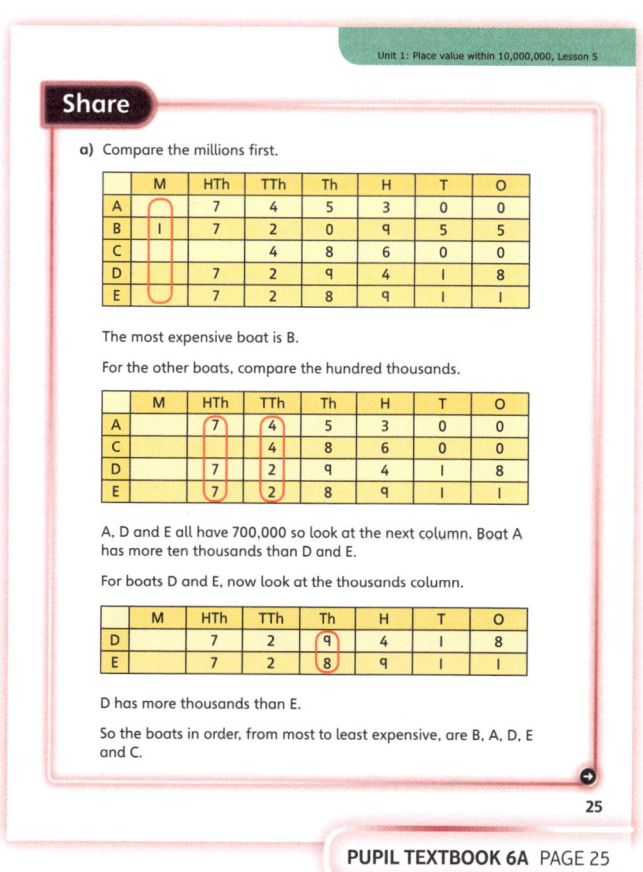

Think together

Whole class teacher led (I do, We do, You do)

ASK

- Question ❶ : *What do the signs < and > mean?*
- Question ❶ : *How can you prove you are correct?*
- Question ❷ : *What order do you have to put the numbers in? How will this change the way you solve the problem?*
- Question ❸ : *Do you agree with Dexter or is there only one solution for each missing digit?*
- Question ❹ : *Why is it difficult to compare the amounts in this question? What can you do to make this easier?*

IN FOCUS Question ❶ gives the opportunity to link the learning in this lesson to children's understanding of the mathematical notation for comparison. Question ❶ d) challenges children's understanding of place value: children may assume that £5,999,999 is bigger because it has more 9s than £7,000,000. Question ❷ will help develop children's flexibility with ordering numbers, ensuring they are able to order in descending order, as well as ascending. As all the numbers are built from different combinations of the same five or six digits, it also challenges their understanding of place value.

STRENGTHEN When working on question ❸, if children find one solution for each digit and do not look for more, ask:
- *Is that the only solution that works?*
- *Can you prove to me that there are no other possible solutions?*

DEEPEN Question ❹ especially challenges children's understanding of place value and the values of numbers, as the highest value amount is written in ones and tenths, while the lowest amount is written using hundreds of thousands. Extend this by asking children to write sets of three numbers, where one number is written in words, one in digits and one in ones and tenths, to give to a partner to order.

ASSESSMENT CHECKPOINT The questions all assess children's ability to compare and order numbers up to 10,000,000. Look for children demonstrating their ability to use the mathematical notation of comparison accurately (< and >). Questions ❸ and ❹ particularly draw out children's fluency and reasoning with place value. Question ❹ also assesses children's ability to understand and interpret the value of numbers written in different ways.

ANSWERS

Question ❶ a): £429,118 < £518,128

Question ❶ b): £392,271 > £392,098

Question ❶ c): £41,510 > £4,151

Question ❶ d): £7,000,000 > £5,999,999

Question ❷ : £320,400, £302,400, £302,040, £32,000

Question ❸ a): The first digit could be 8 or 9; the second digit could be 7, 8 or 9.

Question ❸ b): The first digit could be 0, 1, 2, 3, 4, 5, 6, 7, 8 or 9; the second digit could be 0, 1, 2, 3, 4, 5, 6 or 7.

Question ❹ a): Player 4

Question ❹ b): Player 3

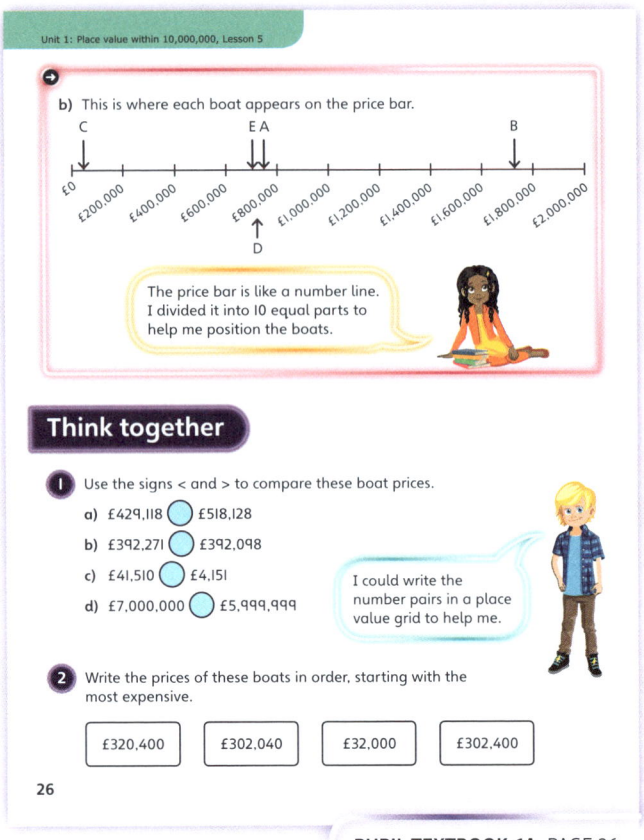

PUPIL TEXTBOOK 6A PAGE 26

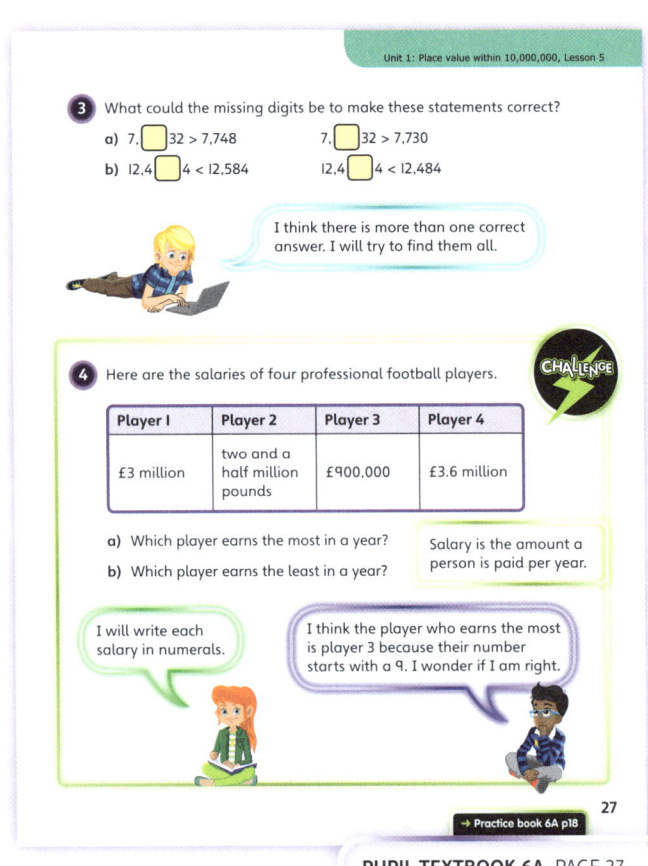

PUPIL TEXTBOOK 6A PAGE 27

Practice

WAYS OF WORKING **WAYS OF WORKING** Independent thinking

IN FOCUS Question **1** offers a pictorial representation of the differences between two numbers, thereby scaffolding children's comparisons and explanations. Question **2** requires children to demonstrate their understanding of the mathematical notation for comparison (<, >). In questions **3** and **5**, children may find it helpful to draw on the previous lesson by imagining they are ordering the numbers on number lines. However, in question **6** children need to demonstrate their fluency and reasoning when comparing numbers up to 10,000,000. Note that question **4** uses the term 'mass'; some children may be more comfortable with the term 'weight' when considering lightest and heaviest, but this should not detract from the task of comparing and ordering.

STRENGTHEN If children are struggling to compare numbers in the abstract, give them place value grids and counters to represent the numbers, as in question **1**. In question **4**, for children struggling to identify the third elephant to be fed, ask:
* *Why is this question trickier than asking for first or fourth place?*
* *How can you write down your thinking and check that you have compared accurately?*

DEEPEN When solving question **7**, deepen children's reasoning by asking:
* *How did you start this problem?*
* *Is there a specific order you have to solve this in? Explain.*

ASSESSMENT CHECKPOINT Questions **1** to **5** assess children's ability to compare and order numbers up to 10,000,000, based on their understanding of place value. Look for confident use of structures and representations to justify children's ideas, as well as accurate use of the < and > signs. In questions **6** and **7**, look for whether children can reason confidently when deciding what the missing numbers should be and whether there is more than one possible solution.

ANSWERS Answers for the **Practice** part of the lesson appear in the separate **Practice and Reflect answer guide**.

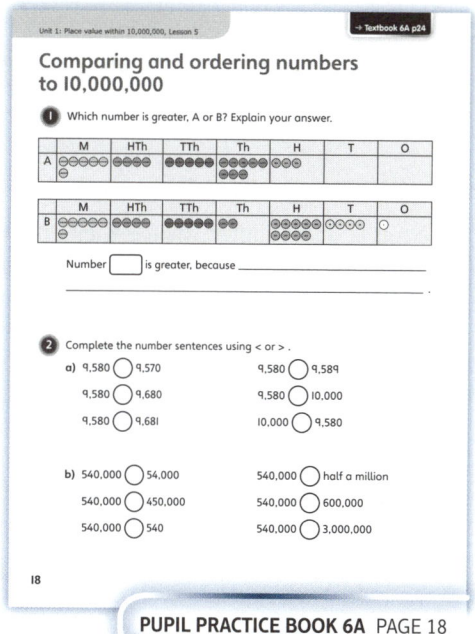

PUPIL PRACTICE BOOK 6A PAGE 18

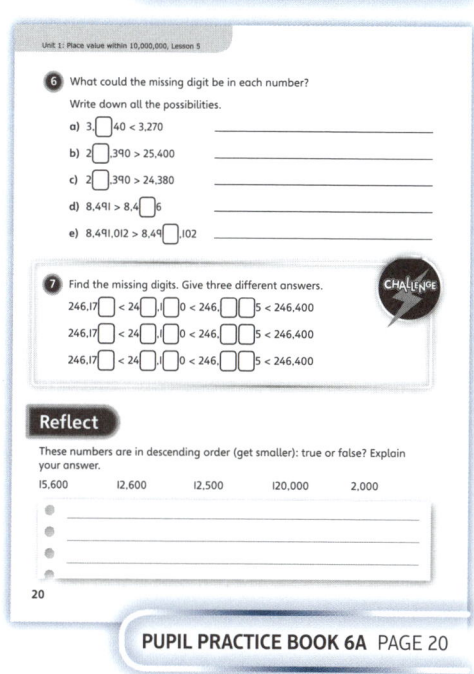

PUPIL PRACTICE BOOK 6A PAGE 19

Reflect

WAYS OF WORKING Independent thinking

IN FOCUS This question requires children to demonstrate their ability to reason by evaluating the place value of digits within numbers in order to accurately order them.

ASSESSMENT CHECKPOINT Look for children's reference to place value or use of a place value grid to prove their ideas and provide a clear explanation of their comparisons. Ensure children understand the meaning of 'descending'.

ANSWERS Answers for the **Reflect** part of the lesson appear in the separate **Practice and Reflect answer guide**.

After the lesson

* How could you apply children's learning to other areas of the curriculum?
* What were the barriers to learning for those children who struggled to make progress?

Rounding numbers

Learning focus

In this lesson, children will use their understanding of place value to help them round numbers up to 10,000,000. They will discuss when rounding is appropriate and which power of 10 to round to in a given context.

Small steps

→ Previous step: Comparing and ordering numbers to 10,000,000
→ **This step: Rounding numbers**
→ Next step: Negative numbers

NATIONAL CURRICULUM LINKS

Year 6 Number – Number and Place Value

Round any whole number to a required degree of accuracy.

ASSESSING MASTERY

Children can recognise and explain how to round numbers fluently. They can apply this understanding to larger numbers, reliably rounding up and down to the nearest 10, 100, 1,000, 10,000, 100,000 and 1,000,000.

COMMON MISCONCEPTIONS

Children may assume that rounding only affects the place value column referenced and those below it (for example, assuming that rounding to the nearest 100 only affects the 100s, 10s and 1s digits). Ask:

- *If you round 1,992 to the nearest 100, what happens?*
- *Can you show me using a picture or resources how the number will change?*
- *What happens when you count on one more 100 from 900?*

STRENGTHENING UNDERSTANDING

Before the lesson, children who may need more help can be encouraged to recap rounding smaller numbers to the nearest 10,000 (or smaller powers of 10). Ask:

- *What does it mean to round a number?*
- *Can you show me how to round x to the nearest y?*

GOING DEEPER

Deepen children's understanding by inviting them to play a 'guess the number' game in pairs. Children could set questions for their partner such as: *I have a 5-digit number that when rounded to the nearest 1,000 is 567,000. What could my number be?*

KEY LANGUAGE

In lesson: rounding/rounded/round up/round down/rounds, maximum, minimum, nearest, tens (10s), hundreds (100s), thousands (1,000s), ten thousands (10,000s), hundred thousands (100,000s)

Other language to be used by the teacher: millions (1,000,000s), greater than (>), less than (<), equal to (=)

STRUCTURES AND REPRESENTATIONS

number line, place value grid, table

RESOURCES

Optional: place value grids, counters

 In the eTextbook of this lesson, you will find interactive links to a selection of teaching tools.

Before you teach

- How fluent are children with the rules for rounding from their work in previous years?

Discover

Rounding numbers

WAYS OF WORKING Pair work

ASK

- Question **1** a): *How will you round this number to the nearest 10,000 and 1,000?*
- Question **1** b): *Which digit do you need to look at?*

IN FOCUS In question **1** a), children have to think about which digit to look at when rounding to different powers of 10 and then decide whether to round the given number up or down. Check whether children recall when to round up and when to round down. This knowledge is then required in question **1** b) when children have to think about what the boundary figures for rounding up and down are.

PRACTICAL TIPS Discuss with children real-life examples of where it may be appropriate to give a rounded estimate, rather than an exact figure. For example, country populations, distances between planets, number of fish in the sea. Children could research and find the rounded estimates for things that interest them and present these to the class. The numbers could then be compared and ordered, linking this lesson to the previous one.

ANSWERS

Question **1** a): The population of Zac's town rounded to the nearest 10,000 is 80,000.
The population of Zac's town rounded to the nearest 1,000 is 76,000.

Question **1** b): Minimum 450,000; maximum 549,999

Discover

1 a) Round the population of Zac's town to the nearest 10,000 and 1,000.

b) The number of termites has been rounded to 500,000 to the nearest 100,000.

What are the minimum and maximum numbers of termites there could actually be?

28

PUPIL TEXTBOOK 6A PAGE 28

Share

WAYS OF WORKING Whole class teacher led

ASK

- Question **1** a): *Did you remember the rules for rounding? What are they?*
- Question **1** a): *How do you know whether to round up or down?*
- Question **1** b): *What is the first number that you would round up to 600,000?*
- Question **1** b): *Why do you think the number has been rounded?*

IN FOCUS Question **1** a) offers the opportunity to recap the rules for rounding. Ensure children recall that they round down if a digit is less than 5 and round up if a digit is greater than or equal to 5. It is important to link these rules to previous learning from this unit regarding place value. In question **1** b), ensure children understand that they need to consider the ten thousands digit and encourage them to think what numbers would round to 400,000, 500,000 and 600,000. Use Ash's comment to prompt children to consider contexts in which rounding is useful (for example, when it is too difficult to count an amount exactly or when a rounded number is easier to understand).

DEEPEN Deepen children's understanding in question **1** a) by encouraging them to consider how context can determine what degree of accuracy to round to. Ask:

- *Which rounded total is more appropriate to use and why?*

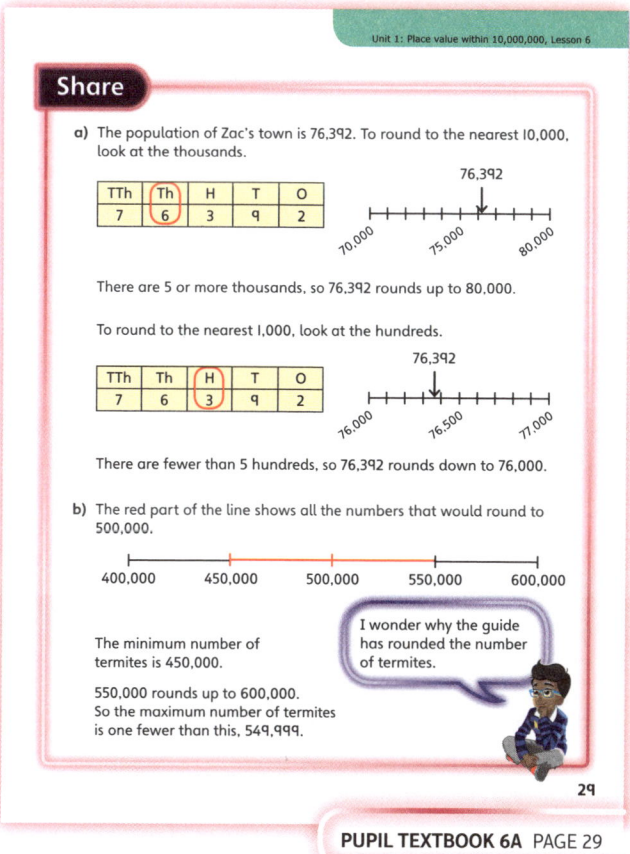

Share

a) The population of Zac's town is 76,392. To round to the nearest 10,000, look at the thousands.

To round to the nearest 1,000, look at the hundreds.

There are 5 or more thousands, so 76,392 rounds up to 80,000.

There are fewer than 5 hundreds, so 76,392 rounds down to 76,000.

b) The red part of the line shows all the numbers that would round to 500,000.

The minimum number of termites is 450,000.

550,000 rounds up to 600,000. So the maximum number of termites is one fewer than this, 549,999.

I wonder why the guide has rounded the number of termites.

29

PUPIL TEXTBOOK 6A PAGE 29

Think together

Whole class teacher led (I do, We do, You do)

ASK

- Question **1** a): *Is there a way of always knowing which column to look at?*
- Question **1** b): *Which column should you look at when rounding to the nearest 100,000? Why?*
- Question **2**: *Is there anything unexpected about some of the numbers that you have rounded to?*
- Question **3**: *Is there a way to work systematically?*
- Question **3**: *Is there anything you can say about your target number before you make it?*

IN FOCUS Questions **1** a) and **1** b) use pictorial representations to scaffold children's understanding of how to round and to aid their verbal reasoning when sharing their explanations. Spend some time discussing question **2** to tackle the misconception that rounding a number to the nearest (for example) thousand will only change the digits in the thousands column and below. Also draw attention to the fact that rounding to two different powers of 10 can sometimes result in the same number.

STRENGTHEN In question **3** a), if children struggle to find numbers that can be made with the digit cards, ask:

- *What can you say about the digits in the thousands and ten thousands columns?*
- *Does it matter what digits are used for the smaller place values? Explain.*

DEEPEN When children have solved question **2** and noticed that some of the solutions are the same, deepen their reasoning and generalisations by asking:

- *Will this happen with every 6-digit number you try? Why?*

ASSESSMENT CHECKPOINT Questions **1** and **2** assess children's ability to round numbers up to 10,000,000 to the nearest 10, 100, 1,000, 10,000 and 100,000. Look for children making valid arguments about place value. Question **3** assesses children's fluency, reasoning and problem-solving with number and rounding; this should be particularly demonstrated in their explanations for part c).

ANSWERS

Question **1** a): 10,559 rounded to the nearest 10 is 10,560.

Question **1** b): 1,556,028 rounded to the nearest 100,000 is 1,600,000.

Question **2**:

Rounded to the nearest...				
100,000	10,000	1,000	100	10
200,000	180,000	180,000	179,900	179,900

Question **3** a): Look for a number that begins 45,*xxx*, 47,*xxx*, 49,*xxx*, 52,*xxx* or 54,*xxx*

Question **3** b): Look for a number that begins 49,5*xx* or 49,7*xx*

Question **3** c): It is not possible. The number would need to be between 49,950 and 50,049, so you would need 2 nines or 2 zeros.

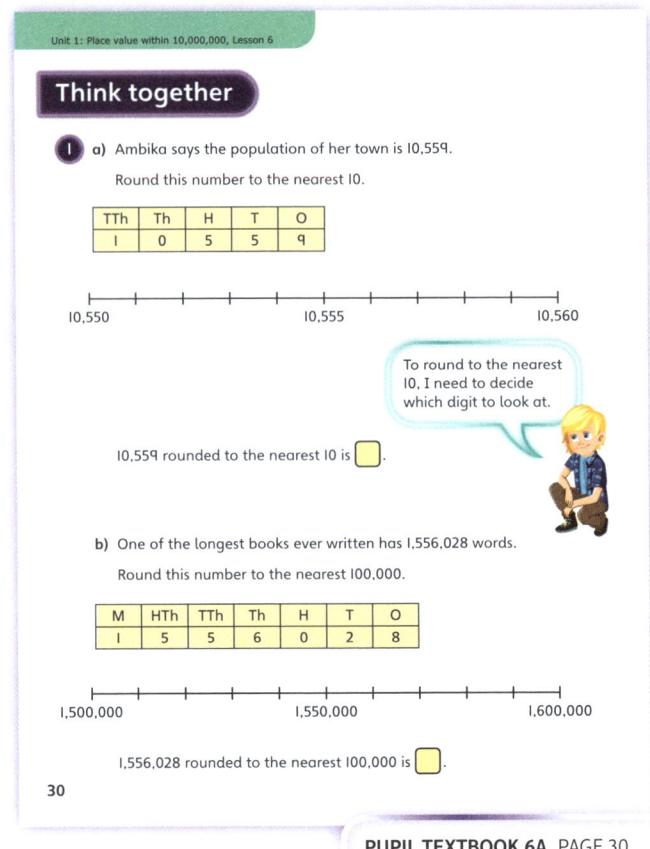

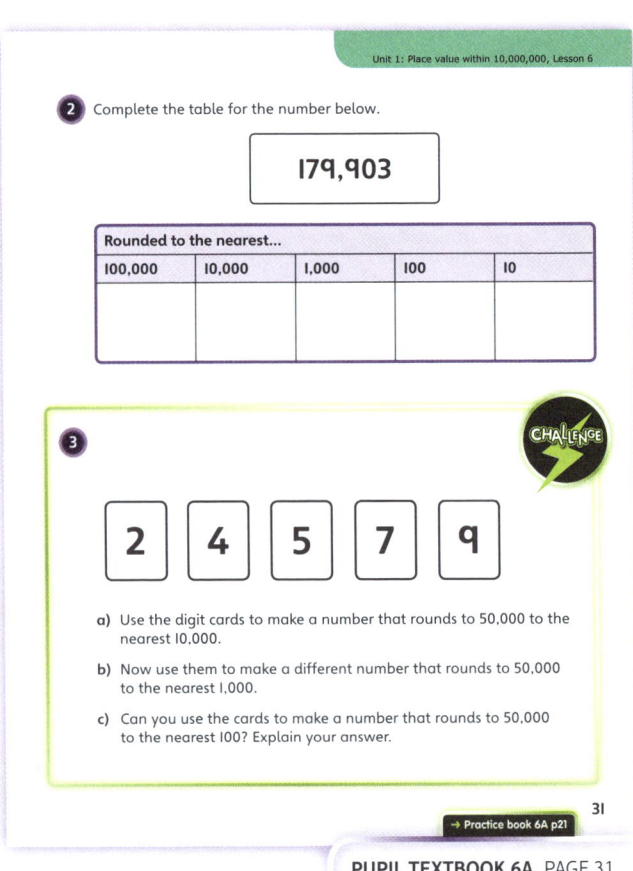

Practice

WAYS OF WORKING Independent thinking

IN FOCUS Question ❶ a) reinforces the rules for rounding by encouraging children to diagnose where another child has made a mistake in their rounding. Question ❶ b) supports the idea of rounding visually by showing why numbers are rounded up or down depending on the proximity to the nearest specified power of 10. Question ❸ offers children the opportunity to spot patterns in rounding and, through this, generalise facts about rounding based on their understanding of number. Questions ❻ and ❼ give children independent opportunity to develop their reasoning and problem-solving with rounding.

STRENGTHEN If children struggle to recognise which digits to round, they may find it helpful to represent numbers using place value grids and counters, as in question ❶. If they find question ❺ difficult, ask:
• *To round to the nearest 1,000, which column of a place value grid would you look at? How does this fact help you to solve the question?*

DEEPEN Question ❼ challenges children to pull together all of their learning from this lesson. Ask:
• *What does each sentence tell you? What facts do you know and what is missing? Is there only one answer to part b)?*

Deepen children's understanding by asking them to write similar sentences of their own to challenge a partner.

ASSESSMENT CHECKPOINT Questions ❶ to ❺ assess children's ability to apply the rules for rounding to numbers up to 10,000,000. Look for children making mistakes with place value, particularly in question ❺. In questions ❻ and ❼, children should demonstrate their ability to reason, using their understanding of place value and rounding to solve the problems.

ANSWERS Answers for the **Practice** part of the lesson appear in the separate **Practice and Reflect answer guide**.

Reflect

WAYS OF WORKING Independent thinking

IN FOCUS This question not only assesses children's understanding and fluency when applying the rules for rounding to numbers but also their ability to represent their thinking in more than one way.

ASSESSMENT CHECKPOINT Children should be able to use concrete, pictorial or abstract representations to demonstrate their conceptual understanding of rounding in more than one way. For example, they may find the difference between the number and 16,000 and 15,000, show it on a number line or explain that 7 hundreds will round the number up.

ANSWERS Answers for the **Reflect** part of the lesson appear in the separate **Practice and Reflect answer guide**.

After the lesson ⏸

• How will you continue to encourage children to see the usefulness of rounding in real-life contexts?
• How did children's reasoning skills improve the mathematics in this lesson? Could you offer more or improved opportunities for reasoning next time you teach this?

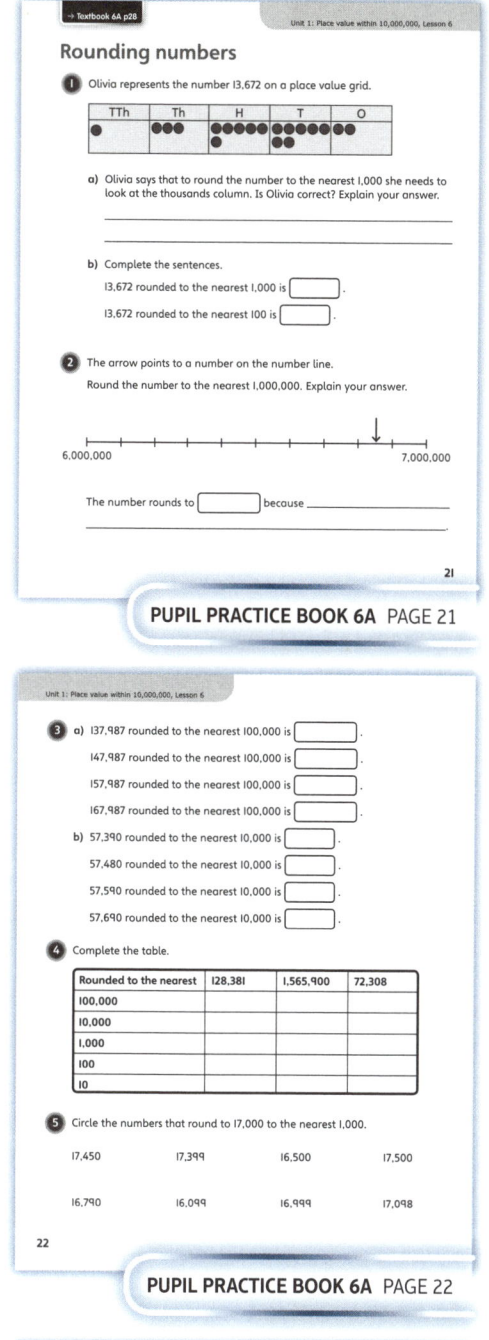

PUPIL PRACTICE BOOK 6A PAGE 21

PUPIL PRACTICE BOOK 6A PAGE 22

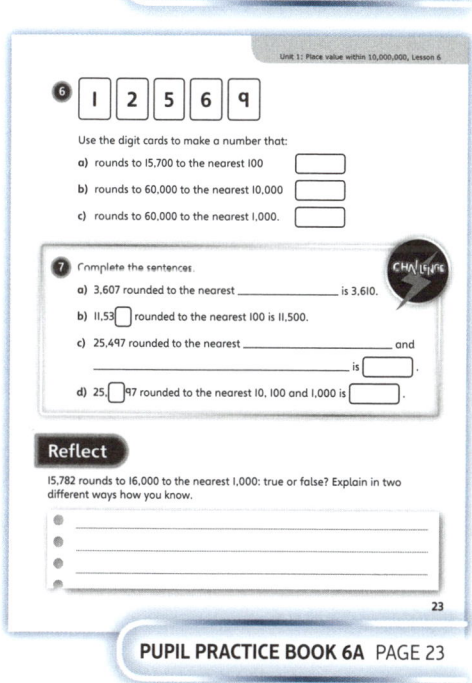

PUPIL PRACTICE BOOK 6A PAGE 23

Negative numbers

Learning focus

In this lesson, children will learn about negative numbers and their relationship with positive numbers. They will use negative numbers in context and use a number line to identify negative numbers and begin calculating with them.

Small steps

→ Previous step: Rounding numbers
→ **This step: Negative numbers**
→ Next step: Problem solving – using written methods of addition and subtraction (1)

NATIONAL CURRICULUM LINKS

Year 6 Number – Number and Place Value

Use negative numbers in context, and calculate intervals across zero.

ASSESSING MASTERY

Children can identify and explain what a negative number is and how negative numbers are similar to and different from positive numbers. They can use negative numbers in context and can reliably find negative numbers on a number line. They can also use a number line to begin calculating with negatives.

COMMON MISCONCEPTIONS

Children may assume that negative numbers work in a similar way to positive numbers. For example, assuming that ⁻5 must be greater than ⁻1 because 5 is greater than 1. Show children a number line and ask:

• *Can you plot the numbers from ⁻10 to 10 on this number line?*
• *Which end of the number line shows the greatest number? Which end shows the smallest?*
• *What can you say about how ⁻5 is different from ⁻1? Explain how you know.*

STRENGTHENING UNDERSTANDING

For children who struggled with finding unmarked intervals on a partially completed number line, it is essential to develop this skill before this lesson begins. Number lines can be linked to bar models. Identify the two known numbers and discuss and identify with children the difference between them. Use this difference as the total bar in a bar model and underneath show bars equal to the number of unlabelled intervals. Ask:

• *What does this bar model show? How will you use it to find how much each unlabelled interval is worth?*
• *What calculation do you need to use? Can you write it?*

GOING DEEPER

Discuss with children how negative numbers are used in real-life contexts (see the **Discover** practical tips section below) and encourage them to use these ideas to create word problems involving negative numbers.

KEY LANGUAGE

In lesson: trial and error, interval, divide/divided, halved/half/half-way, difference

Other language to be used by the teacher: negative, minus, positive

STRUCTURES AND REPRESENTATIONS

number line, bar model, table

RESOURCES

Optional: number lines

 In the eTextbook of this lesson, you will find interactive links to a selection of teaching tools.

Before you teach

• In which real-life contexts may children have experienced negative numbers before?

Discover

WAYS OF WORKING Pair work

ASK

- Question ❶ a): *Can you see how the scale on the thermometer is like a number line?*
- Question ❶ a): *How will you find out what each interval on the thermometer is worth?*
- Question ❶ a): *What is similar and different about where the arrow is pointing to on the thermometer compared with number lines you have seen before?*
- Question ❶ b): *Is ⁻25 °C colder or warmer than ⁻50 °C?*
- Question ❶ b): *Which end of the thermometer shows colder temperatures?*

IN FOCUS Question ❶ a) links the learning in this lesson to children's understanding of number lines from earlier in the unit. They should recognise the scale on the thermometer as a number line. It is important for children to be reminded of how they found the value of each interval on a number line. Question ❶ b) gives children their first opportunity to place a negative number on a number line. Be sure to discuss the number's position in relation to 0.

PRACTICAL TIPS Children could be encouraged to share where they have seen negative numbers before. They could research the different uses for negative numbers (for example, distance below sea level, debits on bank statements, floors below ground in lifts or department stores) and share these ideas with the class.

ANSWERS

Question ❶ a): The temperature in the town in Siberia was ⁻50 °C

Question ❶ b): ⁻25 °C lies half-way between ⁻50 °C and 0.

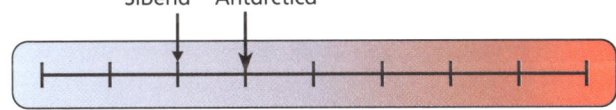

Share

WAYS OF WORKING Whole class teacher led

ASK

- Question ❶ a): *Is Dexter's method the most efficient? Explain.*
- Question ❶ a): *Can you explain Flo's method? Could you use this method to find other temperatures easily along the thermometer?*
- Question ❶ b): *Having found ⁻25 °C, can you estimate where other negative numbers are on the thermometer, such as ⁻10 °C?*

IN FOCUS In question ❶ a), talk though Ash's comment and link it to the bar model shown above the thermometer to help explain how to find the value of each interval. Question ❶ b) is important as it will prompt children to begin to realise that negative numbers work inversely to positives as they see that ⁻50 is smaller than ⁻25.

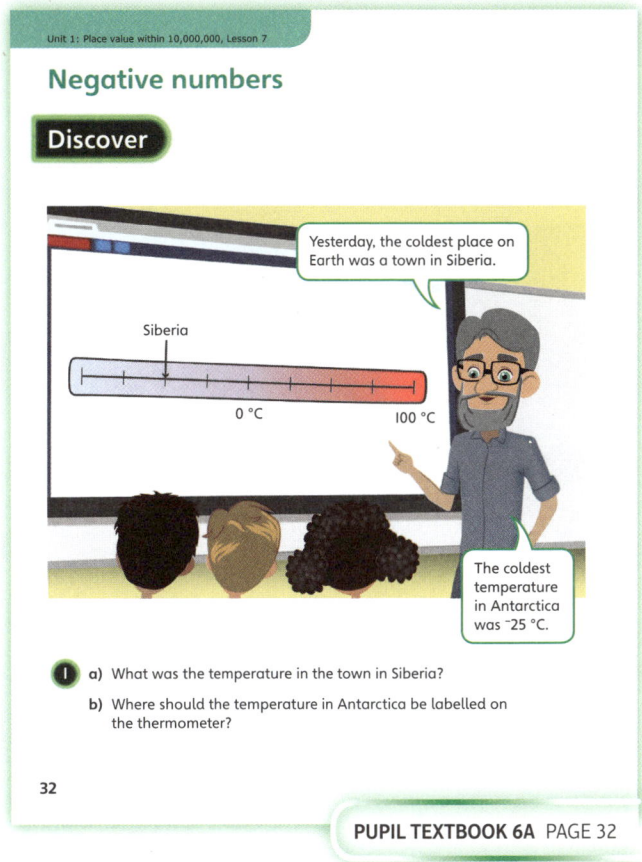

PUPIL TEXTBOOK 6A PAGE 32

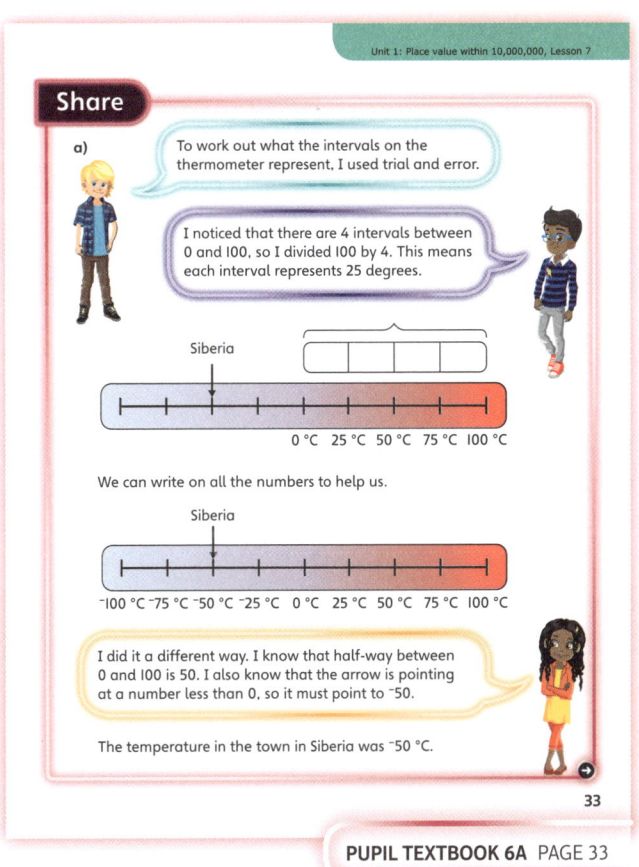

PUPIL TEXTBOOK 6A PAGE 33

hink together

WAYS OF WORKING Whole class teacher led (I do, We do, You do)

ASK

- Question ❶ a): *Will a subtraction help you here?*
- Question ❶ b): *What will you do to find the new temperature?*
- Questions ❷ and ❸: *How will you find out the value of each interval?*
- Question ❸: *What is tricky about this question?*
- Question ❸: *Can you predict whether C is positive or negative? Explain how you know.*

IN FOCUS Question ❶ provides children with further opportunity to work with negative numbers in context. It is also the first time they will consider the difference between a negative number and another given number. Children may count the intervals between two numbers on the number line provided. Look out for any children who get stuck as they count across 0. Question ❶ b) gives children an opportunity to begin adding on to negative numbers. Again, they may use the number line provided, but it is important to discuss how the numbers change when adding on to a negative number, especially when the addition bridges 0. Question ❷ helps children to secure their ability to find the value of intervals on a number line and find numbers greater and less than 0.

STRENGTHEN For children struggling with question ❷, ask:
- *How many intervals are there between 0 and 10?*
- *How will this help you to work out what each interval is worth?*
- *How will what you did for question ❷ a) help you with question ❷ b)?*

DEEPEN Question ❸ deepens children's problem-solving, reasoning and fluency with number lines. Children should recognise that they need to count the intervals between points A and B to work out the value of each interval. Develop their reasoning by asking:
- *How did you find out what each interval was worth?*
- *How was this question different from those you have tackled before? Was it more difficult? Why?*

ASSESSMENT CHECKPOINT Question ❶ assesses children's ability to calculate with negative numbers, using a number line as scaffolding. Children should be able to place positive and negative numbers on the number line and work out difference and addition. Questions ❷ and ❸ assess children's fluency and reasoning with partially completed number lines and negative numbers. They should be able to place numbers even when intervals are unlabelled. Question ❸ also assesses children's problem solving with negative numbers and partially completed number lines.

ANSWERS

Question ❶ a): The difference in temperature between Paris and Moscow was 14 °C.

Question ❶ b): The temperature in Helsinki today is 3 °C.

Question ❷ a): A = ⁻6, B = 9

Question ❷ b): ⁻35 is in the middle of the third interval from left. 8 is at the right-hand end of the seventh interval.

Question ❸: C = ⁻24

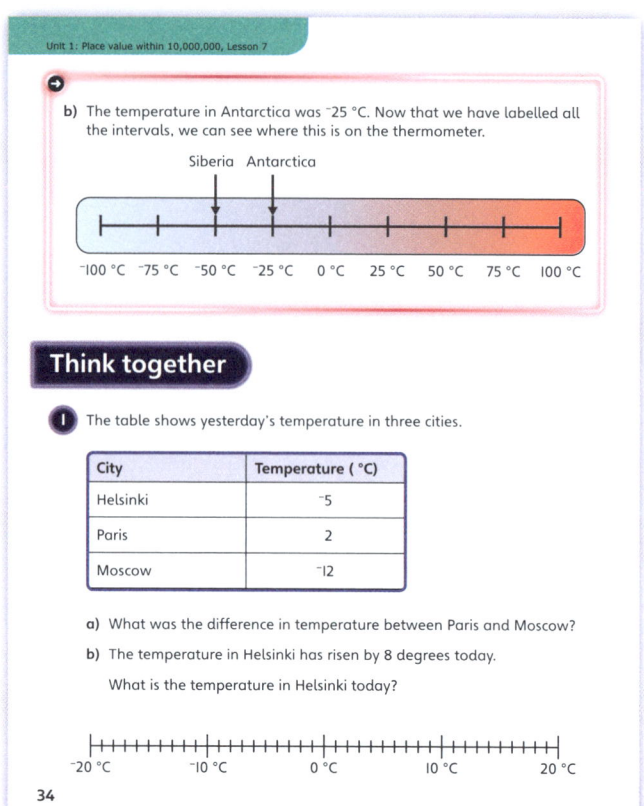

b) The temperature in Antarctica was ⁻25 °C. Now that we have labelled all the intervals, we can see where this is on the thermometer.

PUPIL TEXTBOOK 6A PAGE 34

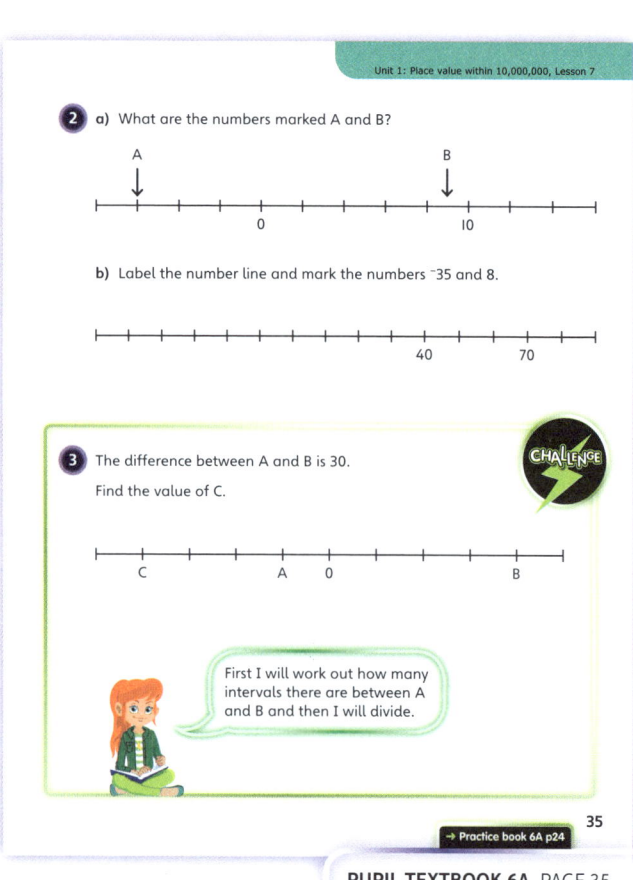

PUPIL TEXTBOOK 6A PAGE 35

Practice

WAYS OF WORKING Independent thinking

IN FOCUS Questions **1**, **2** and **3** offer children the opportunity to develop their ability to calculate with negative numbers in context. It is important for them to see that they will meet negatives in contexts other than temperature. Questions **4**, **5** and **6** develop children's independent fluency and reasoning with number lines that include negative numbers. In question **5**, decimals and fractions are introduced. Look out for children incorrectly assuming that negatives work the same way as positives, for example, thinking that ⁻11.1 is greater than ⁻11.

STRENGTHEN For questions **1**, **2** and **3**, it may help to provide children with a printed number line for them to use to represent their thinking. Ask:
- *Can you complete the number line so it has numbers on it that will help you to solve this question?*
- *How can you use the number line now to help solve this question?*

DEEPEN Question **7** requires children to demonstrate their deep conceptual understanding of what they have learnt about number lines and negative numbers across this unit in order to solve this successfully. Ask:
- *How will you approach this question?*
- *What do you know? What do you not know?*
- *What do you need to find out?*

ASSESSMENT CHECKPOINT Questions **1**, **2** and **3** will demonstrate children's ability to apply their understanding of negative numbers in real-life contexts. They should be able to count forwards and backwards and work out additions and differences involving negative numbers. Questions **4**, **5** and **6** assess children's ability to fluently apply their understanding of number lines and negative numbers when using partially complete number lines. They should be able to use reasoning to work out the values of intervals on number lines and place both positive and negative numbers accurately.

ANSWERS Answers for the **Practice** part of the lesson appear in the separate **Practice and Reflect answer guide**.

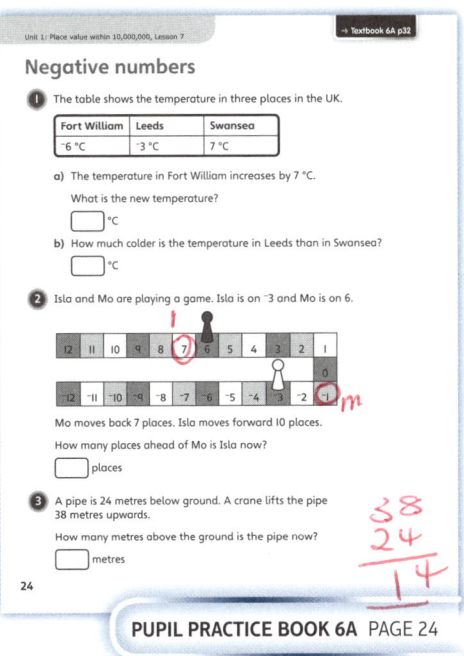

PUPIL PRACTICE BOOK 6A PAGE 24

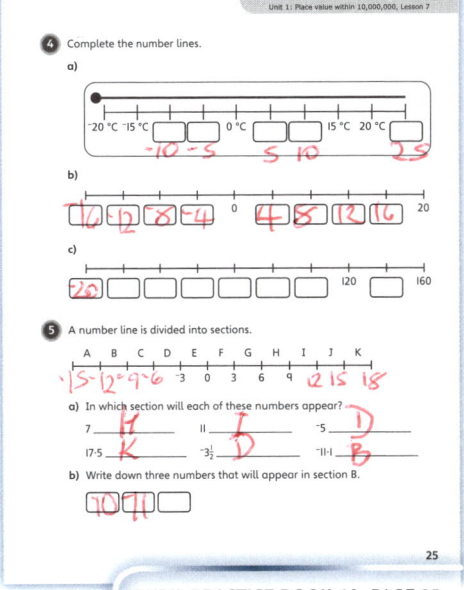

PUPIL PRACTICE BOOK 6A PAGE 25

Reflect

WAYS OF WORKING Independent thinking

IN FOCUS This question requires children to demonstrate their fluency with partially completed number lines and negative numbers. They first need to explain how they work out the value of intervals and then need to show they can count forwards and backwards across 0 on the number line.

ASSESSMENT CHECKPOINT Look for explanations of how children would calculate the value of each interval before working out the value of each letter.

ANSWERS Answers for the **Reflect** part of the lesson appear in the separate **Practice and Reflect answer guide**.

After the lesson

- Could children confidently and fluently explain how to find the value of intervals along a number line?
- How fluent were children's explanations of negative numbers? Could they understand and explain the differences between negative and positive numbers?

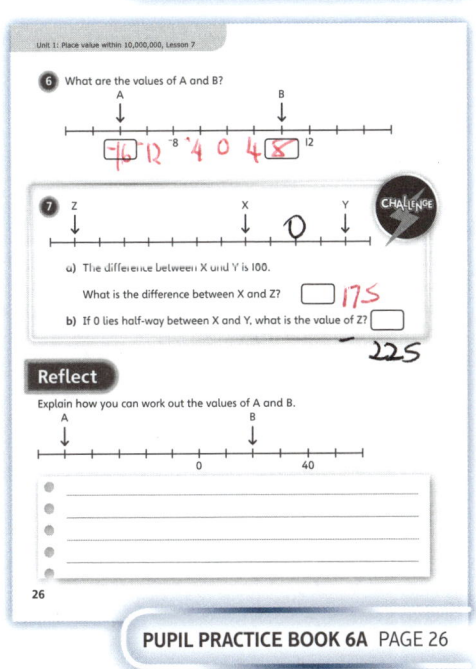

PUPIL PRACTICE BOOK 6A PAGE 26

End of unit check

> Don't forget the *Power Maths* unit assessment grid on p26.

Group work adult led

IN FOCUS

- Question **1** assesses children's recognition of the place value of digits in a large number.
- Question **2** assesses children's recognition of the written vocabulary of number names.
- Question **3** assesses children's understanding of place value and their ability to partition numbers.
- Question **4** assesses children's ability to round numbers.
- Question **5** assesses children's ability to use a number line fluently to identify where numbers are in relation to each other, including calculating the value of intervals and identifying numbers that fall between intervals. Number lines including negative numbers are assessed in question **8**, which is a SATs-style question.
- Question **6** assesses children's understanding of place value and their ability to compare and order numbers. This is also assessed in question **7**, which is a SATs-style question.

ANSWERS AND COMMENTARY

Children who have mastered the concepts in this unit will demonstrate fluent understanding of place value within numbers up to 10,000,000. They will be able to read, write and partition numbers accurately and compare and order numbers up to 10,000,000. They will demonstrate fluency when using a number line and will be able to complete partially completed number lines, confidently working out the value of unlabelled intervals. Children will be able to round up and down numbers up to 10,000,000 to different powers of 10. They will be able to recognise and use negative numbers on number lines and in real-life contexts.

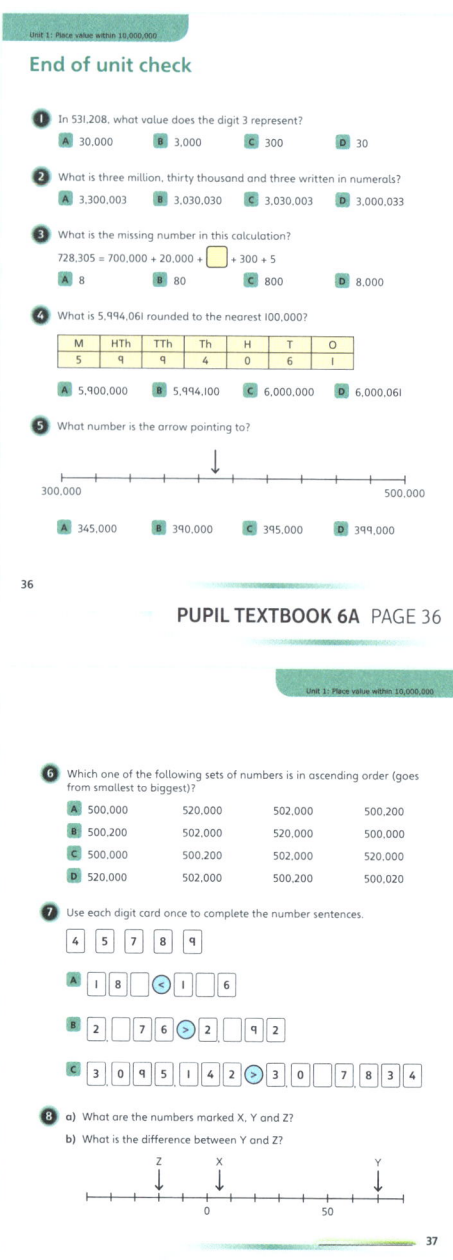

PUPIL TEXTBOOK 6A PAGE 36

PUPIL TEXTBOOK 6A PAGE 37

Q	A	WRONG ANSWERS AND MISCONCEPTIONS	STRENGTHENING UNDERSTANDING
1	A	Incorrect answers suggest children are struggling with place value or have misread the numbers.	Encourage children to use a place value grid. Ask: • *Can you write the number in the place value grid? How can this help you to tell me what each digit is worth? Can you use this to partition the number?* • *If you write all the numbers you are comparing into the place value grid, how can it help you to compare and order them?* To help children with rounding, offer them a number line. Ask: • *Where is the number on the number line? Is it closer to X or Y?* For labelling the intervals, ask: • *What is the difference between the two labelled values? How can this help you to find what each interval is worth?*
2	C		
3	D	A, B or C suggest problems with place value or partitioning.	
4	C	A, B or D suggest problems with rounding or place value.	
5	B	A, C or D suggest difficulties finding the value of unlabelled intervals on a number line and using them to estimate numbers.	
6	C	Incorrect answers suggest children are struggling with place value or comparing and ordering numbers.	
7		Options for missing digits: **a)** 4 or 5, 8; any digit, 9; **b)** first digit > second digit; **c)** any digit except 9. Errors suggest incomplete mastery of topic.	
8		**a)** X = 5, Y = 70, Z = ⁻20, **b)** 90. Errors suggest problems with unlabelled intervals or (errors for Z and b) with negative numbers.	

My journal

WAYS OF WORKING Independent thinking

ANSWERS AND COMMENTARY

Children should be able to demonstrate deep conceptual understanding of number through their reasoning. Children may share ideas such as:

- *I know 130,689 has 3 more hundreds than ten thousands as 6 is 3 more than 3.*
- *I know 6,985,310 is the greatest number less than 7 million I can make, as I used 6 millions and then made sure that the rest of the digits go in descending order so the bigger digits are in the higher place value positions.*
- *I know 60,389 is 10,000 more than 50,389 as the ten thousands digit has increased by 1.*

If children are struggling to begin unpicking the reasoning in the question, ask:

- *What mathematical models might help you in this task?*
- *How would a place value grid help?*

Power check

WAYS OF WORKING Independent thinking

ASK

- *If I gave you a number in the millions, how many different ways could you write it? Could you place it on a number line? How about if the number is negative?*
- *How confident are you with the place value of numbers up to 10,000,000?*
- *How confident are you that you can compare and order numbers up to 10,000,000?*
- *Given a number up to 10,000,000, could you explain to someone how to round it to the nearest x?*

Power puzzle

WAYS OF WORKING Pair work

IN FOCUS Children need to use their reasoning to solve problems involving the numbers they have been studying in this unit. They should demonstrate their fluency with number by trying different solutions and adhering to the given clues. Once they have solved the puzzle, the suggested activity from Sparks will allow children to demonstrably deepen their understanding of the vocabulary and mathematical concepts by writing their own puzzles.

ANSWERS AND COMMENTARY Children should be able to confidently read and interpret the given clues, using their understanding of number and place value. If children are struggling, encourage them to use a place value grid to help them structure their ideas, writing all the possibilities for each place value column and then narrowing them down systematically until they reach the correct solution (5,293,187).

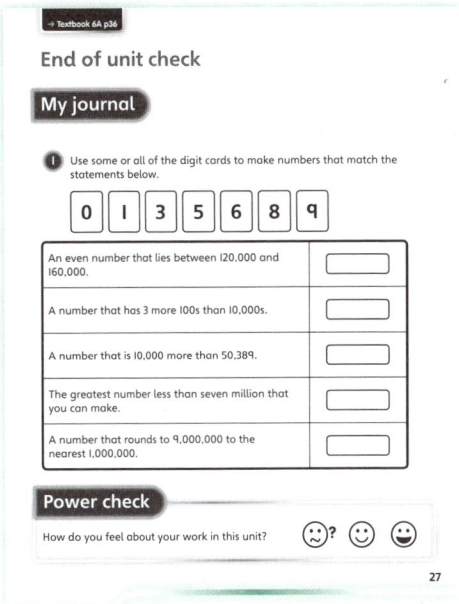

PUPIL PRACTICE BOOK 6A PAGE 27

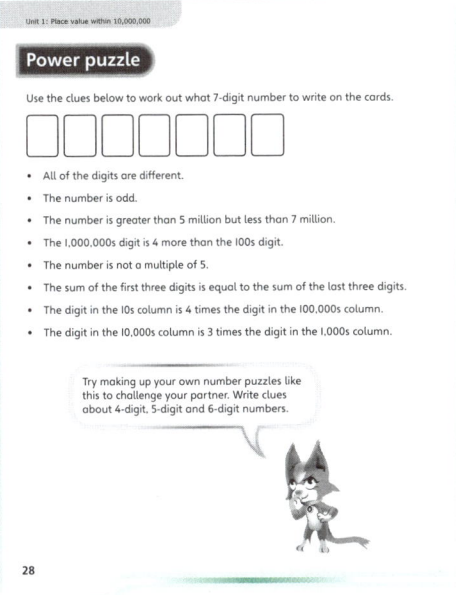

PUPIL PRACTICE BOOK 6A PAGE 28

After the unit

- To what extent did the learning in this unit deepen children's understanding of place value of digits in a number?
- What concrete representations did you make most effective use of in this unit? What representation would you develop the use of next time you teach this? How and why?

Strengthen and **Deepen** activities for this unit can be found in the *Power Maths* online subscription.

Unit 2
Four operations ①

Mastery Expert tip! "Always ask children to explain the links between the written methods they are learning and using. This can really secure conceptual understanding and helps children to work with greater confidence!"

Don't forget to watch the Unit 2 video!

WHY THIS UNIT IS IMPORTANT

This unit allows children to develop fluency with efficient columnar written methods for addition and subtraction, without and with exchanges. They will deepen understanding of the columnar method for multiplication of 4-digit numbers by 1- and 2-digit numbers and develop an understanding of written methods for division. Children will make links to methods they have met before and apply new learning to contextual word problems.

WHERE THIS UNIT FITS

→ Unit 1: Place value within 10,000,000
→ **Unit 2: Four operations (1)**
→ Unit 3: Four operations (2)

This unit builds on children's knowledge of using formal columnar written methods for addition, subtraction, multiplication and division, including with real-life contexts.

Before they start this unit, it is expected that children:
- recognise and understand the symbols for the four operations
- recognise and use written methods for addition and subtraction
- recognise and use written methods for multiplication and division
- understand and explain factors and multiples
- have had experience solving word problems.

ASSESSING MASTERY

Children will demonstrate mastery in this unit by efficiently and fluently solving addition and subtraction calculations using a compact columnar method, with exchanges where necessary, and explaining how this is linked to place value. Children will efficiently and accurately solve multiplication of a 4-digit number by a 1- or 2-digit number using a columnar method. They will be able to use multiple methods to solve division calculations, including short and long division, explaining how their understanding of multiples and factors can help them. They will demonstrate mastery of all written methods by using them fluently to reason and solve problems.

COMMON MISCONCEPTIONS	STRENGTHENING UNDERSTANDING	GOING DEEPER
Children may multiply incorrectly when using short multiplication, for example 2,345 × 6 as '5 × 6', '45 × 6', '345 × 6' and '2,345 × 6'.	Show the numbers in a place value grid and ask what number each digit represents.	Children could write their own story problems for each of the four different operations.
When representing a remainder as a fraction, children may record the fraction upside down, with the remainder as the denominator and the divisor as the numerator.	Ask children to describe how fractions are linked to a division and its remainder and to explain which number will be the numerator and which the denominator, and why.	Make links to known fraction and decimal equivalents and give some examples where children will be able to express the remainder as a fraction and also as a decimal.

WAYS OF WORKING

WAYS OF WORKING

Whole class discussion

Go through the unit starter pages of the **Pupil Textbook**. Introduce the focus and explore different ways of working.

STRUCTURES AND REPRESENTATIONS

Column methods of addition and subtraction: These models are used to enable children to efficiently solve addition and subtraction calculations.

Grid method: The model is used to support children's conceptual understanding of column multiplication. It can also be used inversely to help their conceptual understanding of both short and long division.

	2,000	300	40	5
4	8,000	1,200	160	20

Column method of multiplication: This model is used to enable children to efficiently solve multiplication calculations.

Short division and long division: These models are used to enable children to efficiently solve the division of a number with up to 4 digits by a 1-digit number and by a 2-digit number respectively.

```
      2  3  4  5
  4 | 9 ¹3 ¹8 ²0
```

```
          3  5  6
  24 | 8  5  4  4
     -  7  2  0  0
        1  3  4  4
     -  1  2  0  0
           1  4  4
     -     1  4  4
              0
```

Number line: Number lines are used to help represent division, particularly when teaching about remainders.

Place value grid and counters: This model helps children to recognise the value of each digit in a number and to create and partition numbers.

Bar model: This model is used to represent the solving of division calculations pictorially.

KEY LANGUAGE

There is some key language that children will need to know as part of the learning in this unit.

→ add, subtract, sum, total, difference

→ method, column, columnar

→ multiply, multiplication, product, approximation

→ divide, division, short division, long division

→ factor, multiple, divisor, dividend, remainder

→ inverse grid method

→ fraction, simplify, numerator, denominator

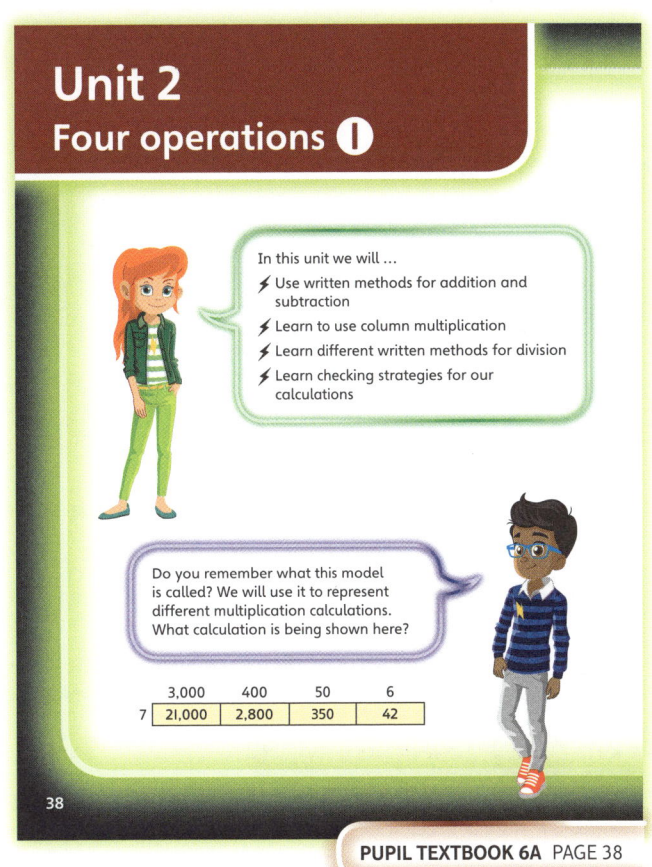

PUPIL TEXTBOOK 6A PAGE 38

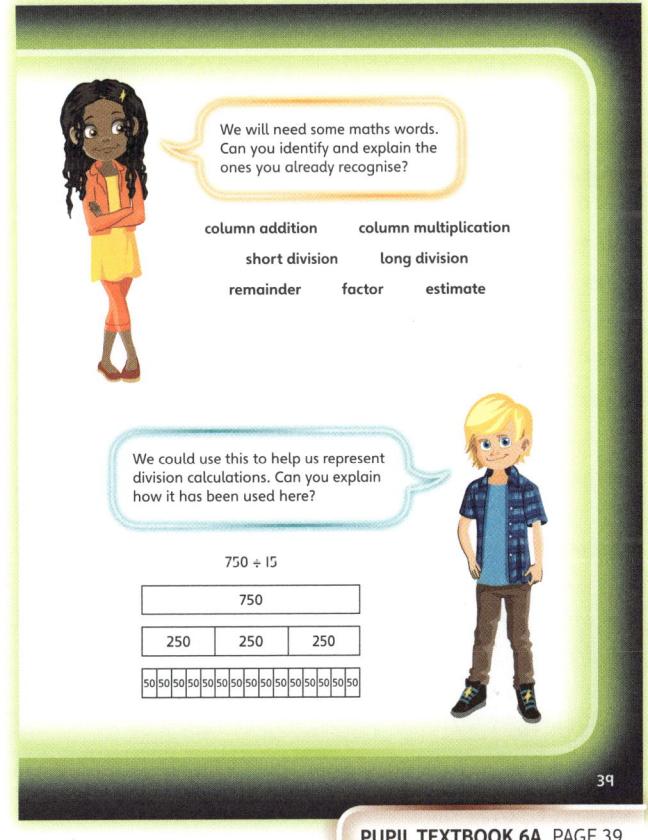

PUPIL TEXTBOOK 6A PAGE 39

Problem solving – using written methods of addition and subtraction ❶

Learning focus

In this lesson children will develop their understanding of the columnar written methods of addition and subtraction where exchanging is not necessary.

Small steps

→ Previous step: Negative numbers
→ **This step: Problem solving – using written methods of addition and subtraction (1)**
→ Next step: Problem solving – using written methods of addition and subtraction (2)

NATIONAL CURRICULUM LINKS

Year 6 Number – Addition, Subtraction, Multiplication and Division

Solve addition and subtraction multi-step problems in contexts, deciding which operations and methods to use and why.

ASSESSING MASTERY

Children can fluently and efficiently use columnar written methods to solve addition and subtraction problems. They can explain why and how these methods work and can represent them clearly.

COMMON MISCONCEPTIONS

Children may confuse the place value headings above the columns. Ask:
• *Can you show the place value headings above each column?*
• *Does each number fit its column heading? Explain.*

STRENGTHENING UNDERSTANDING

Before the lesson, give children concrete opportunities to experience and revise addition and subtraction, such as building numbers with place value counters or base 10 equipment and adding or subtracting by adding or taking away resources, or using money in context through role playing or visiting a shop.

GOING DEEPER

Encourage children to create their own missing number calculations (for example 456,232 + ____ = 563,213) and use them to challenge their partner.

KEY LANGUAGE

In lesson: addition, total, subtraction, method, column, calculate, calculation

Other language to be used by the teacher: difference

STRUCTURES AND REPRESENTATIONS

column addition, column subtraction, number line, place value counters

RESOURCES

Optional: place value counters, printed place value grids, base 10 equipment

 In the eTextbook of this lesson, you will find interactive links to a selection of teaching tools.

Before you teach

• How confident are children with written methods of addition and subtraction?
• Will you need to spend more teaching time on one method than on the others?
• What extra teaching strategies and experiences will you offer children to support each operation?

Discover

WAYS OF WORKING Pair work

ASK

- Question ➊ a): *How will you find out how many runners actually completed the race?*
- Question ➊ a): *What would be the most efficient and accurate way of calculating this difference? Explain.*
- Question ➊ b): *Can you demonstrate Isla's mistake using resources? As a picture? Using abstract methods?*

IN FOCUS Question ➊ a) offers children an opportunity to calculate the difference between two numbers. Encourage them to discuss and decide what would be the most efficient and accurate method for solving this problem.

Question ➊ b) addresses misunderstanding of place value. Encourage children to demonstrate their understanding of the mistake Isla has made using different representations.

PRACTICAL TIPS Children could be encouraged to build the numbers in the picture with place value counters or base 10 equipment to help scaffold their concrete understanding of addition. Discuss with children how they could organise their resources to make the addition and the total clear, moving towards organising them in columns according to place value (as in column addition).

ANSWERS

Question ➊ a): 2,145 runners completed the race.

Question ➊ b): 32,145 + 4,302 = 36,447 is the correct answer.

Isla has tried to use column addition but she has not aligned the digits correctly.

Share

WAYS OF WORKING Whole class teacher led

ASK

- Question ➊ a): *Do any of the representations match how you would have solved the subtraction?*
- Question ➊ a): *Which method is more efficient? Explain.*
- Question ➊ b): *How are the two totals different?*
- Question ➊ b): *What advice would you give to Isla?*
- Question ➊ b): *What representations could you use to help Isla understand her mistake?*

IN FOCUS It will be important to use the multiple representations and methods shown in question ➊ a) to scaffold children's revision of subtraction and to assess their current confidence and understanding. Children should be encouraged to use each of the methods and representations to help secure the links in their mathematical understanding.

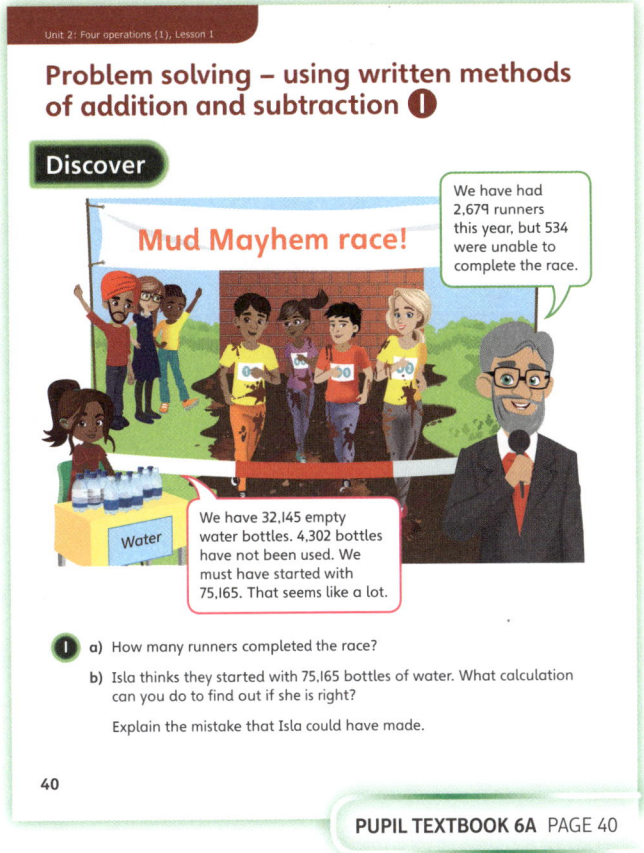

Problem solving – using written methods of addition and subtraction ➊

Discover

We have had 2,679 runners this year, but 534 were unable to complete the race.

Mud Mayhem race!

We have 32,145 empty water bottles. 4,302 bottles have not been used. We must have started with 75,165. That seems like a lot.

Water

➊ a) How many runners completed the race?

b) Isla thinks they started with 75,165 bottles of water. What calculation can you do to find out if she is right?

Explain the mistake that Isla could have made.

40

PUPIL TEXTBOOK 6A PAGE 40

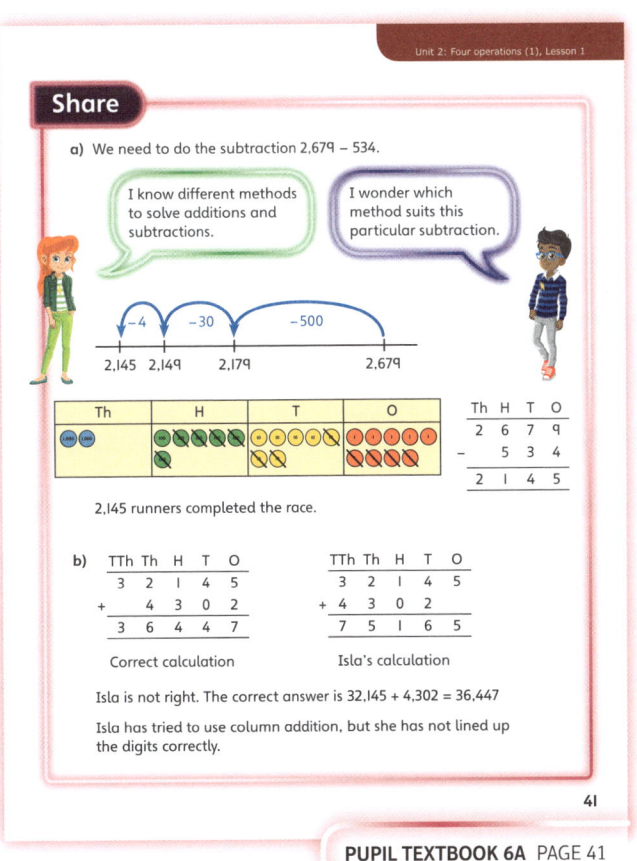

Share

a) We need to do the subtraction 2,679 – 534.

I know different methods to solve additions and subtractions.

I wonder which method suits this particular subtraction.

−4   −30   −500

2,145   2,149   2,179   2,679

2,145 runners completed the race.

Th	H	T	O

	Th	H	T	O
	2	6	7	9
−		5	3	4
	2	1	4	5

b)

	TTh	Th	H	T	O
	3	2	1	4	5
+		4	3	0	2
	3	6	4	4	7

Correct calculation

	TTh	Th	H	T	O
	3	2	1	4	5
+ 4	3	0	2		
	7	5	1	6	5

Isla's calculation

Isla is not right. The correct answer is 32,145 + 4,302 = 36,447

Isla has tried to use column addition, but she has not lined up the digits correctly.

41

PUPIL TEXTBOOK 6A PAGE 41

Think together

WAYS OF WORKING Whole class teacher led (I do, We do, You do)

ASK

- Question **1**: *What operation is needed to solve this question? How do you know?*
- Question **1**: *What is the same/different about each representation? Which is clearer? Which is quicker?*
- Question **2**: *Do all parts of this question require the same operation? Explain how you know.*
- Question **2** b) and **2** c): *Where will you find the information needed to solve this question?*
- Question **3** a): *Is there only one solution to this question? Explain how you know.*

IN FOCUS Question **1** helps children with their conceptual understanding of addition. They will benefit from having the matching resources available to them while they solve the question. Encourage them to make the calculation with place value counters while solving the abstract calculation and discuss what is the same and different about the representations. Question **2** offers the opportunity to solve subtraction and addition calculations in context. Make sure children are aware that they need information from the **Discover** section of the lesson.

STRENGTHEN If children are struggling to decide what operation to use for each part of question **2**, ask:
- *Can you make the problem using resources?*
- *Does the question suggest you will take away from what you have or add to it? Explain how you know.*

DEEPEN If children are quick to solve question **3** b), encourage them to create a similar challenge for their partner. Ask:
- *Can you create an addition or subtraction calculation that could have hidden digits?*
- *Can you create one which is simple and one which is tricky? How are they the same and different?*

ASSESSMENT CHECKPOINT Can children recognise different representations of addition and use them to solve calculations? Do children recognise addition and subtraction calculations in the context of word problems? Do children understand how the columnar methods for addition and subtraction work, and can they use these methods with fluency?

ANSWERS

Question **1**: The marathon runners raised £43,787.

Question **2** a): 1,312 runners finished the marathon.

Question **2** b): 1,061 more runners started the Mud Mayhem race than started the marathon.

Question **2** c): 3,457 runners finished both races.

Question **3** a):

TTh	Th	H	T	O
4	5	7	8	3
− 4	0	0	5	2
	5	7	3	1

Question **3** b): Answers will vary. Look for children recognising that they can cover one digit from each column but when more than one digit is covered in a column it is not possible to be certain what the calculation is.

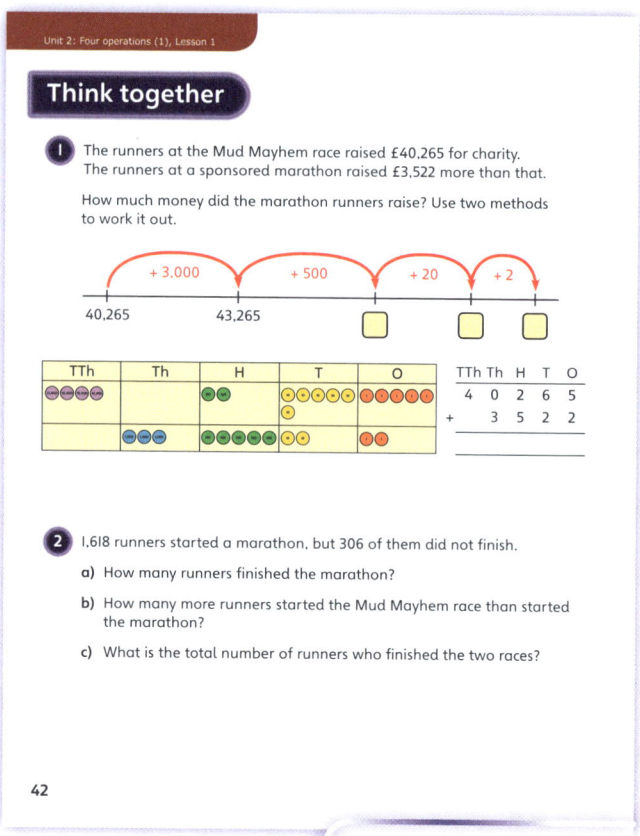

Think together

1 The runners at the Mud Mayhem race raised £40,265 for charity. The runners at a sponsored marathon raised £3,522 more than that.

How much money did the marathon runners raise? Use two methods to work it out.

2 1,618 runners started a marathon, but 306 of them did not finish.

a) How many runners finished the marathon?

b) How many more runners started the Mud Mayhem race than started the marathon?

c) What is the total number of runners who finished the two races?

42

PUPIL TEXTBOOK 6A PAGE 42

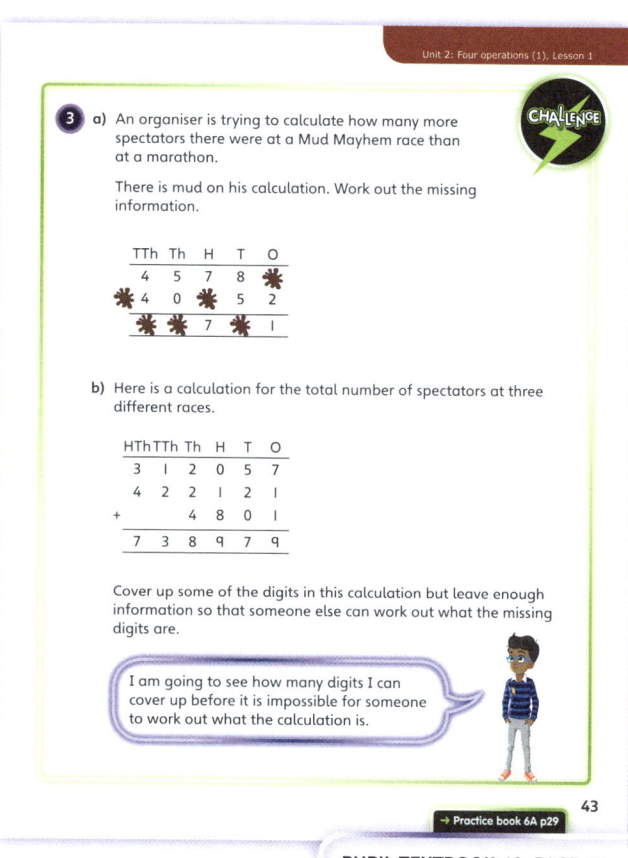

3 a) An organiser is trying to calculate how many more spectators there were at a Mud Mayhem race than at a marathon.

There is mud on his calculation. Work out the missing information.

CHALLENGE

b) Here is a calculation for the total number of spectators at three different races.

Cover up some of the digits in this calculation but leave enough information so that someone else can work out what the missing digits are.

I am going to see how many digits I can cover up before it is impossible for someone to work out what the calculation is.

→ Practice book 6A p29

43

PUPIL TEXTBOOK 6A PAGE 43

Practice

WAYS OF WORKING Independent thinking

IN FOCUS Questions **1** scaffolds children's independent conceptual understanding of addition and subtraction, ensuring they use columnar methods alongside pictorial representations. Question **2** challenges children to identify a calculation using information on a number line.

Question **3** develops children's fluency with calculations that have no context. Encourage children to use their understanding of the representations used in the previous questions to support them in finding the solutions.

In Question **4** children read and understand addition and subtraction in the context of word problems.

Question **5** requires children to demonstrate understanding of written methods by identifying mistakes.

STRENGTHEN If children are struggling to decide how to solve the word problems in question **4**, ask:
• *Can you explain what is happening in the story of the question?*
• *Is something being added or is it being taken away? How do you know?*

DEEPEN Question **7** deepens children's understanding and reasoning with addition and subtraction calculations by requiring them to use their understanding of inverses to solve missing number calculations. Ask:
• *Can you find the missing number with the operation shown? Explain why/ why not.*

ASSESSMENT CHECKPOINT Can children use a formal written method with fluency and link their understanding to pictorial representations? Do they draw out what they know and what they need to find to solve problems?

ANSWERS Answers for the **Practice** part of the lesson appear in the separate **Practice and Reflect answer guide**.

Reflect

WAYS OF WORKING Independent thinking and pair work

IN FOCUS This question will offer two opportunities to assess children's understanding. First, they will need to use their knowledge of inverses to find the missing number in the calculation. Having done this, they will need to demonstrate their ability to communicate this calculation in the form of a word problem. Ask:
• *Check your partner's problem. Is it solvable using the calculation shown?*

ASSESSMENT CHECKPOINT Look for children's ability to use the inverse to solve the missing number problem, potentially using the columnar method to calculate. Children should be able to show fluency with these types of calculation by demonstrating their ability to put the calculation into an appropriate context.

ANSWERS Answers for the **Reflect** part of the lesson appear in the separate **Practice and Reflect answer guide**.

After the lesson

• Are all children sufficiently confident with the columnar methods for both addition and subtraction?
• What support will you offer to children who are still struggling with one or both of the methods?
• How did this lesson develop children's use of mathematical vocabulary?

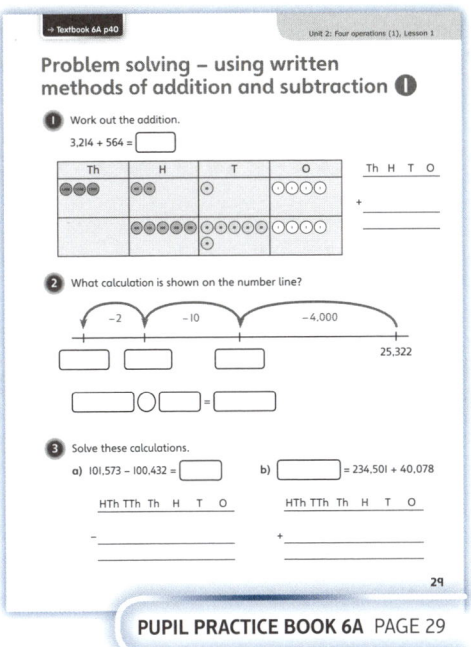

PUPIL PRACTICE BOOK 6A PAGE 29

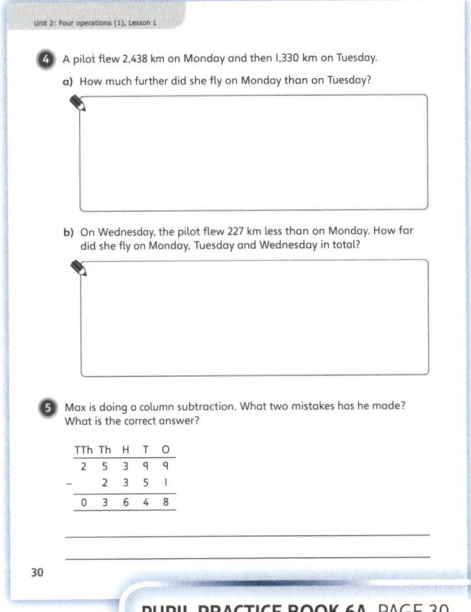

PUPIL PRACTICE BOOK 6A PAGE 30

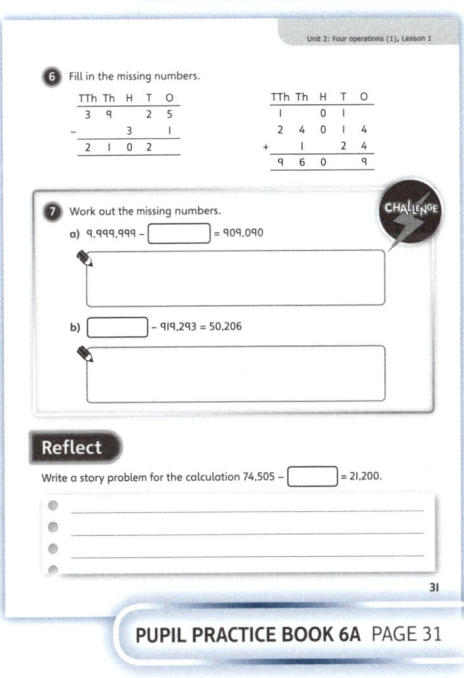

PUPIL PRACTICE BOOK 6A PAGE 31

Problem solving – using written methods of addition and subtraction ②

Learning focus

In this lesson children will develop their understanding of the columnar written methods of addition and subtraction where exchanging is necessary.

Small steps

→ Previous step: Problem solving – using written methods of addition and subtraction (1)

→ **This step: Problem solving – using written methods of addition and subtraction (2)**

→ Next step: Multiplying numbers up to 4 digits by a 1-digit number

NATIONAL CURRICULUM LINKS

Year 6 Number – Addition, Subtraction, Multiplication and Division

Solve addition and subtraction multi-step problems in contexts, deciding which operations and methods to use and why.

ASSESSING MASTERY

Children can fluently and efficiently use columnar written methods to solve addition and subtraction problems, carrying and exchanging where necessary. They can explain why and how these methods work and can represent them clearly.

COMMON MISCONCEPTIONS

Children may exchange to or from the incorrect column. Ask:
- (Addition) *Can you describe the place value of the digits in 12? What column should the 1 digit be in?*
- (Subtraction) *If you exchange from the hundreds column, how many ones will you have? Does this match the number of ones you have recorded?*

STRENGTHENING UNDERSTANDING

Before the lesson, support children in securing understanding of columnar methods with no exchanging – without this conceptual understanding, children will make limited progress in this lesson. Recap concrete and pictorial representations from the previous lesson and give children the opportunity to solve word problems.

GOING DEEPER

Children could create column addition and subtraction calculations that include exchanging. Once they have done this they can cover a digit from each column and challenge a partner to work out what the missing digits are.

KEY LANGUAGE

In lesson: addition, subtraction, difference, method, column, exchange, digit

Other language to be used by the teacher: calculate, calculation

STRUCTURES AND REPRESENTATIONS

column addition, column subtraction, number lines, place value counters, bar models

RESOURCES

Optional: place value grids and counters, base 10 equipment

 In the eTextbook of this lesson, you will find interactive links to a selection of teaching tools.

Before you teach

- How confident are children with the structure and use of columnar methods?
- Are there any misconceptions from the previous lesson that you need to address in this one?

Discover

Independent thinking

ASK

• Questions **1** a) and **1** b): *What calculation will solve this question? Explain how you know.*
• Question **1** a): *How is the addition calculation different from the ones we solved yesterday?*
• Question **1** b): *How is the subtraction calculation different from the ones we solved yesterday?*
• Questions **1** a) and **1** b): *Can you make up another calculation that requires exchanging or carrying?*

IN FOCUS Question **1** a) requires children to find the sum of two numbers with an exchange into the next place value column. Children should make this calculation with base 10 equipment to help them identify this fact.

Question **1** b) requires children to find the difference between two numbers with an exchange from another place value column. Children should make the calculation in a concrete way to help them identify this.

PRACTICAL TIPS Children could research other famous people's birth and death years and use them to work out how many years they lived for by finding the difference. Some will require exchanging while others will not, giving the opportunity to discuss what is the same and different about the calculations. Children could be encouraged to build the numbers using base 10 equipment then demonstrate their calculations using those representations.

ANSWERS

Question **1** a): Queen Elizabeth I died in the year 1603.

Question **1** b): There were 394 years between the beginning of Elizabeth I's reign and the beginning of Elizabeth II's reign.

Share

WAYS OF WORKING Whole class teacher led

ASK

• Questions **1** a) and **1** b): *How did you represent the exchange into another place value column?*
• Questions **1** a) and **1** b): *How did exchanging make the calculation different from the ones we solved yesterday?*
• Question **1** b): *How is exchanging in subtraction different from exchanging in addition?*

IN FOCUS Encourage children to create and compare concrete or pictorial representations of the exchanges in addition and in subtraction. Use these to help children identify similarities and differences and to make links between the ways of exchanging, to develop and secure children's understanding.

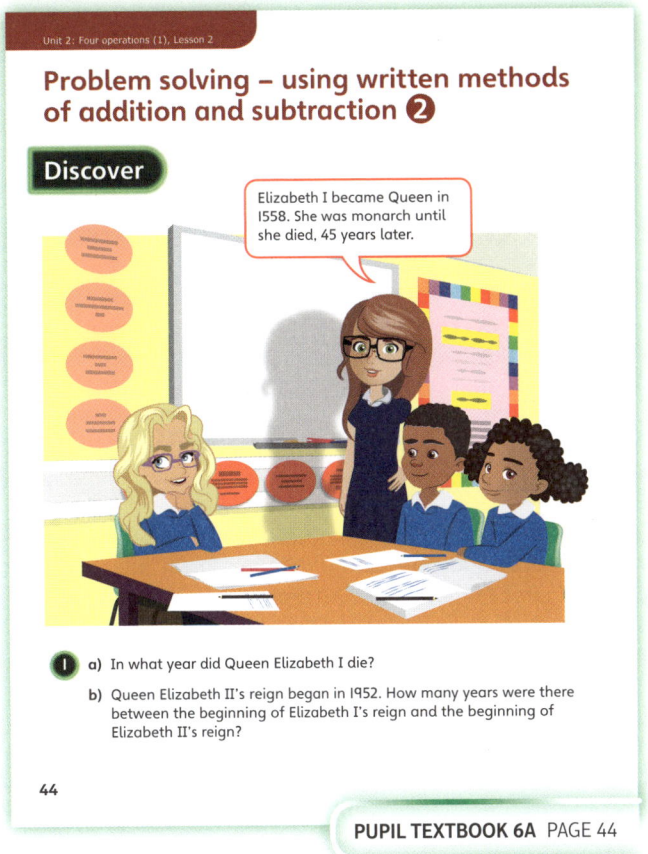

PUPIL TEXTBOOK 6A PAGE 44

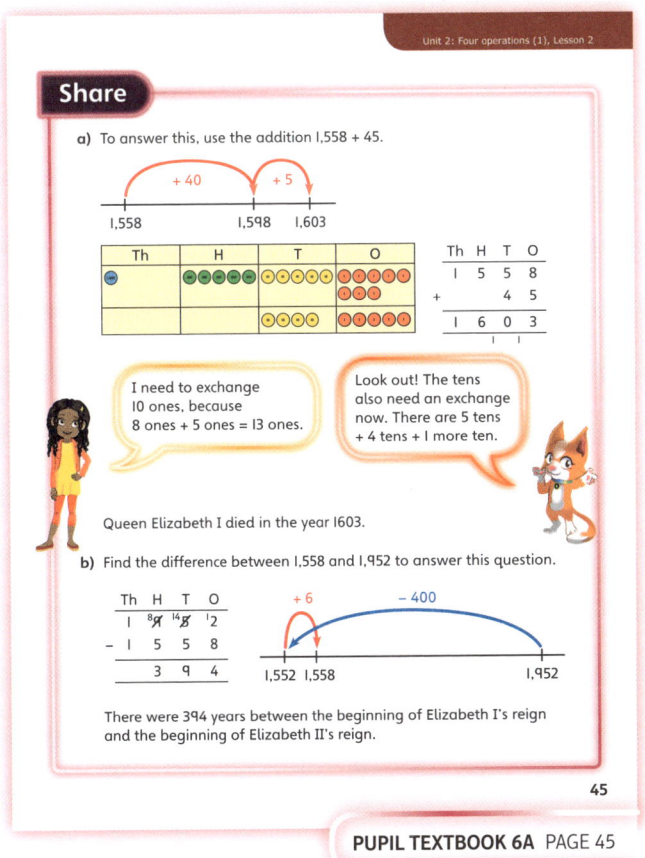

PUPIL TEXTBOOK 6A PAGE 45

Think together

WAYS OF WORKING Whole class teacher led (I do, We do, You do)

ASK

- Question **1**: *How does each representation demonstrate the calculation? Explain.*
- Question **1**: *Which two methods would you use to check your answer? Explain why.*
- Question **2**: *What operation is needed to solve each question? Explain how you know.*
- Question **3**: *Did your subtraction require any exchanges?*
- Question **3**: *Did you create a situation where no exchanges were needed? Can you explain why this was?*

IN FOCUS Question **1** offers children the opportunity to develop their understanding and fluency with subtraction that requires exchanging. It is important to give children concrete experience of these representations, so providing them with a range of manipulatives will be beneficial to their understanding.

Question **2** presents children with a contextual number line (in the form of a timeline). When solving question **2** b) it is possible that some children will be able to surmise the answer by knowing that Elizabeth I was the last Tudor. This will be a good opportunity to discuss how maths is an integral part of the whole curriculum.

STRENGTHEN If children are struggling to solve the problems in question **2**, ask:

- *What do you need to know to be able to work out how long the House of Lancaster ruled for?*
- *Where can you find those facts on the timeline?*
- *Can you make or draw a representation of the calculation to show me how you will solve it?*

DEEPEN Question **3** offers children an opportunity to deepen their understanding of column subtraction and exchanging while generalising about patterns they notice in their solutions. Ask:

- *What happens in every example? Can you explain why?*
- *Did anyone create a situation where they needed no exchanges?*

ASSESSMENT CHECKPOINT Do children know how to use different representations to help them solve addition and subtraction calculations with exchanges?

ANSWERS

Question **1**: Ethelred the Unready reigned for 35 years.

Question **2** a): The rule of the House of Lancaster lasted for 62 years.

Question **2** b): The Tudor reign ended in 1603.

Question **3** a): You will always need to exchange across two columns because in two pairs of digits the number being subtracted will be bigger and in the other two pairs of digits the number you are subtracting from will be bigger.

Question **3** b): Answers will vary. Look for children being able to use specialised examples to begin generalising about the pairs of digits in the calculations they are creating and solving.

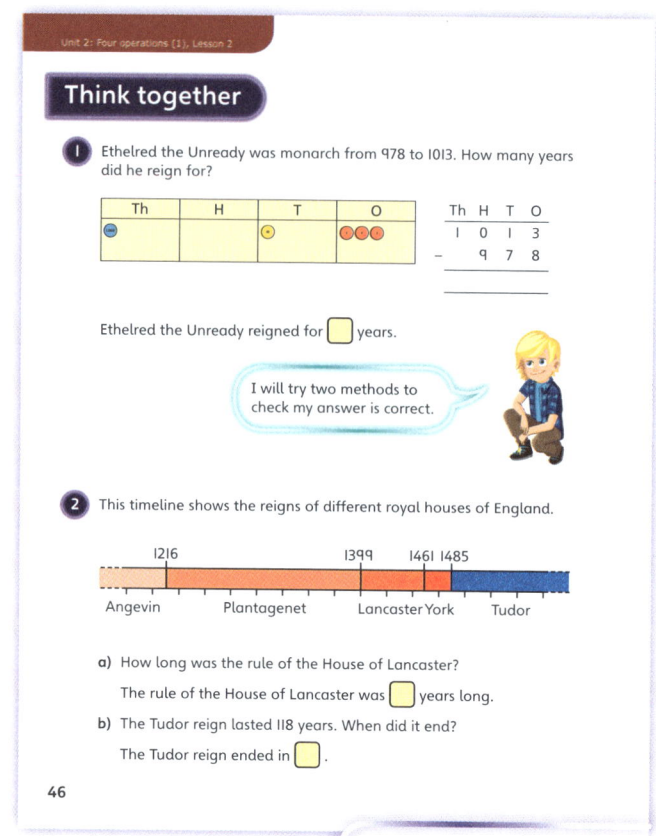

PUPIL TEXTBOOK 6A PAGE 46

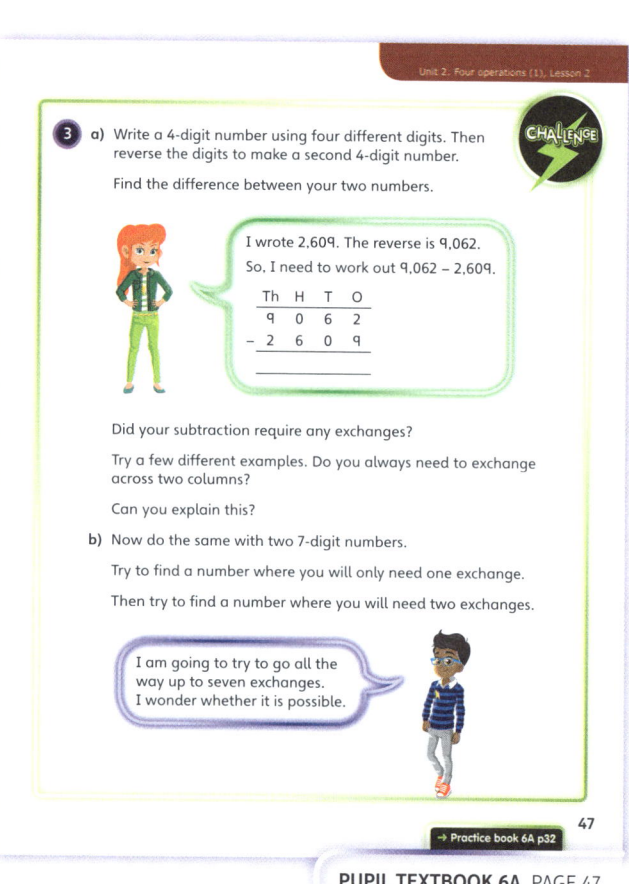

PUPIL TEXTBOOK 6A PAGE 47

Practice

WAYS OF WORKING Independent thinking

IN FOCUS Question **1** provides the opportunity to add and subtract fluently from different starting numbers, exchanging where necessary. Give children the resources to copy the scenario in the pictures and make sure they support this with the formal written method. Question **2** develops children's awareness that a formal written method is not always the most appropriate. The calculation has a difference of 6 which should be easy to solve using mental methods.

STRENGTHEN If children are struggling to set out the column subtractions in question **4**, offer a printed place value grid. Ask:
• *How can you use this to help organise and structure your column method?*

DEEPEN Question **5** deepens children's fluency and problem solving when calculating with addition and subtraction. The question is written in a way that requires some 'untangling'. The bar model will support children with this.

THINK DIFFERENTLY Question **3** offers children the opportunity to think differently as they must interpret the numbers given on the number line in the context of the question, to reason about the value of the third number. They are required to find the sum of all numbers so they must remember to complete this final step of the problem.

ASSESSMENT CHECKPOINT Do children recognise representations of addition and subtraction calculations, and can they solve addition and subtraction calculations that require exchanging, using an efficient columnar method? Can children identify when column methods are appropriate and when there is a more efficient method available? Can children solve problems by using the columnar methods to support and record their working out?

ANSWERS Answers for the **Practice** part of the lesson appear in the separate **Practice and Reflect answer guide**.

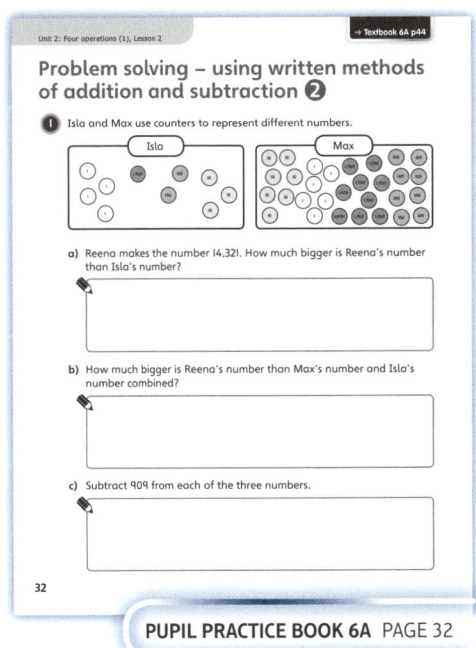

Problem solving – using written methods of addition and subtraction 2

1 Isla and Max use counters to represent different numbers.

a) Reena makes the number 14,321. How much bigger is Reena's number than Isla's number?

b) How much bigger is Reena's number than Max's number and Isla's number combined?

c) Subtract 909 from each of the three numbers.

32

PUPIL PRACTICE BOOK 6A PAGE 32

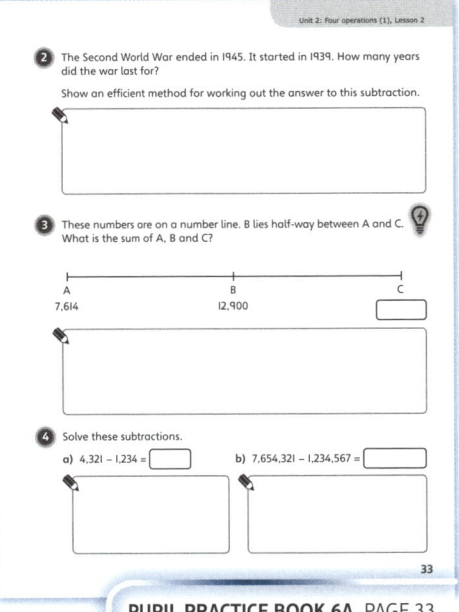

2 The Second World War ended in 1945. It started in 1939. How many years did the war last for?

Show an efficient method for working out the answer to this subtraction.

3 These numbers are on a number line. B lies half-way between A and C. What is the sum of A, B and C?

A
7,614

B
12,900

C

4 Solve these subtractions.
a) 4,321 – 1,234 = [ ]
b) 7,654,321 – 1,234,567 = [ ]

33

PUPIL PRACTICE BOOK 6A PAGE 33

Reflect

WAYS OF WORKING Independent thinking and pair work

IN FOCUS This activity will test children's understanding of a difference. They should recognise that although each number in the subtraction has decreased by 1, the actual difference between the numbers is the same.

Children should recognise that this method will reduce the amount of exchanging required within a written method.

ASSESSMENT CHECKPOINT Look for children's ability to reason about an equivalent difference and their confidence to explain why the difference is the same despite the change in numbers.

ANSWERS Answers for the **Reflect** part of the lesson appear in the separate **Practice and Reflect answer guide**.

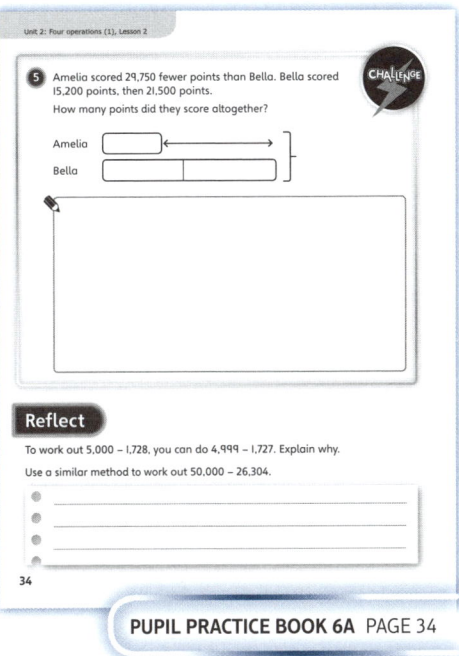

5 Amelia scored 29,750 fewer points than Bella. Bella scored 15,200 points, then 21,500 points.
How many points did they score altogether?

CHALLENGE

Amelia [ ]
Bella [ ]

Reflect

To work out 5,000 – 1,728, you can do 4,999 – 1,727. Explain why.
Use a similar method to work out 50,000 – 26,304.

34

PUPIL PRACTICE BOOK 6A PAGE 34

After the lesson

• Which example of exchanging was trickier for children to grasp: in addition or in subtraction?
• Which concrete or pictorial resource did children most engage with when developing their understanding? How can you use this resource to promote similar engagement in future lessons?

Multiplying numbers up to 4 digits by a 1-digit number

Learning focus

In this lesson children will develop their understanding of the multiplication of 4-digit numbers by 1-digit numbers. They will use multiple representations and methods to solve these calculations.

Small steps

→ Previous step: Problem solving – using written methods of addition and subtraction (2)
→ **This step: Multiplying numbers up to 4 digits by a 1-digit number**
→ Next step: Multiplying numbers up to 4 digits by a 2-digit number

NATIONAL CURRICULUM LINKS

Year 6 Number – Addition, Subtraction, Multiplication and Division

Multiply multi-digit numbers up to 4 digits by a two-digit whole number using the formal written method of long multiplication.

ASSESSING MASTERY

Children can fluently and reliably multiply a 4-digit number by a 1-digit number. They can demonstrate their thinking and understanding through multiple representations and written methods.

COMMON MISCONCEPTIONS

Children may multiply incorrectly when using short multiplication – for example, attempting to solve 2,345 × 6 by calculating '5 × 6', then '45 × 6', then '345 × 6' and finally '2,345 × 6'. Ask:
*   *What do you have if you recombine 5, 45, 345, and 2,345? Is that different from the original calculation?*
*   *What number does the digit '4' represent? How do you know?*

STRENGTHENING UNDERSTANDING

Give opportunities to recap written methods from previous units and years. Show a multiplication and ask:
*   *What is this calculation asking you to do? Can you represent it using resources or with a picture?*
*   *What written method would you use to solve it? Show me.*

GOING DEEPER

Encourage children to write word problems for each other that require them to multiply 4-digit numbers by 1-digit numbers.

KEY LANGUAGE

In lesson: multiplication, total, digit, estimate, rounding

Other language to be used by the teacher: repeated addition, multiply, times, product

STRUCTURES AND REPRESENTATIONS

short multiplication, grid method, place value counters

RESOURCES

Optional: place value counters

 In the eTextbook of this lesson, you will find interactive links to a selection of teaching tools.

Before you teach

*   What methods for solving multiplication calculations are the children fluent with already?
*   How will you use the children's prior learning to engage them and secure progress?

Discover

WAYS OF WORKING Pair work

ASK

- Question ① a): *What number facts could you use to help you estimate the solution to this problem?*
- Question ① b): *How will you go about solving this calculation? What operation could be used?*
- Question ① b): *Can you represent the problem with resources?*

IN FOCUS Question ① a) gives children the opportunity to begin investigating problems that require multiplication to solve. Be aware that neither of the given totals is correct, which may baffle children; encourage them to think about which estimate is more sensible.

Question ① b) gives children their first opportunity to calculate a 4-digit number multiplied by a 1-digit number.

PRACTICAL TIPS Children could role play a travel agency scenario. Give a list of holiday destinations with prices. Children, working in small groups, investigate how much it would cost their group to go on each holiday. What would the price be if two groups went? (Be careful not to let the number of children go over 9.)

ANSWERS

Question ① a): Neither of the totals is correct, but rounding shows that £12,905 is more likely to be correct.

Question ① b): £3,225 × 4 = £12,900. The trip will cost £12,900 for four people.

PUPIL TEXTBOOK 6A PAGE 48

Share

WAYS OF WORKING Whole class teacher led

ASK

- Question ① a): *How have Astrid and Dexter shown that both totals are wrong?*
- Question ① a): *How did you prove they are wrong? Can you make a model or draw a picture to show this?*
- Question ① b): *Can you explain the two different methods shown by the place value counters?*
- Question ① b): *Did you use a written method? Which one?*
- Question ① b): *Can you explain how your method works?*

IN FOCUS When looking at question ① a), discuss why it is important to use the mental methods and number facts demonstrated in Dexter and Astrid's comments when solving any calculation, as they can help children determine whether their solutions are correct or not. Use the pictures of place value counters in question ① b) to develop children's fluency with number and to build their awareness of the commutative nature of multiplication.

Also use question ① b) to show that multiplication is far more efficient than repeated addition.

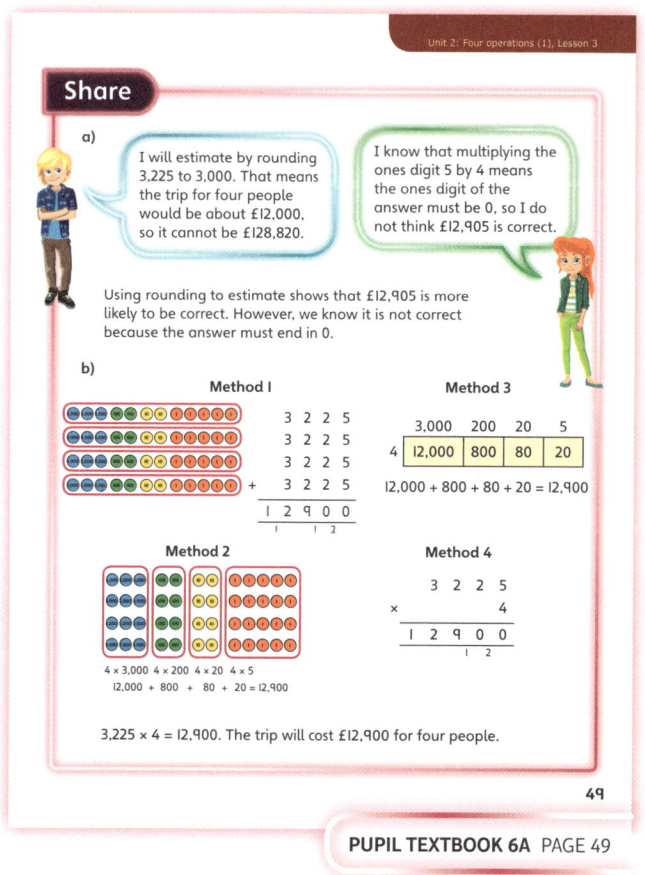

PUPIL TEXTBOOK 6A PAGE 49

Think together

Whole class teacher led (I do, We do, You do)

ASK

- Question **1**: *Can you explain how each method demonstrates the multiplication?*
- Question **1**: *Which method is most efficient? Why?*
- Question **2**: *How can you use the methods shown to help you find the solution for 6 people?*
- Question **3**: *What will you look for to know which digits go where in the calculation?*

IN FOCUS Question **1** helps to secure the link between methods children have learnt previously and short multiplication, which they should recognise as the most efficient method shown. Question **2** requires children to think about the methods used in the previous question and choose the one they prefer to solve this problem.

STRENGTHEN If children are struggling with the abstract methods shown in questions **1** and **2**, ask:

- *Look back at the models in **Share**. How do you think they can help you?*
- *How will you use them to help you solve these calculations?*

DEEPEN To deepen children's investigation of the calculation shown in question **3** a), ask:

- *If the exchanging was moved or taken away, would there be another way of solving this problem? Show me.*

ASSESSMENT CHECKPOINT Do children recognise different multiplication methods, and can they make links with written methods they have learnt about before? Can children use written multiplication methods with fluency to solve problems?

ANSWERS

Question **1**: $2,345 × 4 = 9,380$

The new cost is £9,380.

Question **2** a): $2,865 × 5 = 14,325$ £14,325

Question **2** b): $2,865 × 6 = 17,190$ £17,190

Question **3** a):

TTh	Th	H	T	O
	1	4	5	3
×				2
	2	9	0	6
			1	

Question **3** b): A 4- or 5-digit answer can be made. For example:

TTh	Th	H	T	O
	3	2	1	4
×				5
1	6	0	7	0
1	1		2	

Question **3** c): The greatest answer you can make is:

TTh	Th	H	T	O
	4	3	2	1
×				5
2	1	6	0	5
	1	1		

It is not possible to make a 6-digit answer with the digit cards available.

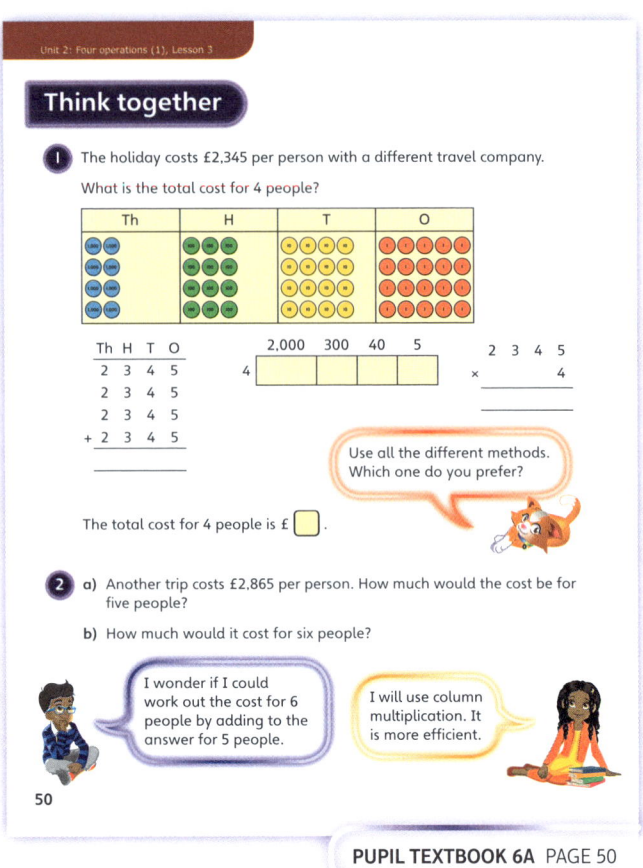

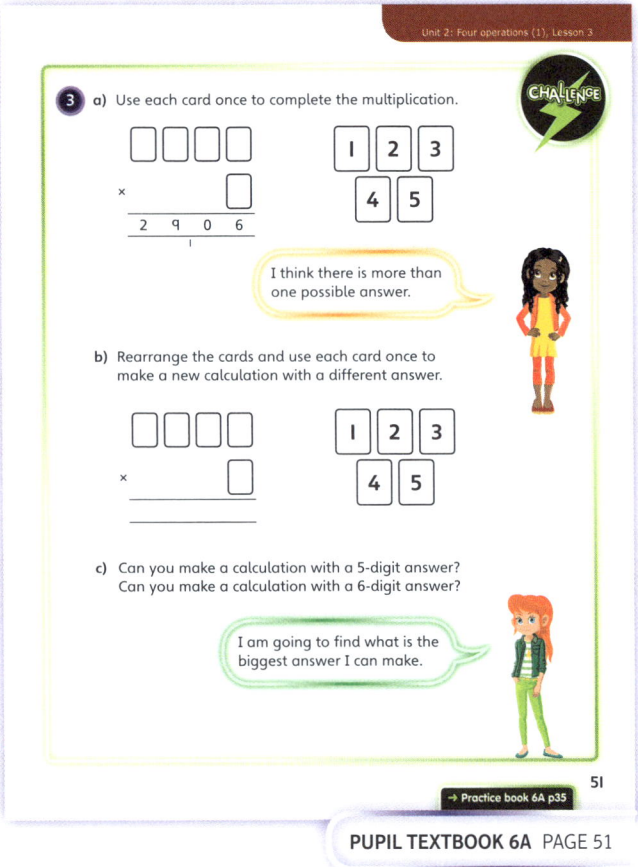

Practice

WAYS OF WORKING Independent thinking

IN FOCUS Questions ❶ a), ❶ b) and ❶ c) give the opportunity to solve multiplication calculations using different methods. This scaffolds understanding through clear partitioning. Children should independently link their understanding of the grid method to short multiplication. Question ❷ links children's understanding of multiplication along a number line with the short multiplication written method.

STRENGTHEN It may help children's understanding in question ❹ if they are able to use resources to create the questions and the calculations. Ask:
• *How could you use resources to represent the problems?*
• *How could you use resources or draw pictures to represent the calculations you need to solve?*

DEEPEN When working on question ❺ b), deepen children's generalisations about the calculations by asking:
• *What is the same/different about the calculations that make the biggest and smallest product?*
• *Is there a way to be sure of making the biggest or smallest product, no matter what digits you use?*

ASSESSMENT CHECKPOINT Can children recognise and interpret different representations of multiplication and link these with the short multiplication written method? Can children solve multiplications using short multiplication with fluency?

ANSWERS Answers for the **Practice** part of the lesson appear in the separate **Practice and Reflect answer guide**.

Reflect

WAYS OF WORKING Independent thinking

IN FOCUS This question will give you the opportunity to assess whether children have been able to identify links between a multiplication method they have learnt before and the method they are currently using. These links will scaffold and cement their understanding of short multiplication so it is important they see and understand them.

ASSESSMENT CHECKPOINT Look for children recognising that short multiplication, like the grid method, uses partitioning when multiplying a number by another number, even though this is less visually obvious than in the grid method.

ANSWERS Answers for the **Reflect** part of the lesson appear in the separate **Practice and Reflect answer guide**.

After the lesson ⏸

• How clearly did your lesson make the links between methods of multiplication?
• What would you do differently next time you teach this, to cement those links even more effectively?
• What percentage of your class is confident and fluent in the use of short multiplication?

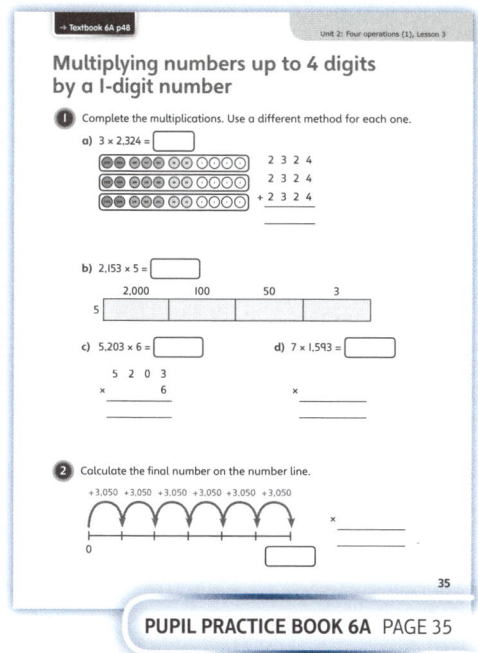

PUPIL PRACTICE BOOK 6A PAGE 35

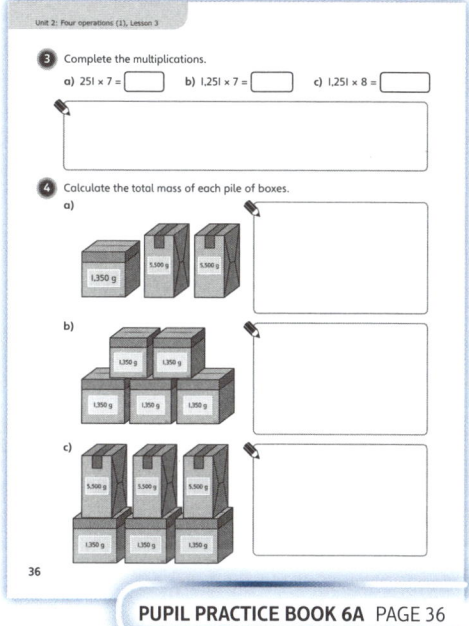

PUPIL PRACTICE BOOK 6A PAGE 36

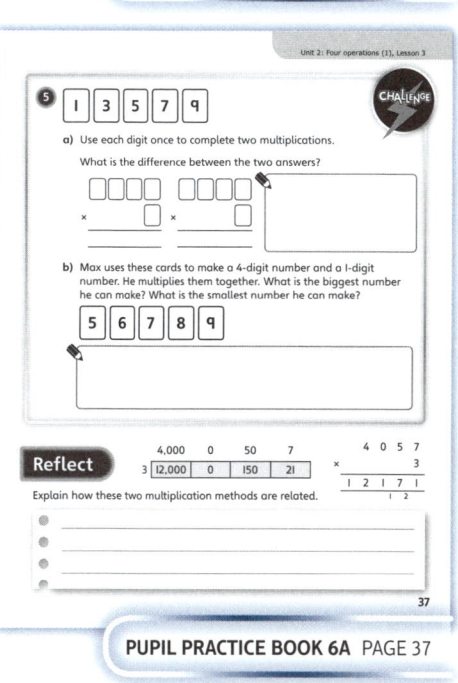

PUPIL PRACTICE BOOK 6A PAGE 37

Multiplying numbers up to 4 digits by a 2-digit number

Learning focus

In this lesson, children will develop their understanding of the multiplication of 4-digit numbers by 2-digit numbers. They will use multiple representations and methods to solve these calculations.

Small steps

→ Previous step: Multiplying numbers up to 4 digits by a 1-digit number

→ **This step: Multiplying numbers up to 4 digits by a 2-digit number**

→ Next step: Dividing numbers up to 4 digits by a 2-digit number (1)

NATIONAL CURRICULUM LINKS

Year 6 Number – Addition, Subtraction, Multiplication and Division

Multiply multi-digit numbers up to 4 digits by a two-digit whole number using the formal written method of long multiplication.

ASSESSING MASTERY

Children can fluently and reliably multiply a 4-digit number by a 2-digit number. They can demonstrate their thinking and understanding through multiple representations and written methods.

COMMON MISCONCEPTIONS

Children may multiply incorrectly when using the column method for multiplication – for example, calculating 345 × 26 as 5 × 6, 40 × 6, 300 × 6, followed by 5 × 2, 40 × 2 and 300 × 2. Ask:

• *What is the place value of each digit in both of the numbers in the calculation?*
• *Is the value of the digit 2 two or twenty? How do you know?*

STRENGTHENING UNDERSTANDING

If children are not confident in using short multiplication to multiply by a 1-digit number, offer opportunities to find the product of a 4-digit and a 1-digit number, using the grid method to support conceptual understanding. Ask:

• *How do the grid method and short multiplication show the multiplication of the 1s by 1s? How do they both show the multiplication of the 10s by 1s?*
• *How are the methods the same and different? How do the similarities help you to use short multiplication?*

GOING DEEPER

Children could write a short multiplication, showing the product of a 4-digit and a 1-digit number. Children could then challenge their partner by hiding one or two of the digits in the calculation. Ask:

• *How can you make the challenge trickier or easier? Explain.*

KEY LANGUAGE

In lesson: multiplication, column method, area, digit

Other language to be used by the teacher: repeated addition, times, multiply, product, calculate, double

STRUCTURES AND REPRESENTATIONS

grid method, long multiplication, short multiplication

RESOURCES

Optional: place value counters

 In the eTextbook of this lesson, you will find interactive links to a selection of teaching tools.

Before you teach

• How confident were children with multiplying by a 1-digit number?
• Were there any misconceptions in the previous lesson that could impact on this lesson? How will you plan for these?

Discover

Pair work

ASK

- Question **1** a): *How can you represent the multiplication you need to solve?*
- Question **1** a): *Does the written method we learnt about in the last lesson work for this calculation? Why?*
- Question **1** b): *Does the coach's method make the calculation simpler? How?*

IN FOCUS Question **1** a) offers children the opportunity to multiply a 4-digit number by a 2-digit number. Discuss how this process is similar to multiplying 4-digit numbers by 1-digit numbers and how it is different.

Question **1** b) develops children's fluency and reasoning by helping them to recognise how factors can be used to simplify a more challenging multiplication.

PRACTICAL TIPS Encourage children to investigate how to represent larger products in different concrete and pictorial ways. Ask:

- *How are the place value counters different when multiplying by a 2-digit number than when multiplying by a 1-digit number?*
- *How could you make this multiplication simpler using your model or drawing?*

ANSWERS

Question **1** a): The athletes in the swimming team swim 25,935 lengths altogether in 21 days.

Question **1** b): $21 \times 1,235 = 3 \times 7 \times 1,235 = 3 \times 8,645 = 25,935$

This method does give the same answer.

Share

WAYS OF WORKING Whole class teacher led

ASK

- Question **1** a): *How do the arrays link with the calculations shown?*
- Question **1** a): *What is different about the written methods shown compared with those we used in the last lesson?*
- Question **1** a): *Did you find the same result? Can you explain how and why?*
- Question **1** b): *How does the grid demonstrate the coach's method?*
- Question **1** b): *Could you have used any other factors to make this calculation simpler?*

IN FOCUS Question **1** a) shows how to record the written method for multiplying a 4-digit number by a 2-digit number. Work through this with children, using the expanded column method to demonstrate how partitioning is used to calculate the product. Link this with the grid method before moving to the compact method, linking back to the previous two methods. Question **1** b) offers children a method of solving this calculation by using two factors of 21 to enable the use of repeated short multiplication with 1-digit numbers (7 first, then 3). Discuss the pros and cons of this method, particularly how it impacts on efficiency.

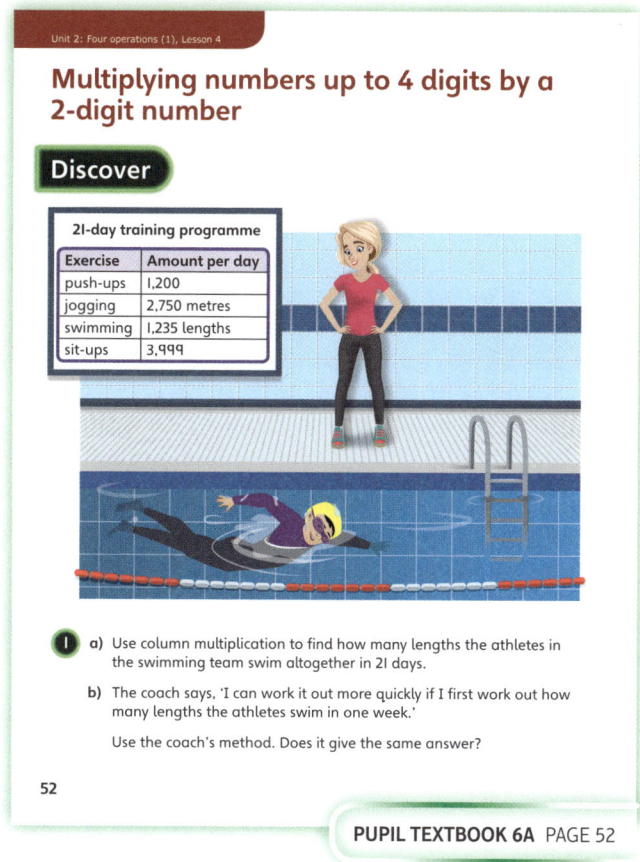

Multiplying numbers up to 4 digits by a 2-digit number

Discover

21-day training programme	
Exercise	**Amount per day**
push-ups	1,200
jogging	2,750 metres
swimming	1,235 lengths
sit-ups	3,999

1 a) Use column multiplication to find how many lengths the athletes in the swimming team swim altogether in 21 days.

b) The coach says, 'I can work it out more quickly if I first work out how many lengths the athletes swim in one week.'

Use the coach's method. Does it give the same answer?

52

PUPIL TEXTBOOK 6A PAGE 52

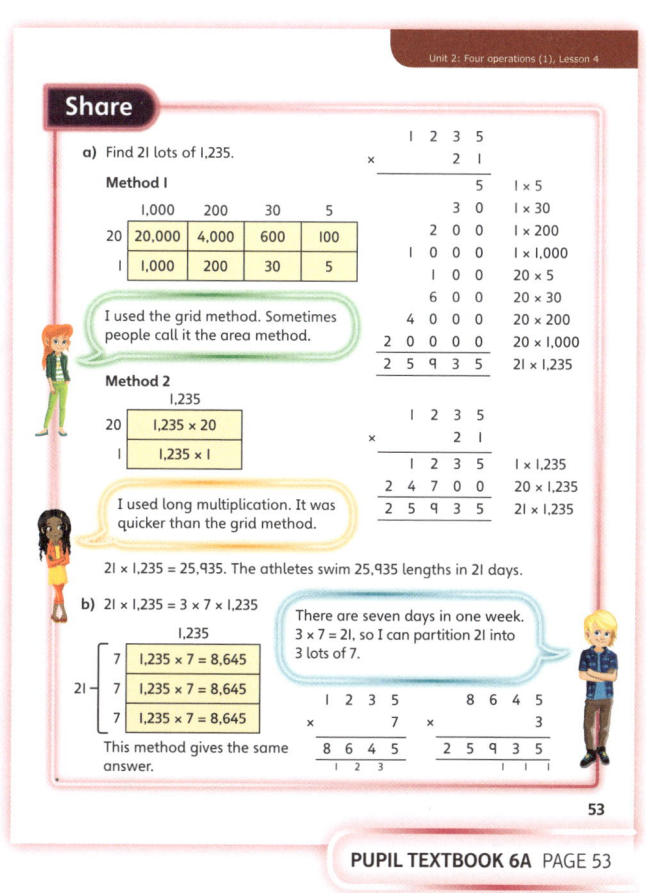

PUPIL TEXTBOOK 6A PAGE 53

Think together

Think together

WAYS OF WORKING Whole class teacher led (I do, We do, You do)

ASK

- Question ❶: *What would be a sensible estimate? How will you round the numbers to help you?*
- Question ❶: *How are the methods linked to each other? How are they the same? How are they different?*
- Question ❷: *Can you explain how Flo's idea makes 3,999 × 35 simpler?*
- Question ❸: *Which method do you think is most effective? Why?*

IN FOCUS Question ❶ demonstrates links between methods children have learnt and are learning. Discuss what is the same and what is different about the methods shown to ensure children can spot where more abstract methods mirror more conceptually explicit methods such as the grid method.

Use question ❷ to develop fluency and reasoning with multiplication calculations. Discuss which calculations are more easily solved using mental methods and which require a written method.

STRENGTHEN If children are struggling to calculate the multiplication in question ❷ using long multiplication, ask:
- *Can you solve this multiplication using the grid method?*
- *How can you use the partitioned multiplication in the grid method to help you with long multiplication?*

DEEPEN Question ❸ deepens children's fluency with different ways of partitioning and representing the same calculation. This will develop number sense and flexibility with numbers. Deepen reasoning by asking:
- *Can you write an explanation of how these methods work? How are they the same and different?*
- *Read your explanation to your partner – do they find your explanation clear and accurate?*

ASSESSMENT CHECKPOINT Can children make links between methods, and do they use understanding of the grid method to secure new understanding of long multiplication? Can children calculate multiplications with fluency, recognising which are more efficiently solved through a mental method and which should be solved using a written method? In Question ❸, look for children demonstrating understanding that the numbers can be partitioned to create simpler multiplications.

ANSWERS

Question ❶: The athletes will swim 29,640 lengths in total.

Question ❷: 139,965

Question ❸: All of Isla's methods produce the same answer.

HTh	TTh	Th	H	T	O
		5	2	0	0
×				2	5
	2	6₁	0	0	0
1	0	4	0	0	0
1	3	0	0	0	0
				1	

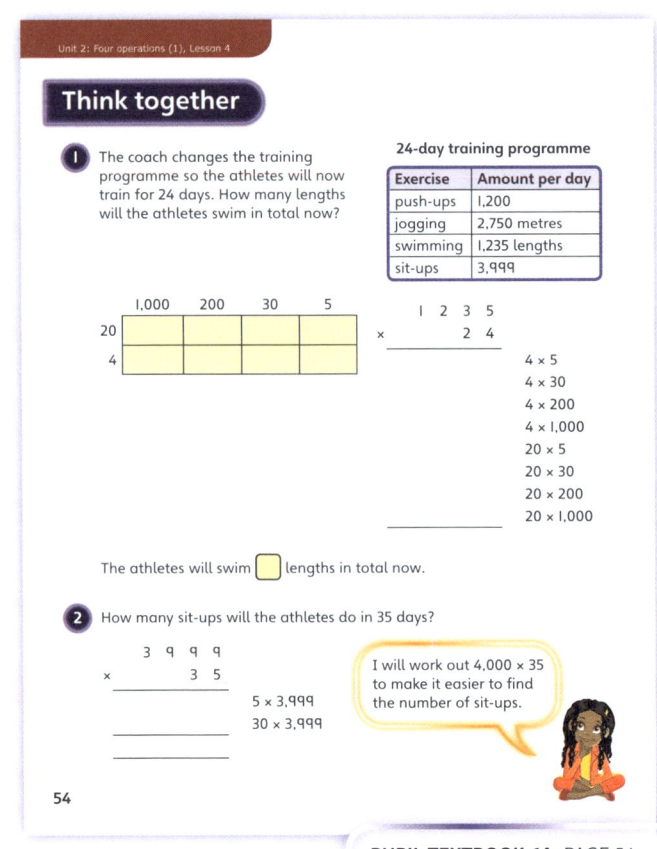

PUPIL TEXTBOOK 6A PAGE 54

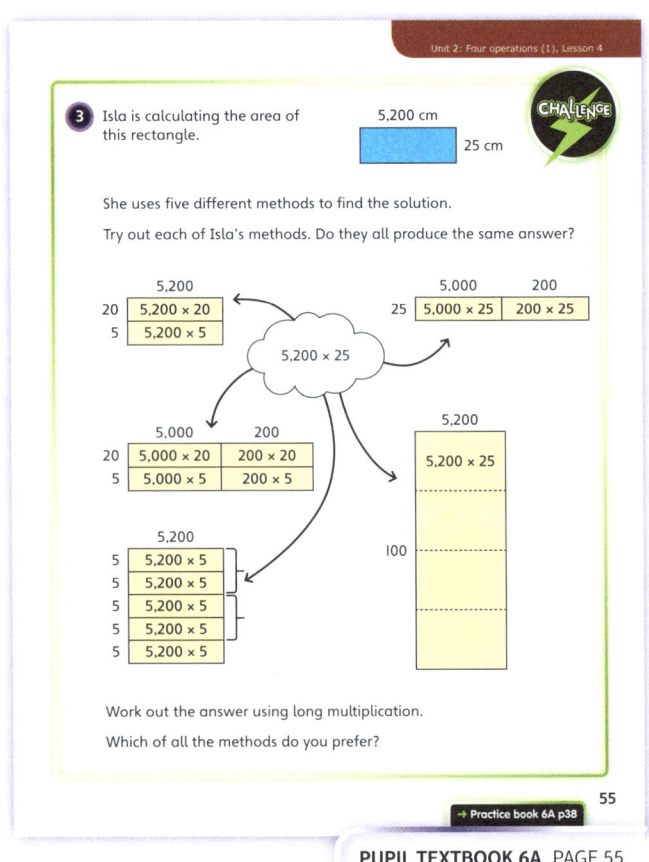

PUPIL TEXTBOOK 6A PAGE 55

Practice

WAYS OF WORKING Independent thinking

IN FOCUS Question ① develops children's ability to recognise and use the links between methods independently.

Question ② allows children to begin using their understanding of long multiplication of 3- and 4-digit numbers in real-life contexts (time).

Question ③ provides another example where factors can be used to complete a long multiplication calculation by breaking it down so that it can be solved in two steps of multiplying by a single digit.

STRENGTHEN If children are not able to answer question ⑤, ask:
- *How could you represent this differently to make it clearer?*
- *Could this be solved through trial and improvement? Is there a more efficient way? Why?*

DEEPEN Question ④ requires children to apply their understanding of multiplication to a problem solving situation. Through rounding and estimating, they should recognise that there is not enough water to fill the pool. They should also be encouraged to use a written method to find out how much water is in the pool now. Ask:
- *How many more buckets does Max need to use? Can you explain how you know it must be more than 100 extra buckets?*

THINK DIFFERENTLY To deepen children's reasoning and fluency with the distributive nature of multiplication in question ③, ask:
- *Does Richard's method work? Can you explain why?*
- *Is there another way you could split the 14 to make two simpler multiplication calculations? How?*
- *Can you create your own calculation where this method could be used to make calculating simpler?*

ASSESSMENT CHECKPOINT Do children make links with previous understanding to fluently and accurately use long multiplication? Are they flexible and fluent in their use of partitioning to make multiplications simpler?

ANSWERS Answers for the **Practice** part of the lesson appear in the separate **Practice and Reflect answer guide**.

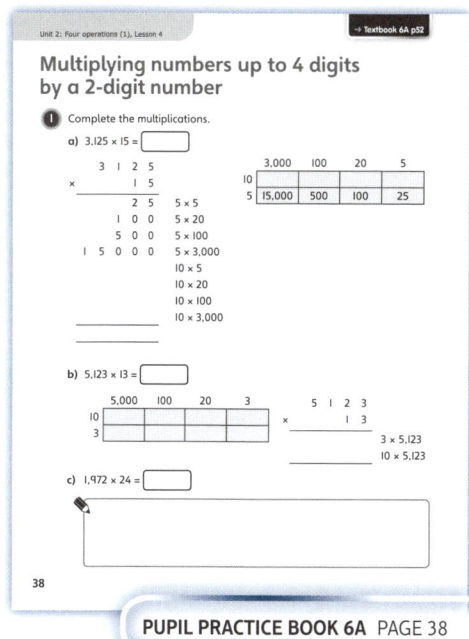

PUPIL PRACTICE BOOK 6A PAGE 38

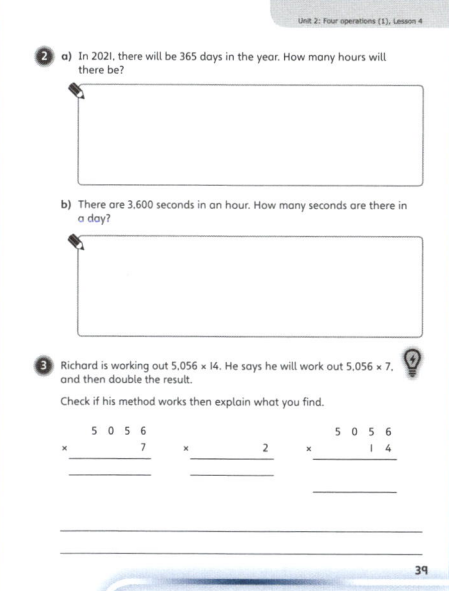

PUPIL PRACTICE BOOK 6A PAGE 39

Reflect

WAYS OF WORKING Independent thinking

IN FOCUS Children may solve this using any of the methods from this lesson. To assess their ability to use long multiplication, ask:
- *Can you represent this with two different methods, one long multiplication and one of your choice? Explain how the two methods are linked.*

ASSESSMENT CHECKPOINT Look for children accurately explaining which product is greater, justifying their solution. Children should find that long multiplication is the most efficient method.

ANSWERS Answers for the **Reflect** part of the lesson appear in the separate **Practice and Reflect answer guide**.

After the lesson ⏸
- Could children fluently and accurately explain the links between the methods shown in this lesson?
- How will you continue to develop the links in children's learning in future lessons?

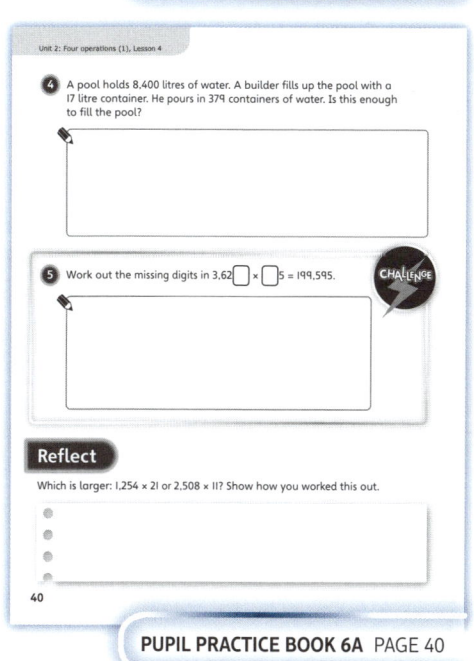

PUPIL PRACTICE BOOK 6A PAGE 40

Dividing numbers up to 4 digits by a 2-digit number ❶

Learning focus

In this lesson, children will develop their understanding of dividing numbers up to 4 digits by 2-digit numbers. They will use multiple representations and methods to solve these calculations.

Small steps

→ Previous step: Multiplying numbers up to 4 digits by a 2-digit number
→ **This step: Dividing numbers up to 4 digits by a 2-digit number (1)**
→ Next step: Dividing numbers up to 4 digits by a 2-digit number (2)

NATIONAL CURRICULUM LINKS

Year 6 Number – Addition, Subtraction, Multiplication and Division

Divide numbers up to 4 digits by a two-digit number using the formal written method of short division where appropriate, interpreting remainders according to the context.

ASSESSING MASTERY

Children can fluently and reliably divide a number with up to 4 digits by a 2-digit number. They can demonstrate their thinking and understanding through multiple representations and written methods.

COMMON MISCONCEPTIONS

Children may have inadequate conceptual understanding of short division, even if they do find the correct answer: for example, thinking that $132 \div 6$ is solved as '6 into 1 goes 0, 6 into 13 goes 2, 6 into 12 goes 2' without an awareness of the place values. Ask

• *Why is your first digit 0 when there are lots of 6s in 100?*

STRENGTHENING UNDERSTANDING

To help children with the inverse grid method, first create a completed grid method calculation together. Ask:
• *What happens if I take away one of the numbers from the outside edge of the grid?*
• *Is there a way of knowing what that number was? What clues are still there that could tell us?*

GOING DEEPER

Encourage children to write real-life division word problems to challenge their partner. Ask:
• *Can you write a problem that has more than one step?*

KEY LANGUAGE

In lesson: division, short division, groups of, hundreds (100s), tens (10s), ones (1s), remainder, digit

Other language to be used by the teacher: divide, area model, written division, shared

STRUCTURES AND REPRESENTATIONS

place value grid and counters, short division

RESOURCES

Optional: place value counters

 In the eTextbook of this lesson, you will find interactive links to a selection of teaching tools.

Before you teach

• How can you use understanding of multiplication to deepen understanding of division?
• How confident are children with division already?
• What misconceptions will you need to plan for?

Discover

WAYS OF WORKING Pair work

ASK

- Question **1** a): *What operation do you need to solve this question? Explain how you know.*
- Question **1** a): *How can you represent your thinking and solution? Can you use more than one method?*
- Question **1** b): *How is this question similar to question **1** a)? How is it different?*
- Question **1** b): *How will you represent your working for this question? How is it similar to your working in question **1** a)? How is it different?*

IN FOCUS In question **1** a) children divide a 3-digit number by a 1-digit number. Encourage them to show their working however they feel most comfortable. Question **1** b) develops children's thinking by extending the calculation to a 3-digit number divided by a 2-digit number. Discuss how their approach is the same as and different from the one used in question **1** a).

PRACTICAL TIPS Ask children to draw the grid method that would link the inverse multiplication (6 × __ = 132) with the division. Discuss what is known and not known. How could you make it easier to find out how many sixes are in 132? Link this to children's fluency with partitioning, as demonstrated in previous lessons on multiplication.

ANSWERS

Question **1** a): 132 bottles of water will last for 22 days.

Question **1** b): The astronauts will eat 21 tubes of fruit puree in one week.

Share

WAYS OF WORKING Whole class teacher led

ASK

- Questions **1** a) and **1** b): *How does the inverse grid method help you to solve the division?*
- Questions **1** a) and **1** b): *How is partitioning used to help complete the inverse grid method?*
- Question **1** a): *Why does the 100s counter get moved into the tens column?*
- Questions **1** a) and **1** b): *Did your method of solving the division calculation match any of those on page 57? How was your method similar or different?*

IN FOCUS It is important at this point to discuss the links between the two methods shown on the page to help children make connections between prior understanding and new learning. When introducing the short division method, ensure children understand what they are actually calculating. This will help them to realise how and why place value fits into the written method and will help secure their conceptual understanding.

PUPIL TEXTBOOK 6A PAGE 56

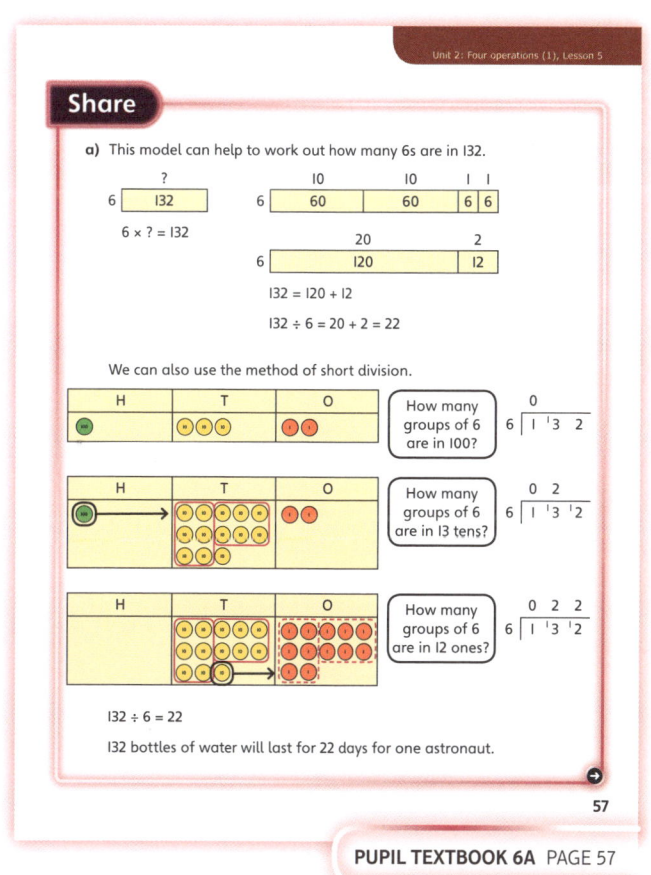

PUPIL TEXTBOOK 6A PAGE 57

Think together

Whole class teacher led (I do, We do, You do)

ASK

- Question ❶: *How can you use the grid method to help you solve the short division?*
- Question ❶: *What is similar about the two methods? What is different?*
- Question ❶: *How many 5s are in 3,350?*
- Question ❷: *How many times do you need to exchange when solving this calculation? Prove it.*
- Question ❸ a): *Are any of these calculations quicker to solve mentally? Why?*
- Question ❸ a): *How will you set out the short division to calculate these divisions?*
- Question ❸ b): *What is the best method to solve these divisions? Can any of the methods from this lesson help?*

IN FOCUS Question ❶ secures children's understanding of using the grid method inversely to solve division calculations, and of using the short division method. Encourage children to solve the calculation using the grid method first, then using short division. Help children to secure their understanding by discussing what is the same and what is different about the two methods. Question ❷ offers the opportunity to calculate the division of a 3-digit number by a 2-digit number. It also provides a good opportunity to discuss how children can use number facts to help them, as this division can be solved easily by seeing it as 56 ÷ 4.

STRENGTHEN If children are not able to solve the calculation in question ❷, ask:
- *Can you represent this calculation using the inverse grid method?*
- *How can you use this representation to help you solve the written calculation?*

DEEPEN Ask children to create their own challenges for their partner like the ones in question ❸ b). Ask:
- *How many digits can you hide before your calculation is no longer possible? Explain why.*

ASSESSMENT CHECKPOINT Can children accurately divide a 4-digit number by a 1-digit number and link the inverse grid method to short division, using both methods? Can children spot when it is possible to use short division to divide a 3-digit number by a 2-digit number and do this accurately? Can children divide a 4-digit number by a 2-digit number, recognising where written calculations are and are not necessary?

ANSWERS

Question ❶: 3,350 ÷ 5 = 670

Question ❷: 560 ÷ 40 = 14

Question ❸ a): 5,050 ÷ 25 = 202

1,770 ÷ 15 = 118

9,840 ÷ 24 = 410

Question ❸ b): 4,6**6**4 ÷ 11 = 4**2**4

2,47**2** ÷ 12 = 20**6**

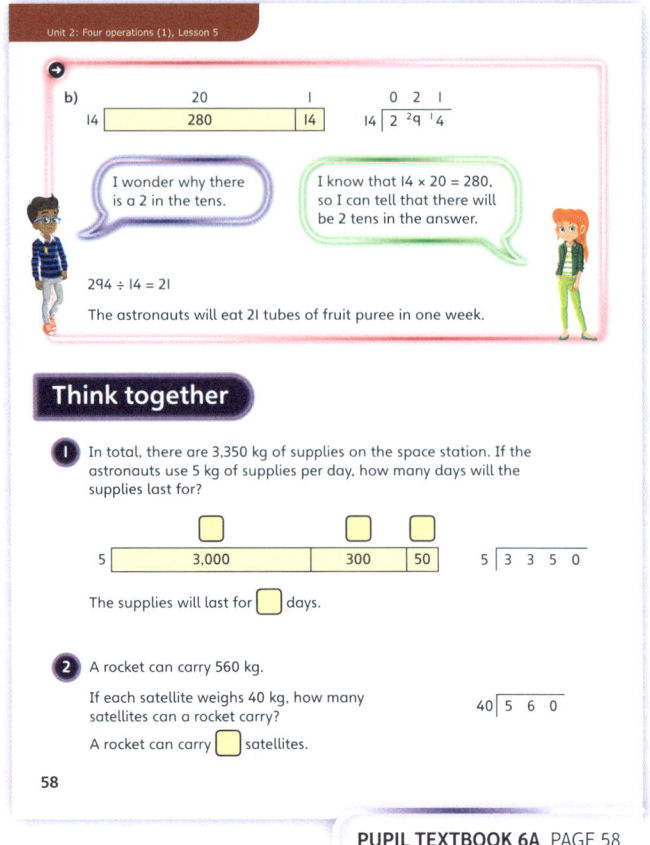

PUPIL TEXTBOOK 6A PAGE 58

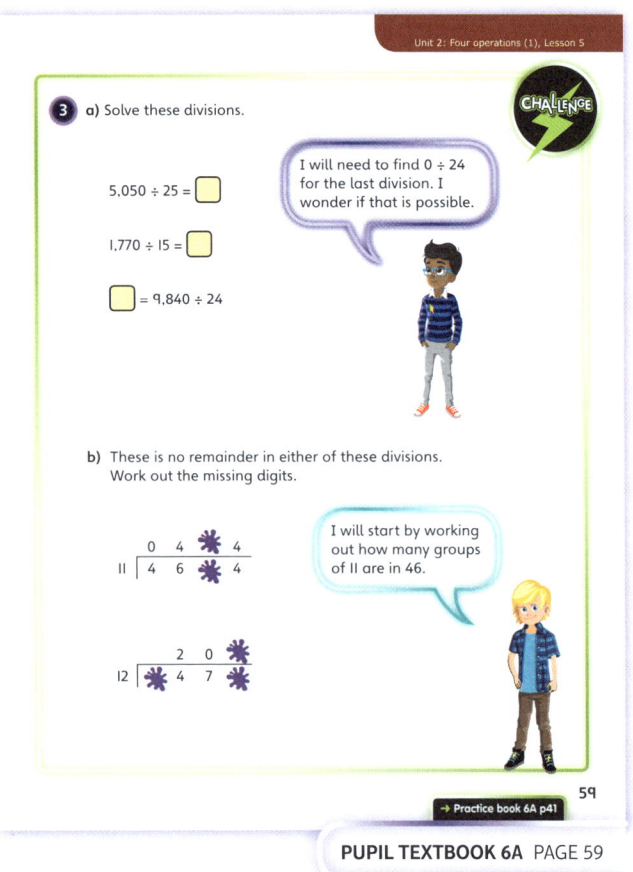

PUPIL TEXTBOOK 6A PAGE 59

90

Practice

WAYS OF WORKING Independent thinking.

IN FOCUS Question **1** scaffolds children's ability to link the inverse grid method with short division. Encourage children to finish the partially completed calculations to help them check and justify their solutions. Question **4** gives children an opportunity to independently use their knowledge of number to solve division calculations. It will be important to discuss with children how, after solving 468 ÷ 9, they can use their knowledge of that calculation to more easily solve 4,689 ÷ 9.

STRENGTHEN If children are struggling with question **3**, ask:
- *What calculation do you need to solve? Explain how you know.*
- *Can you draw the inverse grid method that would match this calculation?*
- *How can the grid method help you show the calculation with short division?*
- *Which method is more efficient? Why?*

DEEPEN If children have solved question **5**, deepen their reasoning by asking:
- *Can you write a clear explanation of how you solved this problem?*
- *How can you make sure your explanation is clear enough so that someone seeing this type of problem for the first time would understand?*

ASSESSMENT CHECKPOINT Can children use the inverse grid method and short division to solve division calculations with fluency? Do they recognise when a calculation can be efficiently solved mentally? Can children solve divisions of up to 4-digit numbers by 1- and 2-digit numbers, recognising where their knowledge of number facts can help them to solve a calculation more efficiently?

ANSWERS Answers for the **Practice** part of the lesson appear in the separate **Practice and Reflect answer guide**.

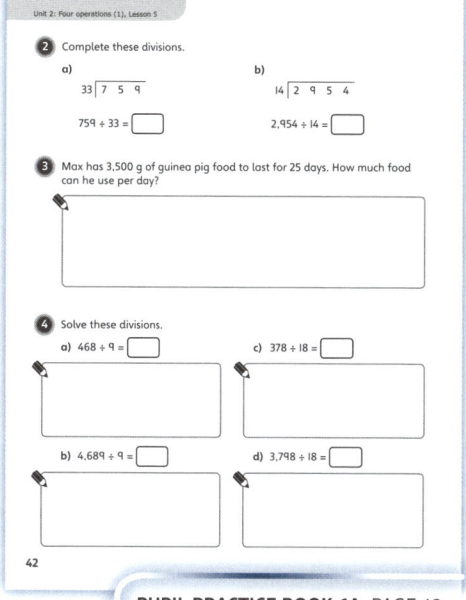

PUPIL PRACTICE BOOK 6A PAGE 41

PUPIL PRACTICE BOOK 6A PAGE 42

Reflect

WAYS OF WORKING Independent thinking

IN FOCUS Children should choose to solve this using the two methods that have been covered in this lesson. To ensure you are able to assess their ability to use the inverse grid method and short division, ask:
- *Can you represent this with both the inverse grid method and short division?*
- *Explain how the two methods are linked.*

ASSESSMENT CHECKPOINT Look for children's fluency in the two methods studied in this lesson. Children should be able to confidently explain the links between the methods.

ANSWERS Answers for the **Reflect** part of the lesson appear in the separate **Practice and Reflect answer guide**.

After the lesson ⏸

- Can children confidently explain how these two methods are linked?
- How can you offer children opportunities to use these methods in other areas of the curriculum?

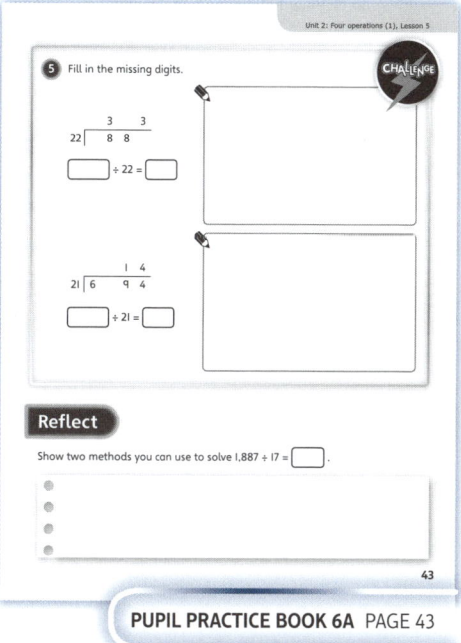

PUPIL PRACTICE BOOK 6A PAGE 43

Dividing numbers up to 4 digits by a 2-digit number ②

Learning focus

In this lesson children will develop their understanding of how 1-digit factors of 2-digit numbers can be used to make the division of numbers with up to 4-digits by 2-digit numbers easier to solve.

Small steps

→ Previous step: Dividing numbers up to 4 digits by a 2-digit number (1)

→ **This step: Dividing numbers up to 4 digits by a 2-digit number (2)**

→ Next step: Dividing numbers up to 4 digits by a 2-digit number (3)

NATIONAL CURRICULUM LINKS

Year 6 Number – Addition, Subtraction, Multiplication and Division

Divide numbers up to 4 digits by a two-digit number using the formal written method of short division where appropriate, interpreting remainders according to the context.

ASSESSING MASTERY

Children can fluently identify the most efficient way to use the single-digit factors of a number to solve challenging division calculations where a number with up to 4 digits is divided by a 2-digit number.

COMMON MISCONCEPTIONS

Children may try to solve a calculation such as $1,260 \div 14$ by partitioning 14, rather than by using its factors. For example, $1,260 \div 10 = 126$, then $126 \div 4 = 31 \cdot 5$; instead of $1,260 \div 2 = 630$, then $630 \div 7 = 90$. Ask:
• *Are 10 and 4 factors of 14? How can you check?*

STRENGTHENING UNDERSTANDING

Children who are not able to find factors of 2-digit numbers may benefit from having the opportunity to use counters to create arrays for different numbers. Ask:
• *Can you use the counters to make an array for 15?*
• *What other arrays are possible? How many factors can you identify?*

GOING DEEPER

Children could create an instructional video, explaining this division method to a lower year group. Ask:
• *How will you make sure your explanation is clear and concise?*
• *What pictures and models will you use and how will you use them?*
• *What vocabulary will you need to include and why?*

KEY LANGUAGE

In lesson: division, divide, factor

Other language to be used by the teacher: group, share

STRUCTURES AND REPRESENTATIONS

bar model, grid method

RESOURCES

Optional: place value counters, multiplication square, counters

 In the eTextbook of this lesson, you will find interactive links to a selection of teaching tools.

Before you teach

• Did any children recognise how they could solve division calculations with these types of number using a similar method in the previous lesson?
• How will you enable these children to deepen their understanding?

Discover

Pair work

ASK

- Question ❶ a): *What do you need to know to be able to solve this question?*
- Question ❶ a): *How will you use the number of people on the ride to work out the solution?*
- Question ❶ a): *How many different ways can you represent this division?*
- Question ❶ b): *How can you use what you have already found out to solve this question more easily?*
- Question ❶ b): *Can you use any factor pairs to make this calculation simpler? Explain.*

IN FOCUS Question ❶ requires children to divide a 3-digit number by two 1-digit numbers. It will be important to discuss how children can go about solving this question – either by solving (750 ÷ 3) ÷ 5 or (750 ÷ 5) ÷ 3, or by solving 750 ÷ 15. Children should be encouraged to discuss which method is easier, which is more efficient, and why.

PRACTICAL TIPS Encourage children to represent the problem using as many representations and methods as they can recall. Offer concrete resources such as place value counters and reminders of the pictorial and written methods they have studied. Ask children to compare and discuss the efficiency of their approaches.

ANSWERS

Question ❶ a): The log flume ran 50 times today.

Question ❶ b): 150 people rode in the log flume boat per hour.

Share

WAYS OF WORKING Whole class teacher led

ASK

- Question ❶ a): *What do the three bar models demonstrate?*
- Question ❶ a): *Is this the only order in which you can solve the divisions? Can you use bar models to show another way?*
- Question ❶ a): *Could you have divided 750 by a 2-digit number and got the same answer?*
- Question ❶ a): *Can you explain the link between your 2-digit number and the two 1-digit numbers?*
- Question ❶ b): *Can you explain how the grid demonstrates the division calculation?*
- Question ❶ b): *How is this division similar to the division in question ❶ a)?*
- Question ❶ b): *What other methods could you have used to answer this question?* [750 ÷ 5, relating this to 75 ÷ 5; or recognising that there must be 10 rides per hour because 50 ÷ 5 = 10 and then multiplying 15 × 10]

IN FOCUS In this part of the lesson, make sure the link between the numbers 3 and 5, and their common multiple 15, is made explicitly clear. Using the bar model shown, make sure children recognise that 750 has been shared into 15 equal groups through the use of its factors, making the division simpler. Encourage children to discuss how this method could help them with other calculations.

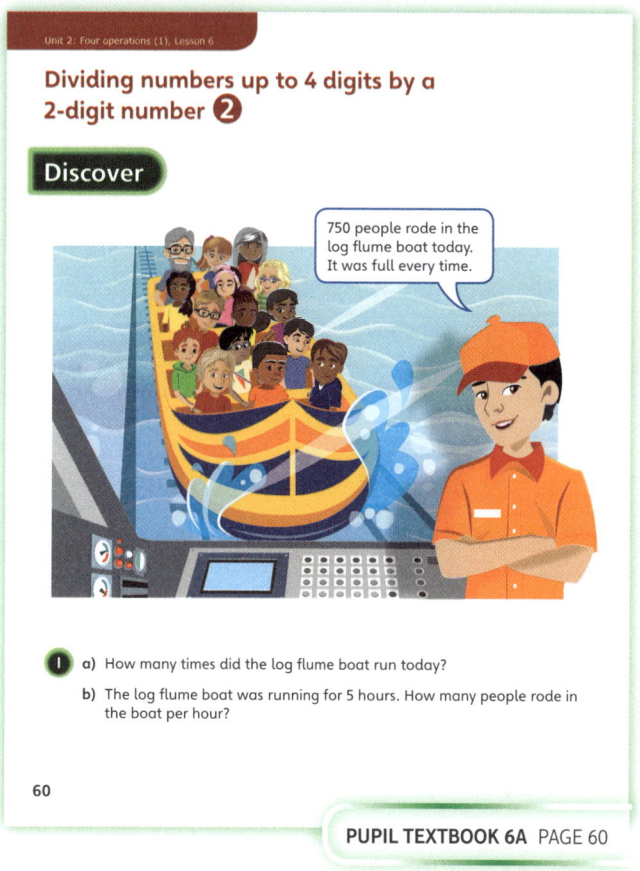

PUPIL TEXTBOOK 6A PAGE 60

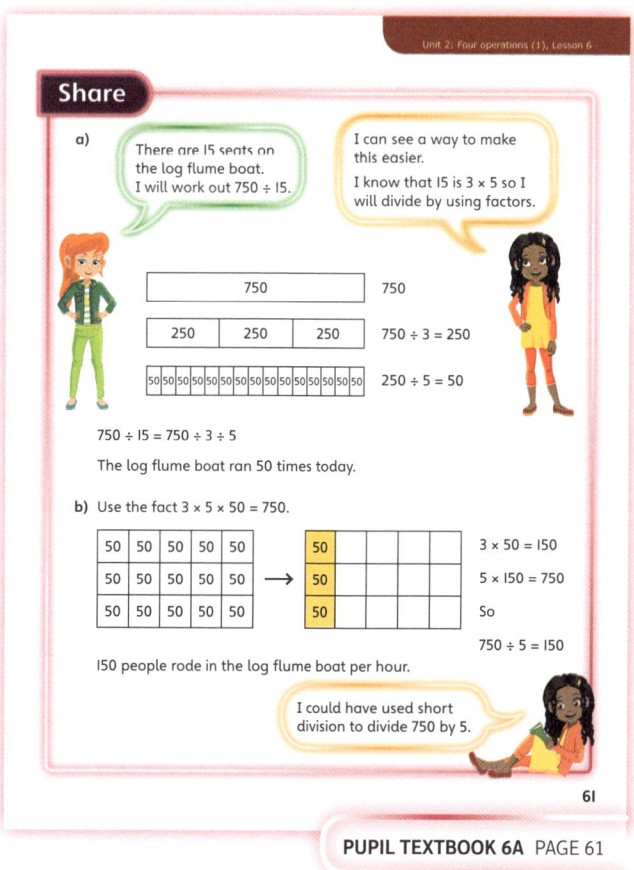

PUPIL TEXTBOOK 6A PAGE 61

Think together

WAYS OF WORKING Whole class teacher led (I do, We do, You do)

ASK

- Question **1**: *What does each box in the grid represent? How are the grids linked?*
- Question **1**: *Which division is it better to solve first, divide by 7 or divide by 2? Why?*
- Question **2**: *What mistake has been made in '5,490 divided by 10 divided by 8'?*
- Question **3** a): *How can you tell if the chains of factors will work? Is there a quick way of knowing?*
- Question **3** b): *Which factors would you choose? Is one method more efficient than the others? Why?*

IN FOCUS Question **1** links children's understanding of the grid representation with their new learning about using 1-digit factors to solve 2-digit divisions. Highlight how these calculations could be approached either way around. Ask whether it is easier to begin by dividing by 2 or by 7, and why.

Question **2** addresses the misconception that partitioning a 2-digit number will result in the correct division being solved. Encourage children to draw the grid representation for both approaches and discuss the solutions.

Question **3** b involves identifying factors of the divisor to solve a division. Emphasise that this strategy requires children to draw on number facts that they already know.

STRENGTHEN If children are struggling to know whether the calculation strings in question **3** a) work or not, ask:

- *Could you draw grids to represent each calculation string?*
- *Now you know that divide by 2, divide by 6 works, does that help you with any other calculations? How?*

DEEPEN Children investigate a calculation such as 765 ÷ 17. The prime number only has 1 and itself as factors. Ask:

- *Can you apply the strategies you have learnt to this calculation? How?*
- *What other calculations could you write that share the same properties as this one?*

ASSESSMENT CHECKPOINT Can children link their abstract understanding of using factors to solve trickier divisions with the pictorial grid representation? Do they recognise what the segments of each grid are worth? Do children understand that partitioning a number will not result in the correct division in the way that finding factors will?

ANSWERS

Question **1**: 90 people ride on the new roller coaster per day.

Question **2**: 5,490 ÷ 6 ÷ 3 solves the division correctly.

$$5,490 ÷ 6 ÷ 3 = 305$$

305 tickets were sold.

Question **3** a): All calculations give the correct answer.

Answers will vary. Look for children referring to their understanding of factors and division to support their explanation.

Question **3** b): 1,800 → ÷6 → ÷4

1,800 → ÷3 → ÷2 → ÷4

1,800 → ÷3 → ÷2 → ÷2 → ÷2

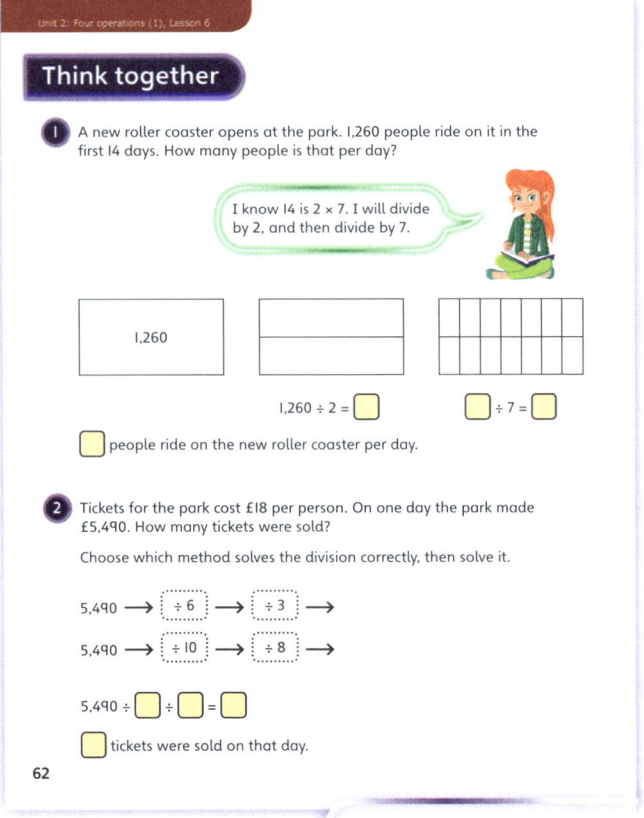

Unit 2: Four operations (1), Lesson 6

Think together

1 A new roller coaster opens at the park. 1,260 people ride on it in the first 14 days. How many people is that per day?

I know 14 is 2 × 7. I will divide by 2, and then divide by 7.

1,260

1,260 ÷ 2 = ☐   ☐ ÷ 7 = ☐

☐ people ride on the new roller coaster per day.

2 Tickets for the park cost £18 per person. On one day the park made £5,490. How many tickets were sold?

Choose which method solves the division correctly, then solve it.

5,490 → ÷ 6 → ÷ 3 →

5,490 → ÷ 10 → ÷ 8 →

5,490 ÷ ☐ ÷ ☐ = ☐

☐ tickets were sold on that day.

62

PUPIL TEXTBOOK 6A PAGE 62

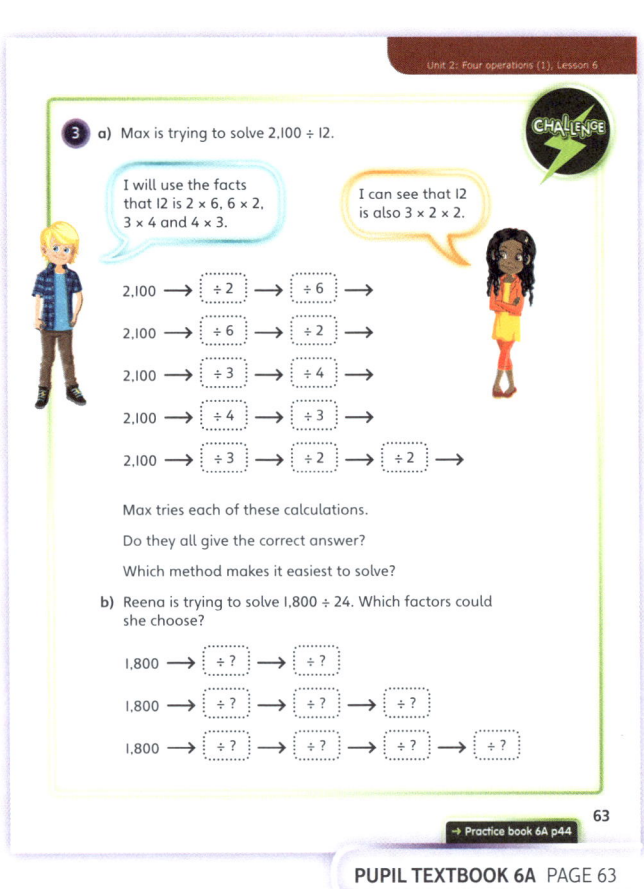

Unit 2: Four operations (1), Lesson 6

3 a) Max is trying to solve 2,100 ÷ 12.

CHALLENGE

I will use the facts that 12 is 2 × 6, 6 × 2, 3 × 4 and 4 × 3.

I can see that 12 is also 3 × 2 × 2.

2,100 → ÷ 2 → ÷ 6 →

2,100 → ÷ 6 → ÷ 2 →

2,100 → ÷ 3 → ÷ 4 →

2,100 → ÷ 4 → ÷ 3 →

2,100 → ÷ 3 → ÷ 2 → ÷ 2 →

Max tries each of these calculations.

Do they all give the correct answer?

Which method makes it easiest to solve?

b) Reena is trying to solve 1,800 ÷ 24. Which factors could she choose?

1,800 → ÷ ? → ÷ ? →

1,800 → ÷ ? → ÷ ? → ÷ ? →

1,800 → ÷ ? → ÷ ? → ÷ ? → ÷ ? →

63

→ Practice book 6A p44

PUPIL TEXTBOOK 6A PAGE 63

Practice

WAYS OF WORKING Independent thinking

IN FOCUS Question **1** a) links children's understanding of the bar model with their new learning in this lesson. Children should be able to use their understanding of the factors of 14 from earlier in the lesson to scaffold their independent work. Question **1** b) moves on from the bar model to the grid representation to represent division calculations.

Question **2** moves on to abstract written calculations. The first two calculations support children's thinking by giving the factors they need to use, but this is withdrawn in the third and fourth calculations.

STRENGTHEN If children are struggling to identify the factors to use to solve the calculations in question **3**, give them a multiplication square and ask:
• *How can you use this to help you find the factors you can use?*

DEEPEN Question **4** b) gives children a good opportunity to begin generalising about the calculations and the properties of the numbers they are working with. Deepen children's reasoning by asking:
• *How can you represent your idea with a picture?*
• *Can you use more than one type of picture to support your thinking?*
• *Explain how your pictures demonstrate your ideas.*

ASSESSMENT CHECKPOINT Can children link pictorial representations with abstract division calculations and explain how a picture represents a calculation? Do they use factors to help them solve divisions of numbers up to 4 digits by a 2-digit number?

ANSWERS Answers for the **Practice** part of the lesson appear in the separate **Practice and Reflect answer guide**.

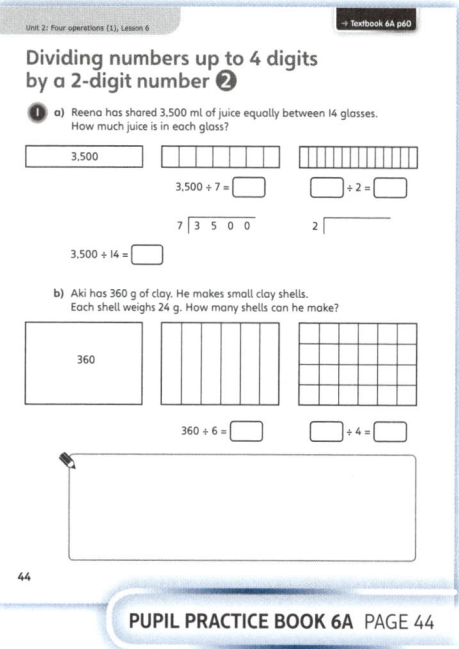

PUPIL PRACTICE BOOK 6A PAGE 44

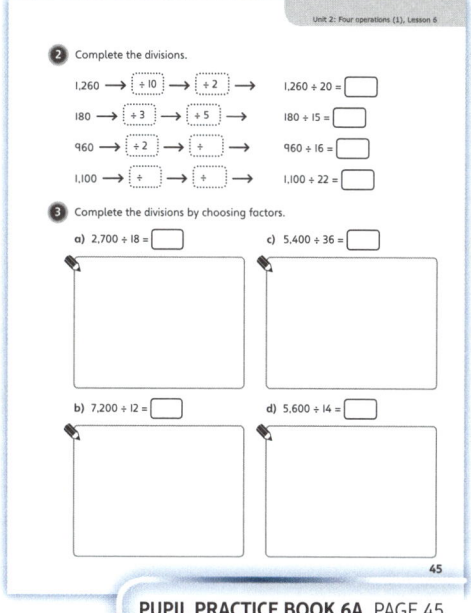

PUPIL PRACTICE BOOK 6A PAGE 45

Reflect

WAYS OF WORKING Independent thinking

IN FOCUS This question will give you a final opportunity to assess children's fluency when using factors to solve the division of numbers up to 4 digits by 2-digit numbers. To encourage children to compare their methods, ask:
• *Did you use the same factors as the person next to you?*
• *Are there other solutions you didn't use? Explain how you know.*

ASSESSMENT CHECKPOINT Look for children recognising that they could solve this by dividing by 10, then by 2 (or vice versa), or by dividing by 5, then by 4 (or vice versa). Children may show this in a pictorial or abstract way.

ANSWERS Answers for the **Reflect** part of the lesson appear in the separate **Practice and Reflect answer guide**.

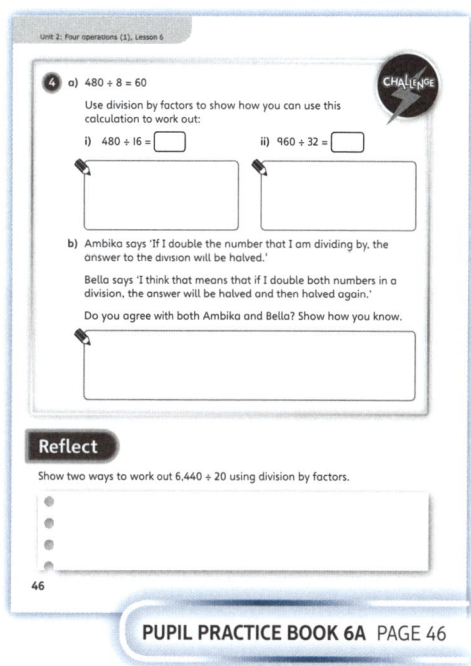

PUPIL PRACTICE BOOK 6A PAGE 46

After the lesson ⏸

• How fluently were children able to identify the factors they could use to solve the divisions efficiently?
• Did any unexpected misconceptions arise in this lesson? How will you plan for them in the future?

Dividing numbers up to 4 digits by a 2-digit number ❸

Learning focus

In this lesson, children will learn long division as a method for solving division calculations where short division is less efficient. They will learn how this method links with those they have already learnt, and how to record it accurately.

Small steps

→ Previous step: Dividing numbers up to 4 digits by a 2-digit number (2)

→ **This step: Dividing numbers up to 4 digits by a 2-digit number (3)**

→ Next step: Dividing numbers up to 4 digits by a 2-digit number (4)

NATIONAL CURRICULUM LINKS

Year 6 Number – Addition, Subtraction, Multiplication and Division

Divide numbers up to 4 digits by a two-digit whole number using the formal written method of long division, and interpret remainders as whole number remainders, fractions, or by rounding, as appropriate for the context.

ASSESSING MASTERY

Children can use their fluency with multiples of 2-digit numbers to accurately divide a number with up to 4 digits by a 2-digit number. They can represent this understanding using an efficient written method.

COMMON MISCONCEPTIONS

When using long division, children may subtract the factor as well as the multiple from the dividend. Ask:
• *What does each of these numbers represent?*
• *Which number should be subtracted from the number you are dividing? Explain how you know.*

STRENGTHENING UNDERSTANDING

Children who find it difficult to secure conceptual understanding of long division may benefit from practising with a 2-digit number divided by a 1-digit number, for example, 28 ÷ 4. Link long division to an array, discussing the multiples children could use to solve the division more quickly. Ask:
• *Can you organise 28 counters into the array that shows this division?*

GOING DEEPER

Children could investigate the statement:
• *Is it always, sometimes or never true that when calculating a division where a number is divided by a 2-digit prime number, short division won't work?*

KEY LANGUAGE

In lesson: group, division, divide, method

Other language to be used by the teacher: multiple, divisor, inverse grid method, approximation, short division, long division, area model, factor

STRUCTURES AND REPRESENTATIONS

grid method, long division

RESOURCES

Optional: place value counters

 In the eTextbook of this lesson, you will find interactive links to a selection of teaching tools.

Before you teach

• How will you show the links between the method in today's lesson and those from previous lessons?
• Are children confident using multiples? Will they need more support or practice to ensure sufficient progress in this lesson?

Discover

WAYS OF WORKING Pair work

ASK

- Question ❶ a): *How could you use rounding to help you approximate an answer?*
- Question ❶ a): *Which method will you use to solve this calculation?*
- Question ❶ b): *What multiplication will you try? Why?*

IN FOCUS Question ❶ a recaps children's previous learning about division and gives them an opportunity to investigate whether the methods they already understand work effectively in this situation. Children may begin to recognise that not all methods are suitable for this calculation. Question ❶ b) reminds children about the use of the inverse operation to check calculations.

PRACTICAL TIPS When solving question ❶ a), encourage children to investigate which of the methods they have learnt about previously works best to solve this calculation. Children should recognise that the methods they already know are less efficient and could result in errors being made.

ANSWERS

Question ❶ a): There are 29 security officers in each group.

Question ❶ b): 29 × 13 = 377

So the answer 29 is correct.

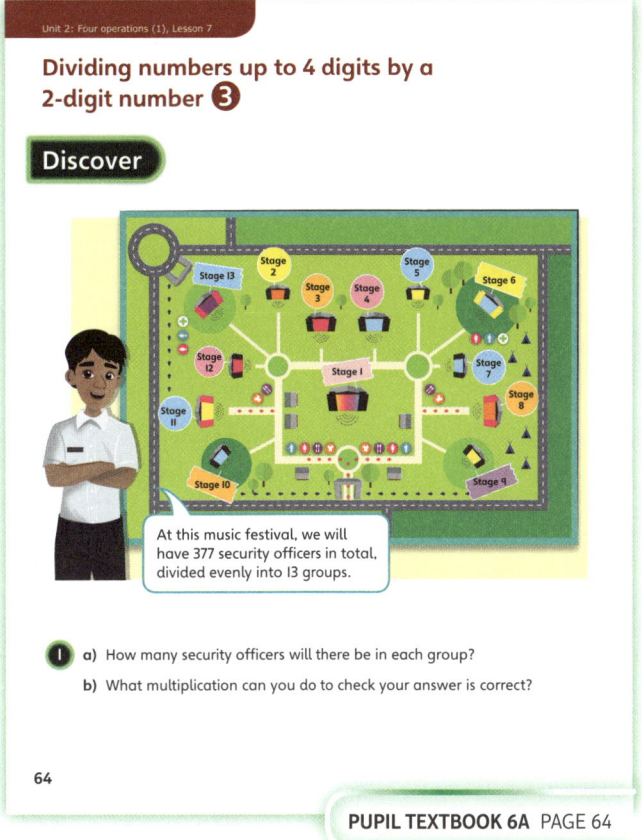

PUPIL TEXTBOOK 6A PAGE 64

Share

WAYS OF WORKING Whole class teacher led

ASK

- Question ❶ a): *How is long division different from short division? Which method is needed here?*
- Question ❶ a): *How do the bar models help with the long division?*
- Question ❶ a): *Why is knowing how to find multiples important when using this method?*
- Question ❶ b): *Did you get back to the number you started with? What does this tell you?*

IN FOCUS At this point in the lesson, it will be important to make the link between the inverse grid method and long division as explicit as possible. Long division is more abstract in its presentation but identical in the thinking that goes on to solve it. Children should be made aware of this to secure their understanding. During class conversation, encourage children to identify the differences between long division and short division.

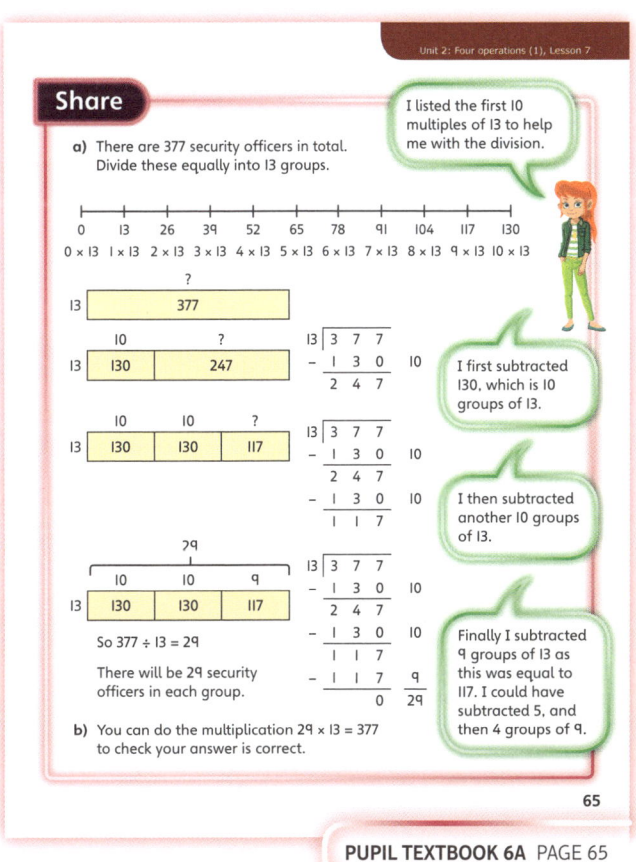

PUPIL TEXTBOOK 6A PAGE 65

Think together

Whole class teacher led (I do, We do, You do)

ASK

- Question **1**: *Describe and explain how the grid method and the long division calculation are linked.*
- Question **1**: *What numbers are missing from the grid method? Explain how you know.*
- Question **2**: *Can you use the grid method to help you complete the long division?*
- Question **3**: *Can you divide 799 by any factors of 17? What numbers multiply together to make 17?*
- Question **3**: *What method can you use when the divisor isn't a neat multiple of other digits?*

IN FOCUS Questions **1** and **2** scaffold children's understanding of the link between the grid method to represent their multiplicative reasoning and the newly learnt method of long division. Discuss these links when solving question **1**, before moving on to question **2** where children are required to apply the methods independently.

STRENGTHEN If children do not know how to approach question **2**, ask them to look at the divisor, 31. Ask:

- *Do you recognise this number from your times-tables? What does that tell you? What method will be needed here?*
- *What do the 6 and the 8 mean? How will you work out how many 31s go into 682? How will you write this down?*

DEEPEN When solving question **3**, encourage children to generalise the type of question Emma is solving, then make up a similar problem for their partner. Ask:

- *What is different about Emma's problem compared with Reena's?*
- *Can you think of other divisors that would require long division?*
- *How could you make up a similar question for your partner, with no remainder? Could you start with the inverse calculation?*

ASSESSMENT CHECKPOINT Do children understand the processes involved when solving a division calculation using long division? Can they explain how to correctly set out the written calculation? Can children compare all the methods they have learnt, recognising where each one is most and least useful?

ANSWERS

Question **1**: There will be 19 balloons in each group.

Question **2**: There are 22 roses in each decoration.

Question **3**: $588 \div 28 = 588 \div 4 \div 7$
$\qquad\qquad = 147 \div 7$
$\qquad\qquad = 21$

$\qquad$ $799 \div 17 = 47$ (using long division)

Look for children who suggest the appropriate method to solve Emma's problem, i.e. long division, and who can explain that, as 17 is a prime number, they cannot use factors of 17 to help.

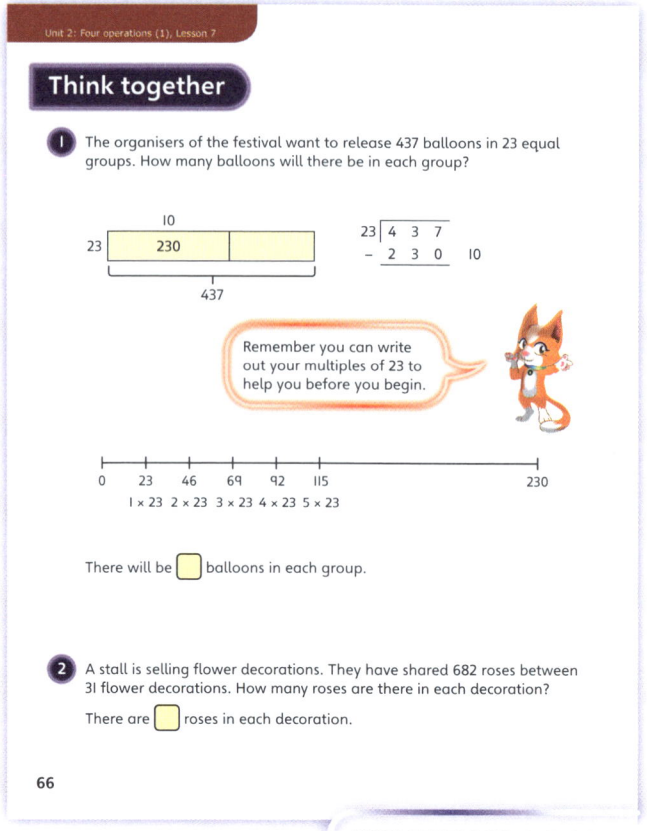

PUPIL TEXTBOOK 6A PAGE 66

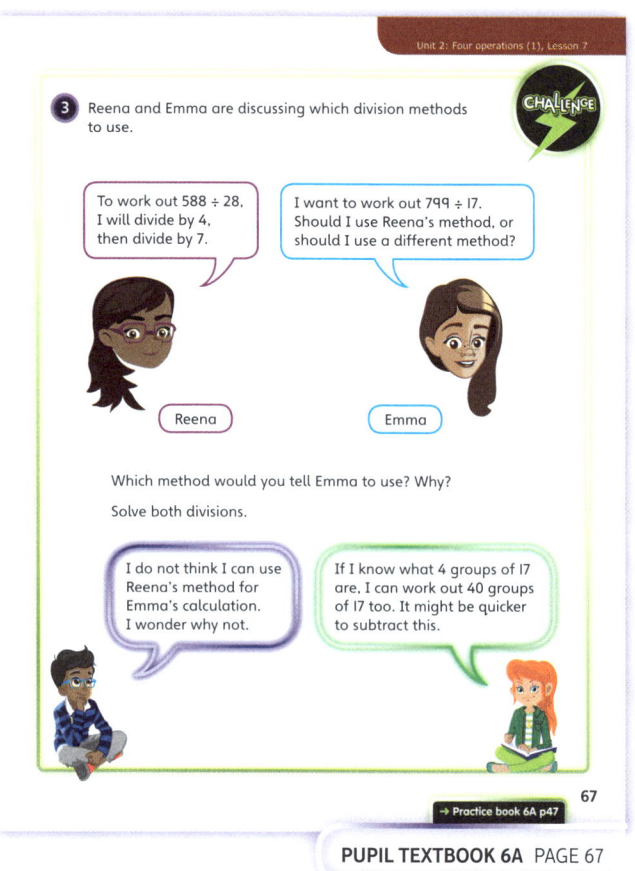

PUPIL TEXTBOOK 6A PAGE 67

Practice

IN FOCUS Question ❶ links children's prior learning with their newly learnt written method of long division. Encourage them to complete each method to help cement their understanding and enable them to visualise the pictorial representation when solving divisions in the future. Where possible, they should use multiples of 10 and 100 first to solve calculations efficiently. For questions ❷ and ❸, children should develop their efficiency by identifying where it is possible to use multiples of 10 to quickly make progress through the division calculations.

STRENGTHEN If children cannot solve the calculations in question ❸, ask:
· *Can you use an inverse grid method? How will that help you?*
· *Can you fit 10 thirteens into 364? How about 20? How can this fact help you?*

DEEPEN Question ❸ helps children to recognise how fluency with doubles and halves can help them to efficiently solve a calculation. Ask:
· *What do you know about the relationship between the answers to the divisions 608 ÷ 16 and 608 ÷ 32?*

THINK DIFFERENTLY Question ❹ requires children to think about the number of groups of the divisor that can be subtracted to fit the given criteria. They should recognise that Mo must subtract some larger groups of the divisor while Olivia subtracts some smaller groups.

ASSESSMENT CHECKPOINT Do children know how to use multiples of 10 to efficiently solve a long division calculation? Can they use their knowledge of multiples, factors and number facts to efficiently and confidently solve calculations? Do children recognise how an understanding of doubling and halving can help them to solve calculations quickly?

ANSWERS Answers for the **Practice** part of the lesson appear in the separate **Practice and Reflect answer guide**.

Reflect

IN FOCUS This question offers an opportunity to assess whether children can confidently use the inverse operation to check the result of a division. They may choose to carry out the actual division using a method of their choice. Children should demonstrate confidence in the method they choose. When children have finished calculating, give them an opportunity to compare their methods. Ask:
· *Have you used the same method to check?*
· *Can you explain which method is more efficient and why?*

ASSESSMENT CHECKPOINT Can children confidently use the methods they have been taught? Can they recognise the most efficient ways of using these methods or the inverse to solve a problem?

ANSWERS Answers for the **Reflect** part of the lesson appear in the separate **Practice and Reflect answer guide**.

After the lesson ⏸

· Can children make the link between the inverse grid method and long division?
· What percentage of the class has mastered long division?
· Are children as confident with this method as they are with short division? How can you tell?

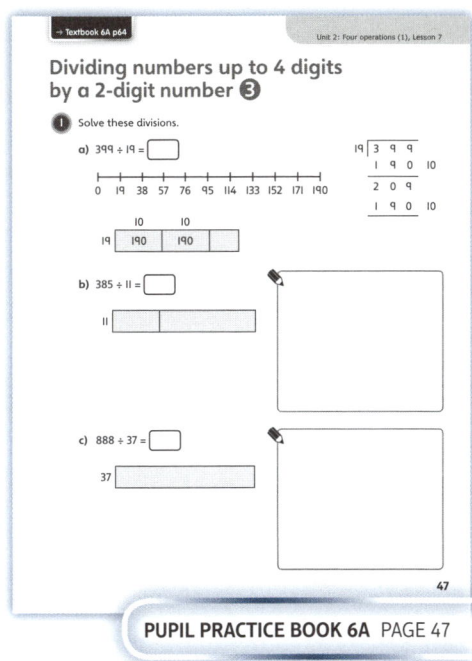

PUPIL PRACTICE BOOK 6A PAGE 47

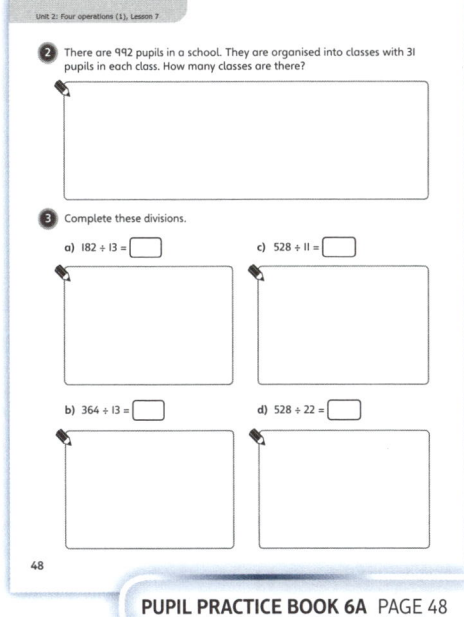

PUPIL PRACTICE BOOK 6A PAGE 48

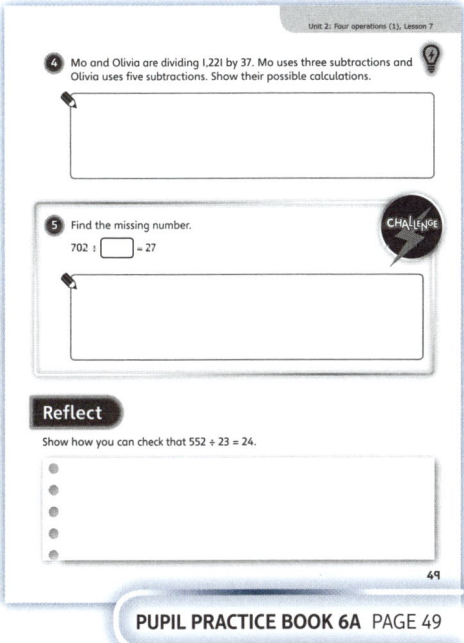

PUPIL PRACTICE BOOK 6A PAGE 49

Dividing numbers up to 4 digits by a 2-digit number ④

Learning focus

In this lesson, children will use the methods they have learnt about in the previous lessons to solve mathematical problems with real-life contexts. They will use different strategies to approximate and check answers.

Small steps

→ Previous step: Dividing numbers up to 4 digits by a 2-digit number (3)

→ **This step: Dividing numbers up to 4 digits by a 2-digit number (4)**

→ Next step: Dividing numbers up to 4 digits by a 2-digit number (5)

NATIONAL CURRICULUM LINKS

Year 6 Number – Addition, Subtraction, Multiplication and Division

Divide numbers up to 4 digits by a two-digit whole number using the formal written method of long division, and interpret remainders as whole number remainders, fractions, or by rounding, as appropriate for the context.

ASSESSING MASTERY

Children can reliably and fluently use the methods for solving division calculations. They can use their knowledge of number to approximate answers and can use various strategies to accurately check working out.

COMMON MISCONCEPTIONS

When using checking strategies, children may assume that, if the original answer does not match the answer to their checking calculation, their solution must be wrong. Ask:

• *Will a checking calculation be exactly the same as your original solution? Why?*

STRENGTHENING UNDERSTANDING

It will be important, before moving into this lesson, to make sure children have a good understanding of all the methods they have been learning about. If children are finding any of the methods challenging, offer them opportunities to practise, outside a real-life context. It will help to use the inverse grid method to secure their conceptual understanding, whatever method they are struggling with.

GOING DEEPER

Children could be encouraged to write their own real-life word problems. Vary the difficulty by asking:

• *Can you write a one- or two-step real-life problem?*
• *Can you write a problem with two division calculations that require the same method to solve?*
• *Can you write a problem with two division calculations that require different methods to solve?*

KEY LANGUAGE

In lesson: division, divide, long division, method

Other language to be used by the teacher: real-life problem, dividend, divisor, quotient, factor, checking strategy, round, inverse, approximate

STRUCTURES AND REPRESENTATIONS

long division, short division, inverse grid method

RESOURCES

Optional: mini whiteboards

 In the eTextbook of this lesson, you will find interactive links to a selection of teaching tools.

Before you teach

• Which method for division are children most confident with? Least confident with?
• How will you use your knowledge of their confidence to direct your teaching in this lesson?

Discover

WAYS OF WORKING Pair work

ASK

- Question **1** a): *What methods for solving division do you know?*
- Question **1** a): *Which method do you predict will work best for this calculation? Why?*
- Question **1** b): *What calculations will you need to use to solve this problem? Why?*
- Question **1** b): *Could you have used the factor method here? Which method do you think is more efficient?*

IN FOCUS Questions **1** a) and **1** b) are division calculations that are solvable using more than one method. Children should be encouraged to consider which method is most efficient and why. Question **2** is a two-step division problem that requires children to use the relationship between days and weeks to help them.

PRACTICAL TIPS When solving the questions in this section of the lesson, encourage children to use all the methods they have been learning about. They could use mini whiteboards to test different methods and identify the most appropriate calculation for each question, justifying their opinions based on their learning in the last three lessons.

ANSWERS

Question **1** a): $2{,}478 \div 21 = 118$

           The food will last for 118 days.

Question **1** b): No. They have enough food for 38 days, which is less than 7 weeks.

Share

WAYS OF WORKING Whole class teacher led

ASK

- Question **1** a): *Which of the methods did you use?*
- Question **1** a): *Was your method the most efficient? Explain how you know.*
- Question **1** a): *Can you explain how each method is linked with the others?*
- Question **1** b): *Why was the digit 3 placed above the line and not 30?*

IN FOCUS At this point in the lesson, it is important to assess children's confidence with all the division methods they have learnt so far.. Any misconceptions, or evidence that they are lacking in confidence with their prior learning, should be tackled now to ensure they are able to access the rest of the lesson. Discuss with children the links between the methods to assess and secure their conceptual understanding.

PUPIL TEXTBOOK 6A PAGE 68

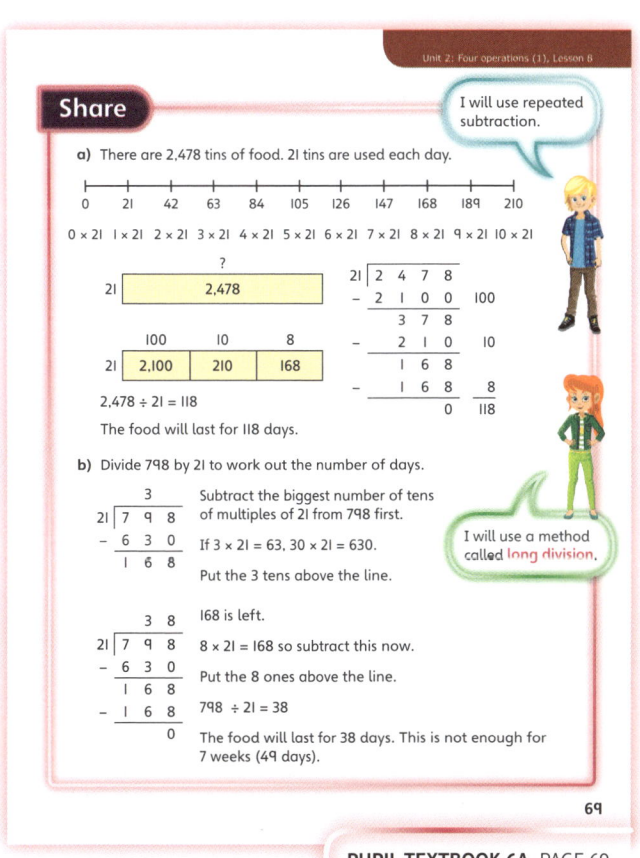

PUPIL TEXTBOOK 6A PAGE 69

Think together

WAYS OF WORKING Whole class teacher led (I do, We do, You do)

ASK

- Question **1**: *What vocabulary in the question lets you know that this requires division to solve?*
- Question **2**: *Do you think Ash is correct to wonder about when to use long division? Explain.*
- Question **3**: *What will you look for in each calculation to decide which method will work best?*
- Question **3**: *Does one method work for more types of calculation than the others?*

IN FOCUS Question **1** allows children to apply their understanding of how their methods for division can be used to solve real-life problems. It will encourage them to consider the links between the different methods of division, recapping and securing their conceptual understanding. Question **2** develops children's ability to independently calculate with the given methods. Like question **1**, it is framed within a real-life context.

STRENGTHEN If children are struggling to complete the long division in question **2**, ask:

- *What multiple has been used already? Can you use it again?*
- *How much is left to share when you have used the multiple?*

DEEPEN Question **3** deepens children's reasoning with the methods they have learnt. Encourage them to find a checking strategy for these calculations. They will need to fluently identify the properties of the calculations that will lead to the use of different checking strategies.

ASSESSMENT CHECKPOINT Can children read, understand and solve a real-life written problem, fluently using the methods of division they have learnt in this unit? Can they assess a division problem and decide which method is most appropriate? Can they work through methods correctly when the scaffolding is removed?

ANSWERS

Question **1**: 812 ÷ 29 = 28. The bags will last for 28 days.

Question **2**: 4,439 ÷ 23 = 193. The centre uses 193 bags of cat litter each month.

Question **3**:

1,890 ÷ 45 = 42	factor method, inverse grid method or long division
1,311 ÷ 23 = 57	inverse grid method or long division
102 = 2,346 ÷ 23	inverse grid method or long division
7,379 ÷ 47 = 157	inverse grid method or long division
101 = 2,525 ÷ 25	factor method, inverse grid method or long division
4,000 ÷ 80 = 50	factor method, inverse grid method or long division

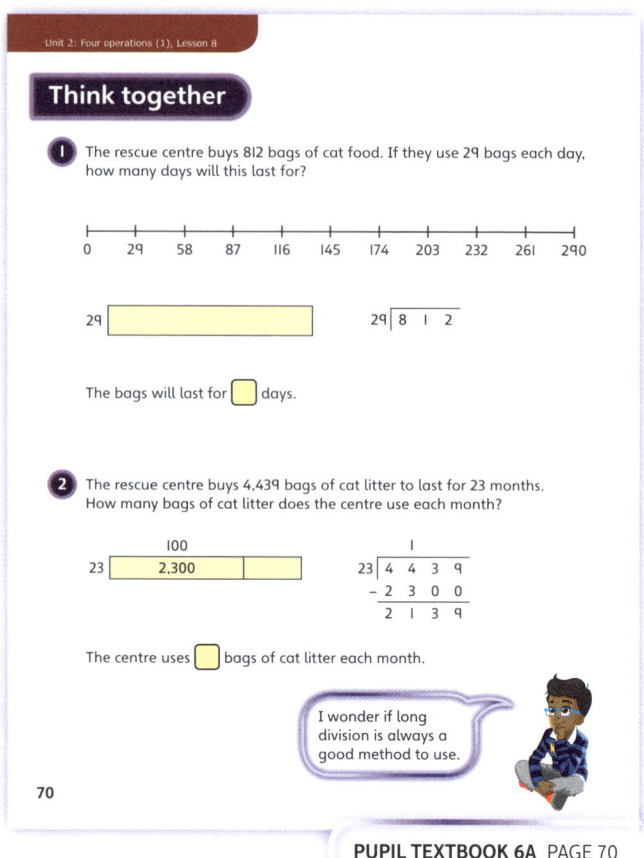

PUPIL TEXTBOOK 6A PAGE 70

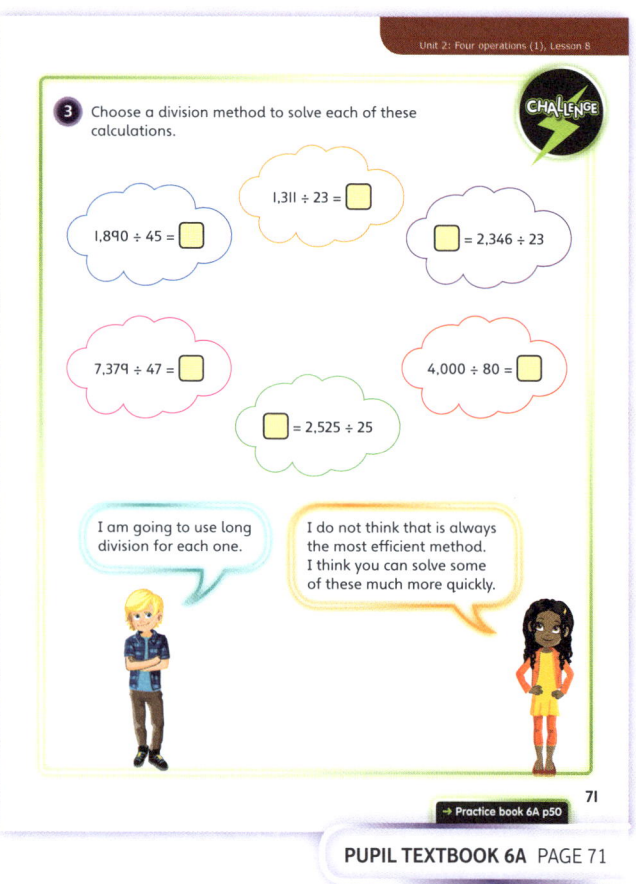

PUPIL TEXTBOOK 6A PAGE 71

Practice

Independent thinking

IN FOCUS Question **1** provides practice in the factor method of division.

Questions **2** and **3** present real-life problems with space for children to apply their chosen method independently. Encourage children to justify the methods they choose to use.

Question **4** reminds children that it is useful to list the first 10 multiples of a divisor and use knowledge of place value when doing long division.

STRENGTHEN To help decide which method is best in questions **3** and **4**, children can look back at the textbook to help them. Ask:

- *Will the calculations needed be similar to any you have done before? How?*
- *Can you use the way you solved earlier calculations to help you?*

DEEPEN Question **5** deepens children's reasoning and use of number facts to solve linked division calculations. Deepen children's generalising about the given calculations and those they create by asking:

- *What is the same and different about the calculations?*
- *How can you use the number facts you have found here to help you with other calculations?*
- *Can you create a mind map for another 4-digit divided by 2-digit calculation and link the facts on it with this one?*

THINK DIFFERENTLY Question **4** requires children to use their understanding of finding 10 times a number to quickly recognise that the tenth multiple of 61 is 610 and not 620. Ask:

- *How can you check where the sequence of multiples has gone wrong?*
- *I think that the sequence in the ones position is correct so the error must be in adding 6 tens each time? How do I know that?*

ASSESSMENT CHECKPOINT Do children select an appropriate method and use it fluently? Can they use appropriate strategies to find approximate answers and check calculations?

ANSWERS Answers for the **Practice** part of the lesson appear in the separate **Practice and Reflect answer guide**.

Reflect

Independent thinking

IN FOCUS This question allows children to recognise and demonstrate where the factor method can and cannot be used.

It will give a final opportunity to assess children's understanding of when the use of long division is appropriate.

ASSESSMENT CHECKPOINT Children should be able to fluently recognise types of calculation where the factor method is, and is not, appropriate. Look for children using knowledge of prime numbers to help them make decisions.

ANSWERS Answers for the **Reflect** part of the lesson appear in the separate **Practice and Reflect answer guide**.

After the lesson

- Are children fluent in determining which method is appropriate for which situation?
- What proportion of the class can reliably pick an appropriate method for each calculation they approach?

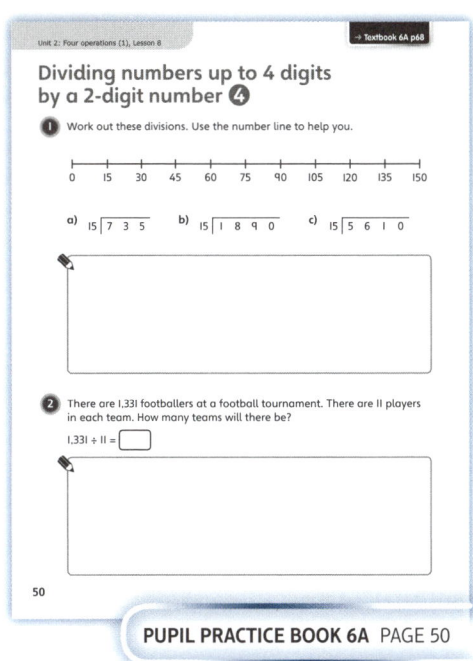

PUPIL PRACTICE BOOK 6A PAGE 50

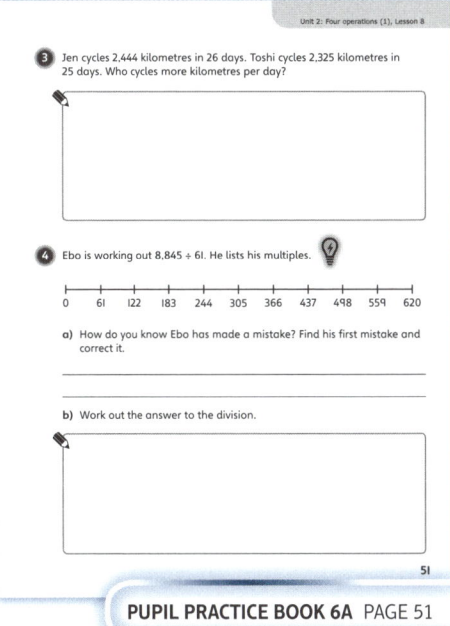

PUPIL PRACTICE BOOK 6A PAGE 51

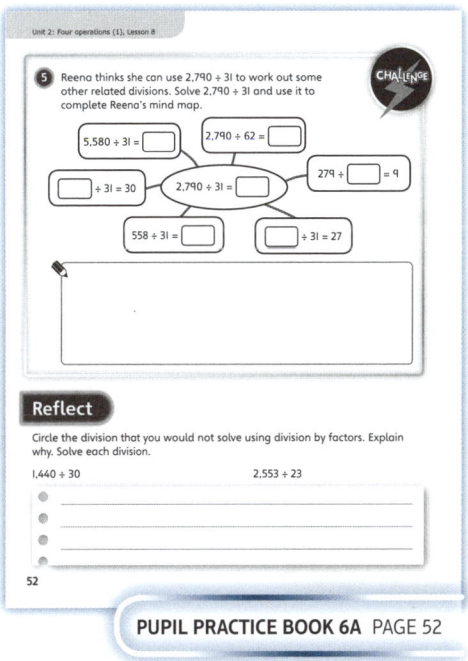

PUPIL PRACTICE BOOK 6A PAGE 52

103

Dividing numbers up to 4 digits by a 2-digit number ⑤

Learning focus

In this lesson, children will develop their understanding of division with remainders. They will learn how the written methods for division they have learnt can represent and solve a division calculation that has a remainder.

Small steps

→ Previous step: Dividing numbers up to 4 digits by a 2-digit number (4)

→ **This step: Dividing numbers up to 4 digits by a 2-digit number (5)**

→ Next step: Dividing numbers up to 4 digits by a 2-digit number (6)

NATIONAL CURRICULUM LINKS

Year 6 Number – Addition, Subtraction, Multiplication and Division

Divide numbers up to 4 digits by a two-digit whole number using the formal written method of long division, and interpret remainders as whole number remainders, fractions, or by rounding, as appropriate for the context.

ASSESSING MASTERY

Children can use their fluency with multiples of 2-digit numbers to divide a number up to 4-digits by a 2-digit number and find a remainder if there is one. They can use an efficient written method to represent this.

COMMON MISCONCEPTIONS

When solving a division with a remainder, children may add the remainder to the factors or multiples in their calculation. Ask:
• *What is the problem with what you have found in your running total?*
• *What do we do with the remainder?*

Children may say that a division 'doesn't work' if they find there is a remainder. Ask:
• *Can you say how many full groups of the divisor fit into the dividend?*
• *What do we call the number that is left over?*

STRENGTHENING UNDERSTANDING

To help children with the idea of remainders, it may be useful to give them a simpler division calculation that they can make with counters or blocks. Using $25 \div 4$ as an example, ask:
• *Can you find 25 counters and share them into groups of 4?*
• *What do you notice about the groups? Is it possible to do this?*
• *What is the leftover counter called?*

GOING DEEPER

Children can write division word problems that have remainders. To deepen this further, children could write a problem where it is necessary to round the remainder up, round it down or leave it as it is.

KEY LANGUAGE

In lesson: remainder, divide, multiple

Other language to be used by the teacher: factor, dividend, divisor, quotient, short division, long division, inverse grid method

STRUCTURES AND REPRESENTATIONS

inverse grid method, short division, long division, factor method, number line

RESOURCES

Optional: mini whiteboards

 In the eTextbook of this lesson, you will find interactive links to a selection of teaching tools.

Before you teach ⏸

• Have children already recognised or found divisions that have remainders?
• How will you ensure children's learning is deepened further in this lesson?

Discover

ASK

- Question ❶ a): *What methods of division do you think would work to solve this problem?*
- Question ❶ a): *Can you represent the calculation in a different way?*
- Question ❶ a): *What is the same and different about this division compared with ones from previous lessons?*
- Question ❶ b): *Can you just double your previous quotient to find the solution to this division? Explain.*
- Question ❶ b): *What methods will you use to solve this division and why?*

IN FOCUS In question ❶ a) children investigate a division calculation where a remainder is left. Encourage them to say what the remainder means in the context of the question – will they need to round it up or down? Question ❶ b) offers children an opportunity to generalise about the properties of numbers and how they affect division with a remainder. They should recognise that, as the dividend doubles, so does the remainder.

PRACTICAL TIPS When introducing this topic, use 2-digit numbers to demonstrate remainders. For example, using 50 ÷ 17, ask children to find 50 counters and share them into groups of 17. Children could then investigate other numbers that result in a remainder when dividing 50.

ANSWERS

Question ❶ a): 100 ÷ 17 = 5 remainder 15. The cicadas will emerge again 5 times over the next 100 years.

Question ❶ b): 200 ÷ 17 = 11 remainder 13. The cicadas will emerge again 11 times over the next 200 years.

Share

ASK

- Question ❶ a): *Can you explain what a remainder is?*
- Question ❶ a): *Which method do you think represents remainders most clearly? Why?*
- Question ❶ a): *Did you use the same methods? How were yours different?*
- Question ❶ a): *Did you find the same solution?*
- Question ❶ b): *Why did Astrid think the solution would be 10? Could there be a remainder of 30? Why?*
- Question ❶ b): *How does the inverse grid method make the division and remainder clear?*

IN FOCUS At this point in the lesson, make sure children are able to recognise what a remainder is and how it can be recorded in different ways. They should also be aware that doubling a dividend does not necessarily mean the quotient will be doubled. This is important as it will help to prevent miscalculations further on in the lesson when children try to use their number knowledge to solve calculations more efficiently.

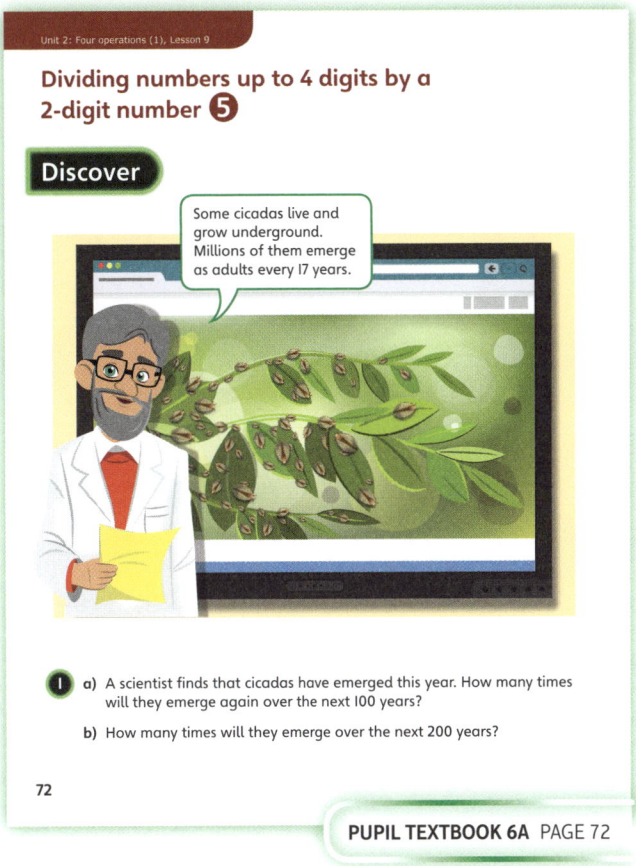

PUPIL TEXTBOOK 6A PAGE 72

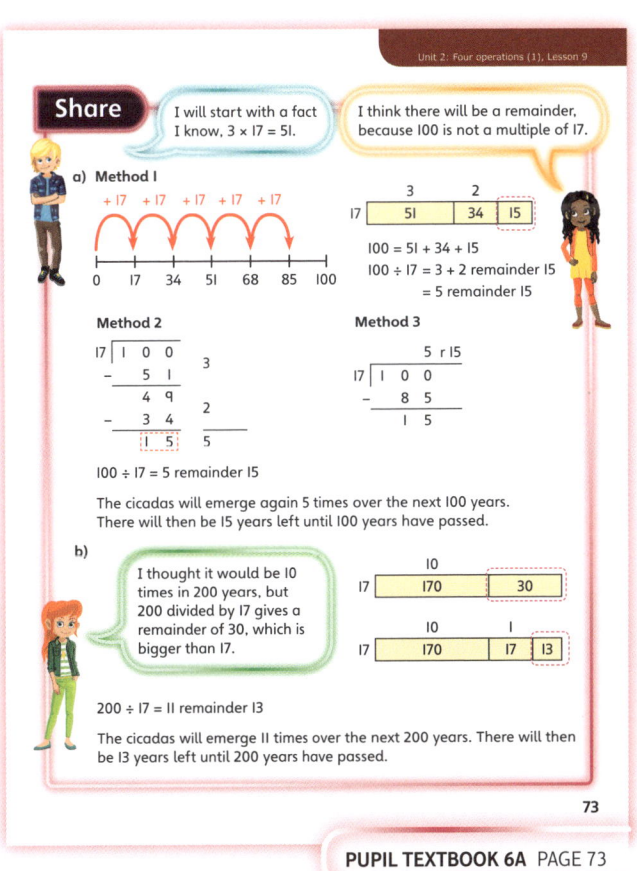

PUPIL TEXTBOOK 6A PAGE 73

Think together

WAYS OF WORKING Whole class teacher led (I do, We do, You do)

ASK

- Question **1**: *How does the number line help you? Which numbers are missing?*
- Question **1**: *Before you start to calculate, why do you know that any remainder must be less than 13?*
- Question **1**: *Is long division a helpful method here?*
- Question **2**: *How can you use this number line to help, when it only goes up to 190?*
- Question **2**: *How can you complete the calculation?*
- Question **3**: *Is Astrid's method correct? How do you know?*

IN FOCUS Question **1** is a contextual problem with a division that results in a remainder. Encourage children to explain the method they use. Question **2** challenges children to complete more than one subtraction in the calculation, as they first subtract 190 in the form of 19×10 and then 114 in the form of 19×6 to identify the remainder as 2. Question **3** a) offers an opportunity to solve a contextual real-life problem. The character comments open conversation about how real-life contexts can change.

STRENGTHEN If children do not recognise how to record the remainders in each calculation, ask:

- *Can you fit any more groups of the divisor into the dividend? Explain.*
- *How can you record the fact that no more groups fit in?*
- *How does the remainder link to the real-life problem? How will you record the answer?*

DEEPEN Question **3** b) deepens children's understanding of remainders and what information a remainder can supply. Ask:

- *What is the same and different about these questions?*
- *How does the question change the way you will use the remainder?*

ASSESSMENT CHECKPOINT Can children solve divisions with methods they have learnt before? Do they recognise where calculations result in a remainder and can they explain how remainders relate to the context of real-life word problems?

ANSWERS

Question **1**: $100 \div 13 = 7$ remainder 9

Question **2**: $306 \div 19 = 16$ remainder 2

Question **3** a): $365 \div 68 = 5$ remainder 25

$\quad\quad\quad\quad 366 \div 68 = 5$ remainder 26

Question **3** b): $365 \div 25 = 14$ remainder 15

It tells you how many days work she has spent working on an unfinished chair.

$365 \div 25 = 14$ remainder 15

If the builder buys 14 packs, he will need 15 more planks, so he must buy 1 more whole pack.

$365 \div 10 = 36$ remainder 5

The bricklayer lays 36 bricks per hour, plus another 5 bricks, over 10 hours.

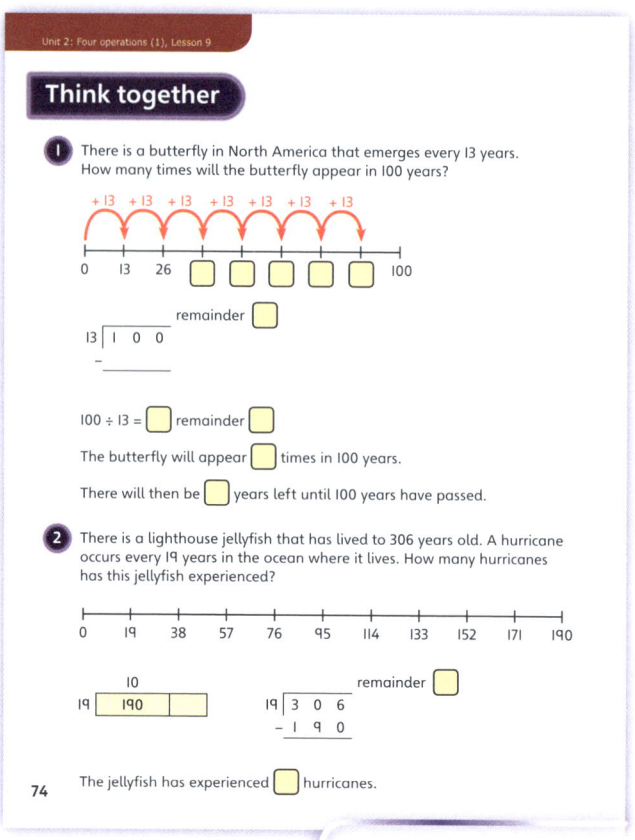

PUPIL TEXTBOOK 6A PAGE 74

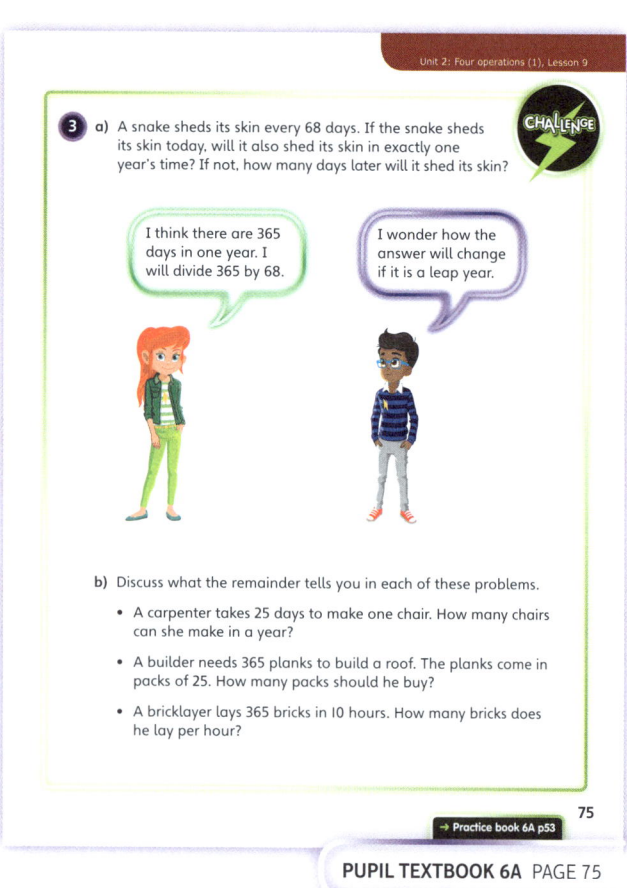

PUPIL TEXTBOOK 6A PAGE 75

Practice

WAYS OF WORKING Independent thinking

IN FOCUS Question ❶ helps reinforce the idea that increasing a divisor will not necessarily have a similar effect on the remainder. Encourage children to explain the differences to help secure their understanding. Question ❷ gives children the opportunity to solve a division calculation using the methods they have learnt. This extends and challenges their understanding as they select methods to apply independently. When children tackle questions ❸, ❹ and ❺, make sure they use their understanding from questions ❶ and ❷ and encourage them to demonstrate their reasoning by showing their working out.

STRENGTHEN If children are convinced that only 13 bags of bird seed are needed in question ❺, ask:
- *Will they be able to make all the bird feeders with only 13 bags? Why?*
- *How many bags should they buy to get the extra 20 kg they need?*
- *What did we have to do with the remainder and why?*

DEEPEN Ask children to create their own word problems involving a remainder. Tell them the remainder should provide useful information, such as 'how many more bags will be needed'.

ASSESSMENT CHECKPOINT Do children demonstrate fluency and reasoning when using known number facts and linking them with division? Can children accurately and fluently solve division calculations that include remainders in multiple representations, including abstract calculations and real-life written problems?

ANSWERS Answers for the **Practice** part of the lesson appear in the separate **Practice and Reflect answer guide**.

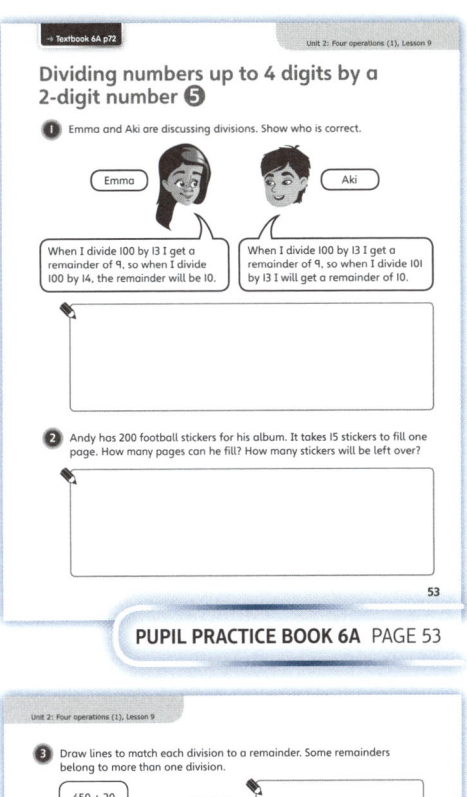

PUPIL PRACTICE BOOK 6A PAGE 53

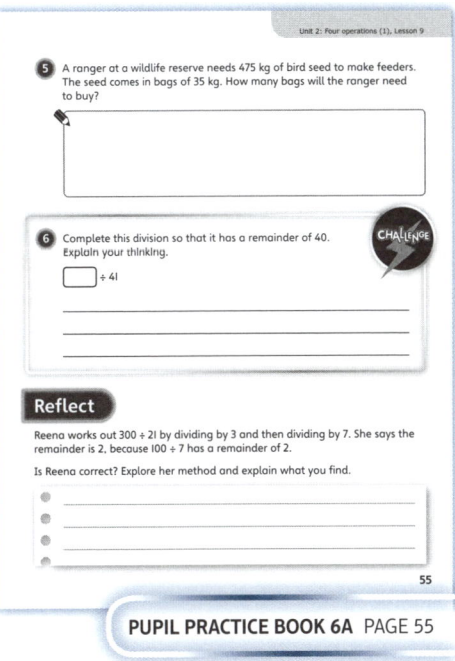

PUPIL PRACTICE BOOK 6A PAGE 54

Reflect

WAYS OF WORKING Independent thinking and pair work

IN FOCUS This question deepens children's understanding of which methods work for divisions with remainders and which do not. Children should recognise and show, possibly with the bar model, that the factor method of division will not find the correct remainder. Ask:
- *Is there a way that you could make the factor method work?*
- *What would you need to multiply your remainder by?*

ASSESSMENT CHECKPOINT Ask children to explain how the factor method could also be used for 450 ÷ 21. Can they explain how to find the correct remainder?

ANSWERS Answers for the **Reflect** part of the lesson appear in the separate **Practice and Reflect answer guide**.

After the lesson

- How confident were children at using written methods to solve division calculations with remainders?
- Were children able to link the context of the written problems to rounding a remainder up or down?

Dividing numbers up to 4 digits by a 2-digit number 6

Learning focus

In this lesson, children will deepen their understanding of remainders and how to represent them. They will learn that representing a remainder as a fraction gives a more accurate answer.

Small steps

→ Previous step: Dividing numbers up to 4 digits by a 2-digit number (5)

→ **This step: Dividing numbers up to 4 digits by a 2-digit number (6)**

→ Next step: Common factors

NATIONAL CURRICULUM LINKS

Year 6 Number – Addition, Subtraction, Multiplication and Division

Divide numbers up to 4 digits by a two-digit whole number using the formal written method of long division, and interpret remainders as whole number remainders, fractions, or by rounding, as appropriate for the context.

ASSESSING MASTERY

Children can reliably solve division calculations with remainders. They can convert a remainder into a fraction using their understanding of divisors.

COMMON MISCONCEPTIONS

Children may record the fraction remainder upside down, recording the remainder as the denominator and the divisor as the numerator. Ask:

• *What do the numerator and denominator represent?*
• *Which number in the calculation represents the size of the equal groups?*

STRENGTHENING UNDERSTANDING

Children may benefit from recapping how to simplify fractions before moving into this lesson. Ask:

• *How can you know if you have found a fraction's simplest form?*
• *Can you explain how to simplify a fraction? Can you use a picture or resources to help you explain?*

GOING DEEPER

Children could write story problems with more than one step and challenge their partner to draw out the steps needed. Ask:

• *Can you write a story that requires a division and another operation?*
• *Can it be solved by doing the calculations either way round? How?*

KEY LANGUAGE

In lesson: fraction, remainder, divide, simplify

Other language to be used by the teacher: division, divisor, dividend, quotient, numerator, denominator, short division, long division, inverse grid method, factor method

STRUCTURES AND REPRESENTATIONS

fractions, short division, long division

RESOURCES

Optional: mini whiteboards

 In the eTextbook of this lesson, you will find interactive links to a selection of teaching tools.

Before you teach

• Are there any misconceptions from the previous lesson on remainders that need to be dealt with before progress can be made in this lesson?
• What structures or representations could you use to help with this?

Discover

Pair work

ASK

- Question **1** a): *How will you solve this question?*
- Question **1** a): *Does this calculation result in a remainder? Can you prove it?*
- Question **1** a): *How many different ways could you represent this calculation?*
- Question **1** a): *Are there any methods you know that don't work with this calculation?*
- Question **1** b): *Does the way in which we record remainders work for this question? How?*
- Question **1** b): *What do we need to do to the remainder to make the solution make sense with the question?*

IN FOCUS Question **1** a) recaps children's learning from the previous lesson on remainders. To ensure quick progress in the lesson, it is important to recap their understanding of the methods they have learnt in this unit. Question **1** b) offers an opportunity to discuss how a remainder can be used, being rounded up, rounded down or left as it is. Recording the remainder as 'remainder 10' would not make sense in the context of the question. Children should discuss how the remainder can be used to solve this problem.

PRACTICAL TIPS To introduce the concept of this lesson, give children a strip of paper measuring 24 cm. They use this as the grid for the inverse grid method with the division 24 ÷ 5. They should split the strip into the necessary segments, including the remainder. Once children have found the remainder, they could fold the remainder box into 5 segments. Discuss how they have now divided the remainder of 4 by 5, which can be shown as a fraction.

ANSWERS

Question **1** a): 1,235 ÷ 25 = 49 remainder 10

Question **1** b): Divide the remainder between the 25 stages; $1{,}235 ÷ 25 = 49\frac{10}{25}$ or $49\frac{2}{5}$

Share

Whole class teacher led

ASK

- Question **1** a): *How does the model help you?*
- Question **1** a): *Which method did you use to solve the question?*
- Question **1** b): *Can you explain Astrid's comment?*
- Question **1** b): *What does the fraction represent?*
- Question **1** b): *Can it be simplified further than $\frac{2}{5}$?*

IN FOCUS At this point in the lesson, it will be important to recap and secure children's understanding that a fraction is a representation of a division and that $\frac{10}{25}$ is the same as 10 ÷ 25. Making sure children understand this will ensure that their conceptual understanding is secured throughout the lesson.

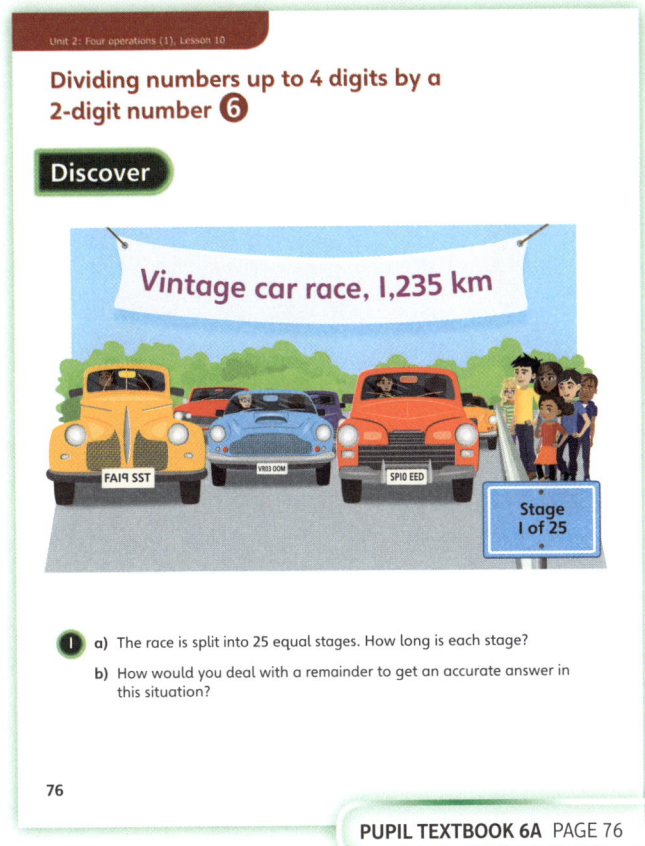

PUPIL TEXTBOOK 6A PAGE 76

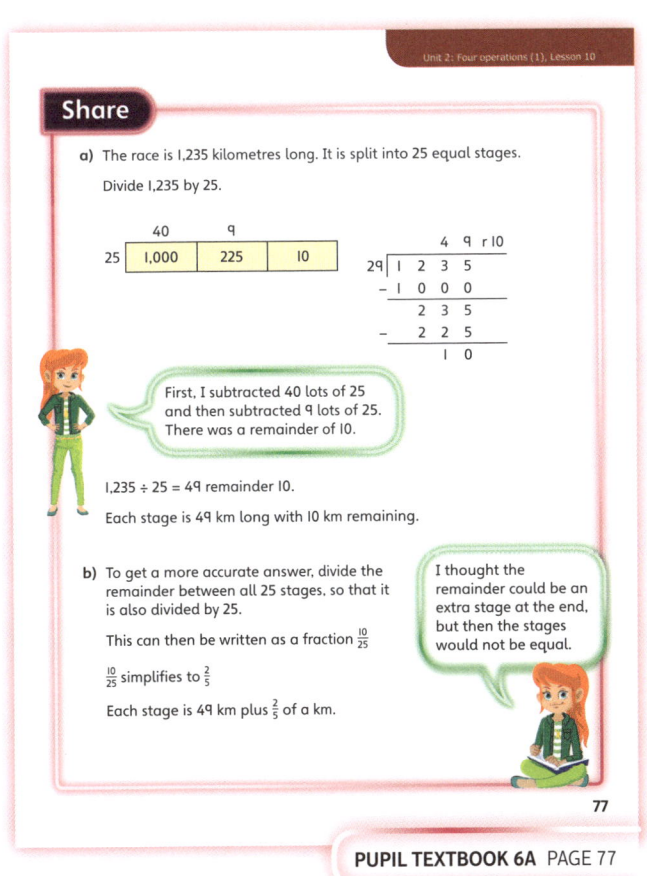

PUPIL TEXTBOOK 6A PAGE 77

Think together

Unit 2: Four operations (1), Lesson 10

Think together

WAYS OF WORKING Whole class teacher led (I do, We do, You do)

ASK

• Question ❶: *How should you use the remainder here?*
• Question ❶: *Can the fraction be simplified?*
• Question ❷: *What is the biggest whole number remainder when the divisor is 50?*
• Question ❷: *How should you use the remainder here?*
• Question ❸: *Has this method made the calculation simpler? How?*
• Question ❸: *How will dividing by 10 affect the remainder?*

IN FOCUS Question ❶ reinforces the use of the remainder within the context of a real-life word question. Discuss the three comments and why each person may think as they do. Question ❸ offers children the opportunity to reason about a remainder, using knowledge of multiples of 50 to make decisions. It is important to discuss how the remainder should be represented. In all questions, clarify whether the fractions can be simplified or not.

STRENGTHEN If children are struggling to convert the remainder to a fraction, ask:
• *How much is left over after solving the division?*
• *What were you dividing the number by?*
• *How can you use a fraction to represent the remainder being divided by the divisor?*

DEEPEN For question ❶, deepen children's understanding and reasoning about the incorrect comments by asking:
• *Can you explain why these comments are incorrect?*
• *What advice would you give each person to help them understand why they are incorrect?*
• *What models or drawings could you use to prove your ideas?*

ASSESSMENT CHECKPOINT Do children understand how remainders can be converted into fractions and how this can be an appropriate way to represent an answer to a division word problem, where a whole number remainder would not make sense? Do children recognise how their knowledge of number facts can be used to solve division calculations more easily?

ANSWERS

Question ❶: Luis is right, though he has not simplified the fraction.

$$1{,}235 \div 40 = 30\frac{35}{40} \text{ or } 30\frac{7}{8}$$

Question ❷: 3,580 litres $\div$ 50 = $71\frac{30}{50}$ or $71\frac{3}{5}$

Question ❸: Ebo needs to multiply his first remainder by 10.

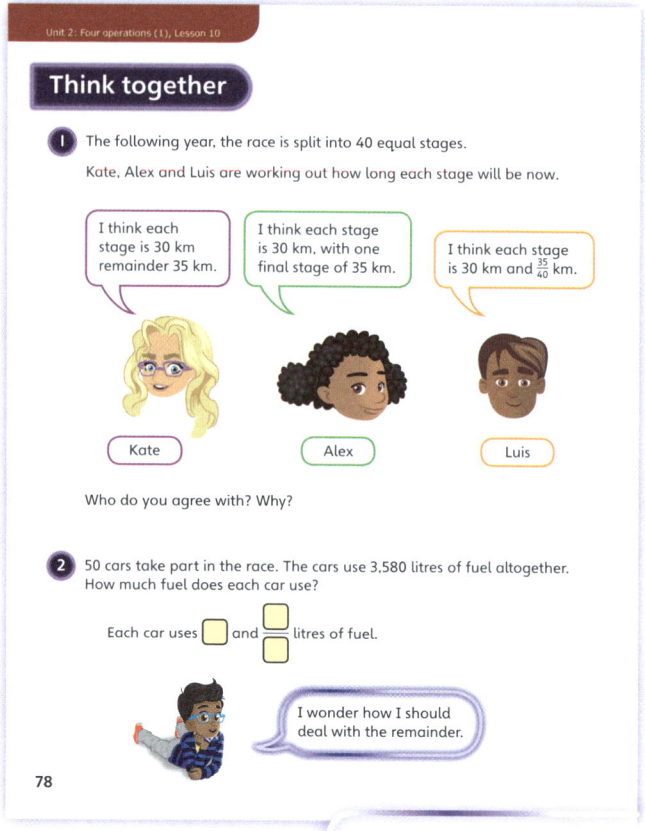

PUPIL TEXTBOOK 6A PAGE 78

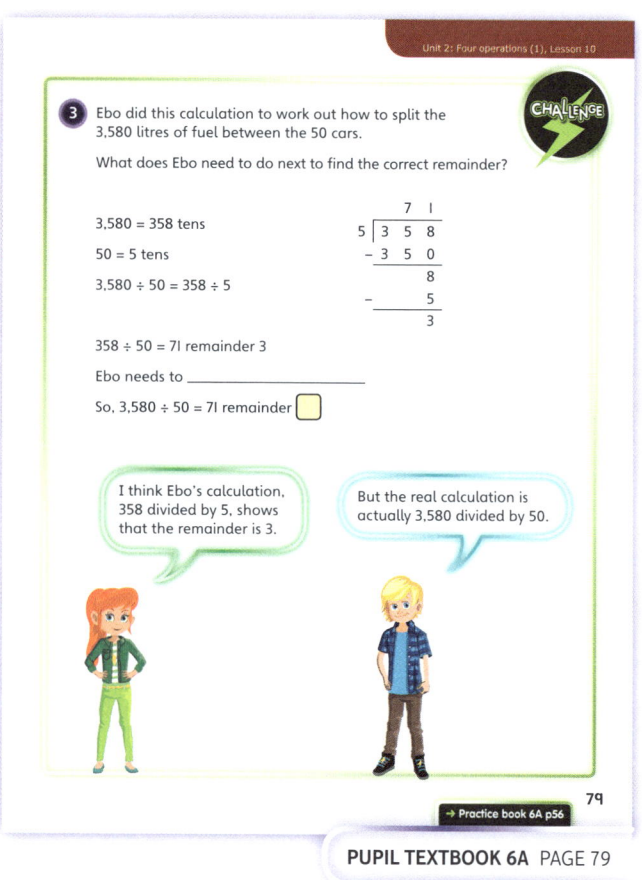

PUPIL TEXTBOOK 6A PAGE 79

Practice

WAYS OF WORKING Independent thinking

IN FOCUS Questions **1** a) and **1** b) challenge children to apply their understanding of remainders in different contexts, starting with a question where it is appropriate to record the remainder as a whole number that is 'left over'. Question **1** c) develops this with a similar context but asking children to convert the remainder into a fraction. Children could discuss how the questions are similar and different and how the fraction represents the answer in question **1** c).

Question **2** helps children to link understanding of number facts with divisions and remainders. Children should share their reasoning and justify their ideas when discussing how they think the calculations are related.

STRENGTHEN If children do not know where to start investigating the possible numbers they could create in question **4**, ask:
- *To get the biggest remainder, does the dividend need to be as big or as small as possible? Why?*
- *Will you need a large or small divisor? Why?*
- *Is there more than one solution? Explain how you know.*

DEEPEN Question **3** deepens children's understanding of using remainders by encouraging them to represent them as a decimal in the context of money. Ask:
- *What is the context of this question?*
- *Can you record the remainder as a fraction? Explain.*
- *How should you record the remainder if you are dealing with pounds and pence?*

ASSESSMENT CHECKPOINT Can children find a remainder when solving a division calculation? Can they accurately convert the remainder to a fraction or a decimal fraction in the context of money? Do children recognise how number facts can help them to solve divisions?

ANSWERS Answers for the **Practice** part of the lesson appear in the separate **Practice and Reflect answer guide**.

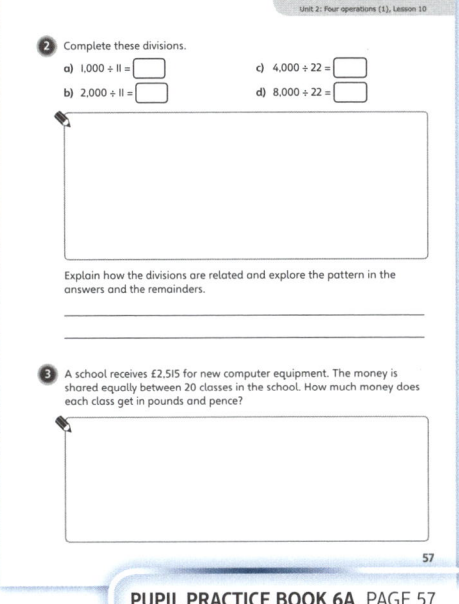

PUPIL PRACTICE BOOK 6A PAGE 56

PUPIL PRACTICE BOOK 6A PAGE 57

Reflect

WAYS OF WORKING Independent thinking and pair work

IN FOCUS Once children have written their story problem, they can challenge a partner to solve it, to check it has all the required features.

ASSESSMENT CHECKPOINT Look for children having written a contextual problem requiring a division that fits the parameters. Children may want to write the abstract calculation first, before putting it into a story.

ANSWERS Answers for the **Reflect** part of the lesson appear in the separate **Practice and Reflect answer guide**.

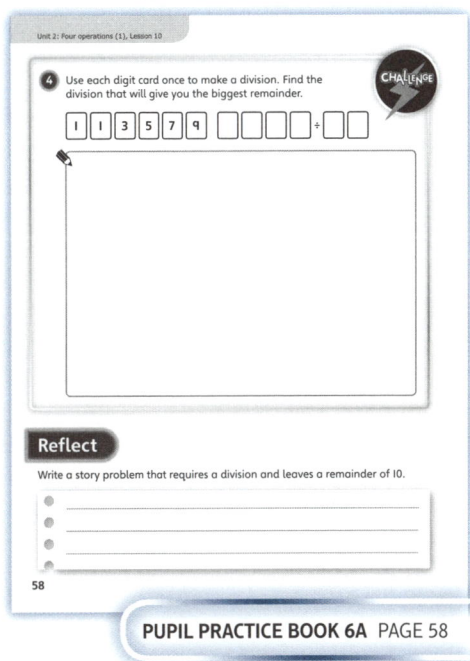

PUPIL PRACTICE BOOK 6A PAGE 58

After the lesson

- Could children clearly explain how the fractions are linked to the remainders they find?
- How could you implement this skill in other areas of the curriculum?

End of unit check

> Don't forget the *Power Maths* unit assessment grid on p26.

WAYS OF WORKING Group work adult led

IN FOCUS These questions focus on knowing and using representations and written methods for the four operations.

- Check that children have fluency when working with number facts and using them to check calculations or solve a missing number problem.
- Children should be able to read, understand and use different written methods for division and show an understanding of how remainders can be used in a contextual division problem.

ANSWERS AND COMMENTARY

Children will be able to efficiently and accurately solve the multiplication of a 4-digit number by a 1- or 2-digit number using a columnar method. Additionally, they will be able to use multiple methods to solve division calculations, including short and long division, explaining how their understanding of multiples and factors can help them. Finally, they will demonstrate mastery of all the written methods through using them fluently to solve real-life word problems.

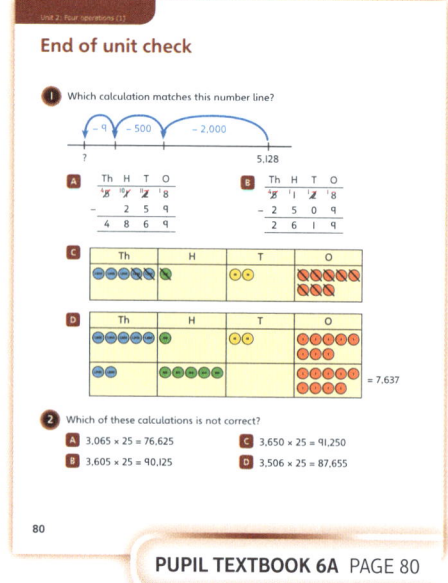

PUPIL TEXTBOOK 6A PAGE 80

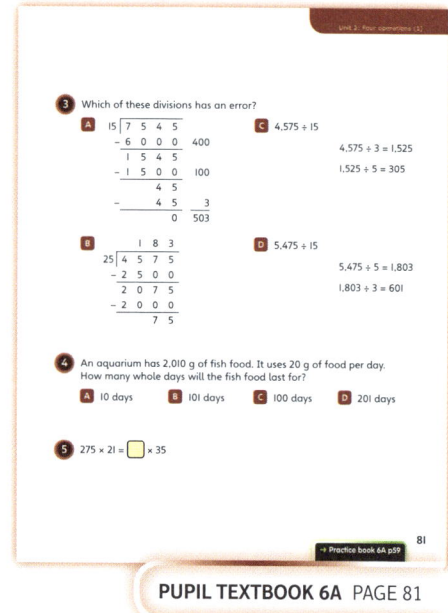

PUPIL TEXTBOOK 6A PAGE 81

Q	A	WRONG ANSWERS AND MISCONCEPTIONS	STRENGTHENING UNDERSTANDING
1	B	A could indicate problems with place value. D may suggest children are unclear about the difference between the representations of addition and subtraction.	When solving calculations, like those in question 2, ask: • *What do we know about the numbers?* • *If we are looking at multiples of 25, what can we say will always be the case about those numbers?*
2	D	Incorrect answers suggest children are not using known number facts to help them.	
3	D	A or C suggests insufficient understanding of long division.	When children are trying to use the factor method to solve a division calculation, ask:
4	C	A may indicate children have mistaken the remainder for the solution. B suggests children were unsure how the remainder fitted the context of the question.	• *What factors of the divisor can you list?* • *Will the dividend easily divide into any of these factors?* • *What will you need to divide the resulting number by? Explain why.* • *Why is it important to factorise and not partition the divisor?*
5	**275 × 21 = 165 × 35** Look for children solving the multiplication first then the division. They may use a written method for the multiplication or multiply by 10, then by 2 and then add 275 to their answer. To solve the division, children may use a written method or the factor method.		

My journal

WAYS OF WORKING Independent thinking

ANSWERS AND COMMENTARY

Children should recognise that they will need to work backwards, finding numbers that divide by 25 to leave a remainder of 10 by adding 10 to a multiple of 25. Following this thought process, they will be able to complete the given calculations. If children do not know where to begin, ask:

- *What do you know about every number that, when divided by 25, leaves a remainder of 10?*
- *How can you use this to begin investigating what numbers can be created?*
- *Could this be solved by trial and improvement? Is that the most efficient way to solve it? Why?*

Power check

WAYS OF WORKING Independent thinking

ASK

- *How confident are you that you could solve a division calculation efficiently and accurately?*
- *What checking methods would you use to check a multiplication calculation?*
- *Do you think you could solve any given word problem? Explain how you know.*

Power puzzle

WAYS OF WORKING Independent thinking

IN FOCUS Use this **Power puzzle** to assess children's ability to accurately subtract a 3-digit number from a 4-digit number. To deepen children's investigation in this **Power puzzle**, ask:

- *What is the longest chain you can make? What is the shortest?*
- *Is there any way of making sure you end up with a long chain of calculations?*

ANSWERS AND COMMENTARY Children should find that, whatever numbers they begin with, they eventually find themselves 'stuck', constantly using and reusing the digits 6, 1, 4, 7. If children do not find this it suggests they need more practice with subtraction.

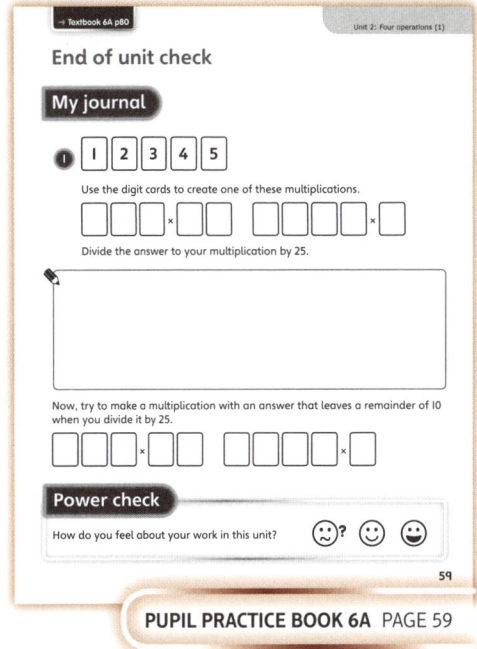

PUPIL PRACTICE BOOK 6A PAGE 59

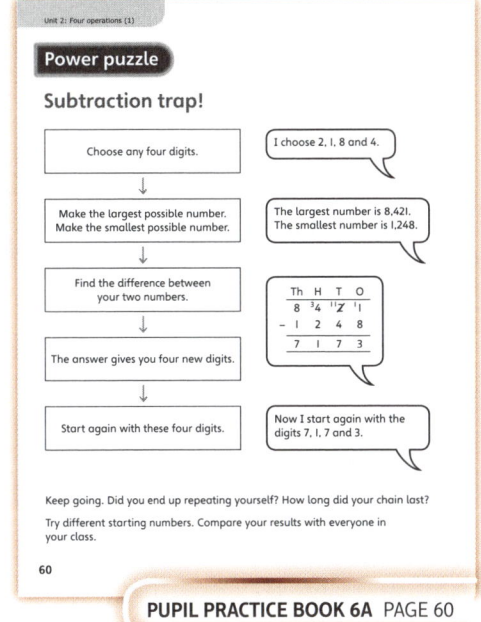

PUPIL PRACTICE BOOK 6A PAGE 60

After the unit ⏸

- Were children more confident with one method than with others? How could you tell?
- If this was the case, how will you continue to offer support in the future to secure their understanding of the other written methods?

Strengthen and **Deepen** activities for this unit can be found in the *Power Maths* online subscription.

Unit 3
Four operations ❷

Don't forget to watch the Unit 3 video!

Mastery Expert tip! "I used as many practical activities and concrete resources as I could in this unit to make sure children engaged with the learning as much as possible. It really brought every lesson alive!"

WHY THIS UNIT IS IMPORTANT

This unit develops children's understanding of how the four operations can be used to manipulate numbers and solve problems. Children begin by learning to recognise and find common factors and multiples, before looking at prime numbers as a special example of numbers with specific factors. Next, children investigate the effects of squaring and cubing, linking this to what they know about the dimensions of the namesake shapes.

After this, children learn about the order of operations, investigating its effect on calculations and considering why it is important to have an agreed order. They then learn how brackets can affect the order of operations. Using these concepts, they complete calculations, solve problems and diagnose mistakes in calculations.

Finally, children learn methods to solve mental calculations with small and large numbers. They consider where mental methods are appropriate and where written methods are appropriate. They also use number facts they already know to solve problems involving related number facts.

WHERE THIS UNIT FITS

→ Unit 2: Four operations (1)
→ **Unit 3: Four operations (2)**
→ Unit 4: Fractions (1)

In this unit, children use their knowledge of the four operations to consider specific properties of numbers. They learn about the order of operations and mental methods, before moving on to work with fractions in Unit 4.

Before they start this unit, it is expected that children:
• are fluent in their multiplication tables
• understand the terms, and are able to find, factors and multiples
• understand and can use the four operations.

ASSESSING MASTERY

Children will demonstrate mastery by fluently finding common factors and multiples of two or more numbers. They will be able to explain how prime numbers differ from other numbers and confidently square and cube numbers. They will also be able to use their understanding of the multiplicative properties of numbers to solve problems and share their reasoning. Children will be able to fluently adhere to the correct order of operations, demonstrating and explaining how brackets can affect this. Finally, they will be able to solve mathematical problems, using efficient mental methods and explaining where written methods are more appropriate.

COMMON MISCONCEPTIONS	STRENGTHENING UNDERSTANDING	GOING DEEPER
Children may confuse the definitions for 'factor' and 'multiple'.	New vocabulary and its meaning should be displayed prominently in the classroom.	Children could investigate the prime factors of different numbers. What patterns can they discover?
Children may muddle the order of operations or neglect to remember how brackets influence how to solve a calculation.	To help children remember the order of operations, they could be encouraged to create a class rhyme or song.	Give children 4–5 random 1-digit numbers. Can they use their numbers and understanding of the order of operations and brackets to create another given number? If not, how close can they get?

Unit 3: Four operations ②

WAYS OF WORKING

Use these pages to introduce the focus to children. You can use the characters to explore different ways of working too!

STRUCTURES AND REPRESENTATIONS

Array: Arrays are a visual representation of multiplication and division. They are an excellent tool for showing equal groups within a number.

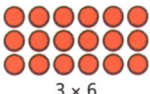

3×6

Sorting circles/diagram: Sorting circles (or sorting diagrams) are used in this unit to organise numbers with certain properties.

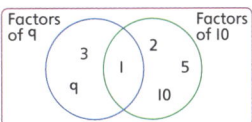

100 square: The 100 square is used in this unit to highlight patterns and relationships between factors and multiples, and to show prime numbers.

Bar model: Bar models enable children to more easily represent a problem. In the context of this unit, they are used to show different types of calculations.

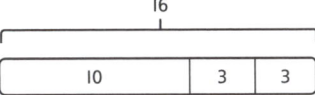

Number line: A number line is a more abstract representation of a sequence of numbers. It is used in this unit to represent different calculations, for example, finding the difference between two numbers.

Part-whole model: Part-whole models help to clearly show the different ways a number can be partitioned.

KEY LANGUAGE

There is some key language that children will need to know as a part of the learning in this unit.

→ factor, common factor
→ multiple, common multiple
→ prime
→ squared (x^2), cubed (x^3)
→ order of operations, brackets
→ inverse operation

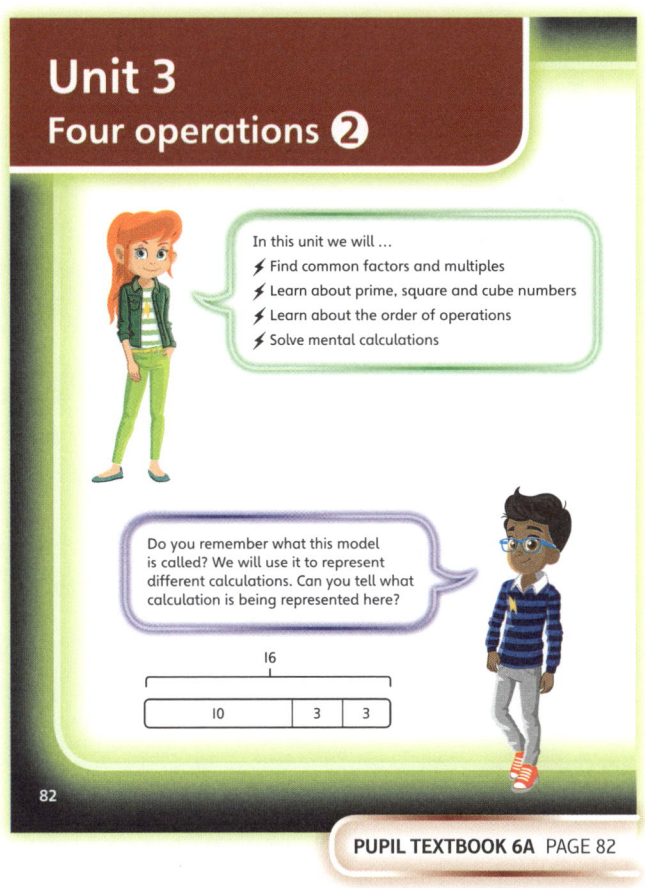

PUPIL TEXTBOOK 6A PAGE 82

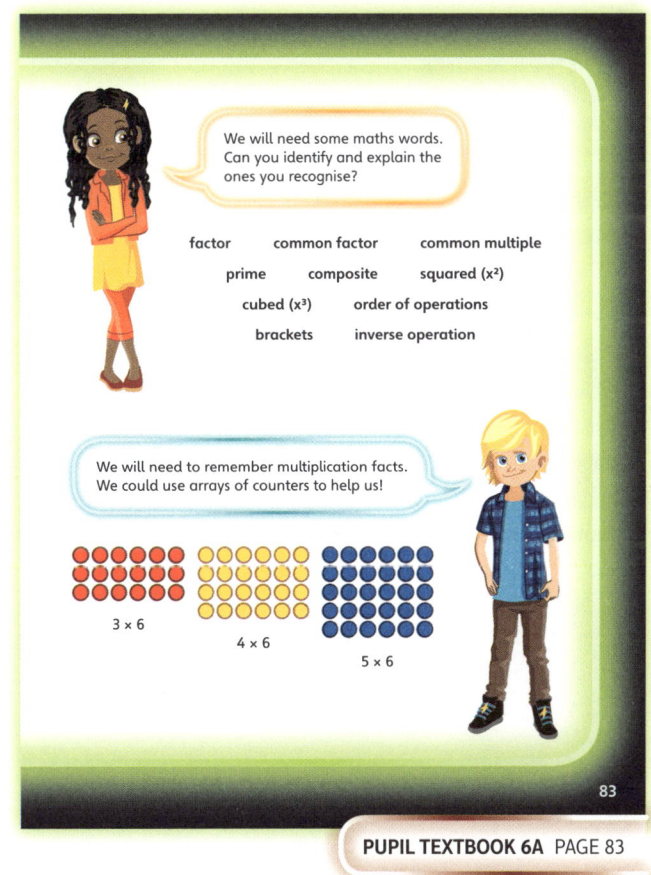

PUPIL TEXTBOOK 6A PAGE 83

115

Common factors

Learning focus

In this lesson, children will develop their understanding of factors and how common factors link two or more numbers. They will use this understanding to find common factors.

Small steps

→ Previous step: Dividing numbers up to 4 digits by a 2-digit number (6)
→ **This step: Common factors**
→ Next step: Common multiples

NATIONAL CURRICULUM LINKS

Year 6 Number – Addition, Subtraction, Multiplication and Division

Identify common factors, common multiples and prime numbers.

ASSESSING MASTERY

Children can, given two or more numbers, find the common factors. They can use this ability to help solve mathematical problems and puzzles reliably, explaining their reasoning and understanding confidently.

COMMON MISCONCEPTIONS

Children may mistakenly look for common multiples of a number instead of factors. Ask:
• *What are factors? What operation can you use to find them?*

Get children to create an array for the number they are finding factors of. Ask:
• *What factors of [number] can you see in this array?*

STRENGTHENING UNDERSTANDING

Before this lesson, give children opportunities to rehearse multiplication facts. For example, use songs and chants, and activities such as 'follow me' cards (for example, *I am 6, who is 3 × 7? I am 21, who is 8 × 12?*).

GOING DEEPER

Encourage children to set challenges for each other, such as:
• *My two numbers share the common factors of 2, 3 and 5. What is my number?*

KEY LANGUAGE

In lesson: factor, common factor, divide, remainder, multiplication, array

Other language to be used by the teacher: multiply, division

STRUCTURES AND REPRESENTATIONS

arrays, sorting circles

RESOURCES

Optional: 'Follow me' cards, counters, multiplication grids

 In the eTextbook of this lesson, you will find interactive links to a selection of teaching tools.

Before you teach

• How confident are children with the recall of multiplication facts? Will this potentially slow the pace of the lesson?
• How will you accommodate children who are less confident with multiplication tables?

Discover

Unit 3: Four operations (2), Lesson 1

WAYS OF WORKING Pair work

ASK

- Question **1** a): *What numbers will split equally into 4 groups?*
- Question **1** b): *What groups will allow for the children and adults to be split equally?*
- Question **1** b): *Can you prove your ideas?*

IN FOCUS Children are required to use knowledge of multiples or divisibility to recognise that they need to find a number that divides into **both** numbers, in order to solve the problem.

PRACTICAL TIPS Children could practically solve the problem posed in the picture, with some children pretending to be adult helpers and others being the children. Change the challenge by varying the numbers used in the question.

ANSWERS

Question **1** a): The adults and children cannot split equally into 4 groups (54 ÷ 4 = 13 with 2 remaining).

Question **1** b): The adults and children could split into 1, 2, 3 or 6 equal groups.

Common factors

Discover

We should divide into groups. We are 24 adults and 30 children.

Treasure Hunt Adventure Land

Let's have 4 groups.

Make sure there are the same number of adults and children in each group.

1 a) Can the adults and children split equally into 4 groups?

   b) What are the equal groups the adults and children could split into?

84

PUPIL TEXTBOOK 6A PAGE 84

Share

WAYS OF WORKING Whole class teacher led

ASK

- Question **1** a): *What other numbers are not factors of 30? How can you prove this?*
- Question **1** b): *How did you show the groups that were possible with those numbers?*
- Question **1** b): *What numbers are 6 and 4 a factor of? Can you prove it?*
- Question **1** b): *Are the common factors of 24 and 30 also factors of any other numbers?*

IN FOCUS To ensure children's concrete understanding of this concept, it is important to link the idea with their experience of arrays. Children should be encouraged to build (for example, using counters) or draw arrays that prove the link between the common factors.

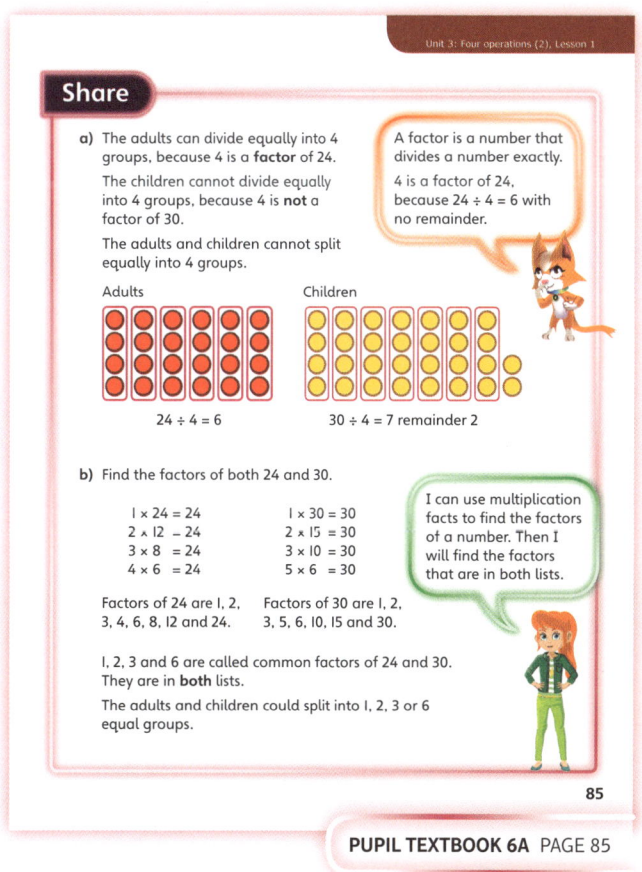

Share

a) The adults can divide equally into 4 groups, because 4 is a **factor** of 24.

The children cannot divide equally into 4 groups, because 4 is **not** a factor of 30.

The adults and children cannot split equally into 4 groups.

A factor is a number that divides a number exactly.

4 is a factor of 24, because 24 ÷ 4 = 6 with no remainder.

Adults   Children

24 ÷ 4 = 6   30 ÷ 4 = 7 remainder 2

b) Find the factors of both 24 and 30.

I can use multiplication facts to find the factors of a number. Then I will find the factors that are in both lists.

1 × 24 = 24	1 × 30 = 30
2 × 12 = 24	2 × 15 = 30
3 × 8 = 24	3 × 10 = 30
4 × 6 = 24	5 × 6 = 30

Factors of 24 are 1, 2, 3, 4, 6, 8, 12 and 24.

Factors of 30 are 1, 2, 3, 5, 6, 10, 15 and 30.

1, 2, 3 and 6 are called common factors of 24 and 30. They are in **both** lists.

The adults and children could split into 1, 2, 3 or 6 equal groups.

85

PUPIL TEXTBOOK 6A PAGE 85

Think together

Whole class teacher led (I do, We do, You do)

ASK

- Question ❶ : *How could you show the children in each group?*
- Question ❷ : *What times-tables knowledge will you need to use to solve this?*
- Question ❸ : *How can you prove that you have found all the common factors?*

IN FOCUS Children should be encouraged to use their fluency with multiplication tables to help solve the problems listed in this part of the lesson. Develop children's ability to work systematically by referring back to Astrid's comment in the previous section. Question ❶ can be used to link children's multiplicative understanding to their new understanding of factors by demonstrating the full multiplication calculations. It may be beneficial to also link them to the inverse division calculations.

STRENGTHEN If children are not yet fluent in the multiplication tables facts, offer multiplication grids for them to use. Ask:

- *How can you use this multiplication grid to help you?*
- *Is it possible to find the common factors using this?*

DEEPEN In question ❸, extend children's reasoning around common factors. Ask: *What other general statements can you make about types of numbers and their common factors? For example, what can you say about the common factors of all even numbers?*

ASSESSMENT CHECKPOINT At this point in the lesson, children should be able to explain what common factors are and how they link two or more numbers. Children should be confident when finding the common factors of two numbers and will be growing in confidence when finding common factors of more than two numbers. Question ❸ will demonstrate these skills in particular; children should be able to explain the generalisations about multiplication facts that will help them find common factors.

ANSWERS

Question ❶ : $1 \times 12 = 12$, $2 \times 6 = 12$, $3 \times 4 = 12$
$1 \times 15 = 15$, $3 \times 5 = 15$
Factors of 12 are 1, 2, 3, 4, 6 and 12.
Factors of 15 are 1, 3, 5 and 15.
The common factors of 12 and 15 are 1 and 3.
The adults and children could split into 1 group or 3 groups.

Question ❷ :

The common factor of 9 and 10 is 1.

Question ❸ a): 1 and 5.

Question ❸ b): 2 and 10.

Question ❸ c): 3, 4, 15 and 20.

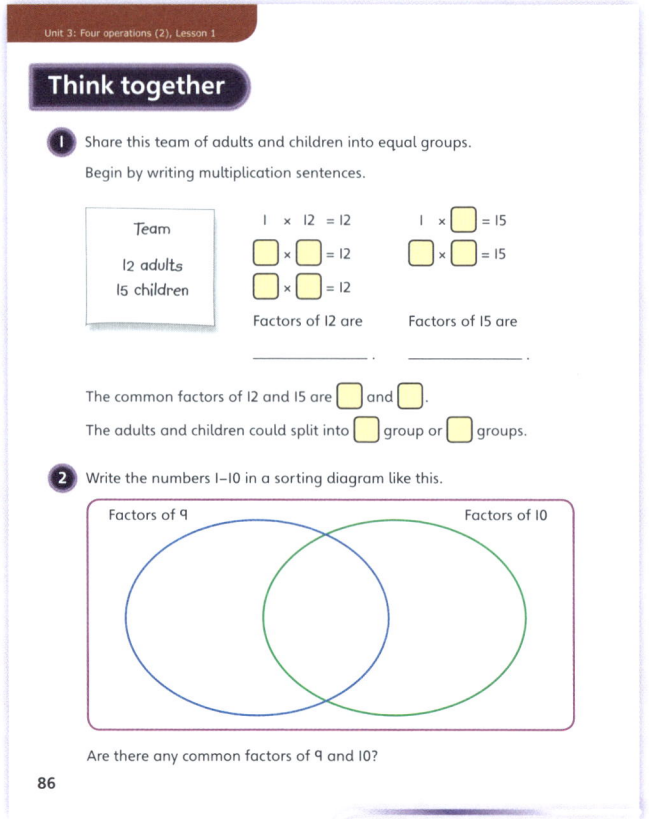

PUPIL TEXTBOOK 6A PAGE 86

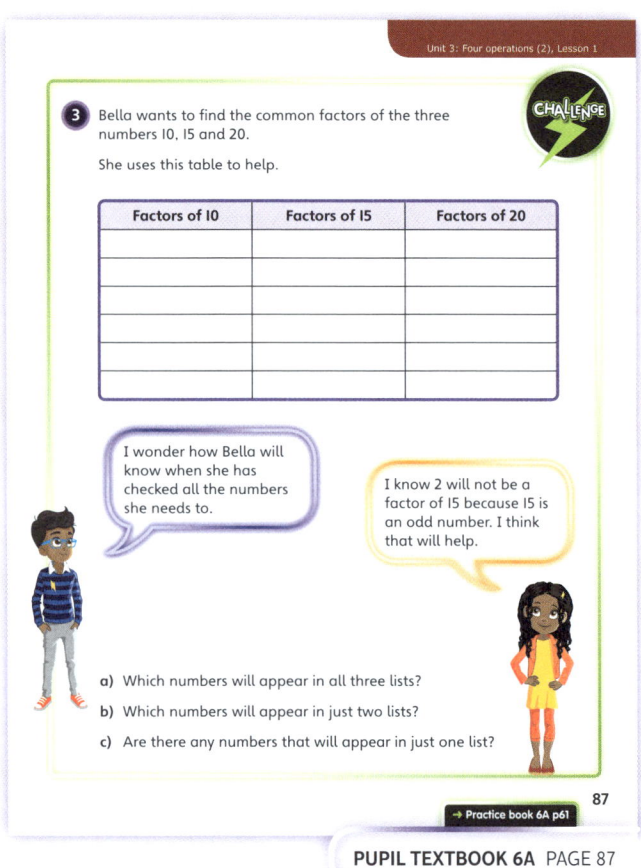

PUPIL TEXTBOOK 6A PAGE 87

Practice

WAYS OF WORKING Independent thinking

IN FOCUS Question ❶ encourages children to use arrays to explain their reasoning and prove their ideas. This will help them to link their concrete understanding with the more abstract ideas presented in the lesson. Children could be encouraged to build and manipulate these arrays to develop their reasoning and explanations. This will be especially useful when approaching question ❶ c).

STRENGTHEN When solving question ❹ b), if children are having trouble getting started, ask:
• *How could you make finding the numbers easier?*
• *What resource might help you to get started finding the numbers that have those factors?*
• *How could you write your findings?*

DEEPEN While solving question ❹ a), deepen children's thinking about the link between numbers by asking:
• *Does knowing the factors of 35 help you to find the factors of 70 more quickly?*
• *Does knowing the factors of 6 help you to find the factors of 60 more quickly? How about 600?*

THINK DIFFERENTLY While solving question ❸, children should be encouraged to provide proof of their ideas. This will be especially appropriate for justifying why 5 is incorrectly placed. Children should be able to prove this through use of an array.

ASSESSMENT CHECKPOINT Question ❸ assesses whether children are able to confidently find common factors of two given numbers. They should be able to use this understanding to identify where someone has made a mistake. Look for children recognising where their generalisations can help them (in this case, for instance, that 5 will be a common factor of all multiples of 10).

ANSWERS Answers for the **Practice** part of the lesson appear in the separate **Practice and Reflect answer guide**.

Reflect

WAYS OF WORKING Independent thinking

IN FOCUS This section assesses children's number sense. They should be able to recognise that, once they have found the factors of both numbers, up to the value of 15, it is not worthwhile going any further as any number bigger than 15 will not be a factor of it.

ASSESSMENT CHECKPOINT Look for children's understanding of the reasoning noted above.

ANSWERS Answers for the **Reflect** part of the lesson appear in the separate **Practice and Reflect answer guide**.

After the lesson

• Were children able to confidently use arrays and their multiplication fluency to clearly explain their reasoning?
• Were there any multiplication tables that children were less confident with? How will you develop their confidence with these in the future?

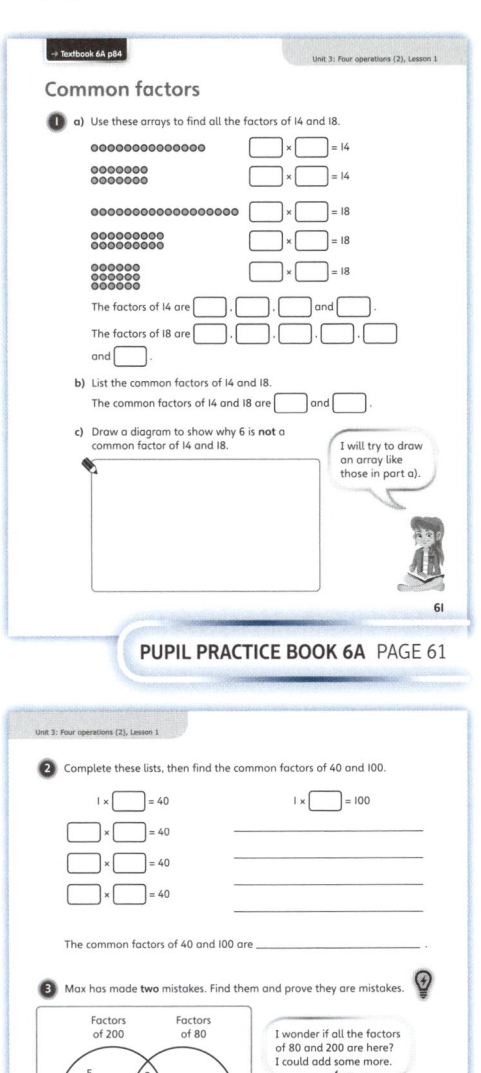

PUPIL PRACTICE BOOK 6A PAGE 61

PUPIL PRACTICE BOOK 6A PAGE 62

PUPIL PRACTICE BOOK 6A PAGE 63

Common multiples

Learning focus

In this lesson, children will develop their understanding of multiples and how common multiples link two or more numbers. They will use this understanding to find common multiples.

Small steps

→ Previous step: Common factors
→ **This step: Common multiples**
→ Next step: Recognising prime numbers up to 100

NATIONAL CURRICULUM LINKS

Year 6 Number – Addition, Subtraction, Multiplication and Division

Identify common factors, common multiples and prime numbers.

ASSESSING MASTERY

Children can, given two or more numbers, find the common multiples. They can use this ability to help solve mathematical problems and puzzles reliably, explaining their reasoning and understanding confidently.

COMMON MISCONCEPTIONS

Children may mistakenly look for common factors of a number instead of multiples. Ask:
• *What are factors? What are multiples? How are they different?*

STRENGTHENING UNDERSTANDING

Children should be given opportunities to develop their fluency with multiplication tables, especially any of those highlighted as areas of development in the previous lesson. Again, give children the opportunity to recite rhymes, songs and chants and play multiplication games.

GOING DEEPER

Children could challenge each other, for example:
• *My two numbers share the common multiples 60 and 75. What numbers could they be?*

KEY LANGUAGE

In lesson: multiple, common multiple, common factor

Other language to be used by the teacher: multiplication

STRUCTURES AND REPRESENTATIONS

100 squares, sorting circles, bar models

RESOURCES

Optional: multiplication grids, hoops, bean bags

 In the eTextbook of this lesson, you will find interactive links to a selection of teaching tools.

Before you teach

• Were there any multiplication tables that needed further input before this lesson?
• How will this influence your teaching?

Discover

Common multiples

WAYS OF WORKING Pair work

Discover

ASK

- Question **1** a): *Can you predict the days that one of the jobs will be done?*
- Question **1** a): *Can you predict the days that both jobs will be done together? Explain how you know.*
- Question **1** b): *What is interesting about the days where more than one job is required?*

IN FOCUS Question **1** a) introduces the concept of common multiples of two numbers. It is important for children to recognise that the common multiples are numbers that feature in both counts. This idea is further developed in question **1** b) with the introduction of a third count.

PRACTICAL TIPS This concept could be introduced in small games outside. For example, a number of hoops could be laid on the ground and children challenged to throw bean bags into the hoops. Give restrictions, such as blue bean bags can only be thrown in every second hoop; red bean bags can only be thrown in every third hoop. Ask:
- *Which hoops will have both colours? Why?*

To engage children further with this concept, it could be introduced with the presentation of a class pet. Give children the instructions as given in the picture and use them to help solve questions **1** a) and b).

ANSWERS

Question **1** a): Lexi will need to do both jobs on day 15 and day 30. (Days that are common multiples of 3 and 5.)

Question **1** b): Lexi will need to do all three jobs on day 30. (Days that are common multiples of 2, 3 and 5)

1 a) On which days will Lexi need to change the bedding and give carrots?

   b) On which days will Lexi need to do all three jobs?

88

PUPIL TEXTBOOK 6A PAGE 88

Share

WAYS OF WORKING Whole class teacher led

ASK

- Question **1** a): *What numbers are multiples of 3? What numbers are multiples of 5?*
- Question **1** a): *Can you find multiples of 5 in the multiples of 3?*
- Question **1** b): *Can you use what you have found to predict all the common multiples?*
- Question **1** b): *If a common multiple is 30, does that mean 300 will be a common multiple too? How about 3,000? Why?*

IN FOCUS In question **1** a) children look at multiples of two numbers and then identify the common multiples between them. A third number is introduced in question **1** b) and children find a common multiple of all three numbers. For both questions, children should be encouraged to use 100 squares to help them identify and write the multiples of each number. Discuss patterns children spot and how these can be used to help predict further common multiples.

Share

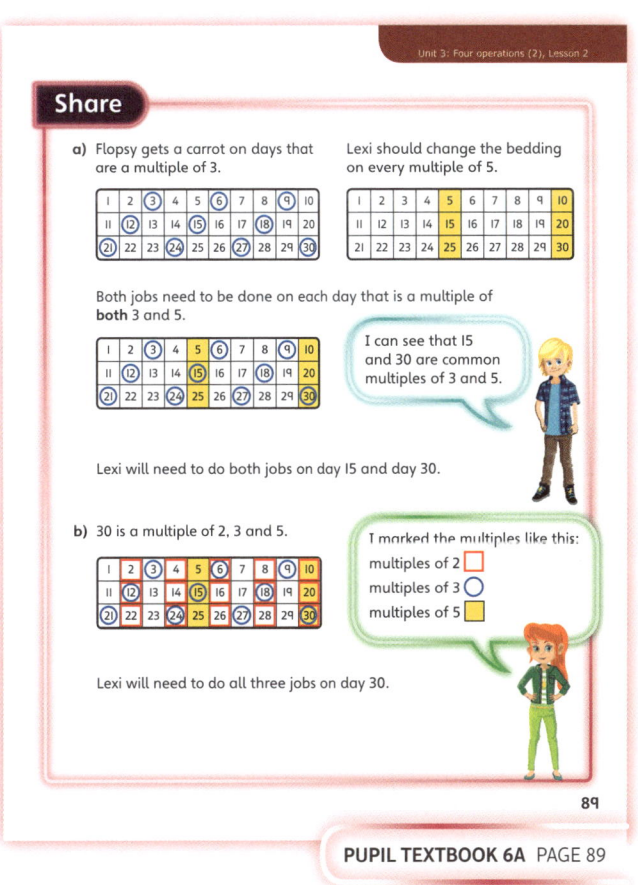

a) Flopsy gets a carrot on days that are a multiple of 3.

Lexi should change the bedding on every multiple of 5.

Both jobs need to be done on each day that is a multiple of **both** 3 and 5.

I can see that 15 and 30 are common multiples of 3 and 5.

Lexi will need to do both jobs on day 15 and day 30.

b) 30 is a multiple of 2, 3 and 5.

I marked the multiples like this:
multiples of 2
multiples of 3
multiples of 5

Lexi will need to do all three jobs on day 30.

89

PUPIL TEXTBOOK 6A PAGE 89

Think together

Whole class teacher led (I do, We do, You do)

ASK

- Question **1** : *How could you write all the multiples of each number?*
- Question **2** : *How will you find the common multiples?*
- Question **3** : *What do you notice about 20 and 100?*
- Question **4** : *Are you able to find all the common multiples? Explain your ideas.*

IN FOCUS Questions **1** and **2** offer children opportunities to find common multiples of different numbers in different contexts. Children should be encouraged to justify their solutions with evidence. While solving question **4**, it is important to ensure children recognise the differences between common factors and common multiples, to avoid confusing the terms in future.

STRENGTHEN To help children begin finding common multiples of the numbers in each question, it may be helpful to give them 100 squares they can colour or write on. Ask: *How can you use this to help you? What patterns can you spot?*

DEEPEN Children could deepen their understanding of common multiples by investigating what happens if they look for common multiples of more numbers. Ask: *What happens if you look for the common multiples of three numbers? Will there be more or fewer common multiples? Why?*

Children could also be encouraged to investigate how much of a difference is made if one of the numbers is increased or decreased by 1. For example, for question **1** ask: *Is there going to be a big change in common multiples if Aki visits his gran every 5 days instead of 4? Explain your ideas.*

ASSESSMENT CHECKPOINT By the end of question **3**, children should be able to find common multiples of two numbers. Use question **4** to assess whether children recognise that, as they are multiplying and looking for larger numbers that their original numbers are factors of, the list of multiples is infinite.

ANSWERS

Question **1** : Common multiples of 4 and 6 are 12, 24, 36, …

They both visit Gran on days 12, 24, 36, 48 (and all following multiples of 12).

Question **2** : The towers could be 60 cm, 120 cm, 180 cm and all subsequent multiples of 60.

Question **3** : They are all the multiples of 100 (as 100 is a multiple of 20).

Question **4** a): Agree with Jamilla.

Question **4** b): All common multiples of 10 and 25 can be described as multiples of 50.

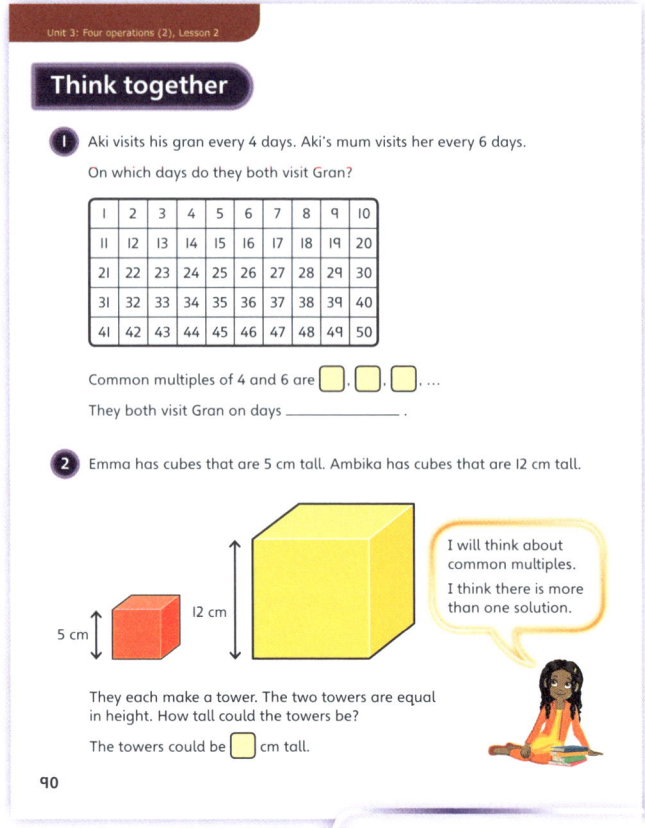

PUPIL TEXTBOOK 6A PAGE 90

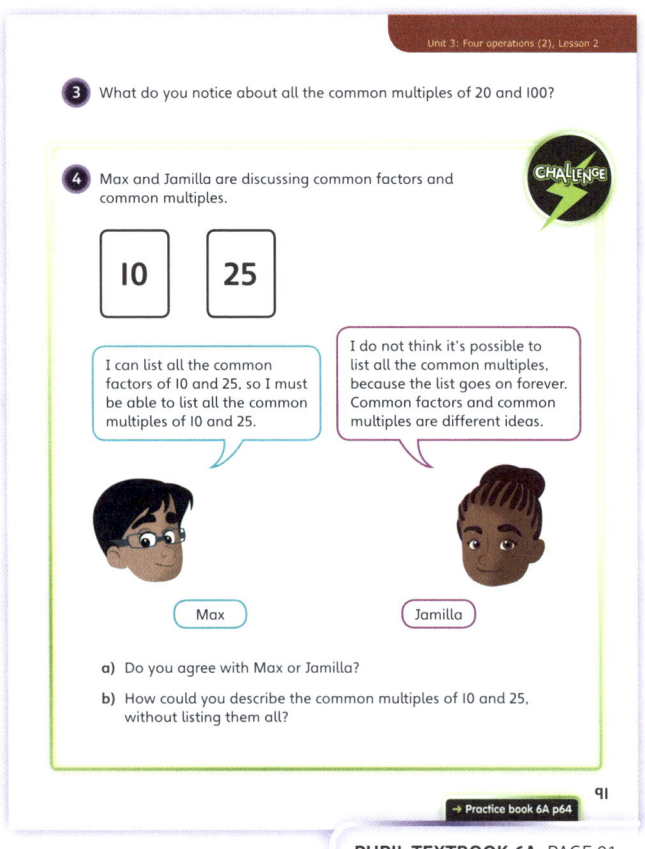

PUPIL TEXTBOOK 6A PAGE 91

Practice

WAYS OF WORKING Independent thinking

IN FOCUS Question **3** assesses whether children can find common multiples of two numbers. It also gives an opportunity to observe whether children are able to make generalisations about the numbers they are finding common multiples of. Ask: *Do your generalisations apply to other numbers? Prove it.*

Question **5** challenges the assumption that merely multiplying the two given numbers together will give the first common multiple. Children should be encouraged to explain why Andy has made this assumption.

STRENGTHEN If children are struggling to place the numbers into the sorting diagram in question **3**, ask:
- *What resource can you use to begin finding multiples of 5?*
- *How can you use it to find common multiples of 4 and 5?*

DEEPEN Building on question **3**, children could be challenged to create a sorting diagram with three circles, each looking for multiples of a different number, less than 10. Can they find the common multiples for these numbers? Can they make any generalisations about the common multiples they find for all three?

ASSESSMENT CHECKPOINT Children should be able to confidently find common multiples of any given numbers. Question **2** assesses this skill and also gives you the opportunity to assess children's ability to begin making generalisations about multiples. Look for children linking their multiplicative understanding to patterns they find in common multiples.

ANSWERS Answers for the **Practice** part of the lesson appear in the separate **Practice and Reflect answer guide**.

Reflect

WAYS OF WORKING Independent thinking

IN FOCUS This question provides a final opportunity to assess whether children can confidently find common multiples of numbers, demonstrating their understanding through the explanation of their reasoning. Ask children to independently find the three common multiples the question asks for. Once they have been given time to do this, they could share their findings with their partner. Ask:
- *Did you both find the same multiples?*
- *Why might they be different?*

ASSESSMENT CHECKPOINT Children should be able to find the common multiples and also recognise that there are an infinite number of possible multiples to pick from.

ANSWERS Answers for the **Reflect** part of the lesson appear in the separate **Practice and Reflect answer guide**.

After the lesson

- Could children recognise and explain the difference between common multiples and common factors?
- Were children able to understand and explain why there are infinite common multiples?
- How could you have made this lesson more practical?

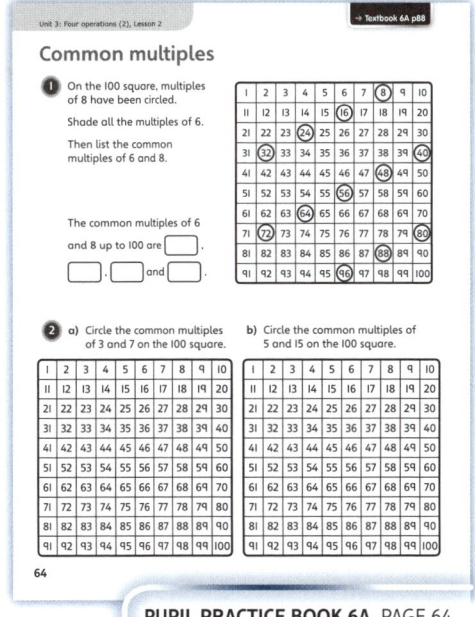

PUPIL PRACTICE BOOK 6A PAGE 64

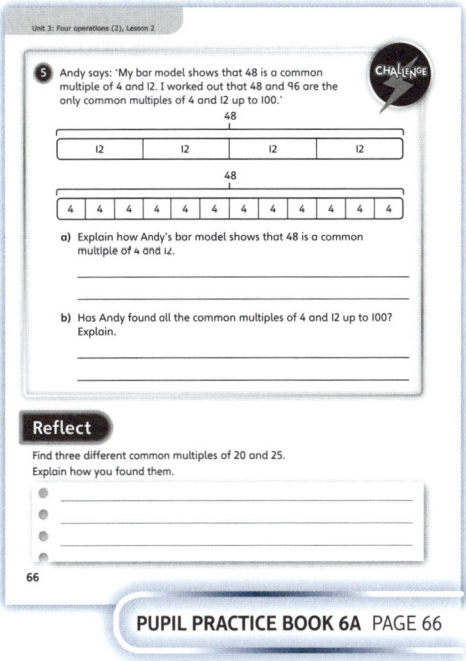

PUPIL PRACTICE BOOK 6A PAGE 65

PUPIL PRACTICE BOOK 6A PAGE 66

Recognising prime numbers up to 100

Learning focus

In this lesson, children will learn to recognise and identify prime numbers. They will explore how these numbers are different from other numbers.

Small steps

→ Previous step: Common multiples
→ **This step: Recognising prime numbers up to 100**
→ Next step: Squares and cubes

NATIONAL CURRICULUM LINKS

Year 6 Number – Addition, Subtraction, Multiplication and Division

Identify common factors, common multiples and prime numbers.

ASSESSING MASTERY

Children can recognise and identify prime numbers, fluently explaining their unique properties. They can explain why 2 is the only even prime number and why 1 is not a prime number.

COMMON MISCONCEPTIONS

Children may assume that 1 is a prime number as it is only divisible by 1 and, therefore, itself. Ask:
• *How many factors does a prime number have?*
• *How many factors does 1 have?*

STRENGTHENING UNDERSTANDING

Arrays will be a powerful tool to help strengthen children's understanding in this lesson. Ask:
• *How many arrays can you make for this number?*
• *How many arrays are possible if the number is prime?*

GOING DEEPER

Children could be encouraged to write each number in question ❸ of the **Think together** section as a product of factors with at least one of the factors as a prime number.

KEY LANGUAGE

In lesson: prime, array, remainder, divide, factor, composite number, reasoning

Other language to be used by the teacher: multiple, multiply, multiplication, division

STRUCTURES AND REPRESENTATIONS

arrays, tables

RESOURCES

Optional: counters

 In the eTextbook of this lesson, you will find interactive links to a selection of teaching tools.

Before you teach

• How will you ensure children are given practical opportunities to investigate prime numbers in this lesson?
• How will you ensure that children's learning from the previous two lessons is drawn on explicitly in this one?

Discover

WAYS OF WORKING Pair work

ASK

- Question ❶ a): *How many factors does 16 have?*
- Question ❶ a): *What happens when one more counter is added? How many factors can you find for your new number?*
- Question ❶ b): *How many factors does 13 have? How is this number similar to 17?*
- Question ❶ b): *Can you find any other numbers that have similar properties?*

IN FOCUS The arrays in the picture show that 16 has several factors. By adding one more counter to the arrays, children should quickly spot that the only arrays without remainders are 1 × 17 and 17 × 1. Children should be encouraged to begin generalising about the numbers mentioned. Can they find any other numbers that have similar properties? Ensuring this is done practically will help secure children's understanding.

PRACTICAL TIPS Children should be given ample opportunities to create the arrays linked to the numbers they are investigating. The scenario shown in the picture can be recreated in the classroom practically using counters. Once these numbers have been explored, can children find other examples where adding one counter creates a number with fewer factors?

ANSWERS

Question ❶ a): Only two different arrays are possible using 17 counters: 1 row of 17 because 17 ÷ 1 = 17 and 17 rows of 1 because 17 ÷ 17 = 1. Isla cannot make more arrays using Aki's counter.

Question ❶ b): 13 and 19 are both prime numbers so you can only make two arrays for each.

Share

WAYS OF WORKING Whole class teacher led

ASK

- Question ❶ a): *How many ways did you try to make an array for 17? How many were successful?*
- Question ❶ b): *What was the same about all the prime numbers? What was different?*

IN FOCUS When looking at arrays for 17, 13 and 19, it is important to ensure that children understand the property that any prime number has exactly two factors: 1 and itself.

DEEPEN It may be that, at this point, children have only found odd numbered primes and so generalise that all primes must be odd. This should be used as an interesting learning point. Ask:

- *Are there any even primes?*

PUPIL TEXTBOOK 6A PAGE 92

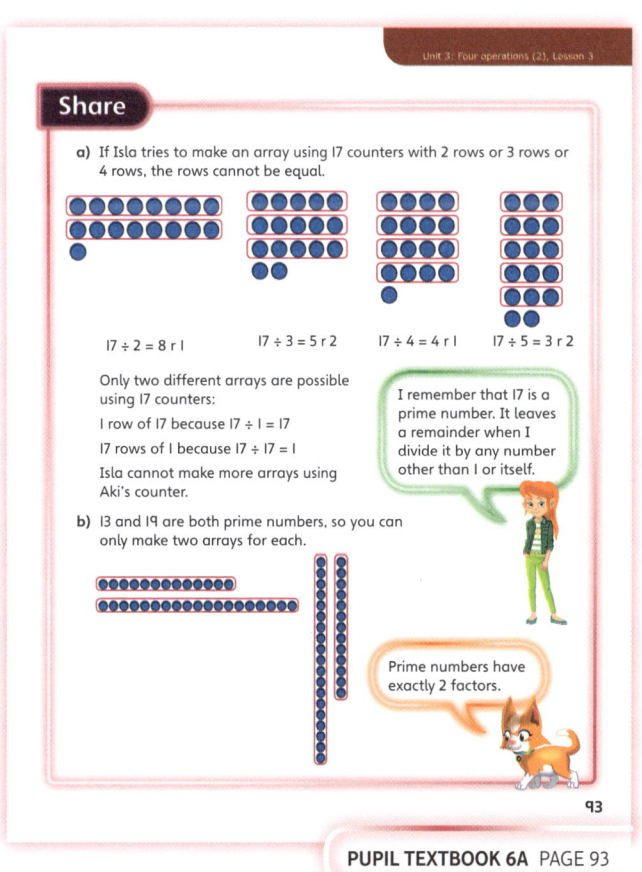

PUPIL TEXTBOOK 6A PAGE 93

Think together

WAYS OF WORKING Whole class teacher led (I do, We do, You do)

ASK

- Question **1** : *Does finding just the factors 1 and the number itself prove that a number is definitely prime?*
- Question **2** : *How could you check that a number is definitely prime?*
- Question **3** : *Are there any ways to quickly judge if a number is prime or not?*

IN FOCUS Question **1** challenges the assumption that if children can prove that a number has the factors 1 and itself, then it must be prime. Be sure to discuss the importance of gathering enough evidence to prove their ideas.

STRENGTHEN For question **3** b), children can be helped to more efficiently identify numbers that are not prime by ensuring they are aware that 2 is the only even prime number. Ask:

- *Are prime numbers more likely to be even or odd? Explain.*
- *How will this help you solve this question more efficiently?*

DEEPEN Question **3** b) provides a good opportunity to make generalisations about numbers with different properties to help children identify primes more efficiently. Children should be encouraged to spot patterns, for example multiples of 5 (except 5 itself) and multiples of 10 are never prime and are easy to identify. Ask: *Can you create some rules for identifying prime numbers?*

ASSESSMENT CHECKPOINT Children should be able to identify the properties of prime numbers and prove a number is prime using resources and arrays. They should be beginning to identify and explain generalisations to help find prime numbers more efficiently.

ANSWERS

Question **1** : Disagree. Mo has not proved that the numbers are definitely prime. While 11 is prime, 21 has the factors 1, 3, 7 and 21.

Question **2** : Alex has circled 39, which is not prime. She has missed 41.

Question **3** a): Bella's method will find out whether or not a number is prime. She can stop at 10, because 10 × 10 is 100 and 100 > 97. So when she gets to 10 she will have found any factor pairs, each of which must contain a number smaller than 10. She will not find any, because 97 is prime.

Question **3** b): Prime numbers: 71, 79

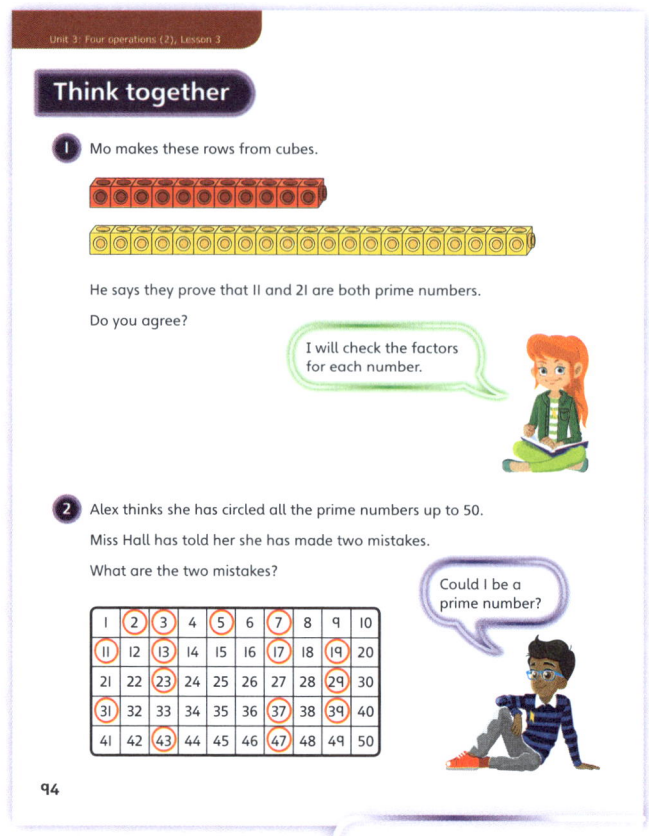

PUPIL TEXTBOOK 6A PAGE 94

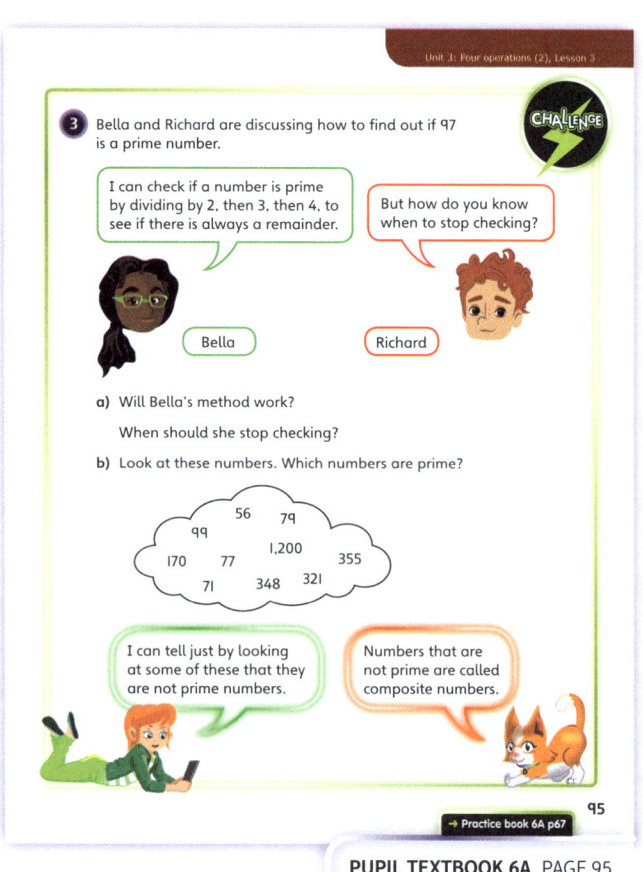

PUPIL TEXTBOOK 6A PAGE 95

Practice

WAYS OF WORKING Independent thinking

IN FOCUS Throughout this part of the lesson, it is important for children to be able to build or draw the arrays to represent the numbers they are dealing with. Questions ❶ and ❷ scaffold children's solutions by offering semi-completed sentences for children to finish. Once they reach question ❸, this support is withdrawn.

STRENGTHEN If children are struggling to find the prime numbers in question ❸, it may be beneficial to recap the generalisations made about the properties of numbers (for example, multiples of 10 are never prime) and display these somewhere prominent in the classroom.

DEEPEN Question ❺ offers children an opportunity to share their reasoning. This activity could be deepened by asking children to investigate if there are any patterns of numbers (for example, 3, 13, 23, 33, 43, 53) where every number is prime. Ask: *Can you find a regular sequence of numbers that are all prime?*

ASSESSMENT CHECKPOINT Children should now be more confident with the properties of numbers and how these can help them identify whether a number is prime or not. They should be able to share their reasoning confidently, using the lesson vocabulary accurately.

ANSWERS Answers for the **Practice** part of the lesson appear in the separate **Practice and Reflect answer guide**.

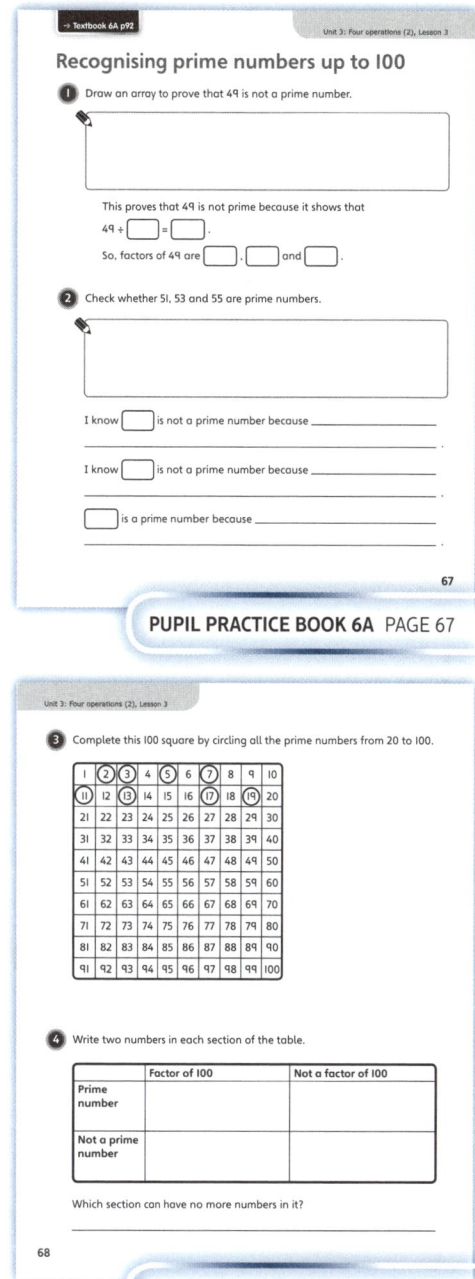

PUPIL PRACTICE BOOK 6A PAGE 67

PUPIL PRACTICE BOOK 6A PAGE 68

Reflect

WAYS OF WORKING Independent thinking

IN FOCUS This question will demonstrate whether children are able to explain how to find out whether a number is prime or not. Children should be given time to formulate and write their proof, which they can then share and discuss with their partner. Ask:
- *Have you investigated the numbers in the same way?*
- *Is one method more efficient than the other? Explain how.*

ASSESSMENT CHECKPOINT Children should be using the concrete or pictorial representations of arrays, coupled with their number knowledge, to identify whether the two numbers are prime.

ANSWERS Answers for the **Reflect** part of the lesson appear in the separate **Practice and Reflect answer guide**.

PUPIL PRACTICE BOOK 6A PAGE 69

After the lesson ⏸

- How was children's ability to generalise developed in this lesson?
- How could you continue to develop children's use of prime numbers in other areas of the curriculum? (For example, possible team groupings in PE – there are 17 people, what equal teams can you make?)

Squares and cubes

Learning focus

In this lesson, children will learn to recognise and identify square and cube numbers. They will explore how these numbers are different from others.

Small steps

→ Previous step: Recognising prime numbers up to 100
→ **This step: Squares and cubes**
→ Next step: Order of operations

NATIONAL CURRICULUM LINKS

Year 5 Number – Multiplication and Division

Recognise and use square numbers and cube numbers, and the notation for squared (^{2}) and cubed (^{3}).

ASSESSING MASTERY

Children can recognise, identify and calculate square and cube numbers, fluently explaining their unique properties. They are able to recognise and use the mathematical notation for squared (^{2}) and cubed (^{3}) and can confidently explain their reasoning using their mathematical understanding.

COMMON MISCONCEPTIONS

Children may mistakenly assume that x^2 means x multiplied by 2, and similarly, x^3 means x multiplied by 3. Ensure children link the terminology of 'squared' and 'cubed' to the properties of their namesake shapes. Ask:

- *To find the area of a square with a side length of 4, would you calculate 4 × 2 or 4 × 4? Why?*
- *How does this relate to 4^2?*

STRENGTHENING UNDERSTANDING

Before beginning this lesson, children could be reminded of their work on area and volume. Give children variously-sized squares and cubes for them to measure and find the area and volume of.

GOING DEEPER

Children could be set 'Always, Sometimes, Never' statements to investigate. For example:
- *When you square an even number, the result is divisible by 4.*

KEY LANGUAGE

In lesson: square, cube, multiplication, multiply (×), array, prime

Other language to be used by the teacher: squared, cubed, multiplied

STRUCTURES AND REPRESENTATIONS

array, 2D square, 3D cube, multiplication grid, sorting circles/diagram

RESOURCES

Mandatory: counters

Optional: multilink cubes, multiplication grids, 100 square

 In the eTextbook of this lesson, you will find interactive links to a selection of teaching tools.

Before you teach

- Are children confident with finding the area of a square and the volume of a cube?
- What resources will you provide to make this link to the properties of shapes explicit?

Discover

WAYS OF WORKING Pair work

ASK

- Question ❶ a): *Can you make a solid cube with 16 small cubes? What regular shape can you make?*
- Question ❶ b): *How many small cubes do you need to make a larger solid cube of length 2?*
- Question ❶ b): *What other amounts make solid cubes?*

IN FOCUS Question ❶ a) requires children to understand the difference between square numbers and cube numbers, and to recognise the common misconception of mistaking one for the other. Children will explore and explain based on the arrangement of 2D square arrays and 3D cube representations. The focus of question ❶ b) is to explore the numerical value of different cube numbers.

PRACTICAL TIPS Children should be encouraged to follow Lee's line of enquiry practically in class. If children are given the opportunity to build the shape being described by Lee, they should be able to explain his mistake more easily.

ANSWERS

Question ❶ a): Lee is incorrect. He cannot make a large solid cube with all 16 cubes.

Question ❶ b): The largest solid cube Lee can make is a 2×2×2 cube using 8 small cubes. Lee would need another 11 small cubes to make a 3×3×3 large solid cube.

PUPIL TEXTBOOK 6A PAGE 96

Share

WAYS OF WORKING Whole class teacher led

ASK

- Question ❶ b): *What numbers of small cubes can you make a square with? How are they similar and how are they different?*
- Question ❶ b): *What numbers of small cubes can you make large solid cubes with? How are they similar and how are they different?*
- Question ❶ b): *Can you find any patterns in the square or cube numbers?*

IN FOCUS In this section, it is important to make explicit the link between the dimensions of the 2D and 3D shapes and finding square and cube numbers.

DEEPEN Give children other numbers to investigate using multilink cubes. Children could begin recording their findings in a systematic way. Their findings could be kept and displayed prominently in the classroom as a learning aid for the rest of the lesson. Make sure to list the numbers with a picture of the corresponding square or cube.

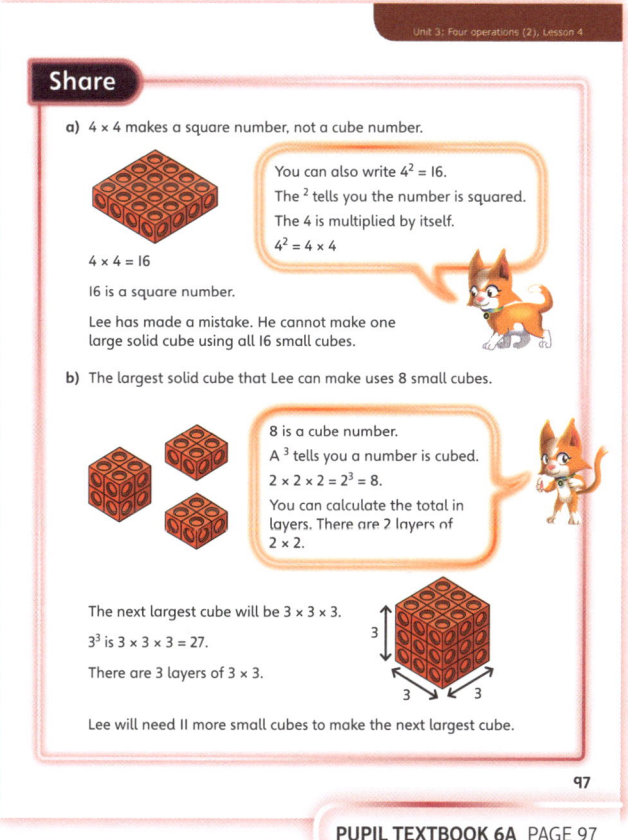

PUPIL TEXTBOOK 6A PAGE 97

Think together

Whole class teacher led (I do, We do, You do)

ASK

- Question ❶ : *How do you know if you need to square or cube a number?*
- Question ❶ : *What do ^{2} ('squared') and ^{3} ('cubed') mean?*
- Question ❷ : *What is different about x^2 and x multiplied by 2?*

IN FOCUS Question ❷ approaches the misconception of reading the square sign as '× 2' and the cube sign as '× 3'. Children should be encouraged to show evidence that this is a misconception by building the shapes to match. When working on question ❸, it is important to discuss why a pattern for cube numbers is not evident on a multiplication grid. Ask:

- *Can you see any cube numbers on the multiplication grid?*
- *What is the same and what is different about square and cube numbers?*
- *Why are there so few cube numbers on the grid?*

STRENGTHEN For question ❶, if children are struggling to match the pictures to the calculations, provide them with the resources necessary to build their own versions. Ask:

- *What multiplication does this array represent? How do you know?*
- *Can you write the multiplication it is representing? Can you find it in the list?*

DEEPEN For question ❸, children could be given a 100 square to investigate whether cube numbers create any patterns on that type of grid. Ask:

- *Do you predict a pattern will be visible on this type of grid?*
- *Explain your prediction.*

ASSESSMENT CHECKPOINT At this point, children should be able to explain what the square and cube signs mean and what calculation they will need to solve when they meet them. Question ❶ assesses whether they are able to link their understanding to the concrete representations of a square and a cube.

ANSWERS

Question ❶ a): $5 \times 5 \times 5 = 5^3 = 125$

Question ❶ b): $4 \times 4 \times 4 = 4^3 = 64$

Question ❶ c): $5 \times 5 = 5^2 = 25$

Question ❶ d): $8 \times 8 = 8^2 = 64$

Question ❷ : Luis has misunderstood the square and cube signs. He has mistaken their meaning as '× 2' and '× 3'. The correct working is $4^3 = 4 \times 4 \times 4 = 64$ and $6^2 = 6 \times 6 = 36$.

Question ❸ : The square numbers appear diagonally downwards from the top left (1–144).

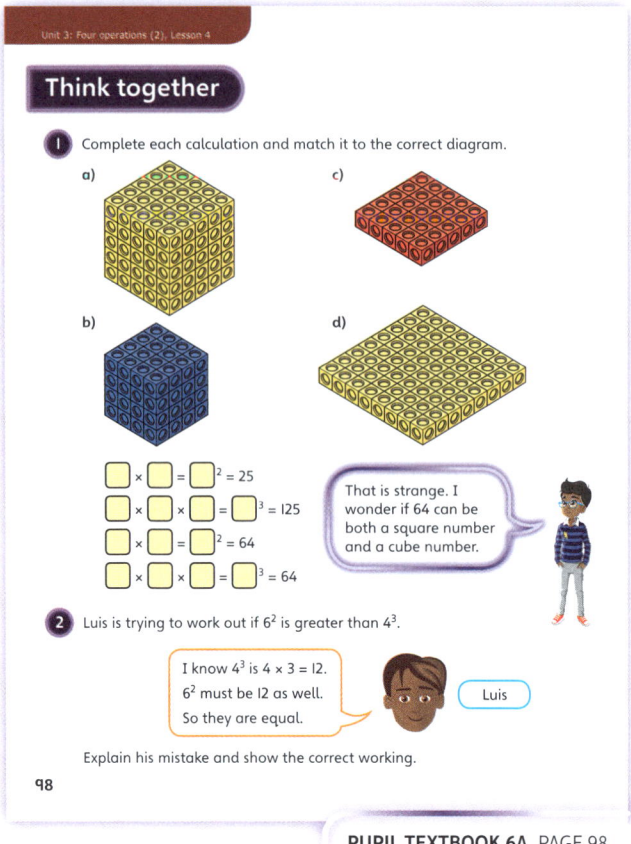

PUPIL TEXTBOOK 6A PAGE 98

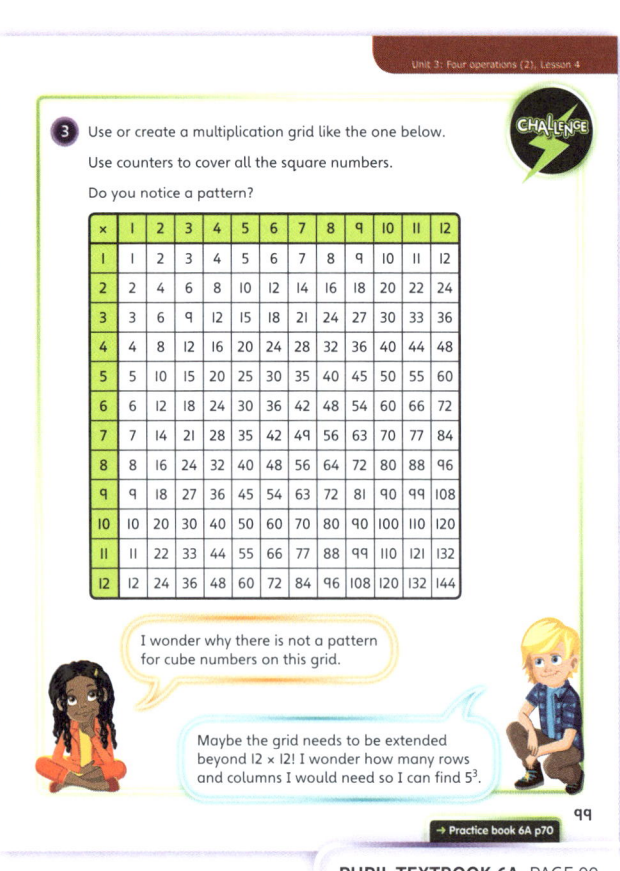

PUPIL TEXTBOOK 6A PAGE 99

Practice

WAYS OF WORKING Independent thinking

IN FOCUS Question ❶ links the pictorial representations with the abstract representations of squared and cubed numbers. If a class record of square and cube numbers has been placed in the classroom from the **Discover** section of the lesson, it would be beneficial to remind children of it for their independent activities.

STRENGTHEN If children are struggling with the pictorial and abstract representation in question ❶, they should be encouraged to make concrete versions of the numbers, for example, using multilink cubes. Likewise, if children are struggling to visualise the cube structure in question ❹, offer them a concrete version of the problem for them to manipulate. This will help them to explain their reasoning more clearly.

DEEPEN In question ❻, if children complete the sorting circles/diagram, they could be challenged to find all numbers that are both squares and cubes. An example of this is 64, which can be found by squaring 8 or cubing 4.

THINK DIFFERENTLY Question ❺ challenges children's assumptions about how the square and cube signs represent numbers that increase alongside the numbers they follow. Children are likely to assume that $30^2 = 90$ is correct, forgetting that the square sign now represents '× 30' not '× 3' as in $3^2 = 9$.

ASSESSMENT CHECKPOINT Children should be able to fluently understand and interpret the square and cube signs, linking them to pictorial and concrete representations of numbers. Questions ❶ and ❷ assess how accurate children are when finding square and cube numbers. Question ❸ assesses whether children can find the result of squaring and cubing numbers and also the number that would need to be squared or cubed to find a particular result.

ANSWERS Answers for the **Practice** part of the lesson appear in the separate **Practice and Reflect answer guide**.

Reflect

WAYS OF WORKING Pair work

IN FOCUS This question assesses whether children understand the two concepts well enough to apply them to previous mathematical knowledge. Children should work with their partner to explain how the mathematical notation has been misused or misinterpreted and devise advice to give Danny. The question highlights three misconceptions; if children are unable to explain where these are, it may indicate that they are liable to make the same mistakes.

ASSESSMENT CHECKPOINT Children should be able to confidently diagnose the misconception and explain where the student has gone wrong. Using their knowledge and understanding of the lesson's concepts, they should be able to give advice on how to correctly solve the problem.

ANSWERS Answers for the **Reflect** part of the lesson appear in the separate **Practice and Reflect answer guide**.

After the lesson

- Were children equally confident with both mathematical concepts?
- If they were weaker in one than the other, how will you support their understanding moving forward?
- How can these concepts be brought into other areas of the curriculum?

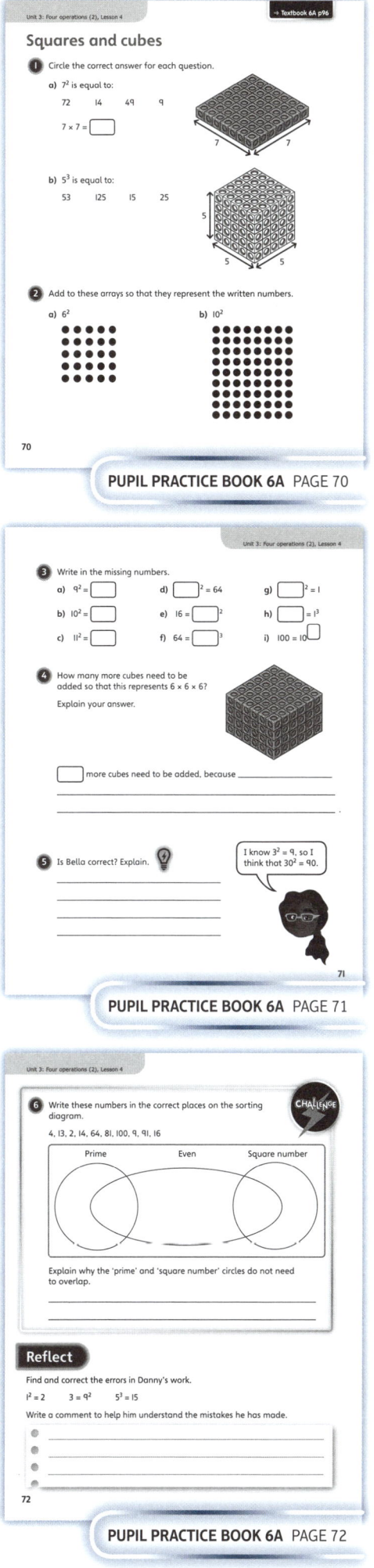

PUPIL PRACTICE BOOK 6A PAGE 70

PUPIL PRACTICE BOOK 6A PAGE 71

PUPIL PRACTICE BOOK 6A PAGE 72

Order of operations

Learning focus

In this lesson, children will learn the correct order of operations and use this to help solve multi-step calculations.

Small steps

→ Previous step: Squares and cubes
→ **This step: Order of operations**
→ Next step: Brackets

NATIONAL CURRICULUM LINKS

Year 6 Number – Addition, Subtraction, Multiplication and Division

Use their knowledge of the order of operations to carry out calculations involving the four operations.

ASSESSING MASTERY

Children can recognise and explain the correct order of operations. They can explain why the order of operations is important and can identify where the order has not been followed.

COMMON MISCONCEPTIONS

Children may be inclined to calculate from left to right, ignoring the order of operations. Have the correct order displayed prominently in the classroom. Ask:
• *Show me how you solved the calculation.*
• *Did you follow the order on the display?*

STRENGTHENING UNDERSTANDING

To help children remember the order of operations, it may be helpful and fun to create a class rhyme or song. It is too early to use 'BIDMAS' (**B**rackets, **I**ndices, **D**ivision and **M**ultiplication, **A**ddition and **S**ubtraction) as children have not yet learnt about brackets.

GOING DEEPER

Children could be given a selection of five numbers and four operations. Ask:
• *How many different solutions can you find using these numbers and operations?*

KEY LANGUAGE

In lesson: order of operations, calculation, addition, subtraction, multiplication, division

Other language to be used by the teacher: calculate, add, subtract, multiply, divide

RESOURCES

Optional: ten frames, counters, multilink cubes, bead strings

 In the eTextbook of this lesson, you will find interactive links to a selection of teaching tools.

Before you teach ⏸

• What real-life contexts could you use for this lesson that will be meaningful to your cohort?

Discover

WAYS OF WORKING Pair work

ASK

- Question **1** a): *Can a single calculation have two solutions?*
- Question **1** a): *How has each child come to their solution?*
- Question **1** a): *How does each child's model demonstrate their thinking?*

IN FOCUS Looking at the picture, children are likely to assume that Ebo is correct as they will be used to solving calculations reading from left to right. It is important at this stage not to give the game away and let children discuss their ideas without any help from you.

PRACTICAL TIPS Children could be encouraged to make their own model of the multi-step calculation using tens frames and counters, multilink cubes or bead strings.

ANSWERS

Question **1** a): Ebo has solved the calculation as (3 + 5) × 2.
Lexi has solved the calculation as 3 + (5 × 2).

Question **1** b): Lexi is correct.

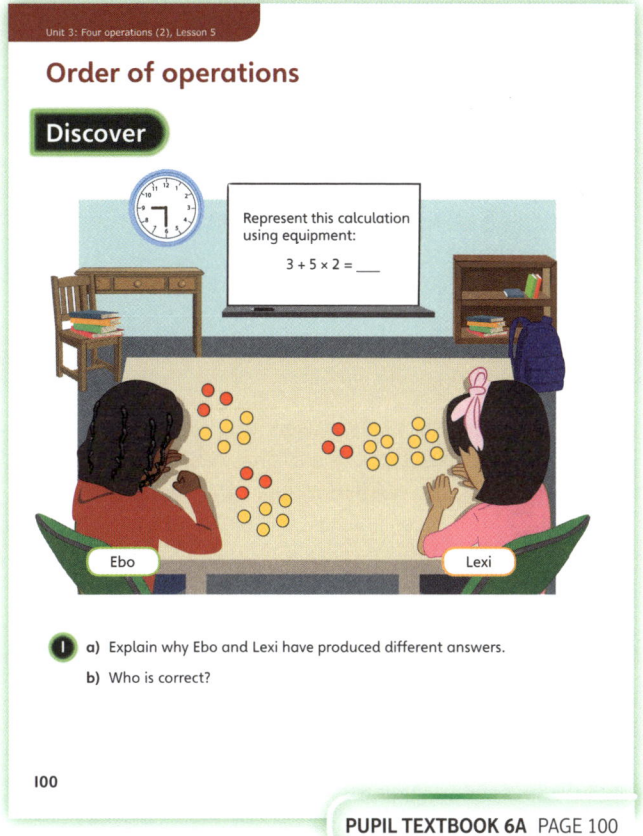

Share

WAYS OF WORKING Whole class teacher led

ASK

- Question **1** b): *Who do you think is correct and why?*
- Question **1** b): *Why do you think it is important to have an agreed order of operations?*
- Question **1** b): *What might happen if you did not agree on the order you solve operations?*

IN FOCUS At this point in the lesson, children are only learning about multiplication and addition. Before moving on, make sure they understand that multiplication is done first, then addition. If children ask about division and subtraction, it would be an interesting opportunity for them to predict where those operations will feature in the order based on what they already know, but these will be covered properly in the next section.

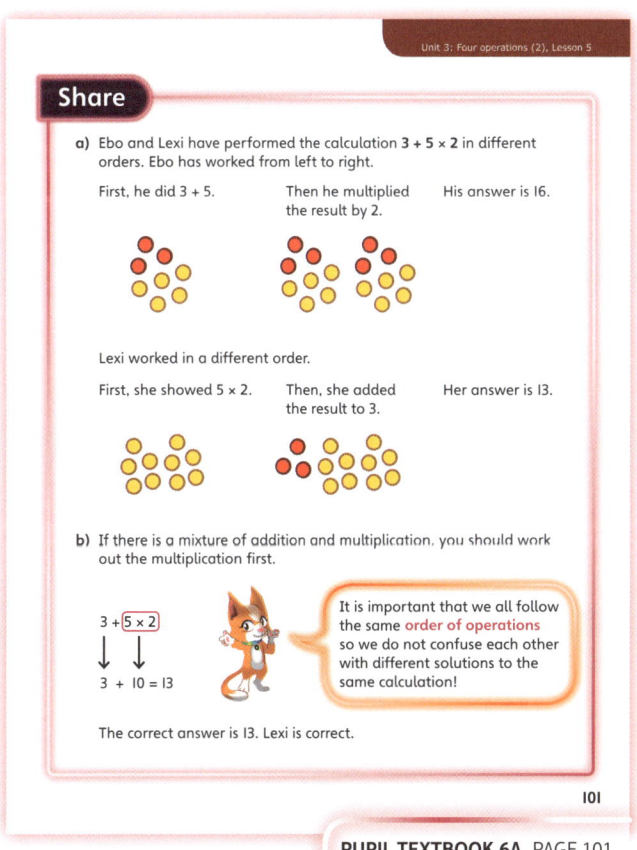

133

Think together

Whole class teacher led (I do, We do, You do)

ASK

- Question **1** : *At what point in the order of operations do you think you should solve subtractions?*
- Question **2** b): *What happens if there are more than two operations?*
- Question **3** b): *What is different about these calculations from those you looked at before?*

IN FOCUS In this part of the lesson, children are introduced to the position of subtraction and division within the order of operations. Question **1** introduces subtraction, while question **3** introduces division. It is important to make sure the position of these calculations is made clear in your teaching, that is, division and multiplication first, then addition and subtraction. Link the additive and multiplicative operations to help children understand why the operations are in the order they are in.

In question **3** b), when children consider Ash's comments, they may conclude that the answer is always the same, regardless of the order in which they work out the multiplication and division. While this is true when the multiplication precedes the division in a written calculation, it is not true if the division precedes the multiplication, as in $10 \div 5 \times 2$. Use this to illustrate that children should work through multiplications and divisions in the order they appear.

STRENGTHEN For question **2** b), if children are struggling to know which multiplication to solve first in the three-part calculation, ask:

- *What happens if you solve the first multiplication first?*
- *What happens if you solve the second multiplication first?*

DEEPEN Children could be given calculations that include all four operations. They could also be given a sequence of three or four numbers (for example, 3 4 5 = 23) and asked:

- *Find the missing operations.*

ASSESSMENT CHECKPOINT At this point in the lesson, children should be able to explain that the operations of multiplication and division are carried out before the operations of addition and subtraction. Children should be able to solve calculations that involve up to three operations and should be able to explain why people may find more than one solution. Question **2** gives you the opportunity to assess children's recognition of the order of multiplication and addition or subtraction operations. Children should recognise in both calculations that the multiplications should be done first.

ANSWERS

Question **1** : $(3 \times 5) - 2 = 13$ is correct

Question **2** a): Solve 25×2 first, then subtract from 100, giving an answer of 50.

Question **2** b): Solve 11×2 and 3×11 first, then add the two results, giving an answer of 55.

Question **3** a): $25 + 100 \div 4 = 50$
$45 = 500 \div 10 - 5$

Question **3** b): Both ways of solving the calculation result in the same solution (10).

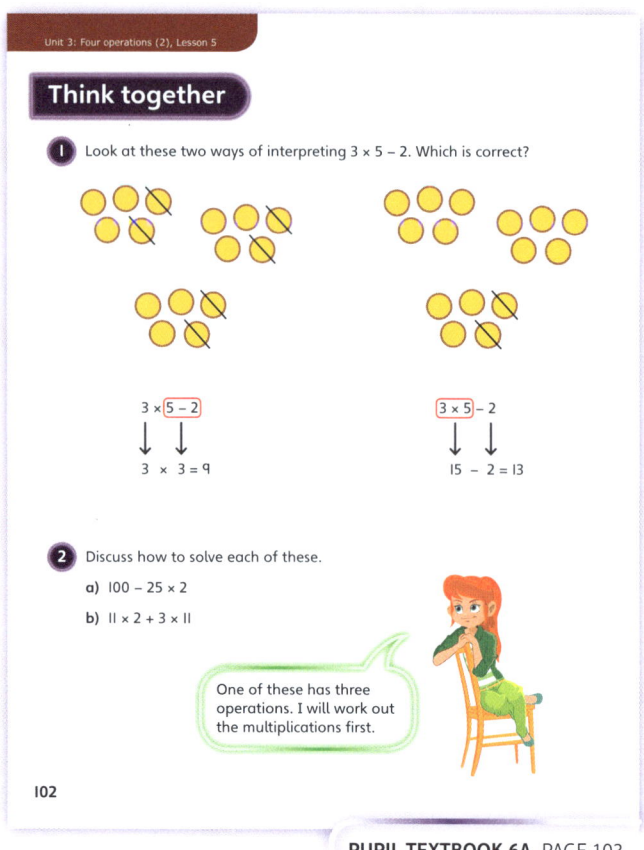

PUPIL TEXTBOOK 6A PAGE 102

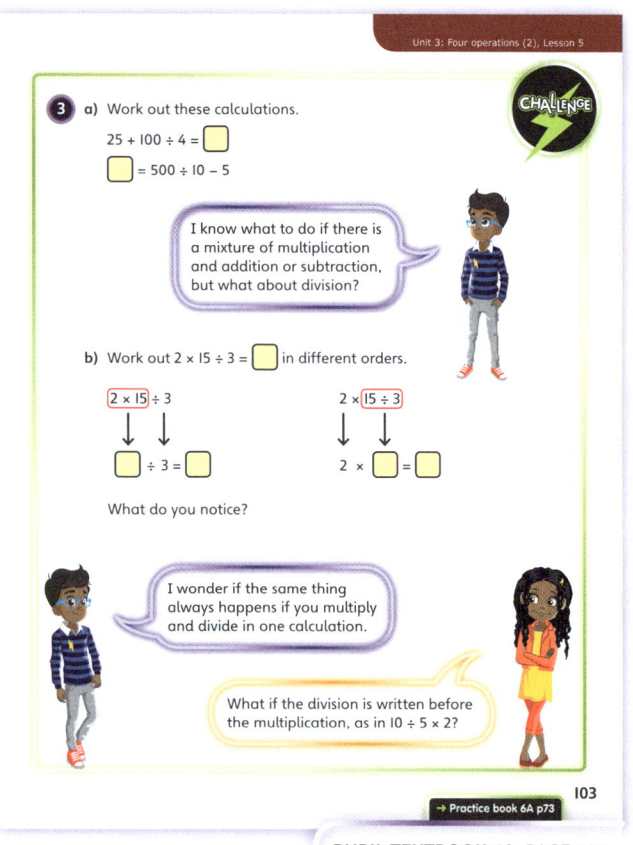

PUPIL TEXTBOOK 6A PAGE 103

Practice

WAYS OF WORKING Independent thinking

IN FOCUS Questions ❶ and ❷ scaffold children's use of the order of operations by linking the abstract to pictures of the concrete representations. This link could be reinforced by offering children the resources pictured to make the calculations themselves. While solving question ❹, it would be beneficial to encourage children to discuss how the calculation and solution change as the operations are altered. Ask:

• How is 30 + 30 ÷ 5 similar to and different from 30 × 30 ÷ 5?

STRENGTHEN For children struggling to find the pairs of numbers in question ❺ b), ask:

• What will the calculation need to total before finally adding 100?
• What will you need to subtract from 100 to make sure that happens?

DEEPEN Question ❹ can be deepened by asking children to suggest other calculations to add to the pairs of calculations, to create three, four or five linked calculations. Ask:

• Are there any links or patterns that you can see in the solutions to your calculations?
• Is there a way of predicting what the solution of your next calculation might be?

ASSESSMENT CHECKPOINT Children should be able to confidently solve a calculation with more than one operation. Question ❸ is a valuable opportunity to assess children's ability to recognise, explain and follow the correct order of operations reliably. Question ❺ assesses whether children can recognise that, by knowing the order of operations, they can work backwards from a number to complete a missing number calculation. Look for children's clarity and confidence when giving explanations linked to their learning earlier in the lesson.

ANSWERS Answers for the **Practice** part of the lesson appear in the separate **Practice and Reflect answer guide**.

Reflect

WAYS OF WORKING Independent thinking, pair work

IN FOCUS Writing their own calculation requires children to demonstrate their grasp of the order of operations without prompts. The activity could be made into a challenge that children set their partner. Once children have designed their calculation, they could share it with their partner. Can they identify all the possible solutions and explain which is the correct one and why? Peer-to-peer feedback will further reinforce children's understanding.

ASSESSMENT CHECKPOINT Look for children's ability to design their own calculations and explain how to solve them correctly. Children should be able to confidently and fluently explain how to use the order of operations to solve their calculations.

ANSWERS Answers for the **Reflect** part of the lesson appear in the separate **Practice and Reflect answer guide**.

After the lesson ⏸

• How confident were children with remembering the order of operations by the end of the lesson?
• Could this lesson have been made more practical? How will you facilitate this next time you teach it?

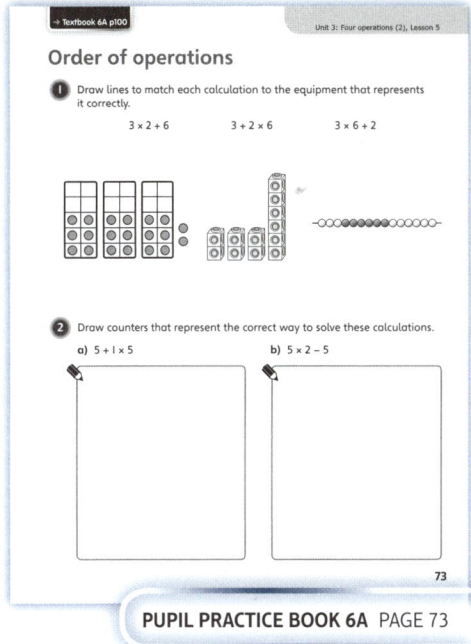

PUPIL PRACTICE BOOK 6A PAGE 73

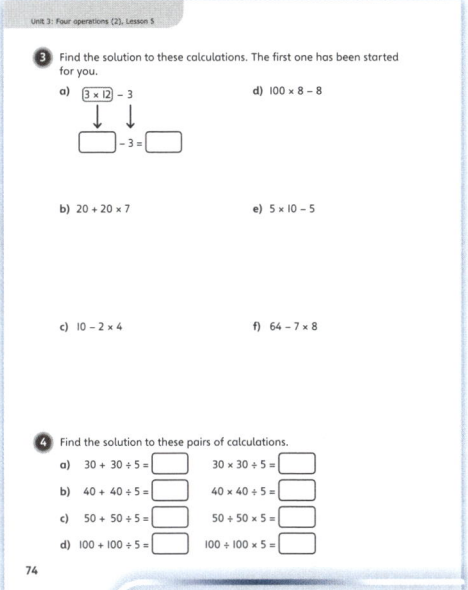

PUPIL PRACTICE BOOK 6A PAGE 74

PUPIL PRACTICE BOOK 6A PAGE 75

Brackets

Learning focus

In this lesson, children will extend their understanding of the order of operations by investigating what effect brackets can have on a calculation.

Small steps

→ Previous step: Order of operations
→ **This step: Brackets**
→ Next step: Mental calculations (1)

NATIONAL CURRICULUM LINKS

Year 6 Number – Addition, Subtraction, Multiplication and Division

Use their knowledge of the order of operations to carry out calculations involving the four operations.

ASSESSING MASTERY

Children can recognise and explain the effect brackets can have on the order of operations. They are able to confidently solve calculations that include brackets by solving what is inside the brackets first.

COMMON MISCONCEPTIONS

Children may ignore the brackets in a calculation and solve it either using the order of operations or ignoring that as well and solving it in the order it is presented. Ask:

- *What was the order of operations you learnt in the last lesson?*
- *What effect do brackets have within a calculation?*

STRENGTHENING UNDERSTANDING

When children are solving questions that require them to add brackets to an already written calculation, it may help to have the calculation written on a large piece of paper and have brackets on cards that can be placed in and around the calculation. This may help children to be more flexible in their approach, as they will be able to easily change the position of brackets without having multiple recorded mistakes across one calculation.

GOING DEEPER

Children could be given a selection of numbers and asked to find a calculation that equals a given (random) number. Ask: *Using these numbers – a, b, c, d, e, f – can you write a calculation that equals x?*

KEY LANGUAGE

In lesson: brackets, multiply, calculation, order of operations

Other language to be used by the teacher: calculate, divide, add, subtract

STRUCTURES AND REPRESENTATIONS

bar models

RESOURCES

Optional: large paper, card, base 10 equipment, counters, multilink cubes, bead strings

 In the eTextbook of this lesson, you will find interactive links to a selection of teaching tools.

Before you teach

- Were there any misconceptions from the previous lesson that will need to be overcome to ensure progress in this lesson?
- How will you integrate this teaching into today's lesson?

Discover

Pair work

ASK

- Question **1** a): *Do you agree with the total the mechanic has come to?*
- Question **1** a): *What do you notice about the calculation she has written?*
- Question **1** a): *What total should the mechanic have found, using the calculation she has written?*
- *Can you show both calculations from questions* **1** *a) and* **1** *b) using a bar model? What is similar and what is different?*

IN FOCUS Question **1** a) recaps children's understanding of the previous lesson. While the mechanic has noted the correct number of tyres, the calculation she has written does not equal 160 when following the order of operations. This will prompt children to see that something else may be needed to show when a calculation should be solved in a different order.

PRACTICAL TIPS Children could create their own versions of the scenario posed in the picture, using toy cars, lorries and trains. Children could be encouraged to see how the calculations stay the same and how they differ depending on the vehicles used.

ANSWERS

Question **1** a): The mechanic's written calculation is incorrect. It gives an answer of 100.

Question **1** b):

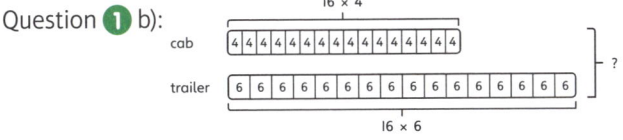

$$16 \times 4 + 16 \times 6 = 160$$

Share

WAYS OF WORKING Whole class teacher led

ASK

- Question **1** a): *What was the problem with the context and the calculation representing it?*
- Question **1** a): *How do the brackets help you make the calculation fit the problem?*
- Question **1** b): *Can you write another calculation that uses brackets to change the order of operations?*

IN FOCUS In question **1** a), it is important to discuss how the brackets have enabled the calculation to reflect the context of the problem. Children should be encouraged to recognise how this can help them to be more efficient, solving what would have otherwise needed to be two separate calculations in one.

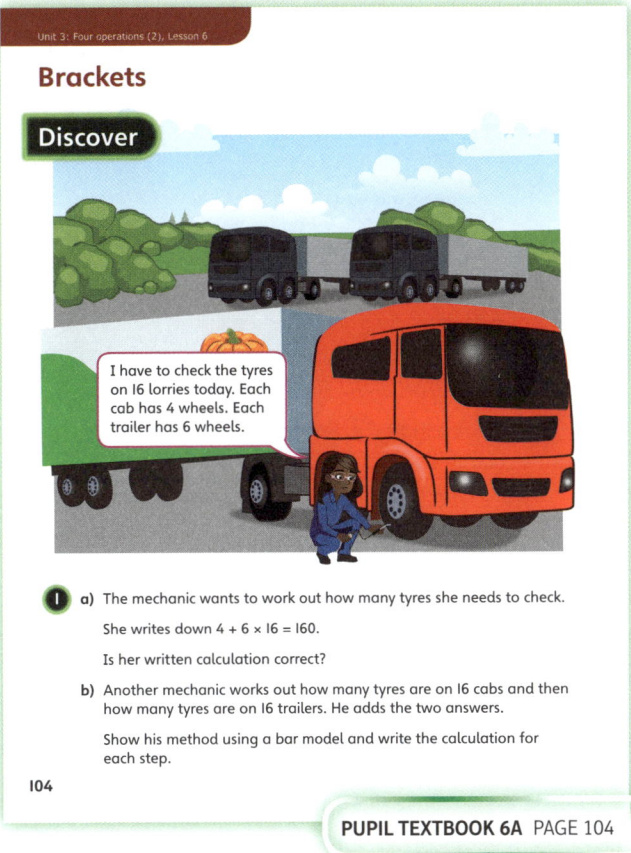

PUPIL TEXTBOOK 6A PAGE 104

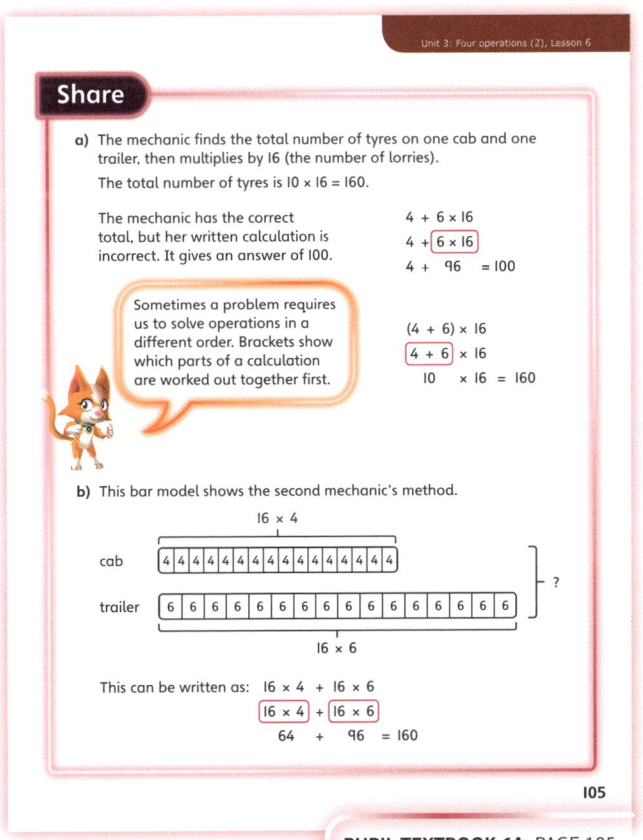

PUPIL TEXTBOOK 6A PAGE 105

Think together

Whole class teacher led (I do, We do, You do)

ASK

- Question **1** : *What do you need to work out first?*
- Question **1** : *Does the calculation you need to do first come first in the order of operations? If not, what do you need to do?*
- Question **2** : *What do the brackets mean in the calculation?*
- Question **3** a): *How many places could brackets be put into the calculation? How many solutions could be found?*

IN FOCUS For question **1**, encourage children to try different possible solutions. Ask:

- *What is the 'story' of your calculation? Does it match the story in the problem?*

After question **1**, contexts are removed so that children think in the abstract.

STRENGTHEN It may help to link the calculations with concrete representations, using base 10 equipment, place value counters, multilink cubes or bead strings. Ask: *What does the concrete representation show? How would you write this as a calculation?*

If children struggle with question **3**, it is important to direct them to Astrid's comment as up until this point they have only experienced brackets around two numbers. This could potentially lead to the misconception that that is the maximum amount seen in any example of brackets. Ask:

- *What is Astrid suggesting?*
- *Do you think she is able to do that? Can you try doing what she suggests? What happens?*

DEEPEN If children solve question **3**, they could be encouraged to continue with Ash and Flo's line of questioning. Ask:

- *Do you predict it is possible to make calculations for all numbers from 1 to 20 using just four 4s? Explain.*
- *Show me how many you can make.*

ASSESSMENT CHECKPOINT Children should now be able to recognise the function of brackets within a calculation and know that whatever calculations are bracketed should be solved first. They should be able to confidently solve calculations with brackets and be more confident at finding where brackets need to be in a calculation, using trial and error. Question **2** assesses children's understanding of how the use of brackets influences their calculations.

ANSWERS

Question **1** : 4 × (£7·50 + £3·50) = £44·00
        4 × £11·00 = £44·00

Question **2** a): (15 − 5) × 3 = 30
        15 − (5 × 3) = 0

Question **2** b): 200 = (15 + 5) × (15 − 5)
        85 = 15 + (5 × 15) − 5

Question **3** a): (4 + 4) × (4 ÷ 4) = 8
        4 + (4 × 4 ÷ 4) = 8
        (4 + 4 × 4) ÷ 4 = 5

Question **3** b): (4 ÷ 4) + (4 ÷ 4) = 2
        (4 × 4) ÷ (4 + 4) = 2
        4 × (4 + 4) − 4 = 28

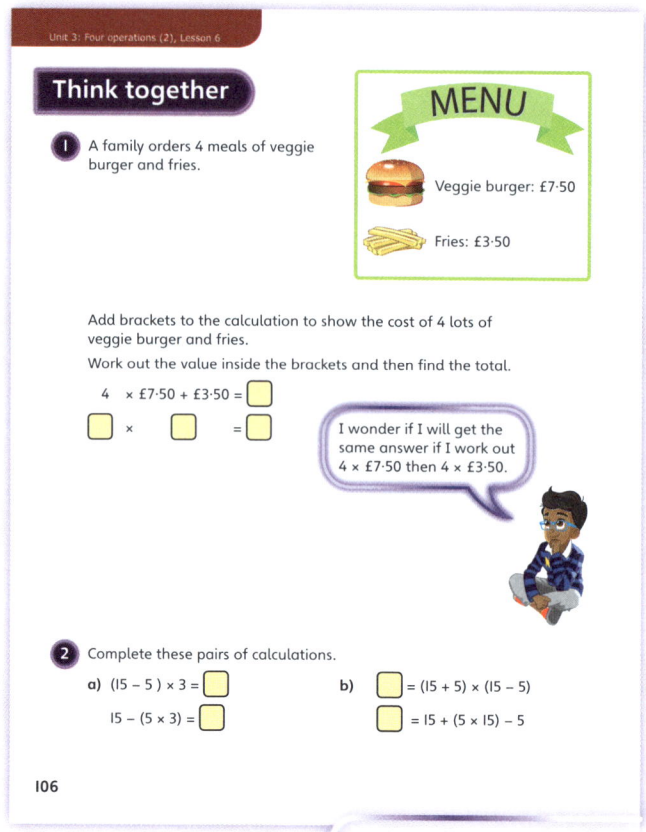

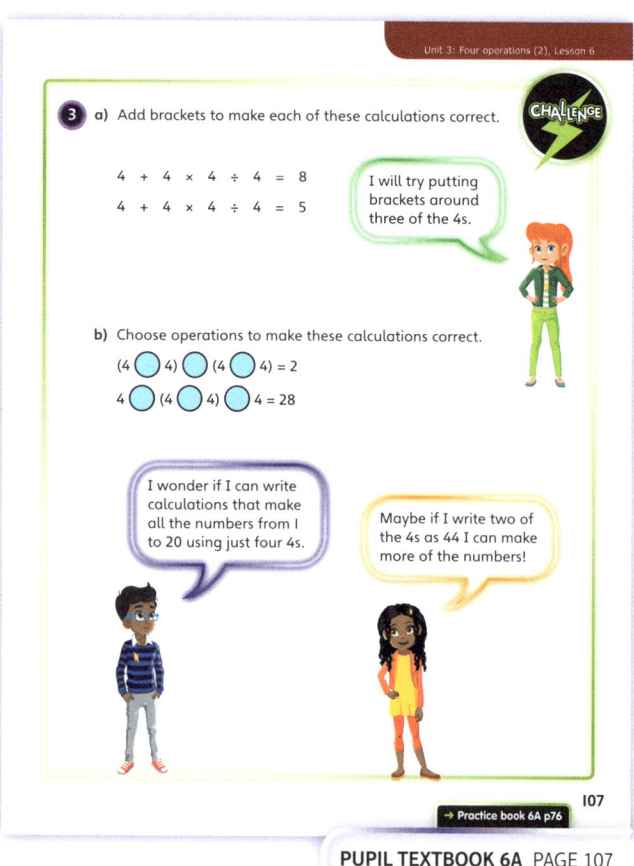

Practice

WAYS OF WORKING Independent thinking

IN FOCUS Question ① links calculations to concrete and pictorial representations. This helps to secure children's understanding of the abstract calculations. If children used equipment earlier in the lesson, it may help them to do so again. Question ③ moves on to word problems, providing a good opportunity for children to apply their learning in context. Children could act out each problem with concrete resources to help them understand the 'story' of the calculation before turning it into a written calculation.

STRENGTHEN If children are finding question ⑤ tricky, ask:
• *What operations could you use?*
• *Is there a way you could write the possible solutions you have tried?*
• *Is one operation more likely to work than another? Why?*

DEEPEN Children's understanding and reasoning could be deepened in question ⑥ by asking:
• *Is there a better operation to use when aiming for smaller results?*
• *Is it possible to get the smallest possible number while using multiply or add? Explain.*
• *Is there more than one way to find the same result?*

ASSESSMENT CHECKPOINT Children should be able to fluently create and solve calculations with brackets. Using their understanding of the function of brackets, they should be able to fluently reason and problem solve, completing partially finished calculations or identifying where mistakes have been made. Question ③ is particularly useful for assessing whether children can recognise how a calculation, taken from a contextual problem, would be presented. Look for children linking their understanding of the problem's 'story' with the order of operations in the calculation.

ANSWERS Answers for the **Practice** part of the lesson appear in the separate **Practice and Reflect answer guide**.

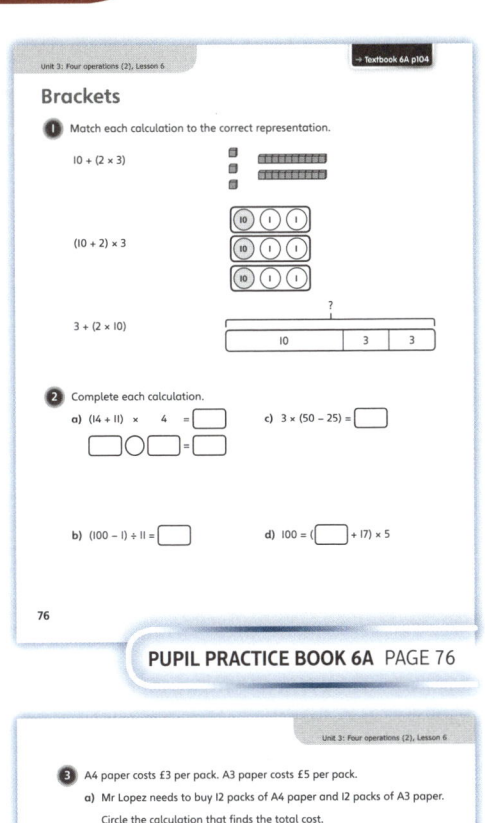

PUPIL PRACTICE BOOK 6A PAGE 76

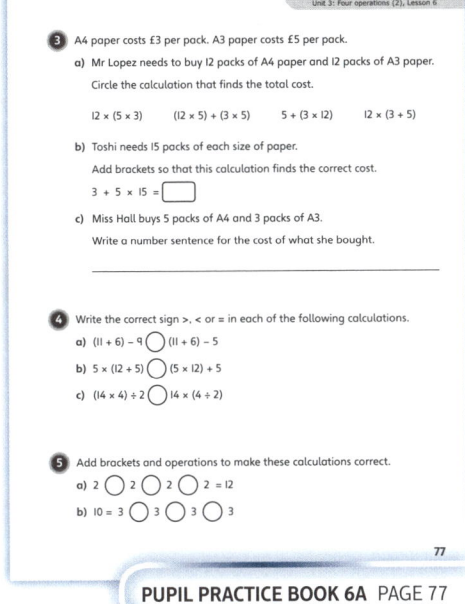

PUPIL PRACTICE BOOK 6A PAGE 77

Reflect

WAYS OF WORKING Independent thinking

IN FOCUS This question offers an opportunity to assess whether children can manipulate calculations confidently using brackets. It also assesses their ability to solve the calculations they create through recognising which calculation equals a greater number.

ASSESSMENT CHECKPOINT Children should recognise that, by placing the brackets around $3 + 4$ in the left-hand calculation, they can multiply 7 by 10. Whereas in the right-hand calculation, if they place the brackets around 10×4, they ensure they only multiply 4 by 10.

ANSWERS Answers for the **Reflect** part of the lesson appear in the separate **Practice and Reflect answer guide**.

After the lesson ⏸

• How flexible were children in their use of brackets?
• Could they use trial and error confidently to achieve a desired result?
• Were children still able to follow the order of operations outside of any brackets or does this need revisiting?

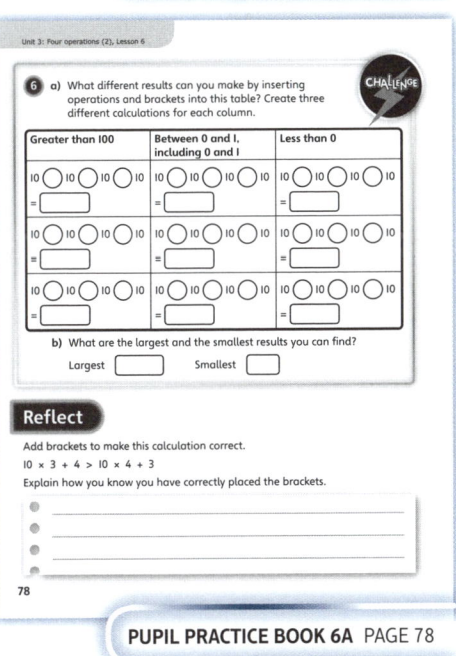

PUPIL PRACTICE BOOK 6A PAGE 78

Mental calculations ❶

Learning focus

In this lesson, children will learn efficient mental methods for solving calculations with smaller numbers, including decimals.

Small steps

→ Previous step: Brackets
→ **This step: Mental calculations (1)**
→ Next step: Mental calculations (2)

NATIONAL CURRICULUM LINKS

Year 6 Number – Addition, Subtraction, Multiplication and Division

Perform mental calculations, including with mixed operations and large numbers.

ASSESSING MASTERY

Children can use efficient mental methods to confidently and fluently solve calculations with smaller numbers, including decimals. They can use these mental methods to help them solve number problems and puzzles, and can describe the methods they used, explaining how they were helpful.

COMMON MISCONCEPTIONS

When children are adding or subtracting by compensating, they may add or take away an insufficient amount at the end of the calculation. For example, when solving 0·99 + 5·98, they may calculate 1 + 6, but then either subtract the wrong amount at the end or not subtract anything at all. Ask:

- *What did you do to the numbers before you solved this mentally?*
- *Have you added back what you subtracted? Have you subtracted what you added?*

STRENGTHENING UNDERSTANDING

Children may benefit from having real-life practice of the concepts covered in this lesson, for example, by running a cake sale. Calculating the amount of ingredients needed to bake multiple cakes and dealing with money in a real context will provide ample opportunity for children to develop confidence with mental methods.

GOING DEEPER

Children could be challenged to investigate how the methods taught in this lesson might transfer to slightly larger numbers, prior to working with thousands and millions in the next lesson. For example, ask:

- *If you can multiply by 9 easily, can you use the same method to multiply by 90 or 900? How about 999? Explain.*

KEY LANGUAGE

In lesson: mental method, calculation, mentally, written method

Other language to be used by the teacher: add, subtract, multiply

STRUCTURES AND REPRESENTATIONS

tables, number lines, column additions, column multiplications, bar models

RESOURCES

Optional: base 10 equipment

 In the eTextbook of this lesson, you will find interactive links to a selection of teaching tools.

Before you teach ⏸

- What mental methods do children already know and use in the classroom?
- Are they more confident with one operation than another?
- How will you support those operations that need more practice?

Discover

Pair work

ASK

• Question ❶ b): *How much does each item cost?*
• *How much does each child spend? How did you work it out?*
• *Is there an easier way to find each total?*

IN FOCUS This part of the lesson provides a good opportunity to assess children's ability to calculate mentally. Make sure children are given time to feed back their ideas into the class discussion to allow you to judge the current class confidence level.

PRACTICAL TIPS Role playing a shop or, if the opportunity is available, a trip to a local shop would give children many chances to work out totals and differences in price mentally.

ANSWERS

Question ❶ a): Holly receives 5p change.

Question ❶ b): Toshi receives £2·03 change.

PUPIL TEXTBOOK 6A PAGE 108

Share

WAYS OF WORKING Whole class teacher led

ASK

• Question ❶ a): *Why is it easier to round 0·99 up to 1 when calculating mentally?*
• Question ❶ a): *What is important to remember when using this method?*
• Question ❶ b): *What are the possible mistakes someone might make? Explain.*
• Question ❶ b): *Could this method be used for other numbers? Explain.*

IN FOCUS While discussing question ❶ a), make sure to point out the pattern in the numbers. Discuss how this can help children at the end to quickly and reliably compensate. Children could be encouraged to discuss how that pattern may change if they compensated numbers such as 0·98 or 0·97.

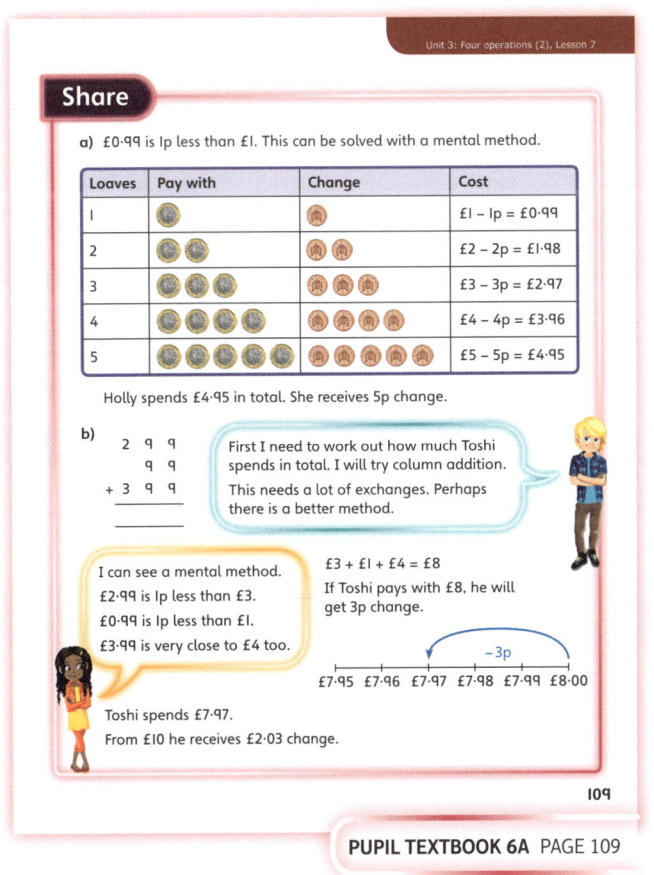

PUPIL TEXTBOOK 6A PAGE 109

Think together

WAYS OF WORKING Whole class teacher led (I do, We do, You do)

ASK

- Question **1** : *How could you multiply by 9 quickly?*
- Question **1** : *What easier number is 9 near?*
- Question **2** b): *How do the brackets help you?*

IN FOCUS Question **1** looks at compensating by multiplying by 10, then subtracting 1 'group' or number to multiply by 9. Discuss with children how this can help them to multiply by other numbers. For example, multiplying by 4 could be achieved by multiplying by 5 and subtracting. Questions **2** and **3** develop children's flexibility and fluency with mental and written methods by asking them to consider which type of method is best suited for each calculation. Encourage children to share their reasoning about the methods they choose.

STRENGTHEN If children are struggling to mentally compensate, it may help to provide them with printed 'parts' of a bar model that would represent the problem they are solving. Children could, for example, lay down 10 parts to represent the simple multiplication, then take one of the parts away to represent the compensation.

DEEPEN While children are solving question **3** a), ask:
- *Do you agree with each character's ideas?*
- *Can you show evidence that their ideas are correct?*
- *Can you show evidence that disproves their ideas?*

ASSESSMENT CHECKPOINT At this point in the lesson, children should be able to recognise compensation as an efficient mental method. They should be able to use this when solving addition, subtraction and multiplication calculations. Question **2** offers an opportunity to assess children's use of both mental methods taught in the lesson. Ask children to explain their method to ensure accurate assessment.

ANSWERS

Question **1** a): Calculating 10 × 45p is quickest.

Question **1** b): £4·05

Question **2** a): 19p + 29p + 39p should be solved as (20p + 30p + 40p) − 3p = 87p

Question **2** b): £10 − (3 × £0·99) should be solved as £10 − (3 × £1) + 3p = £7·03

Question **3** a): 7 × 25 g − 50 g could be solved mentally, as subtracting 50 g can be seen as subtracting 2 × 25 g, giving 5 × 25 g = 125 g.

(14 mm × 5) + (6 mm × 5) could be solved mentally as 70 mm + 30 mm = 100 mm.

10 m − (5 × 95 cm) could be solved mentally by rounding 95 cm up to 100 cm. This would add 5 more lots of 5 cm to the amount subtracted, giving 1000 cm − 500 cm = 500 cm. The final step would be to add on the extra 25 cm that were taken off when 95 cm was rounded up. The final answer is 525 cm.

Question **3** b): Children's own stories.

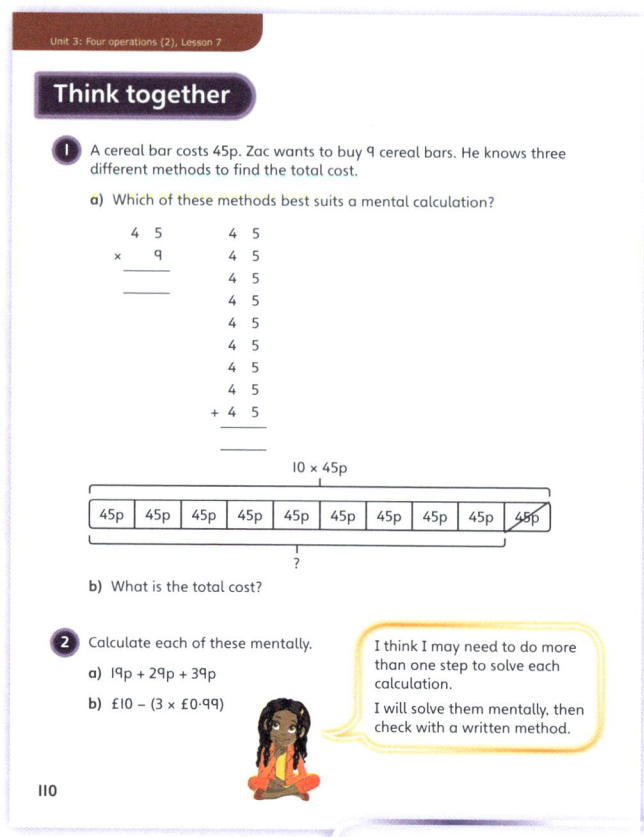

PUPIL TEXTBOOK 6A PAGE 110

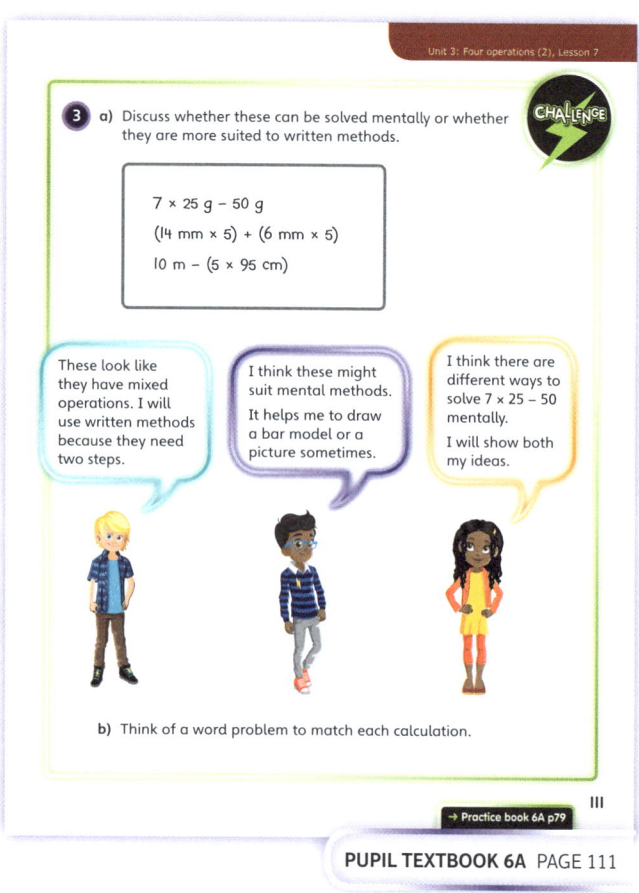

PUPIL TEXTBOOK 6A PAGE 111

Practice

WAYS OF WORKING Independent thinking

IN FOCUS It is important to focus on question ➊ as it links the mental method to the concrete and pictorial representations children will be familiar with. This will help them to visualise each problem and the method they need to use. In question ➌, watch for children who are tempted to use a written method. If you see any child doing this, ask:
- *How would you have solved this if a pencil and paper were not available?*
- *What is similar about your written method and your mental method and what is different?*

STRENGTHEN If children are struggling to solve question ➋, ask:
- *How could you represent each problem?*
- *Could you make finding the totals easier?*
- *How will you make sure you find the correct solution?*

DEEPEN Deepen children's reasoning when solving question ➌ by asking them to write an explanation, in a couple of sentences, about how they approached each calculation. Ask: *Could you have approached the question in a more efficient way?*

THINK DIFFERENTLY Question ➍ encourages children to recognise that the method of compensation can work for numbers other than 9. It also approaches the potential misconception where children add or subtract 1, regardless of the number they are dealing with and its difference between it and the next 10.

ASSESSMENT CHECKPOINT Children should be able to confidently solve mental calculations, using compensation to help them. They should be able to link this to concrete and pictorial representations of their mental methods and use this understanding to help them explain where use of mental methods is appropriate or where written methods are more so. Question ➌ offers an opportunity to assess the different mental methods children are using. Be sure to look at children's jottings as well as discussing their methods with them to best assess their understanding.

ANSWERS Answers for the **Practice** part of the lesson appear in the separate **Practice and Reflect answer guide**.

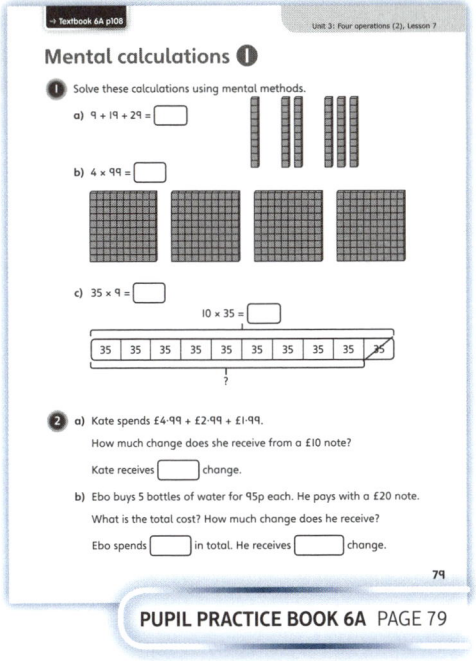

PUPIL PRACTICE BOOK 6A PAGE 79

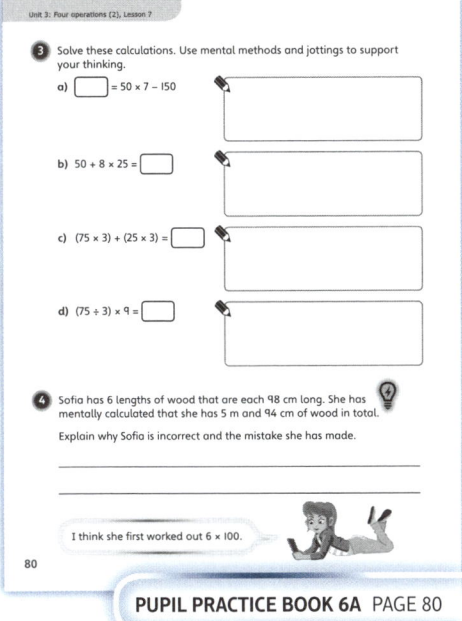

PUPIL PRACTICE BOOK 6A PAGE 80

Reflect

WAYS OF WORKING Independent thinking

IN FOCUS This question assesses whether children can reliably recognise where mental methods may be more appropriate and efficient than written methods. It will give evidence that children's fluency and reasoning with mental and written calculations will enable them to use the mathematics they know flexibly and confidently.

ASSESSMENT CHECKPOINT Children should be able to explain the kinds of clues to look for when deciding if mental methods are appropriate, for example, a number near a multiple of ten.

ANSWERS Answers for the **Reflect** part of the lesson appear in the separate **Practice and Reflect answer guide**.

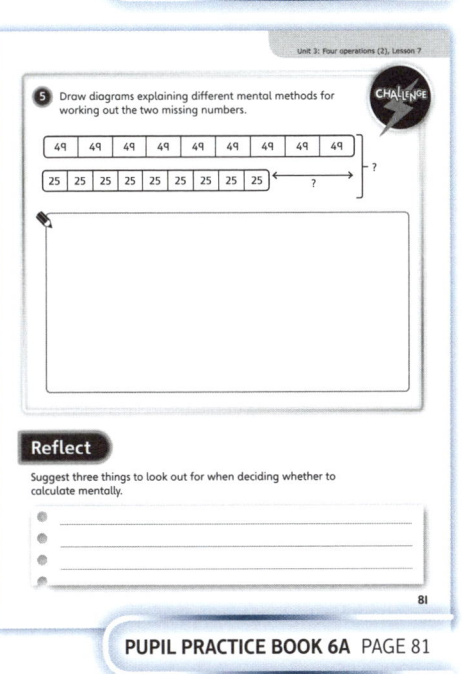

PUPIL PRACTICE BOOK 6A PAGE 81

After the lesson ⏸
- Were children less confident at any particular operation?
- How will you support this in future lessons?
- Could this lesson have been made more practical?

Mental calculations ❷

Learning focus

In this lesson, children will learn efficient mental methods for solving calculations with larger numbers, up to millions.

Small steps

→ Previous step: Mental calculations (1)
→ **This step: Mental calculations (2)**
→ Next step: Reasoning from known facts

NATIONAL CURRICULUM LINKS

Year 6 Number – Addition, Subtraction, Multiplication and Division

Perform mental calculations, including with mixed operations and large numbers.

ASSESSING MASTERY

Children can use efficient mental methods to confidently and fluently solve calculations with larger numbers, up to millions. They can use these mental methods to help them solve number problems and puzzles, and can describe the methods they used, explaining how they were helpful.

COMMON MISCONCEPTIONS

As numbers get larger, children are more likely to confuse the place value of a digit. This may result in incorrectly reading or writing numbers. Ask:

• *Can you identify the place value of each digit in this number?*
• *Having done that, can you read it again?*
• *Does what you have read match what you wanted to write? Explain.*

STRENGTHENING UNDERSTANDING

Before the lesson, it would be beneficial to practise multiplying and dividing larger numbers by 10, 100 and 1,000 through activities such as 'Follow me' cards or bingo games. This will ensure that all children are prepared for the mental methods that will be introduced in this lesson.

GOING DEEPER

Children could be encouraged to design their own word problems using larger numbers. Ask:

• *Can you create a word problem for your partner to solve?*
• *Now can you create a problem that has more than one step and at least two different operations?*

KEY LANGUAGE

In lesson: reduce, column subtraction, mental method, difference, increase, add, subtract, column method, exchange, reasoning, inverse operation, calculation, more than, less than, double, halve, take away

STRUCTURES AND REPRESENTATIONS

column subtractions, place value grids, bar models, number lines, tables

RESOURCES

Optional: 'Follow me' cards, bingo game, number lines, real house sale advertisements

 In the eTextbook of this lesson, you will find interactive links to a selection of teaching tools.

Before you teach

• How confident were children at visualising problems mentally in the previous lesson?
• Does this skill need support before or during this lesson?

Discover

WAYS OF WORKING Pair work

ASK

- Question ❶ a): *What type of calculation is needed to work out a reduction?*
- Question ❶ a): *What is the easiest way of solving each subtraction?*
- Question ❶ a): *How did the numbers change when you solved the subtractions? Explain.*
- Question ❶ b): *Can you change the numbers in any way to make the calculation easier?*
- *Which house has the price that is easiest to calculate with? Explain.*

IN FOCUS Question ❶ a) encourages children to begin considering the most efficient method of solving the given problem. While discussing the question, it would be interesting to consider whether any of the methods children learnt in the last lesson will help them. Can they identify where these methods will be useful and where they will not?

PRACTICAL TIPS This part of the lesson could be easily geared towards the interests of your class, to ensure children are fully engaged. For example, the sale items could be changed to sports cars, jewellery, footballers, breeds of horse and so on. You could use real advertisements from a local newspaper.

ANSWERS

Question ❶ a): This requires two subtractions. Written or mental methods can be used. A mental method works well with these numbers.

Question ❶ b): House A is £800,000 more expensive than house C.

Share

WAYS OF WORKING Whole class teacher led

ASK

- Question ❶ a): *Which method is more efficient in this example: written or mental? Explain.*
- Question ❶ b): *How does looking at 950,000 as 950 thousands make calculating easier?*
- Question ❶ b): *Can you solve 760,000 – 240,000 using this method? Explain.*

IN FOCUS At this point in the lesson, make sure children recognise how their understanding of place value can be very powerful when solving calculations mentally. In question ❶ b), discuss the similarities between 950 – 150 and 950,000 – 150,000; what is similar and what is different about the two calculations? Some children may notice they are dividing the number by 1,000 to make it easier to calculate with mentally.

DEEPEN The discussion about question ❶ b) can be continued to consider how the mental method used in this question can be applied to other numbers.

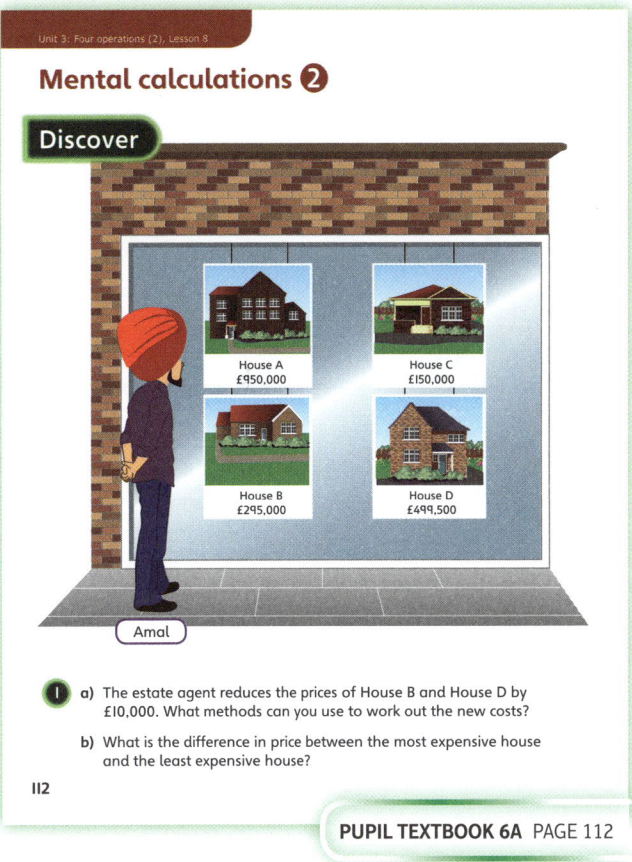

PUPIL TEXTBOOK 6A PAGE 112

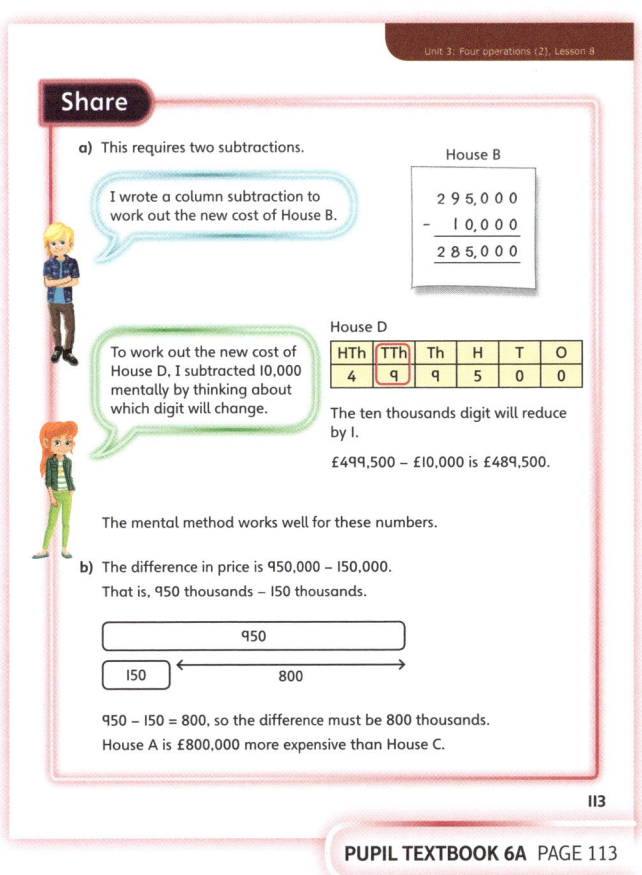

PUPIL TEXTBOOK 6A PAGE 113

Think together

Unit 3: Four operations (2), Lesson 8

Think together

WAYS OF WORKING Whole class teacher led (I do, We do, You do)

ASK

- Question **1** : *Is Dexter's idea a good one? Explain.*
- Question **2** : *What mental method will you use to solve this question?*
- Question **2** : *How is the mental method you are using more efficient than a written one?*
- Question **3** : *How can you use easier number facts to help you with these problems?*

IN FOCUS Use Dexter's comment in question **1** to reiterate the importance of looking at numbers in the thousands as *x* number of thousands. Again, discuss how this can make tackling numbers like those in the **Discover** picture easier. While solving question **2**, discuss with children how visualising the number line can help them solve similar problems. Can they combine this with the method of dividing by 1,000 to solve the problem as efficiently as possible?

STRENGTHEN When solving question **3**, it may help children who are struggling to ask:

- *How could you write each number so it is easier to calculate with?*
- *Can you write the calculation that is being described? How will you solve it mentally?*

DEEPEN Children could be challenged to come up with more than one mental method to solve the calculations in question **4**. Once they have done so, encourage them to think critically about the methods they have come up with. Ask: *Which method is more efficient? Explain.*

ASSESSMENT CHECKPOINT At this point in the lesson, children should be able to mentally calculate addition and subtraction problems. They should be able to explain the mental method they used to do so. Question **4** assesses children's ability to link their learning from this lesson and the last, and their understanding of place value, to efficiently and accurately solve each calculation.

ANSWERS

Question **1** : £75,000

Question **2** : £50,000

Question **3** a): Two hundred and fifty-six thousand.

Question **3** b): 1,450,000

Question **3** c): Fifty thousand

Question **3** d): You need to add 501,000 to 499,000 to make a million.

Question **4** : Look for children using their knowledge and understanding from the **Discover** and **Share** sections to help create mental methods for these calculations:

1,000 − 10 = 990

10,000 − 10 = 9,990

100,000 − 100 = 99,900

10,000,000 − 10,000 = 9,990,000

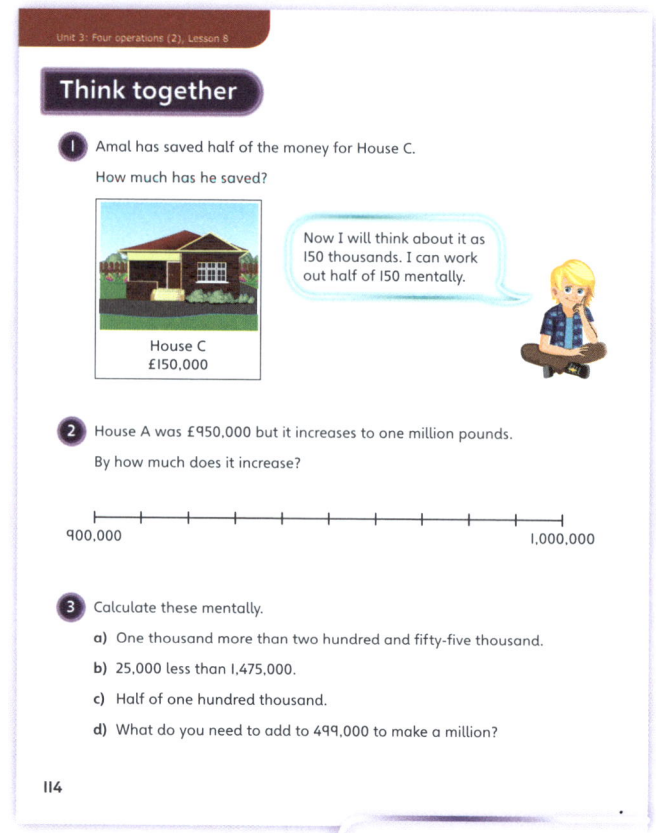

PUPIL TEXTBOOK 6A PAGE 114

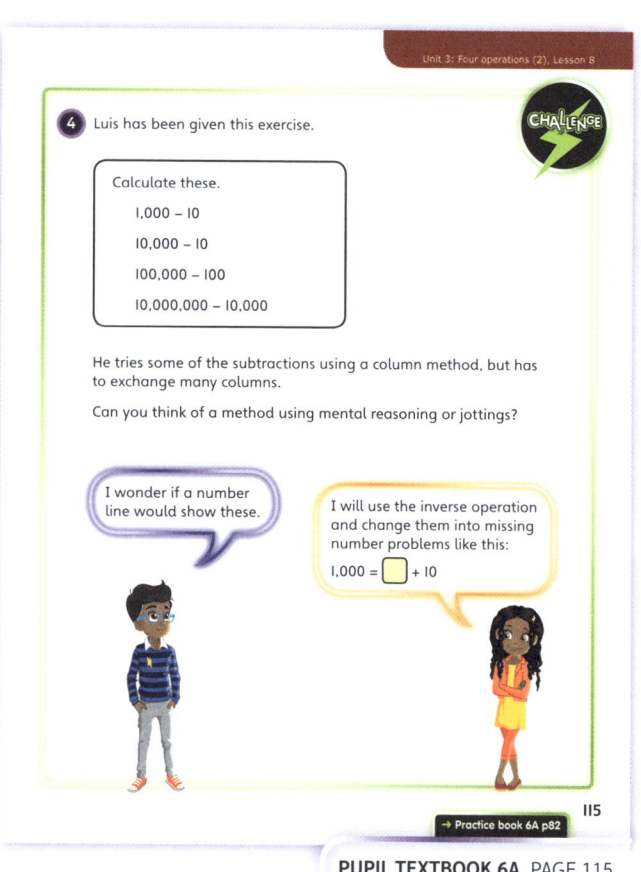

PUPIL TEXTBOOK 6A PAGE 115

Practice

WAYS OF WORKING Independent thinking

IN FOCUS While children are solving the questions in this part of the lesson, it will be interesting and valuable to stop them at regular intervals and discuss how they approached solving each question. Focus on where children have chosen different methods. Ask:
- *Why did you choose to solve it in that way?*
- *Do you think there was a more efficient method? Explain.*
- *Was your method as efficient as your partner's? Explain.*
- *How did you use the mental strategies you have learnt today to help you?*

STRENGTHEN For question ❸, it may help to provide children with a blank number line with divisions drawn on. Ask:
- *How could you use this to represent the problems?*
- *Can you show me a representation of your mental method using this number line?*

DEEPEN Children could be given multiplication or division problems involving large numbers as seen throughout the lesson. Ask: *How can you use the methods you have learnt today to help you solve these calculations mentally?*

ASSESSMENT CHECKPOINT At this point in the lesson, children should be able to confidently solve calculations using larger numbers. They should be able to explain the method they used and how it was efficient at solving the problem they were working on. Question ❻ assesses children's fluency and flexibility with the mental methods covered in this lesson and the last, by asking them to mentally calculate backwards through a problem.

ANSWERS Answers for the **Practice** part of the lesson appear in the separate **Practice and Reflect answer guide**.

Reflect

WAYS OF WORKING Independent thinking

IN FOCUS This question offers the opportunity to assess children's understanding of the mental methods covered in the lesson. If they are able to create questions that can be solved mentally and one that cannot, then they are showing good understanding of the necessary properties of a calculation that can be solved mentally. Once children have written their three calculations, get them to swap with a partner. Ask:
- *Did your partner identify the calculation that you thought could not be solved mentally?*
- *Can they prove to you that they can solve the other two mentally?*

ASSESSMENT CHECKPOINT Look for children's understanding of mental methods through their ability to make a calculation that cannot be solved with them.

ANSWERS Answers for the **Reflect** part of the lesson appear in the separate **Practice and Reflect answer guide**.

After the lesson

- How will you make sure children continue to use and develop mental methods?
- How confident are you that all children were able to fluently solve the problems mentally?

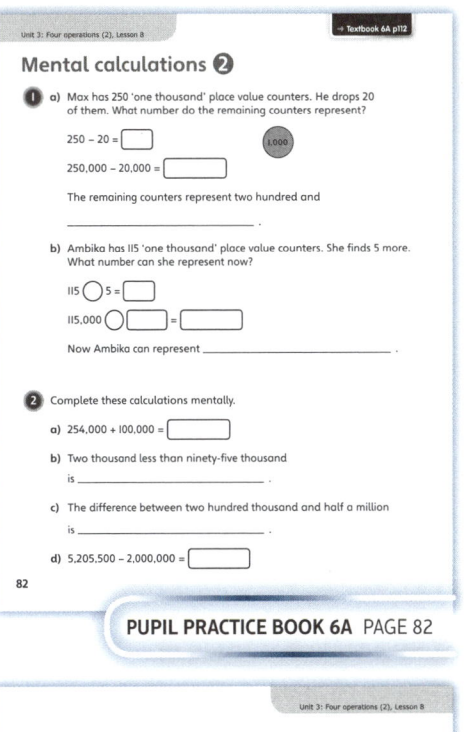

PUPIL PRACTICE BOOK 6A PAGE 82

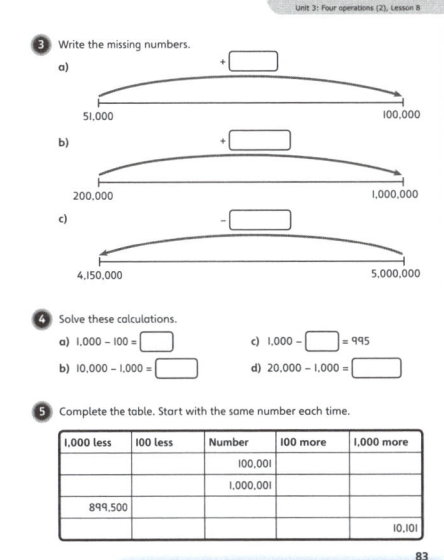

PUPIL PRACTICE BOOK 6A PAGE 83

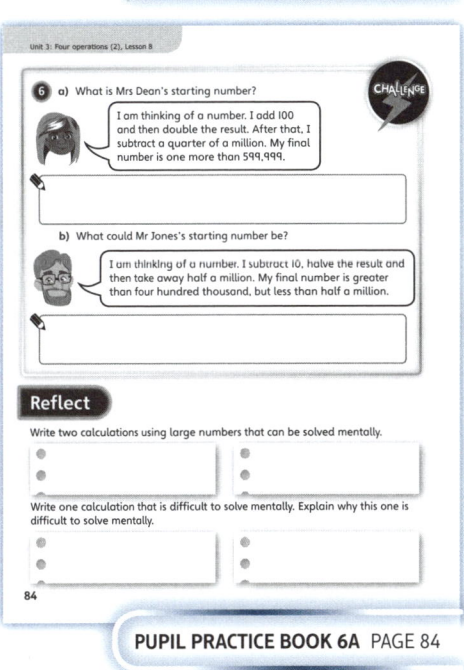

PUPIL PRACTICE BOOK 6A PAGE 84

Reasoning from known facts

Learning focus

In this lesson, children will draw upon their learning throughout the unit to read, understand and solve mathematical puzzles and problems. They will use number facts they know to help them solve more complicated problems.

Small steps

→ Previous step: Mental calculations (2)
→ **This step: Reasoning from known facts**
→ Next step: Simplifying fractions (1)

NATIONAL CURRICULUM LINKS

Year 6 Number – Addition, Subtraction, Multiplication and Division
• Use their knowledge of the order of operations to carry out calculations involving the four operations.
• Solve problems involving addition, subtraction, multiplication and division.

ASSESSING MASTERY

Children can fluently and confidently draw upon their knowledge of number facts to understand and solve mathematical puzzles and problems. They can share their reasoning, confidently expressing their knowledge and understanding of numbers, and use this to help them solve more complicated problems.

COMMON MISCONCEPTIONS

Children may rely too heavily on trial and error, rather than thinking about the number facts they can see in the question and those they know. Ask:
• *Are you solving this problem in the most efficient way?*
• *Are there any clues in the question you could use?*
• *Are there any number facts you already know that would help you to solve this?*

STRENGTHENING UNDERSTANDING

Before teaching this lesson, it may help children to practise calculation strings. For example, $6 \times 3 = 18$, $6 \times 30 = 180$, $6 \times 300 = 1,800$ and so on. Ask: *What other calculations can you find using 6×3?*

GOING DEEPER

Children could be encouraged to create their own versions of the problems found in this lesson. Ask:
• *Using the problems in this lesson as inspiration, can you come up with your own challenges?*
• *Can you create a poster of number challenges that you could challenge teachers with?*

KEY LANGUAGE

In lesson: factor, calculation, product, multiplying, mentally, division, multiplication, multiplied

Other language to be used by the teacher: number fact, inverse

STRUCTURES AND REPRESENTATIONS

bar models, mind maps, part-whole models

RESOURCES

Optional: digit cards, multiplication grids, 100 squares

 In the eTextbook of this lesson, you will find interactive links to a selection of teaching tools.

Before you teach ⏸

• How has your cohort responded to problem-solving activities in the past?
• How will this influence your teaching approach in this lesson?

Discover

Pair work

ASK

- *What do you already know about this calculation?*
- *What do you not know about this calculation?*
- *What numbers could they definitely not be? Why?*

IN FOCUS Discuss with children what they know about the calculation that may help them. This would be a good opportunity to recap some of the themes and concepts covered in this unit. Children should be encouraged to discuss which concepts may help them to solve the problem.

PRACTICAL TIPS To make this part of the lesson more practical, have available lots of different mathematics resources such as digit cards, multiplication grids and 100 squares. These could be resources children have used over the whole unit, but also mixed with some they have not. Discuss which will be helpful and why.

ANSWERS

Question ❶ a: Alex cannot use 1 or 2 to solve the problem.

Question ❶ b: The problem can be solved as
$5 × 6 × 9 = 270$.

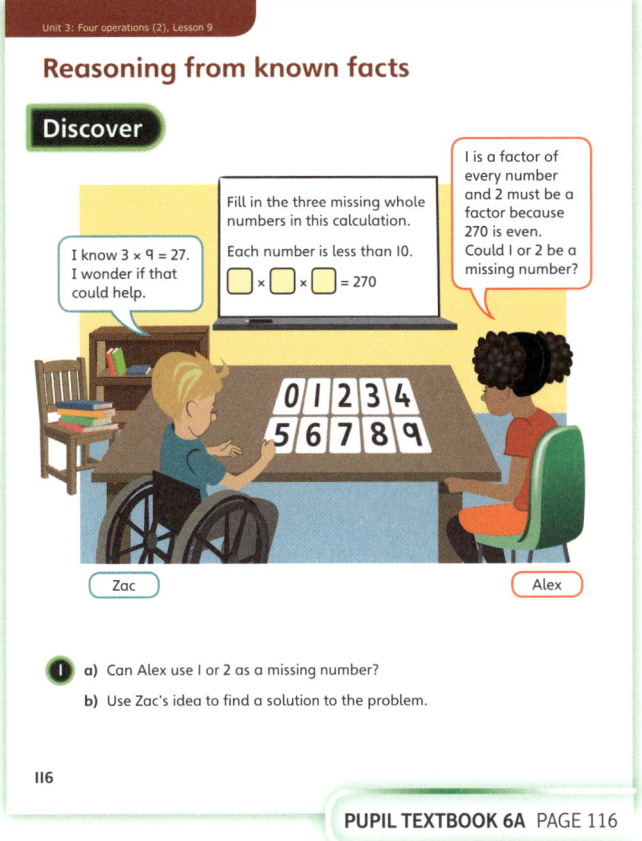

PUPIL TEXTBOOK 6A PAGE 116

Share

Whole class teacher led

ASK

- Question ❶ a): *What other numbers are factors of 270? Explain how you know.*
- Question ❶ a): *Is it essential that every missing number is a factor of 270? Why?*
- Question ❶ b): *What numbers did you rule out as the missing numbers? Why?*
- Question ❶ b): *What was the smallest number that may have worked? Explain your ideas.*

IN FOCUS It is important to point out to children the process behind eliminating some of the numbers from the problem. In question ❶ a), children may not have made the link between how multiplying by 1 or 2 would leave them with a number too great to create with a 1-digit multiplication. In question ❶ b), make sure children see how the known fact $3 × 9 = 27$ leads to the related facts $3 × 90 = 270$ and $30 × 9 = 270$.

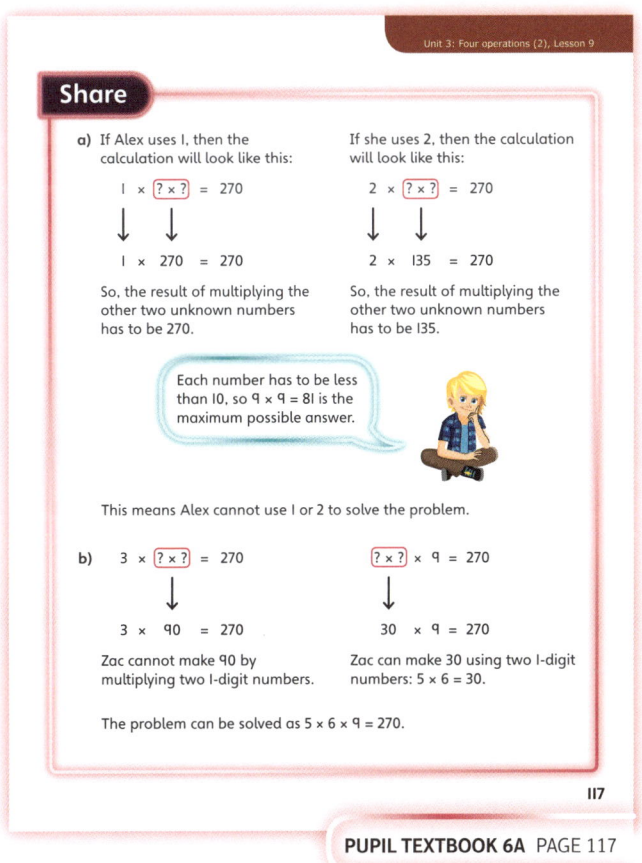

PUPIL TEXTBOOK 6A PAGE 117

Think together

Unit 3: Four operations (2), Lesson 9

Think together

WAYS OF WORKING Whole class teacher led (I do, We do, You do)

ASK

- Question ❶ : *Why is it useful to recognise doubles or multiples of 10?*
- Question ❶ : *How could you show your thinking with resources or as a picture?*
- Question ❷ a): *Are there any other facts you know that are not shown by the diagram? What are they?*
- Question ❷ a): *How could you use the bar model to help you find other multiples of 65?*
- Question ❷ b): *Can you show 66 × 3 using a bar model? How is it similar to and different from the bar model for 3 × 65?*

IN FOCUS The questions in this section help children to see the links between number facts. It is essential to show the different calculations pictorially to help children see how the calculations change and stay the same. Once children are beginning to see the links between the given number facts, they could begin giving their own suggestions. At this point, they might begin to generalise about how this could help them to find any multiple of a given number. While solving question ❷, discuss Flo's comment with children. Ask:

- *What diagrams could you use?*
- *Why is a bar model a useful diagram to use?*

STRENGTHEN If children are struggling to link the number facts in question ❸, they could be encouraged to draw pictorial representations. Ask:

- *Could you show these facts using bar models?*
- *How do your pictures demonstrate links between the number facts?*

DEEPEN Once children have linked the facts in question ❸ and given their justifications, ask:

- *How many other facts can you find that are linked to this division?*
- *Can you give evidence to support your ideas?*

ASSESSMENT CHECKPOINT At this point in the lesson, children should be beginning to more confidently link number facts they know with those they do not. Children should be able to use pictorial representations, such as bar models, to help them explain their reasoning. Question ❷ assesses children's ability to recognise how they can use a known number fact to help them solve a more challenging problem. Expect children to recognise that 4 × 65 requires one more 65, while 66 × 3 requires one more 3.

ANSWERS

Question ❶ a): 6 × 65 = 2 × 195 = 390

Question ❶ b): 65 × 30 = 195 × 10 = 1,950

Question ❷ a): 4 × 65 = (3 × 65) + (1 × 65) = 195 + 65 = 260

Question ❷ b): 66 × 3 = (65 × 3) + (1 × 3) = 195 + 3 = 198

Question ❸ : 170 × 11 = 1,870 is the inverse of
1,870 ÷ 11 = 170
171 × 11 = (170 × 11) + 11 = 1,881
17 × 110 = (170 ÷ 10) × (11 × 10) = 170 × 11 = 1,870
170 × 12 = (170 × 11) + 170 = 2,040

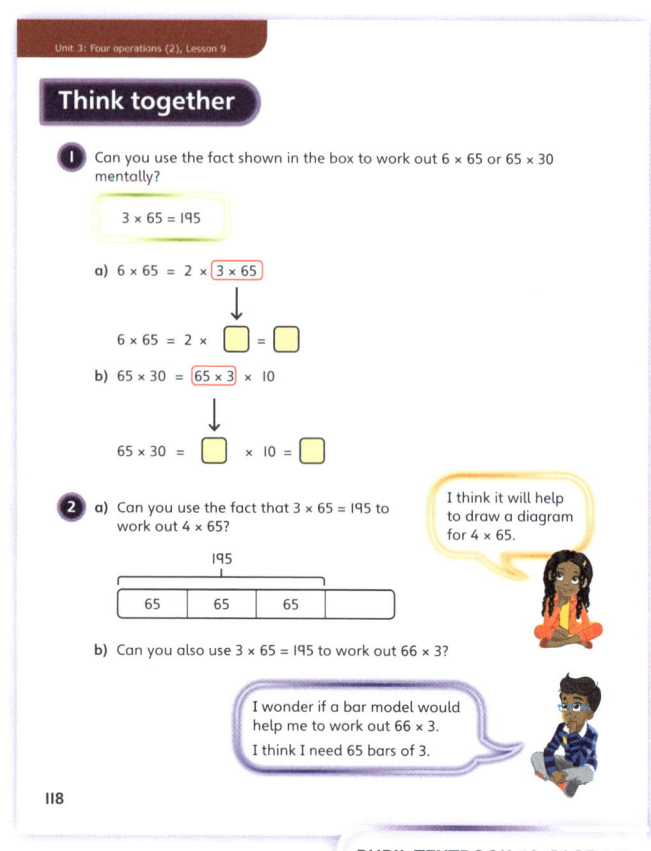

PUPIL TEXTBOOK 6A PAGE 118

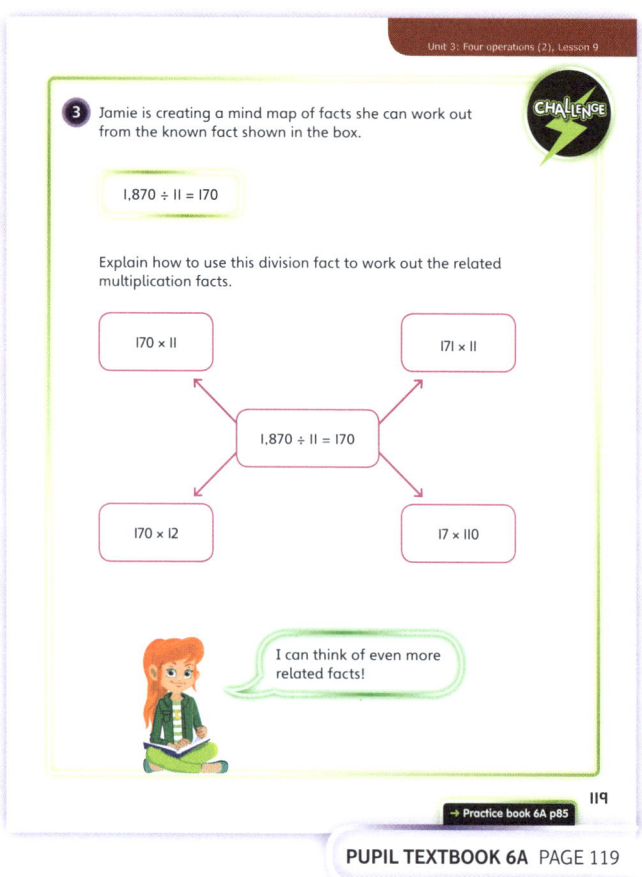

PUPIL TEXTBOOK 6A PAGE 119

Practice

WAYS OF WORKING Independent thinking

IN FOCUS Question **1** scaffolds children's independence when solving the missing number calculations. It is important for them to recognise where the desired product is similar to one they know how to find more easily (for example, 210 being 10 times more than 21). Encourage children to look for similar links to help them find possible numbers to fit the calculations. While solving question **2**, encourage children to complete the given bar models, or draw their own, to help them find and explain the solutions for each calculation.

STRENGTHEN If children are finding it difficult to complete question **2**, ask:
- *Can you see any links between the numbers in each known fact and its related calculation?*
- *Can you describe the links?*
- *How will these help you to solve the problem?*

DEEPEN If children finish the mind map in question **4**, ask:
- *Can you create your own mind map?*
- *What calculation will you put in the middle of your mind map?*
- *How many related facts can you find?*

THINK DIFFERENTLY Question **3** challenges assumptions about multiplying by 100. Children are likely to assume that, as one of the numbers in the calculation has an extra 100, then the answer should be multiplied by 100. The reality is, however, that the calculation will result in 100 more lots of 6, not 100 more lots of 288.

ASSESSMENT CHECKPOINT Children should be fluently using number facts they know to solve mathematical puzzles. They should be able to confidently explain how they use known number facts to solve each problem. Question **1** assesses children's ability to recognise multiplication facts they know, and use their understanding of place value, to solve puzzles.

ANSWERS Answers for the **Practice** part of the lesson appear in the separate **Practice and Reflect answer guide**.

Reflect

WAYS OF WORKING Pair work

IN FOCUS Successfully writing their own number facts will demonstrate that children are able to make links between related calculations. After children have written three other facts on their own, allow them time to share with a partner. Have they found the same facts? Ask:
- *Can you explain the link between the given calculation and your partner's number facts?*
- *Why have they chosen those facts?*

ASSESSMENT CHECKPOINT Children should be able to identify how the three facts link to the original calculation given in the question.

ANSWERS Answers for the **Reflect** part of the lesson appear in the separate **Practice and Reflect answer guide**.

After the lesson

- Where could the problems in this lesson have been given real-life contexts to appeal more to your cohort?
- How resilient were the cohort when they were problem solving?
- If problem solving was a challenge, how will you support and develop this skill in the future?

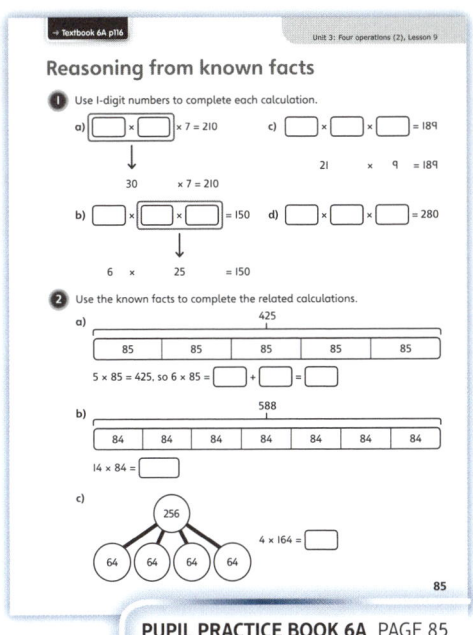

PUPIL PRACTICE BOOK 6A PAGE 85

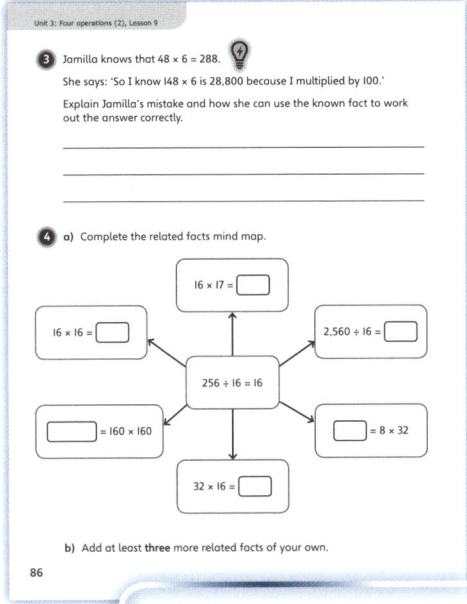

PUPIL PRACTICE BOOK 6A PAGE 86

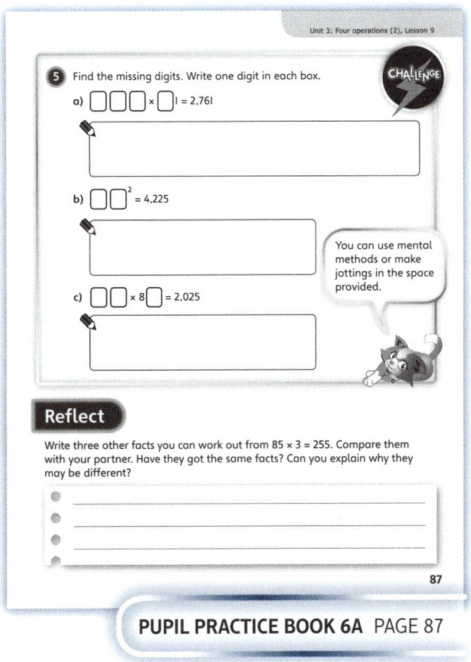

PUPIL PRACTICE BOOK 6A PAGE 87

End of unit check

Don't forget the *Power Maths* unit assessment grid on p26.

WAYS OF WORKING Group work – adult led

IN FOCUS

- Question **1** assesses children's ability to recognise and find common factors of given numbers.
- Question **2** assesses children's understanding of and ability to recognise prime numbers.
- Question **3** assesses children's understanding of cube numbers and how they can be represented.
- Question **4** assesses children's understanding of the order of operations and the effect of using brackets.
- Question **5** assesses children's mental methods and understanding of place value.
- Question **6** is a SATS-style question that assesses children's ability to use a known number fact to solve a more challenging calculation.

ANSWERS AND COMMENTARY

Children who have mastered the concepts in this unit will recognise what a common factor is. They will be able to identify how prime numbers differ from other numbers. They can confidently cube a number, recognising how square and cube numbers can be represented pictorially. Children will be able to fluently adhere to the correct order of operations, including use of brackets. Finally, they will be able to solve problems using efficient mental methods and explain how a known fact can help to solve a related calculation.

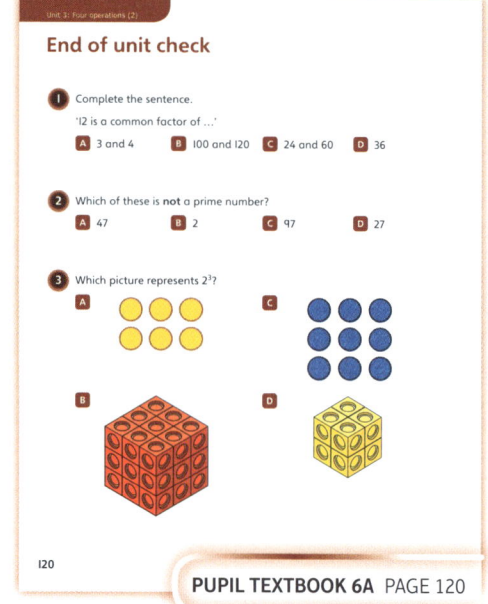

PUPIL TEXTBOOK 6A PAGE 120

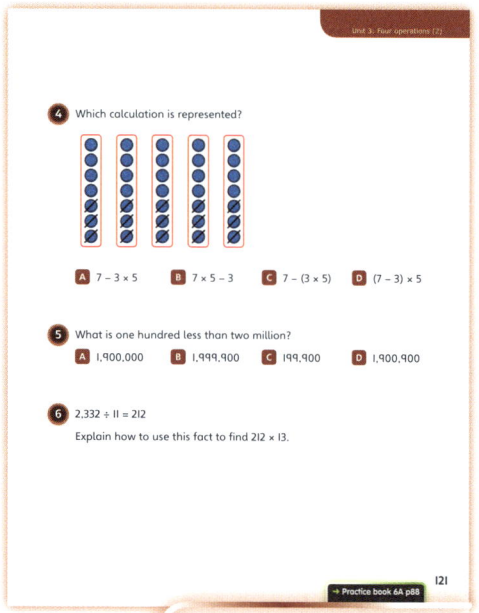

PUPIL TEXTBOOK 6A PAGE 121

Q	A	WRONG ANSWERS AND MISCONCEPTIONS	STRENGTHENING UNDERSTANDING
1	C	A suggests the child has confused factors with multiples. D suggests not understanding the term 'common factor'	For squaring and cubing numbers, ask: *If you were going to find the area or volume of this shape, what calculation would you use?*
2	D	B suggests that the child has not recognised 2 as the only even prime number.	To help children with place value when solving calculations mentally, give them patterns of number facts. For example:
3	D	A suggests the child has interpreted 2^3 as 2 × 3. B indicates cubing 3, not 2. C suggests the child has read 3^2.	$20 - 1 = 19$
4	D	A suggests that the child knows the operations required but is not confident using brackets. B or C indicates the child has not mastered the order in which to write or calculate operations.	$200 - 10 = 190$ $2,000 - 100 = 1,900$ $20,000 - 1,000 = 19,000$
5	B	A, C or D indicates a place value error.	Ask: *What do you notice?*
6		Children should be able to find the related fact 212 × 11 = 2,332 and add two more lots of 212. 2,332 + (2 × 212) = 2,756.	

My journal

WAYS OF WORKING Independent thinking

ANSWERS AND COMMENTARY

Children may write answers such as:
- $3^2 = 3 \times 3 = 9$ and $30^2 = 30 \times 30 = 900$.
- *Because 30 is 3×10, I know that 30^2 could be expressed as $(3 \times 10) \times (3 \times 10) = (9 \times 10 \times 10)$.*
- *Because I know that $30 \times 30 = 900$ then I can just subtract one 30 from that to work out 29×30.*

If children are finding it difficult to unpick the reasoning in the question, ask:
- *Can you show what 3^2 and 30^2 would look like as a picture?*
- *What calculation would you use to find what they are equal to?*
- *What is similar and what is different about the calculations?*
- *Why does 30^2 not equal 90?*

Power check

WAYS OF WORKING Independent thinking

ASK

- *Do you think you could find a prime number?*
- *How confident are you about the order of operations? Do you think you could explain it to someone else?*
- *Did you know what brackets are used for before starting this unit? How confident are you using them now?*

Power puzzle

WAYS OF WORKING Independent thinking

IN FOCUS This **Power puzzle** will assess children's ability to find and manipulate prime numbers. Children should be able to find and use a number of primes to find the given totals. Expect children to recognise where the given numbers are already prime. In these cases, you could ask: *Is it possible to make prime numbers by adding other prime numbers together?*

Sparks offers a suggestion for taking the puzzle deeper by challenging children to find more than one way of adding primes to make a given number.

ANSWERS AND COMMENTARY Children should be able to find the given numbers, possibly through trial and error, using only prime numbers. If they are using numbers that are not prime, it may indicate a misunderstanding of what a prime number is. Ask:
- *What is special about a prime number?*
- *Can you list the first five prime numbers? What is similar or different about them?*
- *How can you show their special properties using a picture or resources?*

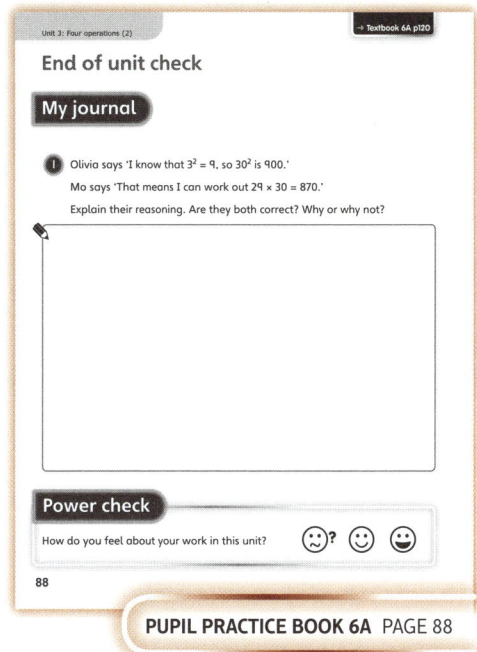

PUPIL PRACTICE BOOK 6A PAGE 88

PUPIL PRACTICE BOOK 6A PAGE 89

After the unit ⏸

- How will you weave the learning from this unit into children's future problem-solving activities? For example, using the order of operations and brackets to more easily generalise about number patterns.
- Did the **End of unit** check show any misconception that the class still has? How will you support and develop this area of learning?

Strengthen and **Deepen** activities for this unit can be found in the *Power Maths* online subscription.

Unit 4
Fractions ①

Don't forget to watch the Unit 4 video!

Mastery Expert tip! "To develop confidence in adding and subtracting fractions and mixed numbers, I encourage children to use fractions in different areas of mathematics. For example, when finding perimeters of shapes, use fractional lengths."

WHY THIS UNIT IS IMPORTANT

This unit is important because it develops children's understanding of fractions, moving on to comparing, adding and subtracting unrelated fractions using common denominators and formal methods. It encourages problem solving of fraction problems while exploring efficient methods.

WHERE THIS UNIT FITS

→ Unit 3 – Four operations (2)
→ **Unit 4 – Fractions (1)**
→ Unit 5 – Fractions (2)

In this unit, children extend their understanding of fractions and mixed numbers by adding and subtracting unrelated fractions using formal written methods involving finding common denominators. Children continue to develop their reasoning and problem-solving skills while exploring efficient methods.

Before they start this unit, it is expected that children:
- can find factors and multiples of numbers using multiplication facts
- can find equivalent fractions and convert between improper fractions and mixed numbers
- can compare and order fractions and add and subtract fractions which have the same denominator.

ASSESSING MASTERY

Children who have mastered this unit will be able to add and subtract fractions and mixed numbers confidently using several formal written methods. They will be able to solve multi-step problems and explain which method is most efficient.

COMMON MISCONCEPTIONS	STRENGTHENING UNDERSTANDING	GOING DEEPER
Children may simplify to find an equivalent fraction but not fully simplify the fraction.	Encourage children to use bar models and to write out factors of the numerator and denominator to identify common factors.	Give children some fractions and ask them to identify which ones have been simplified fully. Encourage them to explain why.
Children may compare the numbers in fractions rather than comparing the overall fractions.	Encourage children to initially compare fractions with the same denominator, so they understand that they can compare the numerators **only if** the denominators are the same.	Encourage children to identify the most efficient way of comparing fractions. This may not always be to find a common denominator, such as $\frac{2}{3}$ and $\frac{16}{17}$ ($\frac{16}{17}$ is closer to 1).
Children may simply add or subtract the numerators and denominators when adding or subtracting fractions.	Use bar models to show that the denominators must be the same before adding/subtracting the numerators. Start with fractions that have the same denominator to clarify why they do not add or subtract the denominators.	Encourage children to add and subtract fractions and mixed numbers using the most efficient method, for example $\frac{1}{2} + \frac{1}{4}$ (could be done mentally) and $5\frac{1}{6} - 4\frac{7}{8}$ (think about adding $\frac{1}{8}$ and $\frac{1}{6}$).

WAYS OF WORKING

Introduce this unit using teacher-led discussion. Allow children time to discuss questions in pairs or small groups and share ideas as a whole class. Children should be encouraged to use representations to visualise the fractions.

STRUCTURES AND REPRESENTATIONS

Number lines: The number line will allow children to see which numbers a fraction sits between and the fractional divisions. They can help children to visualise fractional increases or decreases.

Fraction strips: These models allow children to see how to split up fractions so they can be added or subtracted using common denominators.

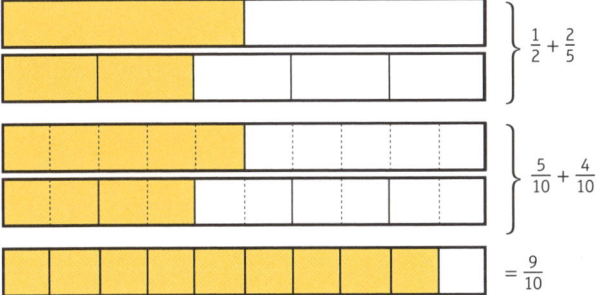

KEY LANGUAGE

There is some key language that children will need to know as part of the learning in this unit.

→ whole, part
→ numerator, denominator, common denominator
→ equivalent
→ simplify, simplest form
→ factor, highest common factor, lowest common multiple
→ compare
→ order, ascending, descending
→ less than, greater than
→ proper fraction, improper fraction
→ mixed number
→ convert

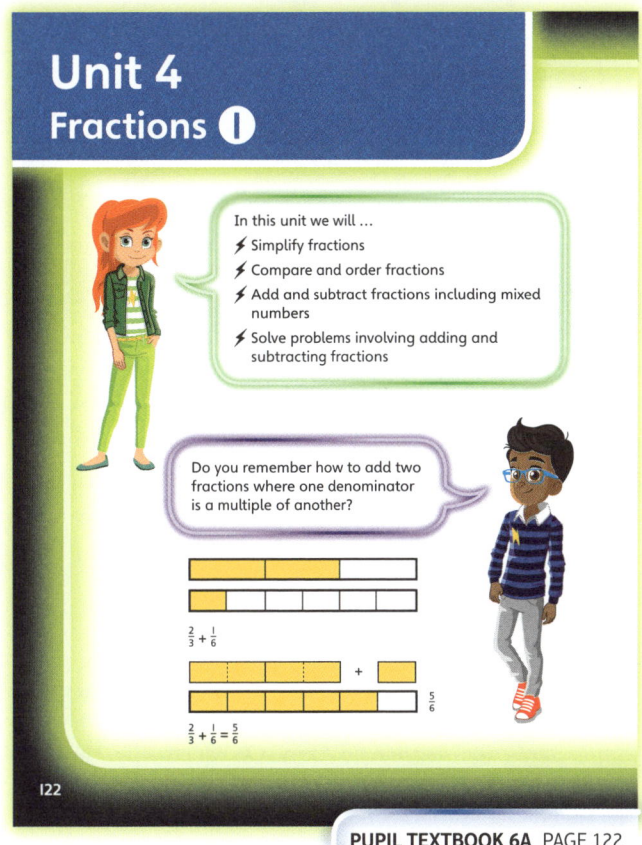

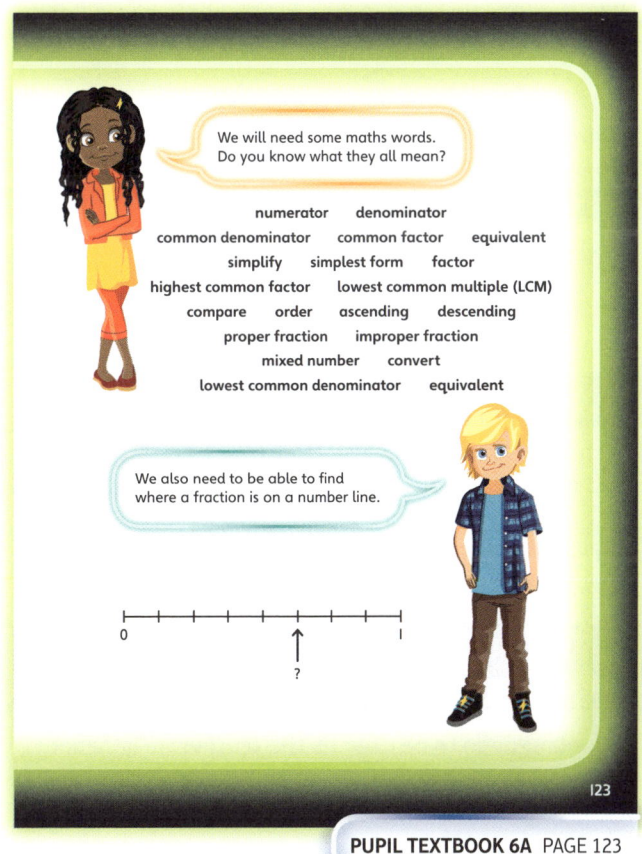

Simplifying fractions 1

Learning focus

In this lesson, children will apply their knowledge of factors to use common factors to simplify fractions.

Small steps

→ Previous step: Reasoning from known facts
→ **This step: Simplifying fractions (1)**
→ Next step: Simplifying fractions (2)

NATIONAL CURRICULUM LINKS

Year 6 Number – Fractions

Use common factors to simplify fractions; use common multiples to express fractions in the same denomination.

ASSESSING MASTERY

Children can confidently simplify fractions by dividing the numerator and denominator by a common factor, showing fluency in using known multiplication facts and finding factors and demonstrating an understanding of the terms simplify, numerator, denominator and equivalent.

COMMON MISCONCEPTIONS

A common misconception among children is that they believe to simplify a fraction you just divide the numerator and denominator by 2; for example, to simplify $\frac{9}{12}$, children may think the answer is $\frac{4\cdot5}{6}$. Conversely, they may realise that the numerator is not divisible by 2 and may incorrectly conclude that the fraction cannot be simplified because of this. To address this misconception, it is important to expose children to problems where the numerator or denominator is not divisible by 2 but the fraction can be simplified. Ask:

* *Are the numerator and denominator odd or even?*
* *If 2 is not a common factor, are there any odd common factors?*

Often, children do not simplify a fraction fully; for example, when simplifying $\frac{12}{18}$, children may reach an answer of $\frac{6}{9}$, without realising that this should be simplified fully to $\frac{2}{3}$. Ensure children understand how to identify when a fraction is simplified fully by asking:

* *Are there any more common factors of the numerator and denominator?*
* *Have you simplified the fraction fully?*

STRENGTHENING UNDERSTANDING

Children who are struggling with simplifying fractions fully should be encouraged to represent the fractions on a bar model. It may be beneficial to practise this with fractions that have only one possible way of being simplified. Encourage children to identify the factors of the numerator and denominator and to consider the common factors before moving on to fractions that can be simplified in several steps.

GOING DEEPER

Children could be encouraged to investigate a range of fractions looking for patterns, for example $\frac{9}{15}, \frac{8}{15}, \frac{9}{16}, \frac{8}{16}$. Ask: *Which of these are simplified fully? How do you know? Can you spot any patterns?*

Not only will this strengthen understanding by helping learners to recognise when a fraction is simplified fully, but it will also deepen learning by developing understanding of how to find factors, using known multiplication facts and drawing on division rules.

KEY LANGUAGE

In lesson: numerator, denominator, **common factor**, simplify, equivalent

Other language to be used by the teacher: fully, factor

STRUCTURES AND REPRESENTATIONS

fraction strips

 In the eTextbook of this lesson, you will find interactive links to a selection of teaching tools.

Before you teach

* Are children confident in finding factors?
* Can children identify the numerator and denominator of a fraction?
* Can children find equivalent fractions?

Discover

ASK

- Question **1** a): *How many cows does Bella get into the pen? What is her score out of? How could you write this as a fraction?*
- Question **1** b): *How many sheep does Lee get into the pen? What is his score out of? How could you write this as a fraction?*

IN FOCUS Question **1** a) introduces children to simplifying fractions where the numerator is a factor of the denominator. Question **1** b) develops this further, requiring children to find a common factor of the numerator and denominator, and simplify the fraction. This is an opportunity to pre-assess children's confidence with multiplication facts and use of factors in this context.

PRACTICAL TIPS Children could be introduced to simplifying fractions in a practical setting, such as playing games in PE. Encourage children to use bar models to understand fractions, and support comparing fractions when simplifying.

ANSWERS

Question **1** a): Bella's score is $\frac{1}{2}$.

Question **1** b): Lee's score is $\frac{3}{4}$.

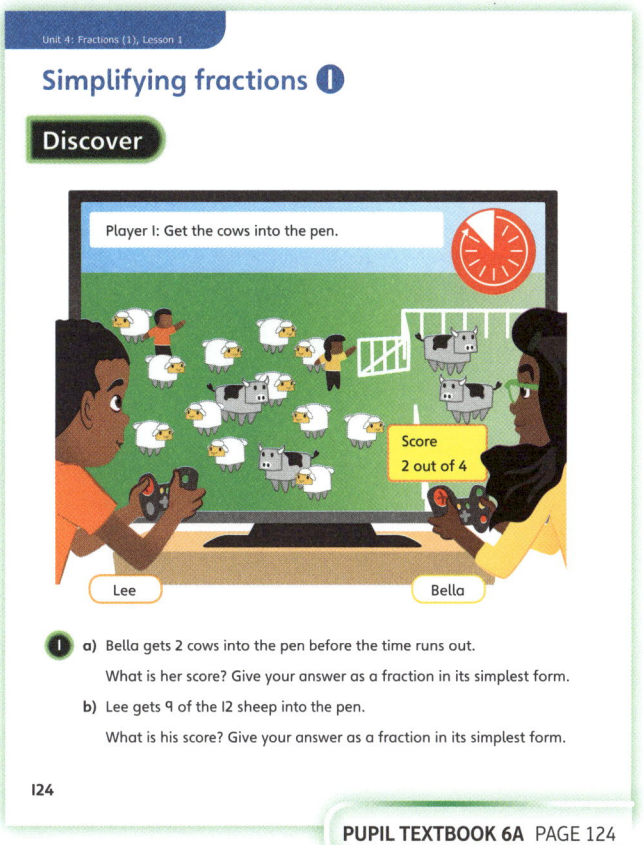

PUPIL TEXTBOOK 6A PAGE 124

Share

ASK

- Question **1** a): *How can you write 2 out of 4 as a fraction?*
- Question **1** a): *What do you think simplified means?*
- Question **1** a): *What common factor could you divide the numerator and denominator by? How do the bar models help?*
- Question **1** b): *Why can't you divide by 2 in this question? How do the bar models help?*

IN FOCUS In question **1** a) the term 'simplify' is introduced, giving an opportunity to explore the term and link this with children's prior knowledge of using factors. Note that the correct terminology is to 'divide the numerator by 2 and the denominator by 2' as opposed to 'divide by 2' which is mathematically incorrect. Encourage children to use fraction strips to help them understand why it is necessary to divide both numerator and denominator by 2. It is important that children realise that the fractions are still the same amount but are written differently because one is in its simplest form. Discuss also that in this instance the numerator is double the denominator and therefore the fraction simplifies to $\frac{1}{2}$.

Question **1** b) requires children to find a different common factor for the numerator and denominator, because it is not possible to divide by 2.

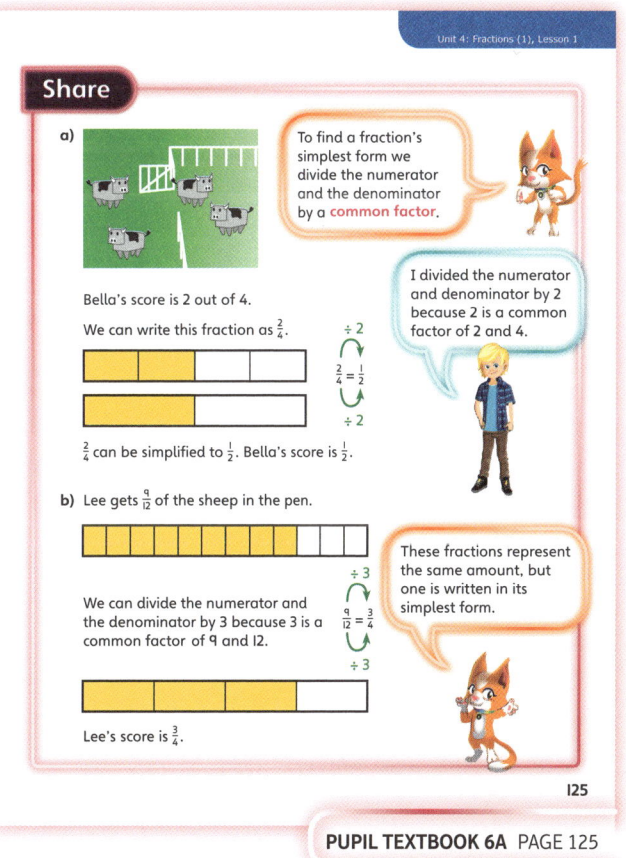

PUPIL TEXTBOOK 6A PAGE 125

Think together

Think together

WAYS OF WORKING Whole class teacher led (I do, We do, You do)

ASK

• Question **1** a): *What is the common factor of 4 and 6? How does the fraction strip help?*
• Question **2**: *What are the common factors of the numerators and denominators? What multiplication facts do you know to help you find the common factors?*
• Question **3**: *What does equivalent mean?*

IN FOCUS Question **1** supports children to develop their understanding of simplifying fractions by dividing both the numerator and the denominator by a common factor, while using fraction strips as scaffolding. In question **1** b), it may be beneficial to discuss how 5 is half of 10 so $\frac{5}{10}$ can be simplified to $\frac{1}{2}$.

Question **2** gives children an opportunity to simplify fractions without using fraction strips as support. Encourage children to discuss what they are dividing the numerator and denominator by, and to use their known multiplication facts and division rules to find common factors.

Question **3** introduces the term 'equivalent', and reduces scaffolding further by requiring children to find equivalent fractions by multiplying by a common factor – in effect, asking them to simplify fractions by working backwards.

STRENGTHEN If children are struggling to simplify fractions, encourage them to draw their own fraction strip to help work out the answer. Some children might struggle to recall times-table facts; in this instance, encourage them to write these down when thinking about common factors and to use their known multiplication facts to link with other facts (for example, doubling to connect the 2, 4 and 8 times-tables).

DEEPEN Question **3** deepens children's understanding of simplifying fractions by linking this knowledge to finding equivalent fractions. This could be explored further by asking children to find other equivalent fractions for each simplified answer. Encourage children to give reasons for their steps and methods.

ASSESSMENT CHECKPOINT Questions **1** and **2** will assess children's ability to simplify fractions using a common factor; question 2 withdraws the pictorial support of bar models. Look for children using multiplication facts to find factors.

Question **3** gives an opportunity to assess children's fluency with simplifying fractions while using their understanding to solve more complex problems involving equivalent fractions.

ANSWERS

Question **1** a): $\frac{4}{6} = \frac{2}{3}$

Question **1** b): $\frac{5}{10} = \frac{1}{2}$

Question **1** c): $\frac{9}{15} = \frac{3}{5}$ (÷ 3)

Question **2** a): $\frac{6}{8} = \frac{3}{4}$

Question **2** b): $\frac{10}{100} = \frac{1}{10}$

Question **2** c): $\frac{7}{7} = \frac{1}{1} = 1$

Question **3** a): $\frac{5}{6} = \frac{10}{12}, \frac{3}{10} = \frac{15}{50}$

Question **3** b): $\frac{15+5}{29+6} = \frac{4}{7}, \frac{6+19}{30} = \frac{5}{6}, \frac{25-4}{28} = \frac{3}{4}, \frac{1+15}{20} = \frac{4}{5}$

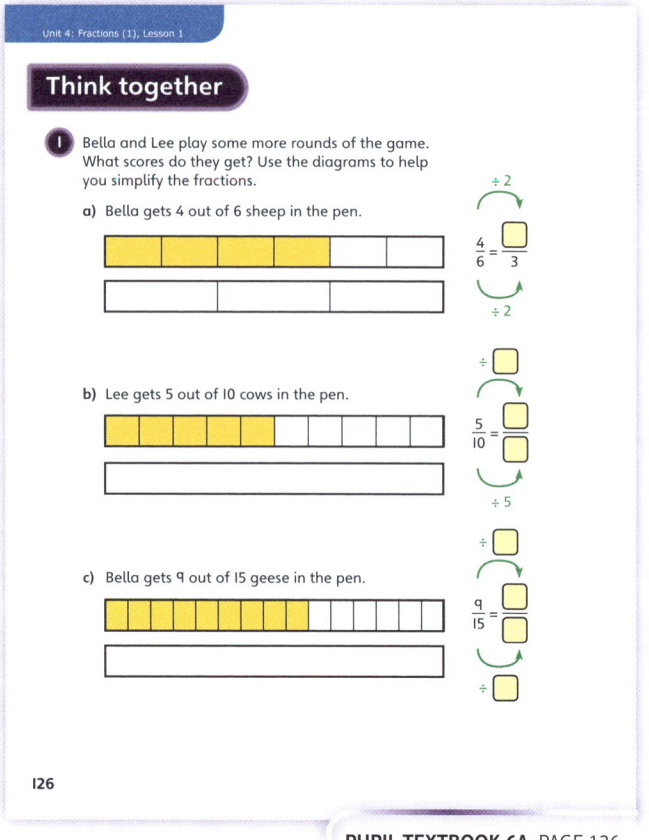

Think together

1 Bella and Lee play some more rounds of the game. What scores do they get? Use the diagrams to help you simplify the fractions.

a) Bella gets 4 out of 6 sheep in the pen.

b) Lee gets 5 out of 10 cows in the pen.

c) Bella gets 9 out of 15 geese in the pen.

126

PUPIL TEXTBOOK 6A PAGE 126

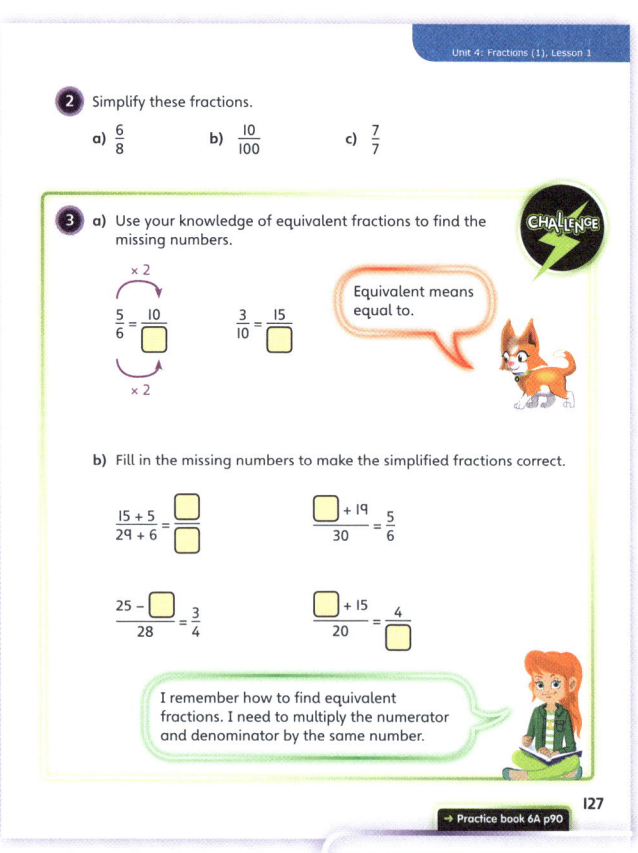

2 Simplify these fractions.

a) $\frac{6}{8}$ b) $\frac{10}{100}$ c) $\frac{7}{7}$

3 a) Use your knowledge of equivalent fractions to find the missing numbers.

CHALLENGE

$\frac{5}{6} = \frac{10}{\square}$ $\frac{3}{10} = \frac{15}{\square}$

Equivalent means equal to.

b) Fill in the missing numbers to make the simplified fractions correct.

$\frac{15+5}{29+6} = \frac{\square}{\square}$ $\frac{\square+19}{30} = \frac{5}{6}$

$\frac{25-\square}{28} = \frac{3}{4}$ $\frac{\square+15}{20} = \frac{4}{\square}$

I remember how to find equivalent fractions. I need to multiply the numerator and denominator by the same number.

→ Practice book 6A p90

127

PUPIL TEXTBOOK 6A PAGE 127

Practice

WAYS OF WORKING Independent thinking

IN FOCUS Questions **1** and **2** consolidate children's understanding of simplifying fractions by dividing the numerator and denominator by a common factor, using limited pictorial representation as support.

In question **3**, children apply their understanding of equivalent fractions and develop a logical chain of reasoning to express $\frac{90}{120}$ in its simplest form.

Question **4** develops children's understanding by linking fractions to pictorial representations (shapes with parts shaded). Encourage children to simplify the fractions first and highlight that the shape is not important but the number of equal parts is.

Question **6** deepens children's conceptual understanding by asking them to identify the fractions that are not equivalent (non-examples).

STRENGTHEN If children are struggling with simplifying fractions in questions **1** and **2**, encourage them to use models and to explain their steps. Ask: *What is a common factor of the numerator and denominator?*

DEEPEN To build on question **2**, give children more fractions that can be simplified using different methods. For example, $\frac{24}{30}$ could be simplified by dividing the numerator and denominator by 2 to get $\frac{12}{15}$, then dividing by 3 to get $\frac{4}{5}$; alternatively, it could be simplified by dividing the numerator and denominator by the highest common factor of 6 to get $\frac{4}{5}$. Ask: *What method did you use? Could you use a different method?*

ASSESSMENT CHECKPOINT Questions **1** and **2** assess children's ability to simplify fractions using a common factor with some pictorial support. Question **5** addresses the common misconception that fractions can be simplified by repeatedly dividing both the numerator and denominator by 2, and simplified fractions can be expressed as decimals. Question **6** assesses children's ability to find and identify equivalent fractions. Question **8** assesses children's fluency in simplifying fractions using common factors and their understanding of finding equivalent fractions using known multiplication facts.

ANSWERS Answers for the **Practice** part of the lesson appear in the separate **Practice and Reflect answer guide**.

Reflect

WAYS OF WORKING Pair work

IN FOCUS This question will assess children's understanding of simplifying a fraction by dividing the numerator and denominator by a common factor. Encourage children to write down and show what they are dividing by and to explain how they know their answer is in its simplest form.

ASSESSMENT CHECKPOINT Children should be able to simplify a fraction fully and to explain confidently the steps in their method. They should have a good understanding of the terms simplify, numerator, denominator and common factor.

ANSWERS Answers for the **Reflect** part of the lesson appear in the separate **Practice and Reflect answer guide**.

After the lesson ⏸

- Are children confident in using the terms 'simplify', 'numerator', 'denominator' and 'equivalent'?
- Can children use known multiplication facts to find common factors and use multiples to find equivalent fractions?

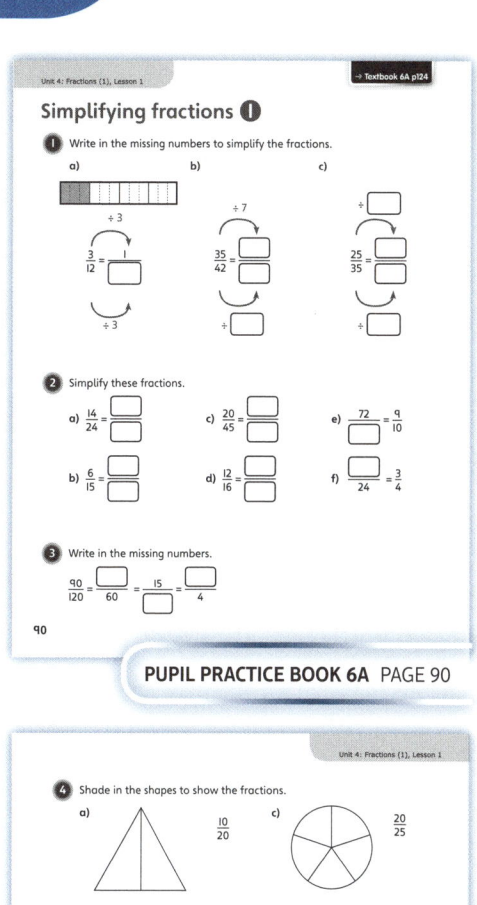

PUPIL PRACTICE BOOK 6A PAGE 90

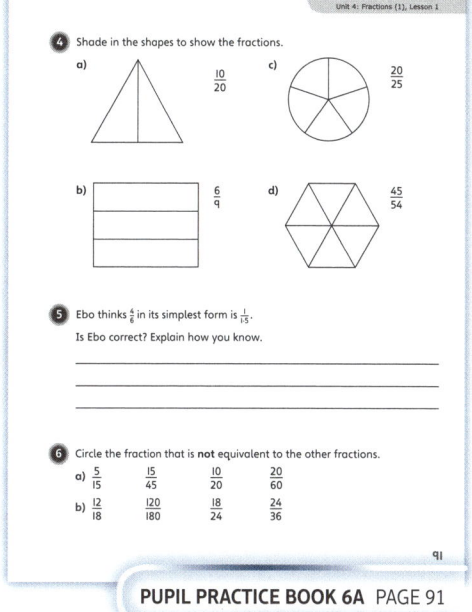

PUPIL PRACTICE BOOK 6A PAGE 91

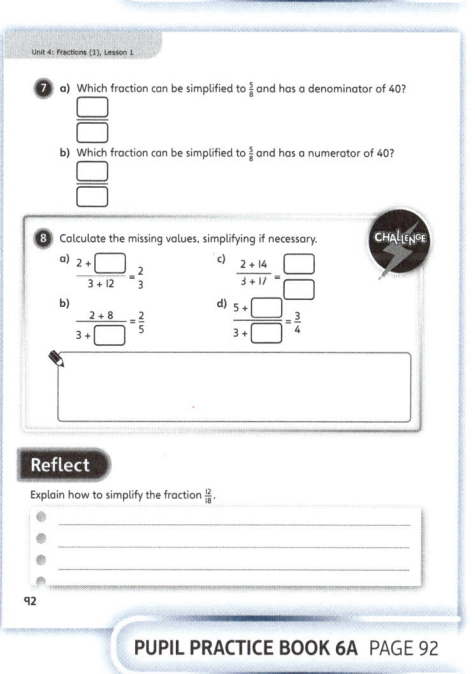

PUPIL PRACTICE BOOK 6A PAGE 92

Simplifying fractions ②

Learning focus

In this lesson, children extend their understanding of simplifying fractions to simplify mixed numbers and improper fractions.

Small steps

→ Previous step: Simplifying fractions (1)
→ **This step: Simplifying fractions (2)**
→ Next step: Fractions on a number line

NATIONAL CURRICULUM LINKS

Year 6 Number – Fractions
• Use common factors to simplify fractions; use common multiples to express fractions in the same denomination.
• Compare and order fractions, including fractions > 1.

ASSESSING MASTERY

Children can confidently fully simplify a fraction including mixed numbers and improper fractions, either by finding and using the highest common factor of the numerator and the denominator, or by using repeated division.

COMMON MISCONCEPTIONS

A common misconception when simplifying a mixed number is to divide all the numbers by a common factor, for example when simplifying $6\frac{8}{10}$ children may correctly identify the common factor as 2 but incorrectly divide all the numbers by 2 to get $3\frac{4}{5}$ when the answer should be $6\frac{4}{5}$. Ensure children see the mixed numbers on a bar model so they comprehend that the whole number stays the same and only the fractional part is simplified.

Another common mistake that children make when simplifying an improper fraction is to convert it to a mixed number without simplifying, or to mix up the numerator and denominator. For example, some children may simplify $\frac{8}{4}$ to $\frac{1}{2}$ instead of $\frac{2}{1}$ or 2. Showing this on a bar model will help children to understand why the fraction simplifies to 2 and not $\frac{1}{2}$.

STRENGTHENING UNDERSTANDING

To strengthen understanding, encourage children to use bar models for pictorial support. Children who struggle to recall times-table facts could be encouraged to write these down when thinking about common factors.

GOING DEEPER

Deepen learning by giving children a fraction or mixed number, for example $\frac{7}{5}$ or $2\frac{1}{3}$, and asking them to find at least three fractions that simplify to the original fraction.

KEY LANGUAGE

In lesson: highest common factor, HCF, simplify, fully, numerator, denominator, factor, common factor

Other language to be used by the teacher: improper fraction, mixed number, equivalent

STRUCTURES AND REPRESENTATIONS

fraction strips

 In the eTextbook of this lesson, you will find interactive links to a selection of teaching tools.

Before you teach

• Can children convert between improper fractions and mixed numbers?
• Can children use the short division method?

Discover

WAYS OF WORKING Pair work

ASK

- Question **1** a): *How many people are there? How many are children? How can you write this as a fraction?*
- Question **1** b): *How many carriages are there? How full is each carriage? How can you write this?*

IN FOCUS Question **1** introduces children to the concept of simplifying fractions and mixed numbers. This builds on their use of known multiplication facts to find common factors of the numerator and denominator.

PRACTICAL TIPS Engage the children by explaining that some people are going on a roller coaster ride. Ask them to visualise the ride and the carriages being full, half full or empty.

ANSWERS

Question **1** a): $\frac{2}{3}$ of the people are children.

Question **1** b): $1\frac{2}{3}$ roller coaster carriages are full.

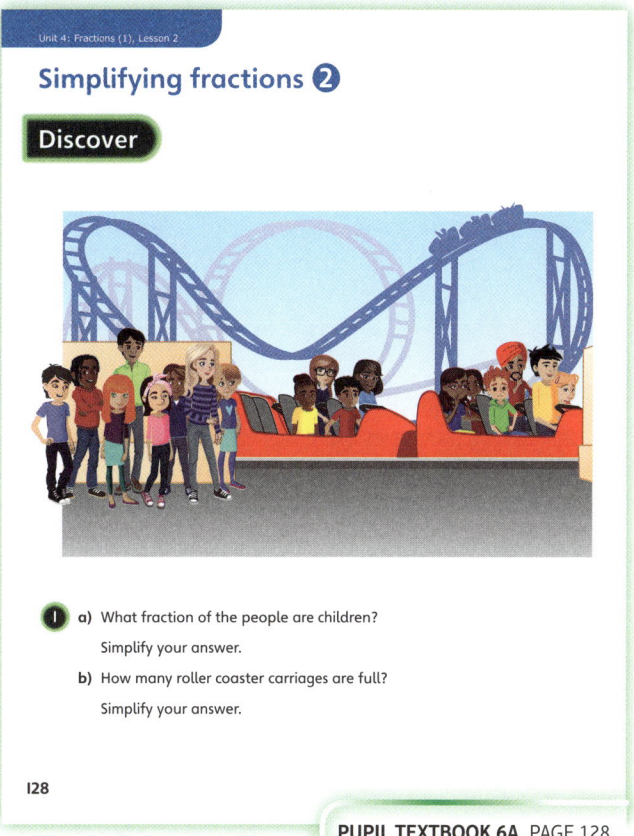

PUPIL TEXTBOOK 6A PAGE 128

Share

WAYS OF WORKING Whole class teacher led

ASK

- Question **1** a): *What are the common factors of 12 and 18? Why do you need to divide **both** the numerator and the denominator by the common factor?*
- Question **1** a): *What if you simplify the fraction using other factors? Is it simplified fully? Will it take more than one step?*
- Question **1** a): *Do both methods give the same answer?*
- Question **1** b): *How full is the first carriage? What number would this relate to?*
- Question **1** b): *How full is the second carriage? How can you write this as a fraction?*
- Question **1** b): *Could you write how full both carriages are as a mixed number?*

IN FOCUS Question **1** a) introduces children to simplifying a fraction by finding the highest common factor of the numerator and denominator, or in steps by finding common factors. Discuss how one way of simplifying $\frac{12}{18}$ is to find the highest common factor of 12 and 18; explore why this is 6 and why children need to divide both the numerator and the denominator by 6. Discuss the other method, of first dividing 12 and 18 by 2 to get $\frac{6}{9}$; explore why this is not simplified fully. Discuss common factors of 6 and 9 and show why it is necessary to divide both the numerator and the denominator again, this time by 3, to get $\frac{2}{3}$.

Question **1** b) requires children to simplify a mixed number. Emphasise that the first carriage is full, so this is 1 whole. Then explore the fraction of the people who are in the second carriage. Use the bar models to support this and to encourage discussion about how to simplify a mixed number. Emphasise that the whole number stays as 1: only the fractional part needs to be simplified.

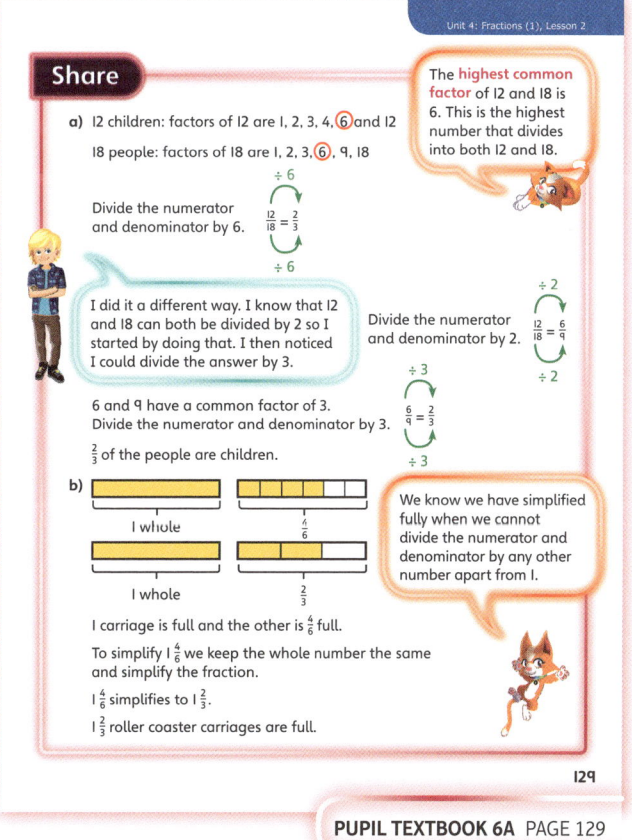

PUPIL TEXTBOOK 6A PAGE 129

Think together

WAYS OF WORKING Whole class teacher led (I do, We do, You do)

ASK

- Question ❶ b): *How can you convert an improper fraction to a mixed number? Can you simplify further? What are the common factors of the numerator and denominator?*
- Question ❸: *Which fraction can be simplified? What are the common factors of the numerator and denominator?*

IN FOCUS Question ❶ reinforces simplifying a mixed number and an improper fraction, reducing scaffolding in part b). In part a), encourage children to find the common factors of the numerator and denominator and to use the bar model to see why the whole number, 3, stays the same.

In question ❸, discuss the different ways in which $\frac{288}{160}$ can be simplified and which way might be most efficient. Encourage children to use the short division method.

STRENGTHEN Encourage children to identify the factors of the numerator and denominator and to consider common factors before trying to simplify, writing them down if necessary.

DEEPEN Question ❷ can be extended by giving children a fraction or mixed number and asking them to consider and discuss in their own words all the different ways in which it can be simplified.

ASSESSMENT CHECKPOINT Question ❷ is an opportunity to assess children's understanding of the different methods for simplifying fractions and to uncover any misconceptions. Look for children understanding that the whole numbers are never divided and recognising the need to divide by common factors until the fraction is simplified.

Question ❸ gives an opportunity to assess children's ability to simplify large improper fractions. Look for children repeatedly simplifying using common multiples until they are confident the fractions are fully simplified.

ANSWERS

Question ❶ a): $\frac{8}{10} = \frac{4}{5}$, so $3\frac{8}{10} = 3\frac{4}{5}$

Question ❶ b): $\frac{25}{10} = 2\frac{5}{10} = 2\frac{1}{2}$

Question ❷:
- Olivia has divided all the numbers by 8 but she should have left the whole number as 16 so her method is incorrect.
- Lexi has kept the whole number as 16 and has divided the numerator and denominator by the highest common factor to simplify the fraction fully. Her method is correct.
- Isla has kept the whole number as 16 but has divided the numerator and denominator by 4: she has not fully simplified the fraction so her method is incorrect.
- Amelia has divided all the numbers by 2, then by 2 again, then by 2 again. This is correct for the fraction part but she should have left the whole number as 16 so her method is incorrect.
- Emma has kept the whole number as 16 and has divided the numerator and denominator by 2 then by 4. This simplifies the fraction fully so her method is correct.

Question ❸: $\frac{288}{160} = \frac{9}{5}$ so both fractions are the same (highest common factor is 32).

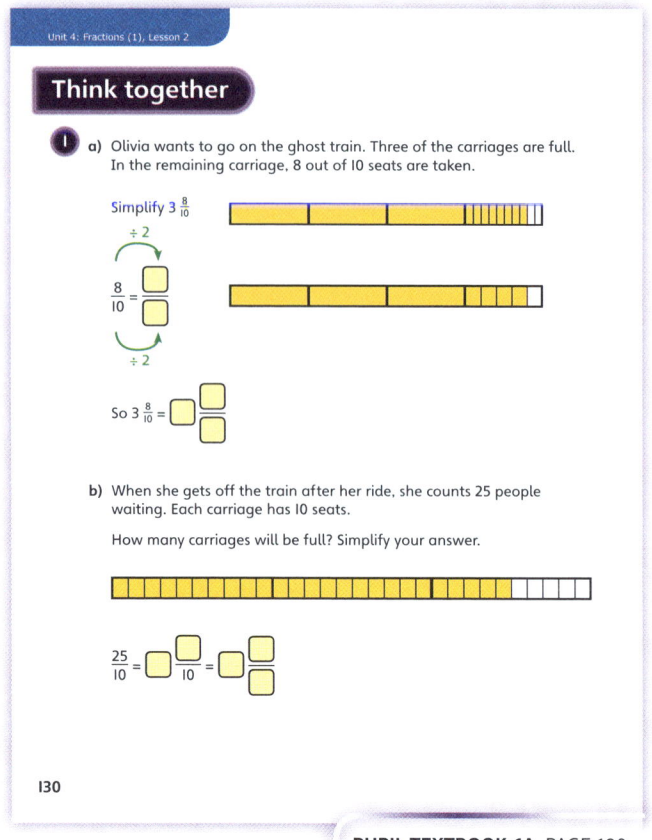

PUPIL TEXTBOOK 6A PAGE 130

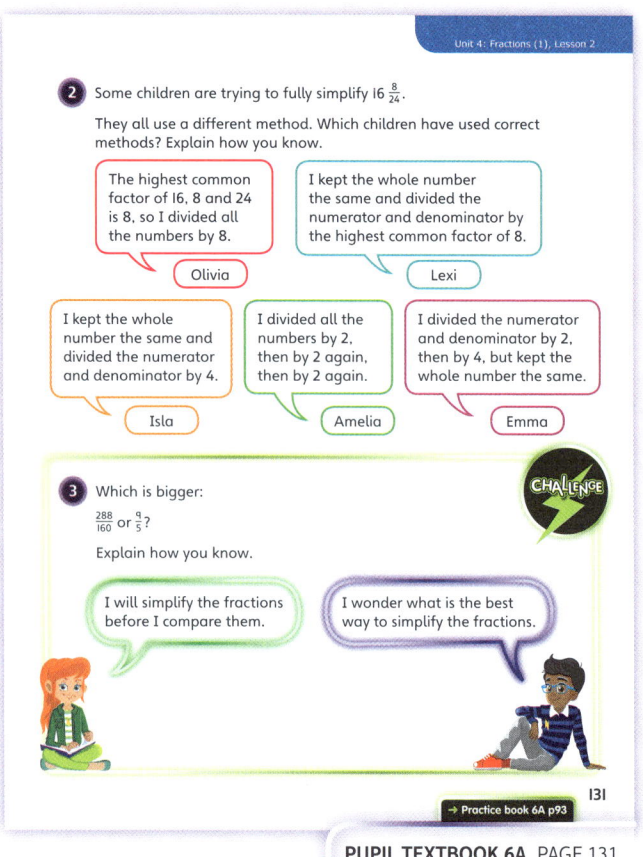

PUPIL TEXTBOOK 6A PAGE 131

Practice

WAYS OF WORKING Independent thinking

IN FOCUS Question **1** and **2** aim to consolidate children's understanding of simplifying a mixed number and to address the misconception of dividing all the numbers by the common factor.

Question **5** requires children to spot the 'odd one out' by finding, in part a), the fraction that is not equivalent. Part b) involves identifying which fractions do not simplify to a whole number. Look for children who convert $\frac{20}{40}$ to 2, because this will indicate more support is needed.

STRENGTHEN Strengthen learning by encouraging children to simplify in multiple steps if they are struggling to find the highest common factor. Ask: *What number do you know goes into the numerator and denominator? How do you know? Can you simplify further?*

DEEPEN Question **7** can be explored further by asking children to write their answers as mixed numbers.

ASSESSMENT CHECKPOINT Question **4** will assess children's ability to fully simplify a proper fraction, an improper fraction and a mixed number in abstract form. Look for children confidently using the lowest common multiple (LCM; concept not formally introduced until lesson 4) to simplify.

Question **7** will assess children's ability to simplify large improper fractions. Look for children repeatedly simplifying using common multiples until they are confident that they have fully simplified the fractions.

ANSWERS Answers for the **Practice** part of the lesson appear in the separate **Practice and Reflect answer guide**.

Reflect

WAYS OF WORKING Pair work

IN FOCUS This **Reflect** activity involves children explaining the steps for simplifying a mixed number. Encourage children: to write down and show what they are dividing the numerator and denominator by; to explain why the whole number stays the same; and to explain how they know the fraction part is in the simplest form.

ASSESSMENT CHECKPOINT This activity assesses children's understanding of simplifying a mixed number; it particularly aims to highlight the misconception of dividing all the numbers by the common factor. Look for children confidently explaining the different steps and reliably using multiples to simplify, repeatedly simplifying the fraction until it is simplified fully while not simplifying the whole number.

ANSWERS Answers for the **Reflect** part of the lesson appear in the separate **Practice and Reflect answer guide**.

After the lesson ⏸

- Can children fully simplify a proper fraction, an improper fraction and a mixed number?
- Can children confidently use repeated division to fully simplify a fraction?
- Do children understand that in a mixed number the whole number does not need to be simplified?

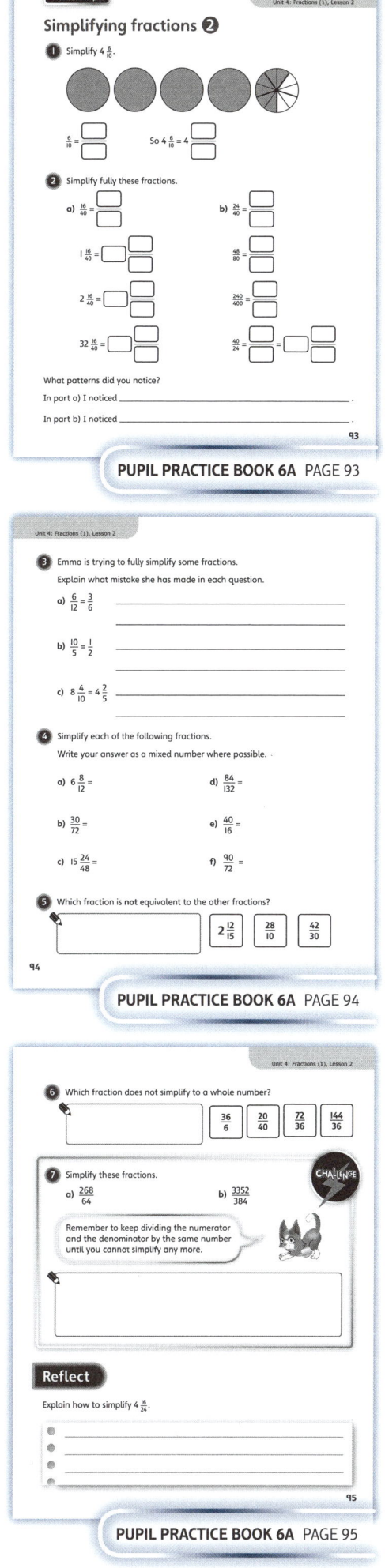

PUPIL PRACTICE BOOK 6A PAGE 93

PUPIL PRACTICE BOOK 6A PAGE 94

PUPIL PRACTICE BOOK 6A PAGE 95

Fractions on a number line

Learning focus

In this lesson, children use their understanding of fractions to count up and down on a number line, place missing fractions on a number line and find missing numbers in a fractional sequence.

Small steps

→ Previous step: Simplifying fractions (2)
→ **This step: Fractions on a number line**
→ Next step: Comparing and ordering fractions (1)

NATIONAL CURRICULUM LINKS

Year 6 Number – Fractions

Compare and order fractions, including fractions >1.

ASSESSING MASTERY

Children can reliably count up and down fractional increases or decreases on a number line, find missing fractions in a sequence and place missing fractions on a number line, including lines where the fractional divisions are not marked.

COMMON MISCONCEPTIONS

Children may count up or down by the wrong amount when counting on a number line, for example to find 1 more than $5\frac{1}{7}$ they may get an answer of $5\frac{2}{7}$ as they have counted up to the next fractional division instead of counting up 1 whole. Some children may become confused when counting beyond 1. For example, when counting in steps of $\frac{2}{3}$, they may say: $\frac{2}{3}$, $1\frac{2}{3}$, $2\frac{2}{3}$, $3\frac{2}{3}$ etc, instead of bridging the whole number correctly. Ask:
• *What fractional divisions is the number line divided into?*

Children may struggle to place numbers on a number line where the fractional divisions are not marked, for example, on a number line that is divided into fifths they may struggle to show $\frac{1}{2}$. Reason with children where the fraction lies, for example $\frac{1}{2}$ would lie exactly in the middle.

Another common misconception is that children often think a pattern cannot be going up or down by the same amount if the fractions have different denominators. For example, children may think the sequence $2\frac{3}{4}$, $3\frac{1}{2}$, $4\frac{2}{8}$, 5 is incorrect because it has quarters and halves and eighths in it. Ask:
• *How are quarters, halves and eighths related?*

STRENGTHENING UNDERSTANDING

Children who are struggling to place fractions on a number line can be encouraged to strengthen their understanding by counting up or down in fractions where the numerator is 1 (unit fractions) on number lines where all the numbers are labelled, before moving on to counting up or down in non-unit fractions using number lines where some of the numbers are missing. Encourage children to fill in all the unknown numbers on a number line before working out the missing numbers in a pattern.

GOING DEEPER

Children can be encouraged to build on previous lessons and deepen understanding of using number lines for fractions, by simplifying fractions before placing them on the number line. For example, children could be asked to draw a number line from 2 to 4 that is divided into thirds, and then place on it numbers such as $2\frac{3}{9}$, $3\frac{8}{12}$, $2\frac{11}{33}$.

Encourage children to use their knowledge of equivalent fractions when placing fractions on a number line. For example, ask children to draw a number line from 5 to 7, divided into twelfths, and then place numbers on it such as $5\frac{2}{3}$, $6\frac{3}{4}$, $5\frac{1}{2}$ and $6\frac{5}{6}$.

KEY LANGUAGE

In lesson: number line, divide, fraction, greatest, more, less, gaps

Other language to be used by the teacher: division, pattern, simplify, equivalent

STRUCTURES AND REPRESENTATIONS

number lines with fractional divisions

 In the eTextbook of this lesson, you will find interactive links to a selection of teaching tools.

Before you teach

• Can children divide up a number line to show fractions?
• Are children confident in simplifying fractions and finding equivalent fractions?
• Can children read mixed numbers from a number line?

Discover

PUPIL TEXTBOOK 6A PAGE 132

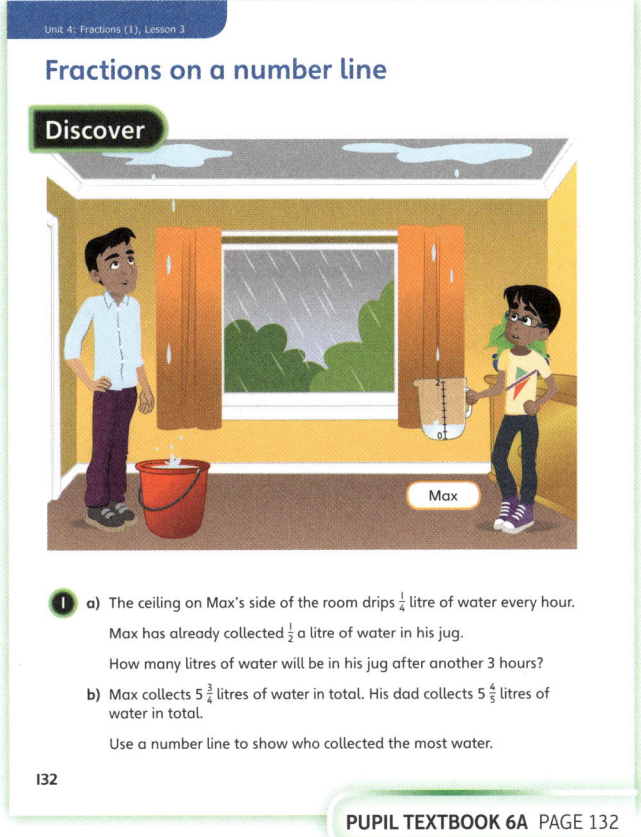

WAYS OF WORKING Pair work

ASK

• Question ① a): *What could help you to calculate how many litres of water will be in Max's jug after 3 hours?*
• Question ① b): *Can you place the numbers on the number line to help you compare them?*

IN FOCUS Question ① a) introduces children to the concept of counting forwards in jumps of $\frac{1}{4}$ using a number line for pictorial support. Question ① b) develops this further by introducing mixed numbers on a number line and using the number line to make comparisons.

PRACTICAL TIPS Ensure children understand the context of the question. It may be beneficial to have measuring jugs for the children to handle, ideally split into fractions, so they can see that there is a number line on the jugs.

ANSWERS

Question ① a): There are $1\frac{1}{4}$ litres of water in Max's jug after another 3 hours.

Question ① b): Max's dad collected the greatest amount of water throughout the day.

Share

PUPIL TEXTBOOK 6A PAGE 133

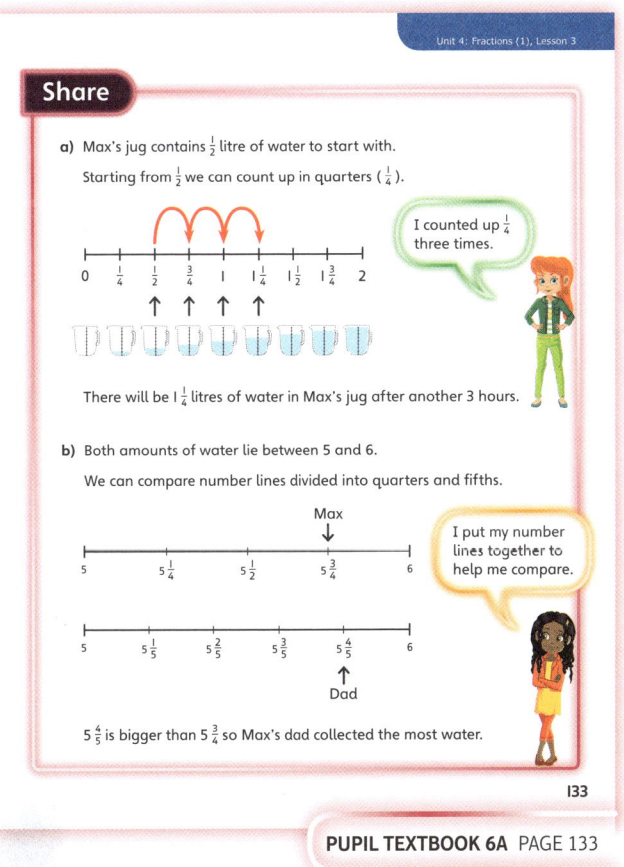

WAYS OF WORKING Whole class teacher led

ASK

• Question ① a): *Where do you need to start on the number line? How could you work out how many litres are in Max's jug after 1 hour? After 2 hours? After 3 hours?*
• Question ① a): *What operation do you need to do with $\frac{1}{4}$? How many times?*
• Question ① b): *Which whole numbers should you place at the start and end of the number line?*
• Question ① b): *Why is it important to use number lines the same length when making comparisons?*

IN FOCUS With question ① a), it will be important to ensure that children are comfortable using the number line; make sure they know where to start on the line and how many times to add on $\frac{1}{4}$. Discuss with children why they have added on $\frac{1}{4}$ three times to get an answer of $1\frac{1}{4}$ litres of water.

Question ① b) develops the concept of representing fractions on a number line by introducing mixed numbers. Again, ensure that children can use the number line to show which whole numbers the two amounts of water lie between. Encourage children to reason that when they are dealing with quarters ($5\frac{3}{4}$) they need to divide the line into 4 equal parts, but when dealing with fifths ($5\frac{4}{5}$) they need to divide it into 5 equal parts. This will provide an opportunity to discuss how it is important for the number lines to be the same length when making comparisons.

Think together

WAYS OF WORKING Whole class teacher led (I do, We do, You do)

ASK

- Question **2** c): *Is $\frac{2}{6}$ a simplified fraction? Can it be simplified to help you? Does simplifying a fraction change the size of the jumps on the number line?*
- Question **3**: *What is the number line divided into between the whole numbers?*

IN FOCUS Question **2** b) could be broken down into counting in halves and then counting in quarters. Encourage children to notice that every other quarter is on the halves number line. For question **2** c), encourage children to notice that jumps of $\frac{2}{6}$ are the same as jumps of $\frac{1}{3}$.

In question **3**, look out for the common error of counting on $\frac{1}{5}$ (one division on the line) instead of 1 whole.

Question **4** a) requires children to place numbers on a number line where the denominators are different. Encourage them to divide the line into 10 equal parts and then use their knowledge of equivalent fractions to mark on $2\frac{1}{2}$. Question **4** b) gives children an opportunity to explore missing numbers in a pattern when given two of the numbers.

STRENGTHEN If children are struggling, encourage them to write any unknown whole numbers onto the number line before trying to work out any missing fractions. Encourage children to draw arrows between numbers to represent the jumps.

DEEPEN Question **3** can be explored further by asking children to place other numbers on the number line, where they will have to use their knowledge of equivalent fractions, for example place $3\frac{7}{8}$, $3\frac{1}{2}$ and $3\frac{3}{4}$ on a number line.

ASSESSMENT CHECKPOINT Question **2** assesses children's ability to work out fractional intervals on a number line in order to label the line. Question **3** assesses children's ability to use a number line to calculate a fractional increase or decrease.

ANSWERS

Question **1**: $2\frac{2}{3}$, 3, $3\frac{1}{3}$, $3\frac{2}{3}$, 4, $4\frac{1}{3}$, $4\frac{2}{3}$. There are $4\frac{2}{3}$ litres of water in the bath after 8 seconds.

Question **2** a): Missing numbers are $7\frac{2}{5}$, $7\frac{3}{5}$, 8, $8\frac{1}{5}$, $8\frac{3}{5}$, $8\frac{4}{5}$

Question **2** b): $11\frac{1}{4}$, $11\frac{1}{2}$, $11\frac{3}{4}$, 12, $12\frac{1}{4}$, $12\frac{1}{2}$, $12\frac{3}{4}$, 13, $13\frac{1}{4}$, $13\frac{1}{2}$, $13\frac{3}{4}$, 14, $14\frac{1}{4}$, $14\frac{1}{2}$, $14\frac{3}{4}$, 15

Question **2** c): Jumps show the following numbers:
1, $1\frac{2}{6}$, $1\frac{4}{6}$, 2, $2\frac{2}{6}$, $2\frac{4}{6}$, 3

Question **3** a): The arrow is pointing to $5\frac{4}{5}$.

Question **3** b): $6\frac{1}{5}$

Question **3** c): $5\frac{3}{5}$

Question **3** d): $6\frac{4}{5}$

Question **4** a):

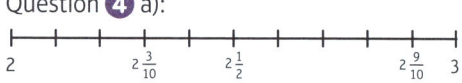

Question **4** b): $2\frac{5}{7}$, $3\frac{4}{7}$, 4, $4\frac{6}{7}$ $(+\frac{3}{7})$

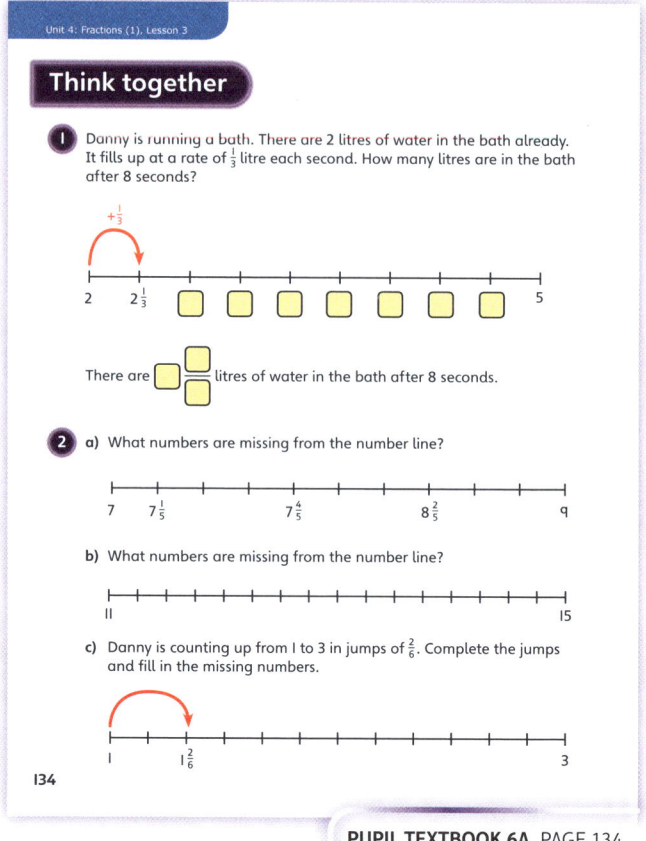

PUPIL TEXTBOOK 6A PAGE 134

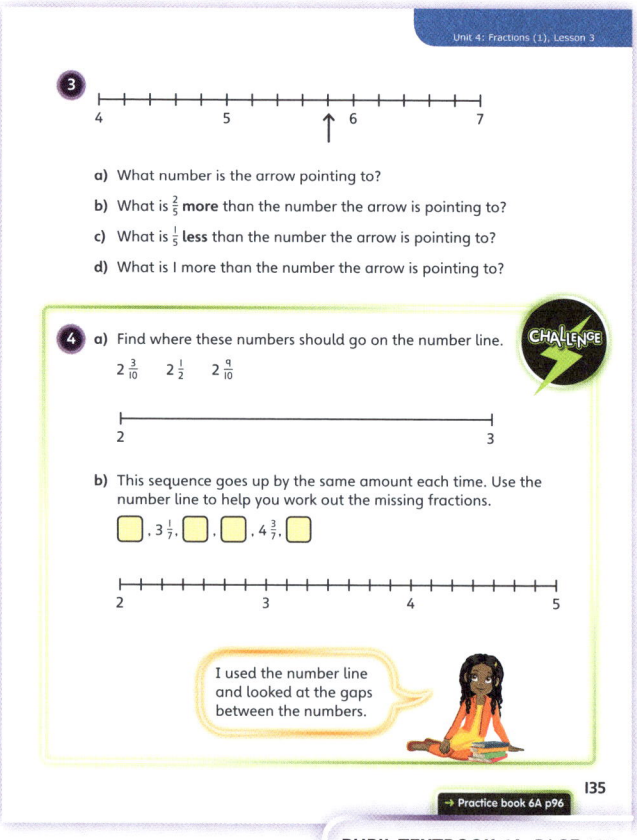

PUPIL TEXTBOOK 6A PAGE 135

Practice

WAYS OF WORKING Independent thinking

IN FOCUS Question ② develops counting backwards on a number line in fractional and whole number jumps. Even in part b), ensure that children count back in fifths rather than 1 whole, to address the common misconception of counting in whole numbers rather than fractions.

Question ③ introduces children to placing numbers on a number line where the fractional divisions are not indicated or use different denominators. Encourage children to use their prior understanding to change the improper fractions to mixed numbers and simplify as needed.

Questions ④ and ⑤ allow learners to practise finding missing numbers in a pattern. Question ⑤ addresses the misconception that a pattern cannot be going up or down by the same amount if the fractions have different denominators. Encourage children to use equivalent fractions to count up/down and to simplify fractions if needed.

In question ⑥, encourage children to look at the numbers before attempting to put them on the number line and to consider whether they need to convert the improper fraction into a mixed fraction; they should notice that $7\frac{5}{10}$ can be simplified to $7\frac{1}{2}$, so there are only three different denominators.

STRENGTHEN If children struggle with placing fractions on a number line, give them separate number lines divided into different amounts. Ensure these number lines are the same length to aid recognition.

DEEPEN To build on question ⑥, use number lines with only the start and end points marked. Deepen learning further by giving children a mixture of mixed numbers and improper fractions, some of which may need simplifying – for example, a number line with only 3 and 5 marked, for placing numbers such as $3\frac{2}{5}$, $4\frac{1}{2}$ and $\frac{76}{20}$.

ASSESSMENT CHECKPOINT Questions ⑤ and ⑥ assess children's ability to find missing fractions in a sequence using a number line. Look for children confidently using mixed numbers and improper fractions, converting between these and simplifying.

ANSWERS Answers for the **Practice** part of the lesson appear in the separate **Practice and Reflect answer guide**.

Reflect

WAYS OF WORKING Pair work

IN FOCUS Encourage children to write all the unknown numbers on the number line and to explain how they know their answer is correct. Children may need support with identifying where an eighth lies on a number line that is divided into quarters.

ASSESSMENT CHECKPOINT Look for children fluently explaining where a fraction lies on a number line, reliably converting between mixed numbers and improper fractions as needed.

ANSWERS Answers for the **Reflect** part of the lesson appear in the separate **Practice and Reflect answer guide**.

After the lesson ⏸

- Can children use a number line to calculate a fractional increase or decrease?
- Can children find missing numbers in a sequence?
- Can children place missing numbers on a number line, including lines where the fractional divisions are not marked?

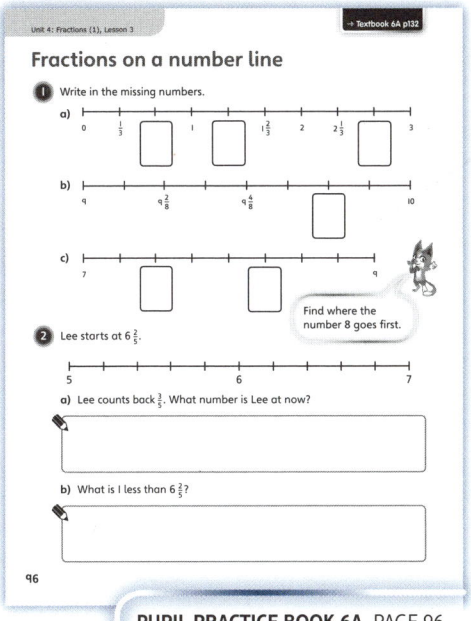

PUPIL PRACTICE BOOK 6A PAGE 96

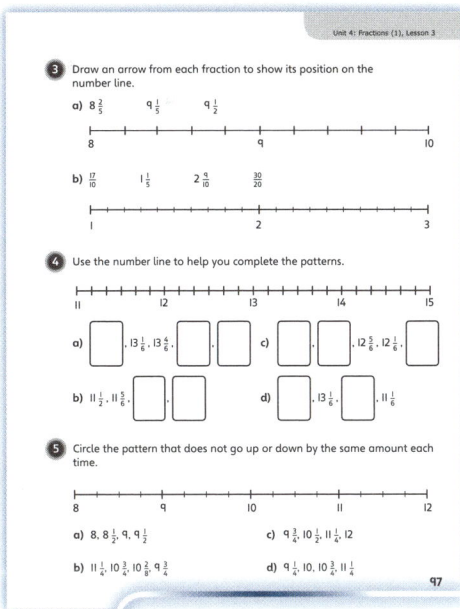

PUPIL PRACTICE BOOK 6A PAGE 97

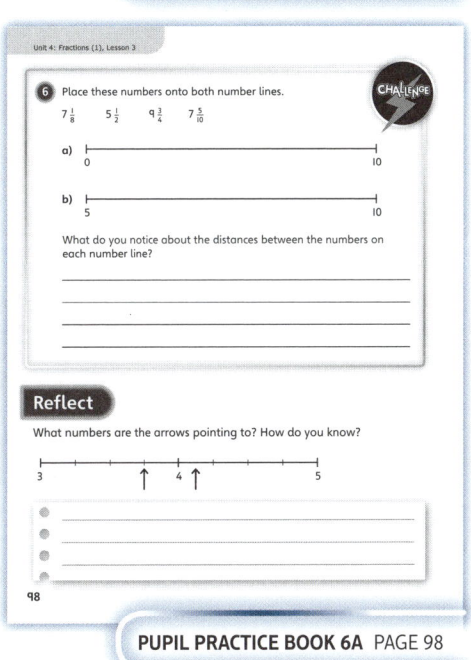

PUPIL PRACTICE BOOK 6A PAGE 98

Comparing and ordering fractions

Learning focus

In this lesson, children will use their understanding of fractions to develop their ability to compare and order fractions by making the denominators the same and comparing the numerators.

Small steps

→ Previous step: Fractions on a number line
→ **This step: Comparing and ordering fractions (1)**
→ Next step: Comparing and ordering fractions (2)

NATIONAL CURRICULUM LINKS

Year 6 Number – Fractions
· Compare and order fractions, including fractions >1.
· Use common factors to simplify fractions; use common multiples to express fractions in the same denomination.

ASSESSING MASTERY

Children can confidently compare and order more than two fractions by finding the lowest common multiple (LCM) and comparing the numerators.

COMMON MISCONCEPTIONS

Children may focus on the numbers in the fractions and compare these, rather than looking at the overall fractions. For example, when comparing $\frac{2}{3}$ and $\frac{3}{5}$ children may think $\frac{2}{3}$ is bigger because 3 and 5 are bigger than 2 and 3. Ask:
· *How can you compare the fractions, not just the digits?*
· *Do you need to find the LCM of the denominators?*
· *How can you use equivalent fractions to compare the fractions?*
· *Would a fraction strip help?*

Encourage children to realise that to compare and order fractions, they need to consider the value of the fractions rather than the individual digits, using the LCM of the denominators and equivalent fractions to make the comparison. Support children with representations on a bar model to secure understanding and avoid this misconception.

STRENGTHENING UNDERSTANDING

Children who are struggling should be encouraged to consolidate learning by comparing and ordering fractions that have the same denominator. Encourage them to show these fractions on a bar model and use the numerators to compare the fractions. Move on to compare fractions where one of the denominators is a multiple of the other, such as $\frac{1}{2}$ and $\frac{3}{4}$ or $\frac{2}{3}$ and $\frac{5}{12}$, so they can easily find the common denominator.

GOING DEEPER

Children could be challenged to find missing fractions: give children some fractions and ask them to find the missing fraction so that the fractions are in order, for example $\frac{1}{8}, \frac{1}{4}, \frac{1}{2}, \_\_, \frac{7}{8}$.

KEY LANGUAGE

In lesson: lowest common multiple, LCM, greater than, less than, order, denominator, numerator, equivalent

Other language to be used by the teacher: ascending, inequality symbol, comparison

STRUCTURES AND REPRESENTATIONS

fraction strips

 In the eTextbook of this lesson, you will find interactive links to a selection of teaching tools.

Before you teach

· Can children compare and order fractions that have the same denominator?
· Can children find the LCM of two or more numbers?
· Can children find equivalent fractions?

Discover

WAYS OF WORKING Pair work

ASK

- Question **1** a): *How many people wear glasses in each group? Out of how many? How can you write this as a fraction?*
- Question **1** a): *Could you use a model to help you compare the fractions?*
- Question **1** b): *Which fraction do you need to compare $\frac{2}{3}$ with, if you are looking to find the greatest fraction of people who wear glasses?*

IN FOCUS Question **1** a) and b) will introduce children to comparing fractions. In question **1** a) one denominator is a multiple of the other (related fractions) whereas in question **1** b) the denominators are unrelated, hence, children are required to use the LCM to find equivalent fractions to compare.

PRACTICAL TIPS This situation could be recreated in a class setting using children from the class. Children could be asked to line up in groups, with children wearing glasses on one side and children not wearing glasses on the other side, and then to recreate a bar model. However, it may be best to do this after children have tried the problem themselves, to reinforce the concepts introduced.

ANSWERS

Question **1** a): $\frac{3}{4}$ is bigger than $\frac{5}{8}$ so group A has a greater fraction of people who wear glasses.

Question **1** b): $\frac{3}{4} > \frac{2}{3}$ so group A has the greatest fraction of people who wear glasses.

Share

WAYS OF WORKING Whole class teacher led

ASK

- Question **1** a): *What do you need to do to the denominators in order to compare the fractions?*
- Question **1** b): *Which fractions do you need to compare now? Why don't you need to compare all three?*
- Question **1** b): *How can you compare these fractions now? What is the lowest common multiple of 4 and 3?*

IN FOCUS Ensure that children realise they need to make the denominators the same in order to compare the fractions. Question **1** a) introduces children to comparing related fractions. Question **1** b) then gives children a third, unrelated, fraction and asks them to compare it with the first two fractions, using the LCM to find equivalent fractions. Explore with children why they need to make the denominators the same and how to do this.

In question **1** b), guide the children to decide which fractions they need to compare and ensure they understand why it is not necessary to compare all three. Emphasise the need to find the LCM of 4 and 3, and remind children to use the inequality symbol.

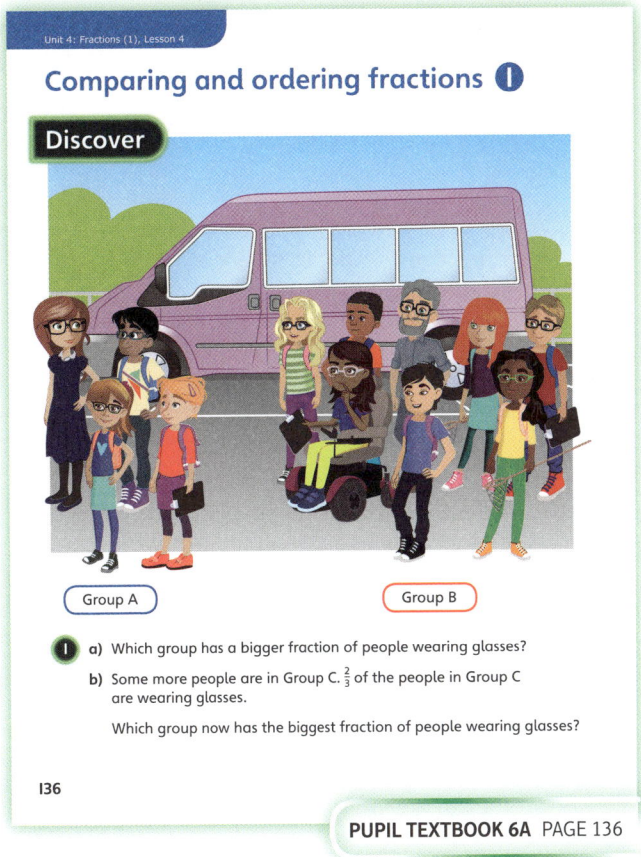

Comparing and ordering fractions ❶

Discover

Group A   Group B

❶ a) Which group has a bigger fraction of people wearing glasses?

b) Some more people are in Group C. $\frac{2}{3}$ of the people in Group C are wearing glasses.

Which group now has the biggest fraction of people wearing glasses?

136

PUPIL TEXTBOOK 6A PAGE 136

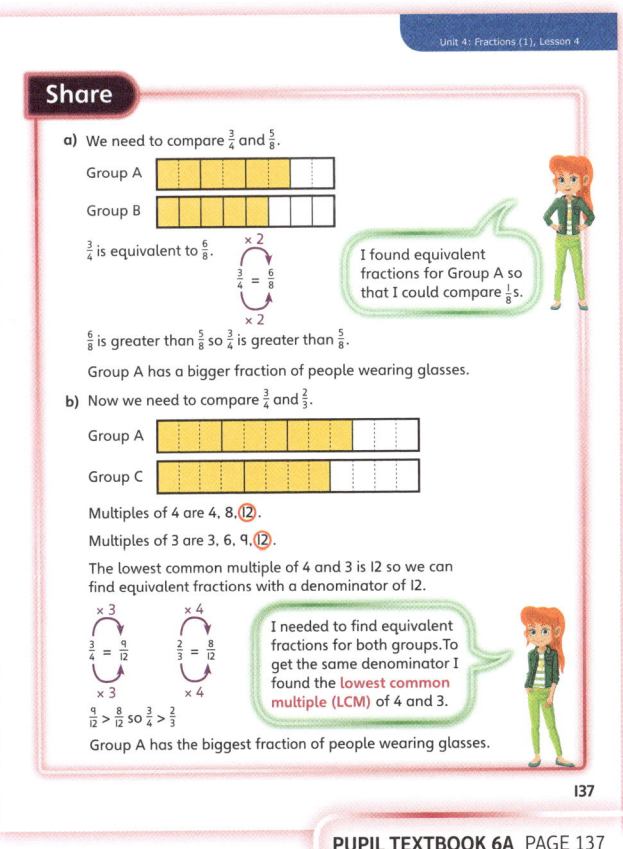

Share

a) We need to compare $\frac{3}{4}$ and $\frac{5}{8}$.

Group A

Group B

$\frac{3}{4}$ is equivalent to $\frac{6}{8}$. $\frac{3}{4} \overset{\times 2}{\underset{\times 2}{=}} \frac{6}{8}$

I found equivalent fractions for Group A so that I could compare $\frac{1}{8}$s.

$\frac{6}{8}$ is greater than $\frac{5}{8}$ so $\frac{3}{4}$ is greater than $\frac{5}{8}$.

Group A has a bigger fraction of people wearing glasses.

b) Now we need to compare $\frac{3}{4}$ and $\frac{2}{3}$.

Group A

Group C

Multiples of 4 are 4, 8, ⑫.

Multiples of 3 are 3, 6, 9, ⑫.

The lowest common multiple of 4 and 3 is 12 so we can find equivalent fractions with a denominator of 12.

$\frac{3}{4} \overset{\times 3}{\underset{\times 3}{=}} \frac{9}{12}$ $\frac{2}{3} \overset{\times 4}{\underset{\times 4}{=}} \frac{8}{12}$

*I needed to find equivalent fractions for both groups. To get the same denominator I found the **lowest common multiple (LCM)** of 4 and 3.*

$\frac{9}{12} > \frac{8}{12}$ so $\frac{3}{4} > \frac{2}{3}$

Group A has the biggest fraction of people wearing glasses.

137

PUPIL TEXTBOOK 6A PAGE 137

Think together

Whole class teacher led (I do, We do, You do)

ASK

- Question **1**: *What part of the fraction helps you to compare once the denominators are the same?*
- Question **2**: *How can you find the lowest common multiple of the denominators?*

IN FOCUS Question **1** gradually reduces scaffolding to enable children to work out the answer without the support of a fraction strip. Children should be encouraged to use inequality symbols when making comparisons and to use the abbreviation LCM.

Question **2** asks children to order three or more fractions, and moves more into the abstract. Discuss Astrid's suggestion – would this be a good way to answer the question? Encourage children to find the LCM of the denominators and then look at the numerators to order the fractions.

STRENGTHEN To support understanding, children can represent the fractions on a bar model. Encourage children to write out the multiples of the denominators so they can find the LCM.

DEEPEN Explore question **1** further by grouping children in the class and finding the fractions of children who wear glasses, have a pet, etc. Deepen understanding of finding the LCM with problems that require them to do more than simply multiply the denominators, for example comparing $\frac{5}{6}$ and $\frac{3}{4}$, where the LCM is 12 not 24 (6 × 4 = 24).

Deepen learning in question **3** by challenging children to explore Flo's point and find all the possible answers, encouraging them to use improper fractions where appropriate.

ASSESSMENT CHECKPOINT Question **1** assesses children's ability to compare two fractions by finding the LCM. Questions **2** and **3** develop this further, providing an opportunity to assess children's ability to compare three or more fractions by finding the LCM and comparing the numerators.

ANSWERS

Question **1** a): The LCM of 6 and 3 is 6; $\frac{2}{3} = \frac{4}{6}$; $\frac{5}{6} > \frac{4}{6}$ so $\frac{5}{6} > \frac{2}{3}$
Class A has a greater fraction of children with brown hair.

Question **1** b): The LCM of 2 and 8 is 8, $\frac{1}{2} = \frac{4}{8}$; $\frac{1}{2} > \frac{3}{8}$; Class A has a greater fraction of children with a pet.

Question **1** c): The LCM of 5 and 3 is 15; $\frac{3}{5} = \frac{9}{15}$; $\frac{2}{3} = \frac{10}{15}$; $\frac{3}{5} < \frac{2}{3}$; Class B has a greater fraction of girls.

Question **2** a): $\frac{1}{2}, \frac{2}{3}, \frac{5}{6}$

Question **2** b): $\frac{5}{12}, \frac{1}{2}, \frac{3}{4}, \frac{5}{6}$

Question **2** c): $\frac{3}{10}, \frac{4}{5}, \frac{49}{50}, \frac{99}{100}, \frac{4}{4}$

Question **3** a): a) 5 or greater; b) 7 or greater; c) 1;
d) 3 or greater in first box, 1, 2, 3 or 4 in second box; e) 3 or greater;
f): Various answers – for example, 4 in 1st box and 3 in 2nd box.

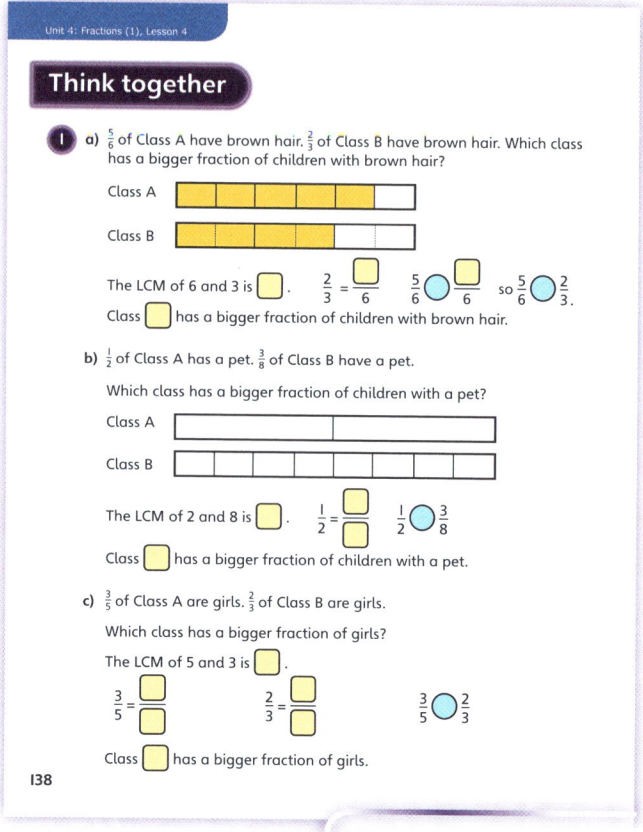

PUPIL TEXTBOOK 6A PAGE 138

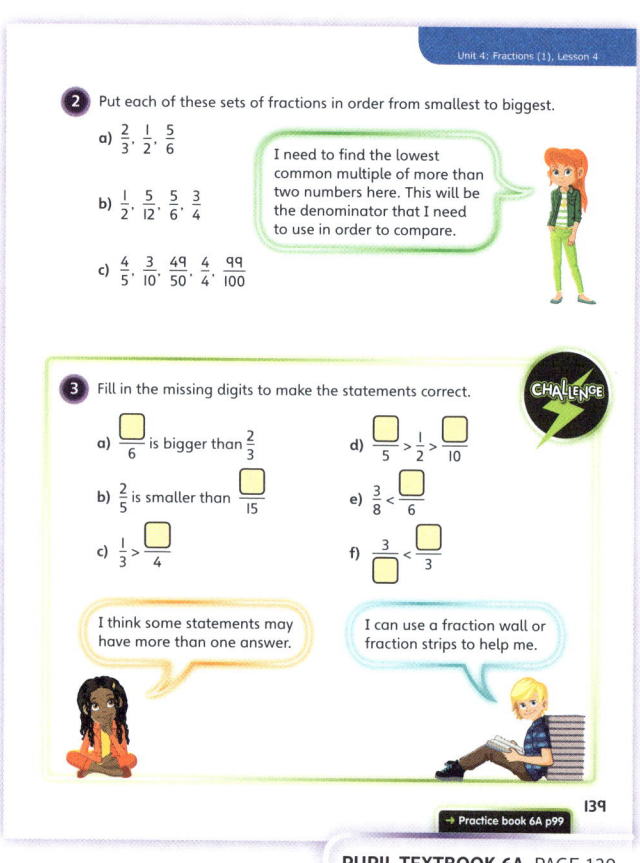

PUPIL TEXTBOOK 6A PAGE 139

Practice

WAYS OF WORKING Independent thinking

IN FOCUS Question ❶ aims to consolidate children's understanding of comparing fractions by finding the LCM. The question progresses from support and structure for question ❶ a) – where one denominator is a multiple of the other – to requiring children to follow and write out the method independently by question ❶ d).

Question ❸ uses a visual representation of fractions and asks children to identify fractions and then put them in order. This procedural variation ensures they can apply the skill of comparing fractions to a new context. Encourage children to find the LCM and not to rely on the diagrams to order the fractions. This is further developed in Question ❹, which moves to the abstract with no pictorial representations given.

STRENGTHEN When comparing fractions, encourage children to represent the fractions on a bar model. If necessary, use the structure from question ❶ for other questions. Ask: *How can you represent the fractions on a bar model? How many parts does each model need to be divided into? What part of the fraction does this relate to?*

DEEPEN Deepen learning in question ❻ by exploring with children whether they can come up with more than one answer.

THINK DIFFERENTLY Question ❺ develops children's understanding of comparing fractions in a problem-solving context. Encourage children to explain their answers, to develop their written reasoning.

ASSESSMENT CHECKPOINT Question ❶ provides an opportunity to assess children's ability to compare two fractions by finding the LCM.

Questions ❹ and ❻ develop this further and will assess children's ability to compare and order more than two fractions by finding the LCM. Look for children confidently using known multiplication facts to find multiples and demonstrating understanding that the LCM is not always the denominators multiplied together.

ANSWERS Answers for the **Practice** part of the lesson appear in the separate **Practice and Reflect answer guide**.

Reflect

WAYS OF WORKING Pair work

IN FOCUS This **Reflect** activity provides an opportunity to check children's understanding of comparing fractions where the numerators are the same, and encourages children to write an explanation in their own words. Discuss which method is most efficient when comparing fractions that have the same numerator. Children should be able to explain that if the numerators are the same, the smaller the denominator, the bigger the fraction.

ASSESSMENT CHECKPOINT Look for children confidently explaining how to compare fractions that have the same numerators.

ANSWERS Answers for the **Reflect** part of the lesson appear in the separate **Practice and Reflect answer guide**.

After the lesson ⏸

- Can children compare fractions that have different denominators?
- Can children find the LCM?
- Do children understand that to order fractions they need to find the LCM and compare the numerators?

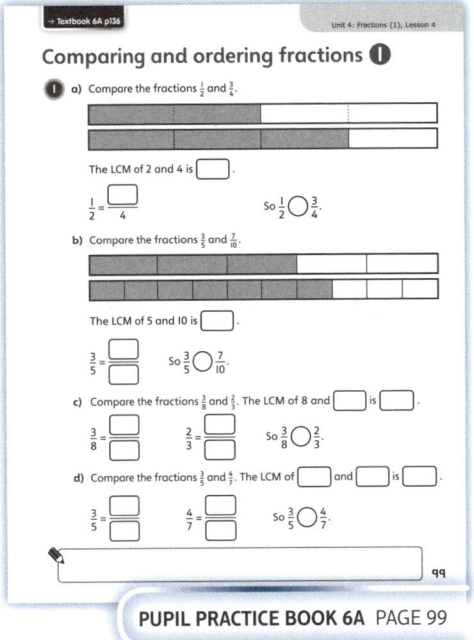

PUPIL PRACTICE BOOK 6A PAGE 99

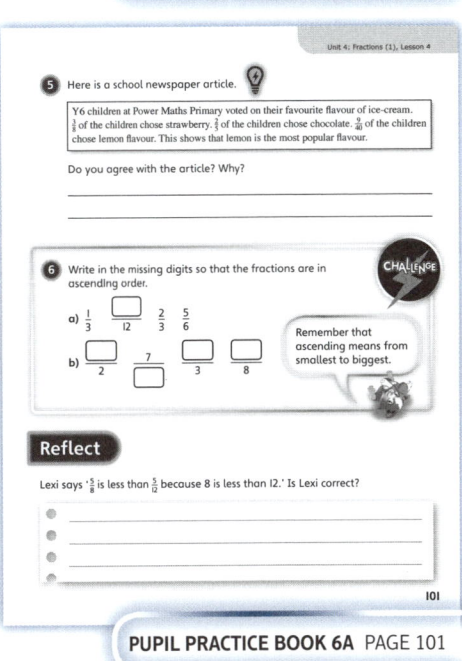

PUPIL PRACTICE BOOK 6A PAGE 100

PUPIL PRACTICE BOOK 6A PAGE 101

Comparing and ordering fractions ②

Learning focus

In this lesson, children develop their understanding of comparing and ordering mixed numbers and improper fractions by converting between improper fractions and mixed numbers and using a common denominator.

Small steps

→ Previous step: Comparing and ordering fractions (1)
→ **This step: Comparing and ordering fractions (2)**
→ Next step: Adding and subtracting fractions (1)

NATIONAL CURRICULUM LINKS

Year 6 Number – Fractions
• Compare and order fractions, including fractions >1.
• Use common factors to simplify fractions; use common multiples to express fractions in the same denomination.

ASSESSING MASTERY

Children can reliably compare and order mixed numbers and improper fractions, confidently converting between mixed numbers and improper fractions and finding common denominators while showing fluency in using inequality symbols.

COMMON MISCONCEPTIONS

Children may focus on the digits in the fractions and compare these rather than looking at the overall mixed number or improper fraction. For example, when comparing $1\frac{2}{3}$ and $\frac{11}{7}$ children may think $\frac{11}{7}$ is greater because 11 and 7 are greater than 1, 2 and 3. Encourage children to understand that to compare and order mixed numbers and improper fractions, it is important to consider the size of the fractions rather than the individual digits in the fractions. Ask:
• *Would converting the improper fraction to a mixed number help?*
• *How can you make the denominators the same?*
• *What is the lowest common multiple of the denominators?*

Another misconception is that children may compare the fraction parts without considering the whole number parts, for example, when comparing $5\frac{1}{5}$ and $4\frac{2}{3}$ children may think $4\frac{2}{3}$ is greater because $\frac{2}{3}$ is greater than $\frac{1}{5}$. To overcome this stumbling block, encourage children to look at the whole number first when comparing mixed numbers.

STRENGTHENING UNDERSTANDING

Children who are struggling to compare and order mixed numbers and improper fractions should firstly recap comparing and ordering proper fractions by finding a common multiple, as in the previous lesson. Children may need a detailed recap on converting improper fractions to mixed numbers to support their understanding.

GOING DEEPER

Children could be encouraged to solve missing number problems involving fractions to deepen understanding. Give children some missing number problems and ask them to find the missing number so that the statement is correct, for example, ask children to find the missing number in the number sentence $\frac{\square}{7} > 1\frac{3}{4}$.

KEY LANGUAGE

In lesson: lowest common denominator, denominator, greater than

Other language to be used by the teacher: numerator, equivalent, mixed number, improper fraction, less than

STRUCTURES AND REPRESENTATIONS

fraction strips, number lines

 In the eTextbook of this lesson, you will find interactive links to a selection of teaching tools.

Before you teach

• Can children compare and order proper fractions?
• Can children find the LCM of two or more numbers?
• Can children convert between improper fractions and mixed numbers?

Discover

Pair work

ASK

• Question **1** a): *What is the same and what is different about the fractions?*
• Question **1** a): *What do you need to look at to compare the fractions?*
• Question **1** b): *How can you compare an improper fraction and a mixed number more easily?*

IN FOCUS This question introduces children to the comparison of mixed numbers and improper fractions in the context of feeding animals. It will be important to ensure that children are confident with converting between mixed and improper fractions.

PRACTICAL TIPS The context in this lesson is feeding animals. This could be made into a practical activity, where children discuss how much their animals eat and compare their fractions, perhaps creating a poster to illustrate the results.

ANSWERS

Question **1** a): $1\frac{2}{3} > 1\frac{3}{5}$ so no, Bella's cat does not eat more than Jamie's cat each day.

Question **1** b): $\frac{11}{7} < 1\frac{3}{5}$ so no, Ebo's cat does not eat more than Bella's.

PUPIL TEXTBOOK 6A PAGE 140

Share

Whole class teacher led

ASK

• Question **1** a): *Do you need to compare the whole numbers? Why?*
• Question **1** a): *What is the LCM of 3 and 5?*
• Question **1** b): *What type of fraction is $\frac{11}{7}$? How can you compare these numbers?*
• Question **1** b): *Why can you just compare the fraction parts?*

IN FOCUS Question **1** a) introduces children to comparing mixed numbers, where the whole number is the same for each mixed number and the fractions are unrelated. Encourage children to use the fraction strips to support their reasoning but ensure they do not rely on these to solve the problem. Emphasise that because both cats eat one whole pouch a day it is only necessary to compare the fraction parts.

Question **1** b) is an opportunity to compare an improper fraction with a mixed number. It may be worth discussing an alternative approach – to convert the mixed number to an improper fraction before comparing the improper fractions to find the answer.

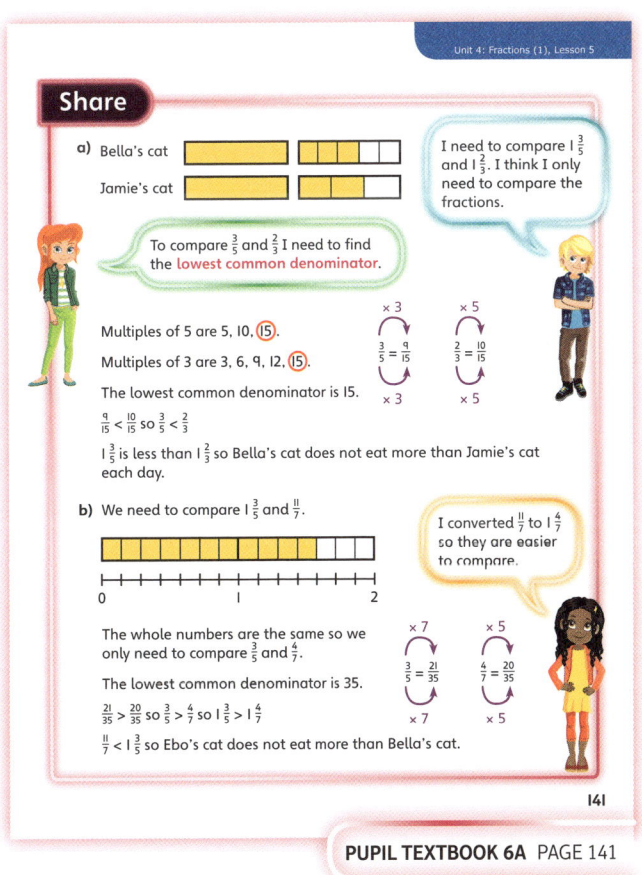

PUPIL TEXTBOOK 6A PAGE 141

Think together

WAYS OF WORKING Whole class teacher led (I do, We do, You do)

ASK

- Question **1**: *Can you find the common denominators? What do you compare once the denominators are the same? Which inequality sign will you use?*
- Question **2**: *Can you convert the improper fraction to a mixed number? Can you find the common denominators of all the fractions? Would simplifying any of the fractions help find a common denominator?*
- Question **3**: *Does comparing the whole numbers help? Can you make the denominators on some of the fractions the same to help you compare them? How can you make the fraction on the final card greater than $2\frac{3}{5}$?*

IN FOCUS Question **1** progresses from finding an equivalent fraction with support, to converting an improper fraction to a mixed number, and then to finding an equivalent fraction with reduced support.

Question **2** asks children to compare and order four numbers with no support. Encourage them to convert the improper fraction to a mixed number. Can they simplify one of the fractions to help find a common denominator?

STRENGTHEN Encourage children to represent the fractions on a bar model. Ask them to change any improper fractions to mixed numbers and to write out the multiples of the denominators so they can find the common denominator of the fractions.

DEEPEN Question **3** can be explored further by giving children another mixed number or an improper fraction and asking them to decide which of the fraction cards are greater/smaller.

ASSESSMENT CHECKPOINT In question **1**, check whether children rely on the fraction strips to make their comparisons or use the models for support to find the lowest common denominators.

Questions **2** and **3** will assess children's ability to compare more than three mixed numbers or improper fractions in abstract form with no pictorial support. Children should fluently convert between mixed numbers and improper fractions and find the lowest common denominator to compare fractions, while confidently using inequality symbols. Look for children being able to explain their steps and their reasoning.

ANSWERS

Question **1** a): $\frac{3}{4} = \frac{9}{12}$, so $2\frac{3}{4} = 2\frac{9}{12}$ and $2\frac{3}{4} > 2\frac{7}{12}$; Lexi's hamster eats more each week.

Question **1** b): $\frac{27}{8} = 3\frac{3}{8}$; Roxy eats $3\frac{3}{8}$ bags of carrots per week.

Question **1** c): $3\frac{1}{2} = 3\frac{4}{8}$; $3\frac{3}{8} < 3\frac{4}{8}$; Mia the horse eats more carrots per week.

Question **2**: $5\frac{10}{21}$ is the largest.

In ascending order: $\frac{36}{7}, 5\frac{2}{7}, 5\frac{6}{14}, 5\frac{10}{21}$

Question **3**: $4\frac{1}{8}, \frac{21}{4}, 2\frac{2}{3}$ are all greater than $2\frac{3}{5}$.
To make the last card greater, the numerator would need to be 13 or greater.

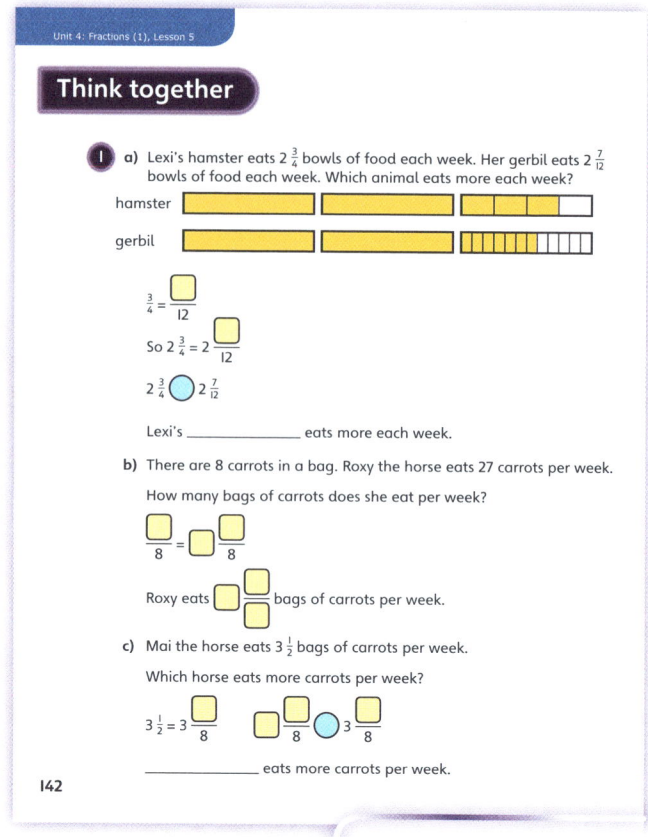

PUPIL TEXTBOOK 6A PAGE 142

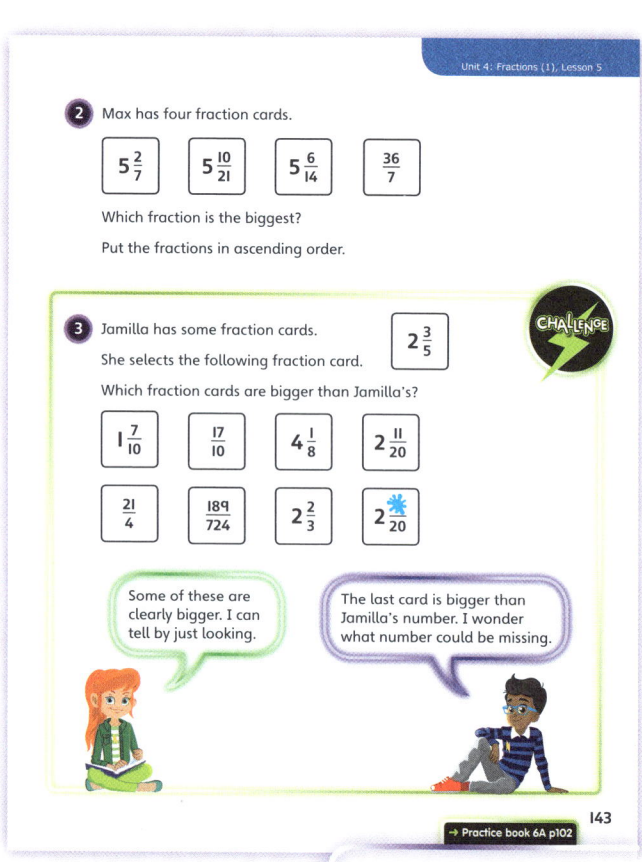

PUPIL TEXTBOOK 6A PAGE 143

Practice

WAYS OF WORKING Independent thinking

IN FOCUS Question **1** develops children's understanding of comparing two mixed numbers or two improper fractions by finding a common denominator, while using prompts to encourage children to look for the lowest common multiple of the denominators.

In question **2** b), children will compare two mixed numbers that have different whole numbers; this is designed to address the common misconception of focusing on the fractions while ignoring the whole numbers.

STRENGTHEN If children are finding it difficult to compare the improper fractions with mixed numbers, remind them to convert the improper fractions to mixed numbers first and compare the whole numbers. Ask: *How can you convert the improper fractions? How can you make the denominators the same? Would using a model help?*

DEEPEN Question **4** could be extended to deepen learning by encouraging the children to order groups of fractions, where some of the digits are missing, for example, order from smallest to biggest $\frac{10}{3}$, $\Box\frac{\Box}{8}$, $3\frac{7}{12}$.

Question **5** can be explored further by challenging children to find a number that is closer to 4 than those listed.

THINK DIFFERENTLY In question **5** there is no prompting or visual representation and children are required to interpret the question and realise that they are being asked to compare the fractions using the lowest common denominator. Encourage children to explain their answer.

ASSESSMENT CHECKPOINT Question **1** will assess children's ability to compare a combination of mixed and improper fractions with prompting. Look for children showing fluency with finding the LCM and confidently using inequality signs.

In questions **4** and **5**, look for children being able to explain their steps and their reasoning.

ANSWERS Answers for the **Practice** part of the lesson appear in the separate **Practice and Reflect answer guide**.

Reflect

WAYS OF WORKING Pair work

IN FOCUS This **Reflect** activity gives children an opportunity to describe the steps needed to compare a mixed number and improper fraction, encouraging children to use their own method and write their explanation in their own words.

ASSESSMENT CHECKPOINT Children can confidently explain how to compare a mixed number and an improper fraction, breaking this into the steps of converting the improper fraction to a mixed number (or vice versa), finding the lowest common denominator to compare fractions, and using inequality symbols in their final answer. Look for children confidently explaining their steps and their reasoning in their own words.

ANSWERS Answers for the **Reflect** part of the lesson appear in the separate **Practice and Reflect answer guide**.

After the lesson ⏸

- Can children compare and order improper fractions?
- Can children compare and order mixed numbers?

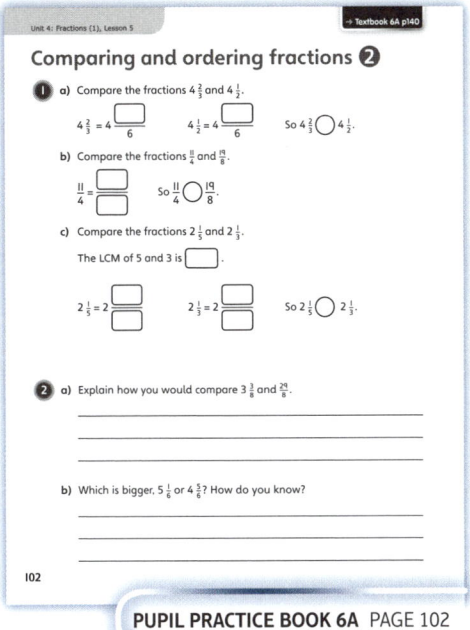

PUPIL PRACTICE BOOK 6A PAGE 102

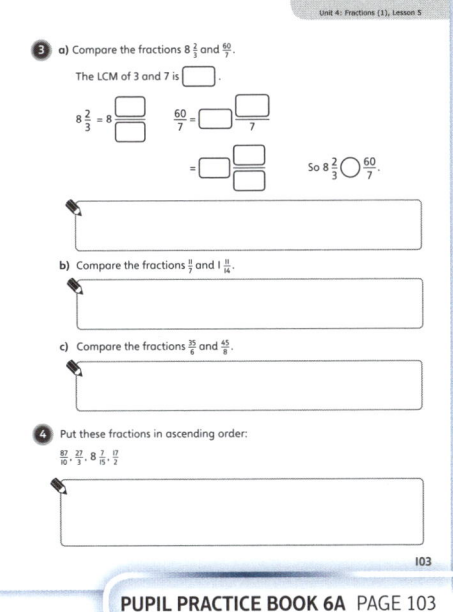

PUPIL PRACTICE BOOK 6A PAGE 103

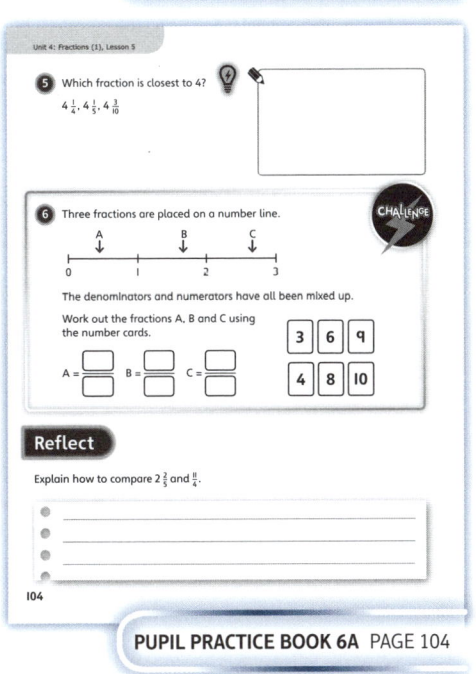

PUPIL PRACTICE BOOK 6A PAGE 104

Adding and subtracting fractions ❶

Learning focus

In this lesson, children link their prior knowledge of finding equivalent fractions with common denominators to adding and subtracting fractions where the answer is between 0 and 1.

Small steps

→ Previous step: Comparing and ordering fractions (2)
→ **This step: Adding and subtracting fractions (1)**
→ Next step: Adding and subtracting fractions (2)

NATIONAL CURRICULUM LINKS

Year 6 Number – Fractions

Add and subtract fractions with different denominators and mixed numbers, using the concept of equivalent fractions.

ASSESSING MASTERY

Children can fluently add and subtract fractions by using a common multiple to create equivalent fractions with a common denominator and then adding and subtracting the numerators, explaining the steps in their own words.

COMMON MISCONCEPTIONS

Children may simply add or subtract the numbers in the fractions, for example, children may give an answer of $\frac{2}{7}$ to the calculation $\frac{1}{3} + \frac{1}{4}$. It will be beneficial to use bar models to secure understanding and avoid this misconception, and to encourage children to understand that they can only add and subtract fractions that have the same denominator.

Another common mistake is that children may find a common denominator but forget to convert the numerators; for example, children may rewrite the calculation $\frac{3}{4} + \frac{1}{5}$ as $\frac{3}{20} + \frac{1}{20}$ to get an answer of $\frac{4}{20}$. Ask:

- *Can you find the lowest common multiple of the denominators?*
- *What do you do to the denominator and the numerator?*
- *Can you write down the equivalent fractions before carrying out the calculation?*

STRENGTHENING UNDERSTANDING

If children are struggling with adding and subtracting fractions, strengthen understanding by adding and subtracting fractions with the same denominator using bar models. Next, encourage children to add and subtract fractions where one of the denominators is a multiple of the other, so they can easily find the common denominator. Explain that the denominators need to be the same when adding and subtracting fractions and encourage children to write out the equivalent fractions before adding or subtracting.

GOING DEEPER

Children can be encouraged to solve number problems finding missing fractions, for example, $\frac{\Box}{\Box} + \frac{2}{5} = \frac{24}{35}$.

KEY LANGUAGE

In lesson: common denominator, equivalent, difference, lowest common denominator

Other language to be used by the teacher: numerator

STRUCTURES AND REPRESENTATIONS

bar models, number lines

RESOURCES

Optional: cooking cups, fraction strips

 In the eTextbook of this lesson, you will find interactive links to a selection of teaching tools.

Before you teach

- Can children add/subtract fractions that have the same denominator?
- Can children find the lowest common multiple of two or more numbers?
- Can children find equivalent fractions?

Discover

WAYS OF WORKING Pair work

ASK

- Question ❶ a): *What fraction of the hay does Hattie eat in the morning? What fraction of the hay does she eat in the evening? How can you work out how much Hattie eats in the whole day? What calculation do you need to do?*
- Question ❶ b): *How much does Hattie eat in a day? How can you work out $\frac{1}{4}$ less? What calculation do you need to do?*

IN FOCUS Question ❶ introduces children to adding and subtracting fractions, linking with their knowledge of finding equivalent fractions to convert two fractions so they have a common denominator before completing the calculation.

PRACTICAL TIPS There are many ways in which fractions can be added in a practical context, for instance, cooking using cups. Encourage children to use fraction strips so they have pictorial representations for support.

ANSWERS

Question ❶ a): Hattie eats $\frac{5}{6}$ of a bale of hay in a day.

Question ❶ b): Molly eats $\frac{7}{12}$ of a bale of hay.

Share

WAYS OF WORKING Whole class teacher led

ASK

- Question ❶ a): *How can you add $\frac{2}{3}$ and $\frac{1}{6}$? How do the bar models help?*
- Question ❶ a): *Can you find a common denominator? Will finding the lowest common multiple help?*
- Question ❶ a): *What fraction is equivalent to $\frac{2}{3}$ with a denominator of 6? Why is it necessary to add the numerators and keep the denominator as 6?*
- Question ❶ b): *If Molly eats $\frac{1}{4}$ less, what calculation do you need to do? What is the common denominator? How much does Molly eat in a day?*

IN FOCUS With this question, it is important that children recognise the need to make the denominators the same using the LCM and then to add or subtract the numerators. Encourage children to use the bar models so they have a visual representation of why they need to make the denominators the same in order to add or subtract the fractions. Explore with children how to make the denominators the same using the bar models, and highlight the conversion calculations shown. It will be beneficial to discuss the additions and subtractions again, referring to the bar models to give pictorial support.

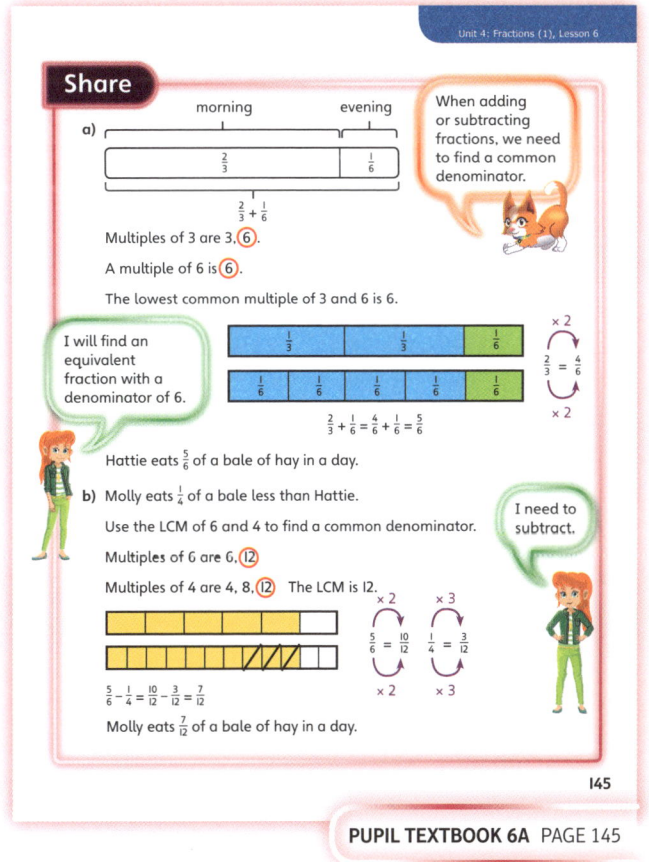

177

Think together

WAYS OF WORKING Whole class teacher led (I do, We do, You do)

ASK

- Question 1: *Can you find the lowest common multiple of each number? What part of the fraction do you add or subtract when adding and subtracting fractions?*
- Question 2: *Can you find the lowest common multiple of the denominators?*
- Question 3: *Which questions involve addition? Which questions involve subtraction?*

IN FOCUS Questions 1 and 2 develop the concept of adding and subtracting fractions using the LCM to find common denominators. The questions progress from fractions where one denominator is a multiple of the other, with support, diagrams and a context, to adding and subtracting fractions in an abstract way with prompting.

Question 3 introduces problem solving in the form of missing number pyramid problems. In these pyramids, the two fractions at the bottom are added together to give the fraction at the top. Children are required to decide which digits are needed to make the addition pyramid correct. In some puzzles, children are required to work backwards and use the inverse operation of subtraction; explore with the children which pyramids involve subtraction.

STRENGTHEN To strengthen understanding, encourage children to represent the calculations with bar models. Clearly break down the calculations into steps, by asking: *What do you need to do first? What do you need to make the same before you can add or subtract?*

DEEPEN Question 2 can be explored further to deepen learning by giving children two more fractions and asking them to find the total and the difference. This will reinforce the importance of subtracting the smaller fraction from the larger fraction to find the difference.

ASSESSMENT CHECKPOINT Questions 1 and 2 assess children's ability to add and subtract fractions using the LCM to find a common denominator and adding/subtracting the numerators with prompting. Children who rely on the bar models to make the calculations may need more support.

Questions 2 and 3 assess children's ability to add and subtract fractions in the abstract. Children who accurately complete question 3 are likely to have mastered this topic.

ANSWERS

Question 1 a): Hector eats $\frac{7}{8}$ of a bale of hay in a day.

Question 1 b): Scoobie eats $\frac{1}{9}$ of a bale of hay.

Question 2: $\frac{1}{6} + \frac{3}{8} = \frac{13}{24}$

Question 3 a): $\frac{13}{20}$

Question 3 b): $\frac{1}{6}$

Question 3 c): For example: $\frac{1}{3}$ and $\frac{23}{24}$; other answers such as $\frac{2}{3}$ and $\frac{31}{24}$ etc.

Question 3 d): $\frac{4}{5}$ in second row; $\frac{9}{100}$ in left-hand box of bottom row; $\frac{71}{100}$ in right-hand box of bottom row. (Children may give equivalent fractions.)

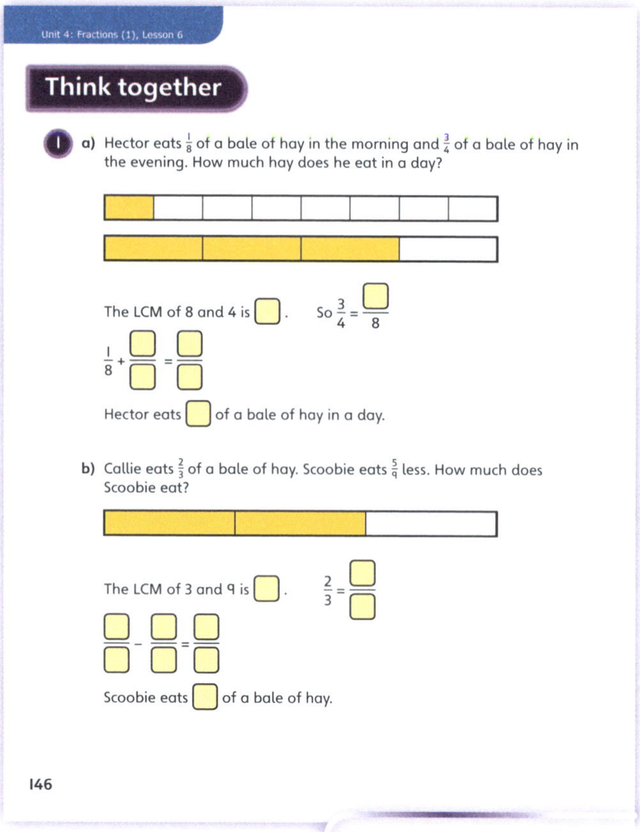
Unit 4: Fractions (1), Lesson 6

Think together

1 a) Hector eats $\frac{1}{8}$ of a bale of hay in the morning and $\frac{3}{4}$ of a bale of hay in the evening. How much hay does he eat in a day?

The LCM of 8 and 4 is ☐. So $\frac{3}{4} = \frac{☐}{8}$

$\frac{1}{8} + \frac{☐}{☐} = \frac{☐}{☐}$

Hector eats ☐ of a bale of hay in a day.

b) Callie eats $\frac{2}{3}$ of a bale of hay. Scoobie eats $\frac{5}{9}$ less. How much does Scoobie eat?

The LCM of 3 and 9 is ☐. $\frac{2}{3} = \frac{☐}{☐}$

$\frac{☐}{☐} - \frac{☐}{☐} = \frac{☐}{☐}$

Scoobie eats ☐ of a bale of hay.

146

PUPIL TEXTBOOK 6A PAGE 146

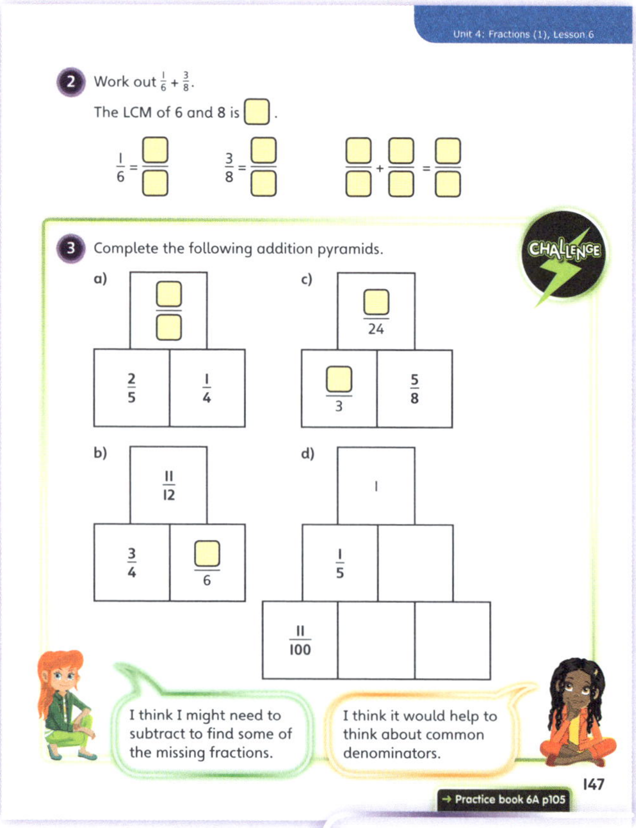
Unit 4: Fractions (1), Lesson 6

2 Work out $\frac{1}{6} + \frac{3}{8}$.

The LCM of 6 and 8 is ☐.

$\frac{1}{6} = \frac{☐}{☐}$ $\frac{3}{8} = \frac{☐}{☐}$ $\frac{☐}{☐} + \frac{☐}{☐} = \frac{☐}{☐}$

3 Complete the following addition pyramids.

CHALLENGE

147

→ Practice book 6A p105

PUPIL TEXTBOOK 6A PAGE 147

Practice

WAYS OF WORKING Independent thinking

IN FOCUS Questions ❶ and ❷ develop children's understanding of adding and subtracting fractions using the LCM and equivalent fractions.

Question ❸ addresses the misconception of adding the numerators together and then adding the denominators together, rather than using the LCM to find a common denominator and then using equivalent fractions.

Questions ❹ and ❺ introduce children to adding and subtracting fraction problems in abstract form and then in words, with no scaffolding. Encourage children to show their method for these questions.

STRENGTHEN If children are struggling with Questions ❹ and ❺, or with adding and subtracting fractions in general, encourage them to represent the fractions on a bar model and to work through the steps systematically, writing down the multiples to find the LCM.

DEEPEN Question ❺ can be explored with more worded questions, for example, 'I think of a fraction and add $\frac{3}{4}$. My answer is $\frac{11}{12}$. What fraction was I thinking of?' Children could also work in pairs, setting each other questions.

THINK DIFFERENTLY Question ❻ requires children to add fractions and to decide if the answer makes 1 whole. Explore with children what the fraction will need to look like if it is over 1 whole.

ASSESSMENT CHECKPOINT Questions ❶ and ❷ assess children's ability to add and subtract fractions using the LCM to find equivalent fractions. It will be useful to observe whether children rely on bar models to make the calculations, or if they are struggling to find the LCM and use equivalent fractions; if so, they may need more support.

Question ❸ provides an opportunity to assess children's understanding of the steps needed to complete the calculations. This is an important question dealing with a common misconception and will indicate if children need more support.

ANSWERS Answers for the **Practice** part of the lesson appear in the separate **Practice and Reflect answer guide**.

Reflect

WAYS OF WORKING Pair work

IN FOCUS This **Reflect** activity addresses the common misconception of finding the LCM of the denominators but leaving the numerators the same. Encourage children to identify and correct the mistake.

ASSESSMENT CHECKPOINT Children should be able to explain that the denominators have been changed to a common denominator but the numerators have incorrectly been left the same. Look for children who can identify the mistake and can also correctly calculate the answer using the LCM to find equivalent fractions before adding.

ANSWERS Answers for the **Reflect** part of the lesson appear in the separate **Practice and Reflect answer guide**.

After the lesson ⏸

- Can children convert two fractions to equivalent fractions with the same denominator?
- Can children add and subtract fractions that have different denominators, without support such as bar models?
- Can children explain the steps for adding and subtracting fractions in their own words?

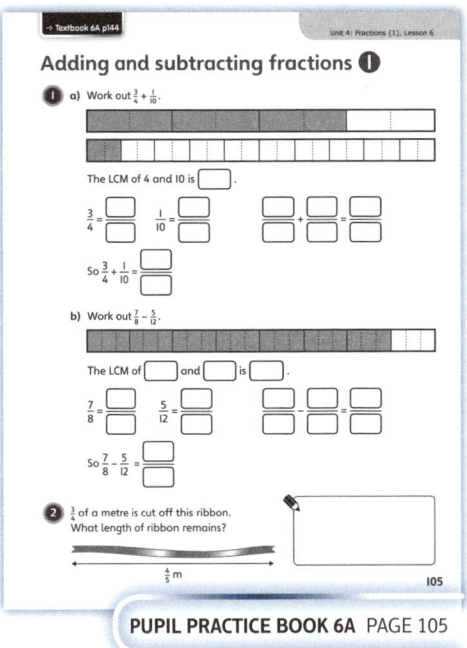

PUPIL PRACTICE BOOK 6A PAGE 105

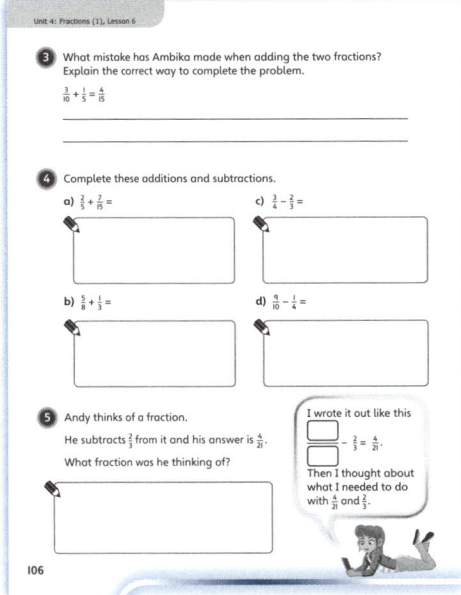

PUPIL PRACTICE BOOK 6A PAGE 106

PUPIL PRACTICE BOOK 6A PAGE 107

Adding and subtracting fractions ②

Learning focus

In this lesson, children understand how to add and subtract mixed numbers where the fractional answer is between 0 and 1 and does not cross the whole.

Small steps

→ Previous step: Adding and subtracting fractions (1)
→ **This step: Adding and subtracting fractions (2)**
→ Next step: Adding fractions

NATIONAL CURRICULUM LINKS

Year 6 Number – Fractions

Add and subtract fractions with different denominators and mixed numbers, using the concept of equivalent fractions.

ASSESSING MASTERY

Children can fluently add and subtract mixed numbers using equivalent fractions by adding/subtracting the wholes and adding/subtracting the fractional parts, while explaining the steps clearly in their own words.

COMMON MISCONCEPTIONS

Children may simply add or subtract all the numbers – for example, in the calculation $3\frac{2}{5} + 1\frac{1}{3}$ children may give an answer of $4\frac{3}{8}$. Use bar models to address this misconception and to secure understanding. Encourage children to break down the method into steps by reasoning that to add and subtract mixed numbers they need to add/subtract the wholes and add/subtract the fractions separately. Ensure children remember that fractions need to have the same denominator before they can be added/subtracted. Ask:

• *What steps do you need to take to find the answer?*
• *How can you add/subtract the wholes and the fractions?*
• *What do you need to do to the fractions before you can add/subtract them?*

STRENGTHENING UNDERSTANDING

Children who are struggling to add and subtract mixed numbers should be encouraged to revisit adding and subtracting fractions with different denominators. When moving on to mixed numbers, encourage children to use bar models so they can visualise why they add/subtract the wholes and parts separately.

GOING DEEPER

Deepen learning using missing number problems where children are asked to find missing fractions. For example, $7\frac{24}{35} - \frac{\square}{\square} = 3\frac{2}{5}$. Children could also be encouraged to add more than two fractions but it will be necessary to ensure the fractions do not cross the whole.

KEY LANGUAGE

In lesson: wholes, parts, lowest common multiple (LCM), subtract, difference

Other language to be used by the teacher: denominator, numerator, common denominator, fractional parts

STRUCTURES AND REPRESENTATIONS

fraction strips, bar models

 In the eTextbook of this lesson, you will find interactive links to a selection of teaching tools.

Before you teach

• Can children find equivalent fractions using lowest common multiples?
• Are children confident in using mixed numbers?
• Can children add/subtract fractions with different denominators?

Discover

WAYS OF WORKING Pair work

ASK

- Question **1** a): *How can you work out how far Amelia has cycled in total?*
- Question **1** b): *What calculation do you need to do to find how much more?*

IN FOCUS Questions **1** a) and b) develop the concept of adding or subtracting fractions using mixed numbers.

Children should start to see that to add and subtract mixed numbers they can add/subtract the wholes and add/subtract the fractions.

PRACTICAL TIPS The context of Amelia and her dad cycling could be introduced in PE or as an out of classroom activity at the park. Encourage children to think about the distances they travel (involving fractions) and add these together or subtract to find the difference.

ANSWERS

Question **1** a): Amelia cycles $4\frac{11}{15}$ miles.

Question **1** b): Amelia cycles $2\frac{1}{15}$ more miles on Saturday.

Adding and subtracting fractions ❷

Discover

1 a) On Saturday, Amelia cycles $3\frac{2}{5}$ kilometres with her dad.

On Sunday, she cycles $1\frac{1}{3}$ kilometres.

How many kilometres does Amelia cycle in total?

b) How many more kilometres does Amelia cycle on Saturday than on Sunday?

148

PUPIL TEXTBOOK 6A PAGE 148

Share

WAYS OF WORKING Whole class teacher led

ASK

- Question **1** a): *What do the fraction strips show you? What do you need to find to add fractions with different denominators? What is the common denominator?*
- Question **1** b): *What operation do you need to use? How can you subtract the numbers?*

IN FOCUS Question **1** a) requires children to add together mixed numbers using a common denominator. Break this down into adding the wholes and then adding the fractions. Reinforce previous learning by reminding children that a common denominator is required before they can add the fractional parts.

Question **1** b) involves subtracting mixed numbers. Explore with children how finding out 'how many more' leads to finding the difference and carrying out a subtraction calculation. Encourage children to use the comparison bar model to identify what they are looking for. Some children may struggle with the fact that they are adding numbers together even though they are working on a subtraction calculation; use the diagrams to show why this is and to strengthen their understanding.

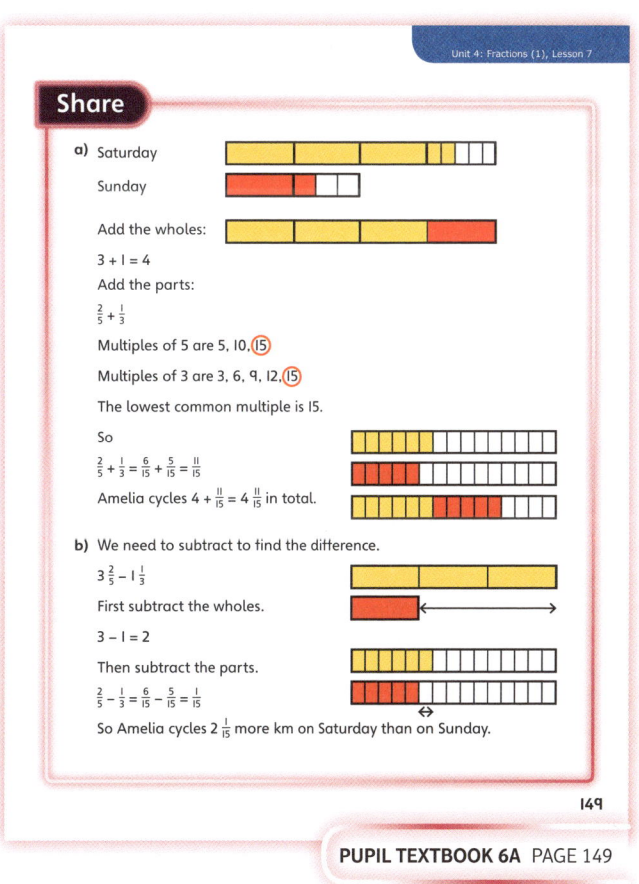

Share

a) Saturday

Sunday

Add the wholes:

$3 + 1 = 4$

Add the parts:

$\frac{2}{5} + \frac{1}{3}$

Multiples of 5 are 5, 10, 15

Multiples of 3 are 3, 6, 9, 12, 15

The lowest common multiple is 15.

So

$\frac{2}{5} + \frac{1}{3} = \frac{6}{15} + \frac{5}{15} = \frac{11}{15}$

Amelia cycles $4 + \frac{11}{15} = 4\frac{11}{15}$ in total.

b) We need to subtract to find the difference.

$3\frac{2}{5} - 1\frac{1}{3}$

First subtract the wholes.

$3 - 1 = 2$

Then subtract the parts.

$\frac{2}{5} - \frac{1}{3} = \frac{6}{15} - \frac{5}{15} = \frac{1}{15}$

So Amelia cycles $2\frac{1}{15}$ more km on Saturday than on Sunday.

149

PUPIL TEXTBOOK 6A PAGE 149

Think together

WAYS OF WORKING Whole class teacher led (I do, We do, You do)

ASK

- Question ❶: *Can you add/subtract the wholes and the parts? Can you find the LCM of each denominator?*
- Question ❷: *What are the steps you need to take to find the answers? How can you find the LCM of each denominator? What do you need to do next?*
- Question ❸: *What calculation is needed to find the length of the second train track? What calculation should you use to work out the total length of both train tracks?*

IN FOCUS   Question ❶ requires children to add and subtract mixed numbers. Questions ❶ a) and ❶ b) have structured support with a diagram, which encourages children to add/subtract the wholes and the parts separately.

Question ❷ links adding and subtracting two mixed numbers with puzzle solving: children have to create the questions from the diagrams with no structure or support given. Encourage children to reason whether to add or subtract and to show how they got their answers.

Question ❸ introduces problem solving and involves both addition and subtraction of mixed numbers. Children will need to first find the length of the second track, then find the total length of both tracks. Encourage them to use the correct units (metres) in their answer.

STRENGTHEN   To support understanding, children can represent the mixed numbers on a bar model or number line. Encourage them to write down their calculations and to do them in stages, i.e. adding or subtracting the wholes, then adding or subtracting the parts; and finding equivalent fractions before trying to add or subtract the fractions.

DEEPEN   Question ❸ can be extended by giving children a third length of track and asking them to find the new total, for example, a third track is $\frac{2}{3}$ metre longer than the second one. This track is put together with the other two tracks. How long is the train track now?

ASSESSMENT CHECKPOINT   Question ❶ will assess children's ability to add and subtract mixed numbers while gradually reducing support and structure. Look for children clearly explaining the steps needed and confidently using equivalent fractions to find the answers, even in an abstract form.

Questions ❷ and ❸ assess children's ability to add and subtract mixed numbers in problem-solving contexts with no support or structure. Look for children who can fluently find a common denominator and then complete the calculations, while clearly explaining the steps.

ANSWERS

Question ❶ a): $2\frac{1}{4} + 2\frac{3}{8} = 4\frac{5}{8}$ so Luis walks $4\frac{5}{8}$ miles.

Question ❶ b): $5\frac{1}{2} - 3\frac{2}{5} = 2\frac{1}{10}$ so Jamie swims $2\frac{1}{10}$ more lengths than Ambika.

Question ❷ a): $9\frac{1}{6} + 7\frac{5}{8} = 16\frac{19}{24}$

Question ❷ b): $4\frac{2}{3} - \frac{2}{7} = 4\frac{8}{21}$

Question ❸: The new train track is $4\frac{9}{10}$ metres long.

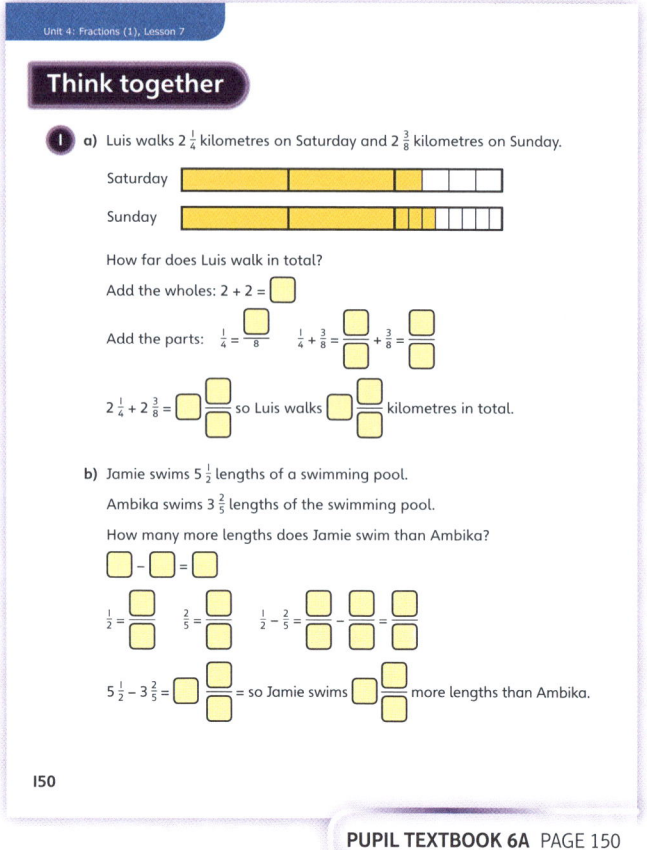

PUPIL TEXTBOOK 6A PAGE 150

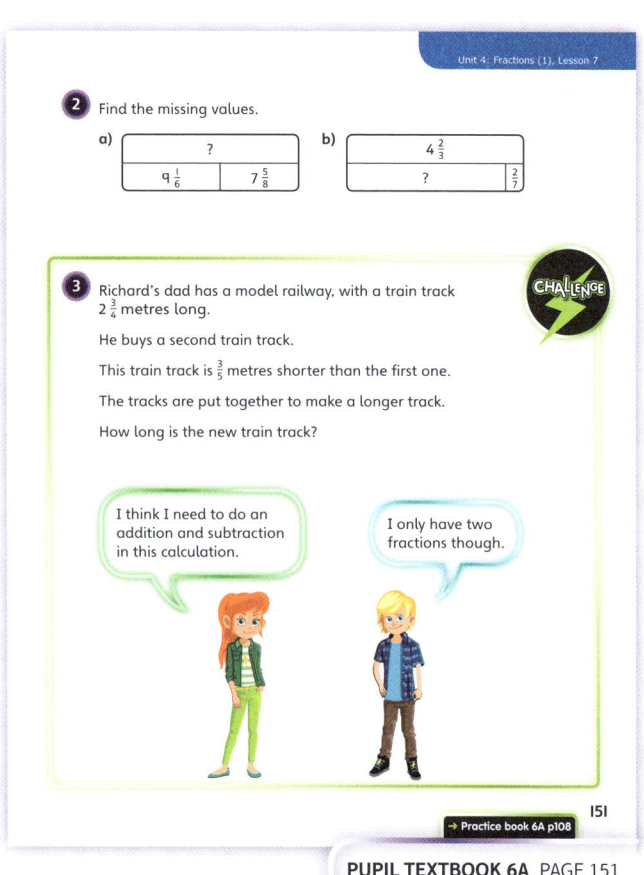

PUPIL TEXTBOOK 6A PAGE 151

Practice

WAYS OF WORKING Independent thinking

IN FOCUS Questions ❶ and ❷ consolidate children's understanding of adding and subtracting mixed numbers. Question ❶ uses prompting and pictorial representations for support, while question ❷ has no structure. Encourage children to write down their calculations and explain their steps. Question ❸ requires children to add and subtract fractions in a context. Encourage children to use their answer from part a) to work out the answer to part b). Question ❹ develops children's understanding of comparing fractions, requiring them to make the biggest total possible from two fractions. It may be useful to revisit and reinforce the comparing fractions concepts from Unit 4, Lesson 4. Question ❺ involves problem solving. Children are given the answer and one number and asked to find the other number. Encourage them to write the numbers in a number sentence.

STRENGTHEN To strengthen understanding, encourage children to represent the fractions on bar models as needed. Refer children to the structure in question ❶ to organise and carry out their calculations. If children are struggling with question ❺, encourage them to write the numbers on a part-whole model to give visual support.

DEEPEN Question ❸ can be explored further by asking children how much water will be left in the bucket after 3 minutes.

ASSESSMENT CHECKPOINT Questions ❶ and ❷ provide an opportunity to assess children's ability to add and subtract mixed numbers while gradually reducing support and structure. Look for children clearly explaining the steps needed, confidently using equivalent fractions to find the answers and showing fluency when working abstractly. Questions ❸ and ❹ assess children's ability to add and subtract mixed numbers in problem-solving contexts with no support or structure. Children should be able to find the common denominator and then complete the calculations in wholes and in parts, understanding when to add and subtract.

ANSWERS Answers for the **Practice** part of the lesson appear in the separate **Practice and Reflect answer guide**.

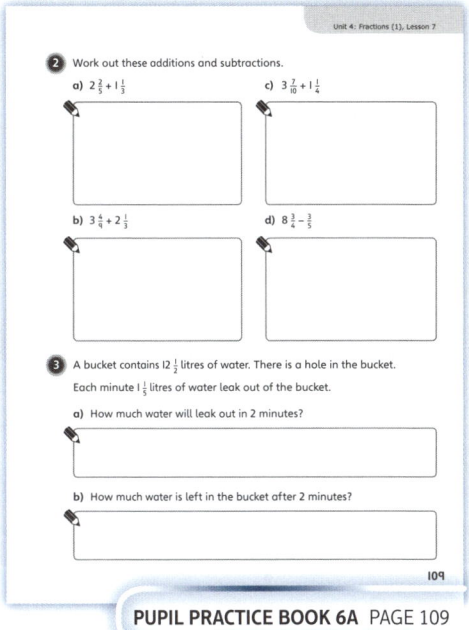

PUPIL PRACTICE BOOK 6A PAGE 108

PUPIL PRACTICE BOOK 6A PAGE 109

Reflect

WAYS OF WORKING Pair work

IN FOCUS This **Reflect** activity gives children an opportunity to describe the steps needed to subtract mixed numbers. Encourage children to explain how they know whether Ebo is correct and ask them to show the answer.

ASSESSMENT CHECKPOINT This activity assesses children's ability to identify the calculation required to solve a problem involving adding or subtracting mixed numbers. Children should be able to confidently explain that this problem requires subtraction, and to describe how to complete it.

ANSWERS Answers for the **Reflect** part of the lesson appear in the separate **Practice and Reflect answer guide**.

After the lesson ⏸
- Can children add and subtract the fractional parts of mixed numbers using equivalent fractions?
- Do children need support adding and subtracting mixed numbers?

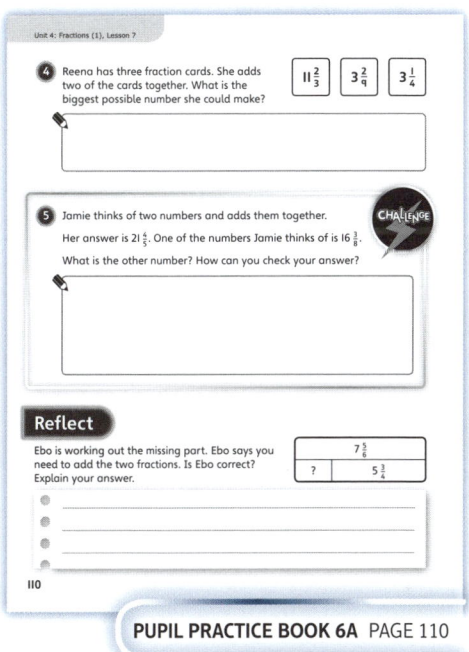

PUPIL PRACTICE BOOK 6A PAGE 110

Adding fractions

Learning focus

In this lesson, children extend their knowledge of adding mixed numbers and fractions by using two methods to add mixed fractions where the fractional answer is greater than 1.

Small steps

→ Previous step: Adding and subtracting fractions (2)
→ **This step: Adding fractions**
→ Next step: Subtracting fractions

NATIONAL CURRICULUM LINKS

Year 6 Number – Fractions

Add and subtract fractions with different denominators and mixed numbers, using the concept of equivalent fractions.

ASSESSING MASTERY

Children can fluently add and subtract mixed numbers either by adding/subtracting the wholes and fractional parts or by converting to improper fractions and adding these, while clearly explaining the steps for both methods.

COMMON MISCONCEPTIONS

Children may add the wholes and add the fractions separately but then fail to convert the fractional answer to a mixed number; for example, when adding $1\frac{2}{3}$ and $2\frac{3}{4}$ children may correctly do $1 + 2 = 3$ and $\frac{2}{3} + \frac{3}{4} = \frac{17}{12}$ but then incorrectly write the final answer as $3\frac{17}{12}$. Use bar models or fraction strips to provide pictorial support to secure this understanding, while encouraging children to realise that $\frac{17}{12}$ is an improper fraction requiring conversion to a mixed number. Ask:
• *What type of fraction is your answer?*
• *Do you need to convert this further?*

STRENGTHENING UNDERSTANDING

When adding mixed numbers, strengthen understanding by showing children the numbers on a bar model so they can see why they can add the wholes and add the parts and why they can convert to improper fractions and add these.

GOING DEEPER

Deepen learning using diagrams of quadrilaterals and triangles that have mixed numbers as their lengths, asking children to find the perimeter of each shape.

Encourage children to solve missing number problems that involve working backwards, such as $\square - 3\frac{2}{5} = 6\frac{3}{4}$.

KEY LANGUAGE

In lesson: wholes, parts, mixed number, improper fraction, convert, add

STRUCTURES AND REPRESENTATIONS

fraction strips

 In the eTextbook of this lesson, you will find interactive links to a selection of teaching tools.

Before you teach

• Can children add fractions with different denominators?
• Can children convert between mixed numbers and improper fractions?
• Can children add mixed numbers where the fractional parts total less than 1?

Discover

WAYS OF WORKING Pair work

ASK

- Question ❶ a): *How can you work out how many carrots the farmer has harvested so far?*
- Question ❶ a): *What calculation do you need to do?*
- Question ❶ b): *How can you work out how many carrots the farmer has harvested now?*
- Question ❶ b): *What calculation do you need to do?*

IN FOCUS Question ❶ develops the concept of adding or subtracting fractions by introducing calculations where the fractional answer is greater than 1. An extra conversion step is required to complete the calculation.

PRACTICAL TIPS Adding mixed numbers can be introduced in a practical setting in the classroom – for instance, by adding fractions of distance between local places of interest. Encourage children to use pictorial supports such as bar models to give a visual representation.

ANSWERS

Question ❶ a): The farmer has harvested $4\frac{1}{4}$ tonnes of carrots so far.

Question ❶ b): The farmer has harvested $5\frac{1}{20}$ tonnes, so he has harvested enough.

Adding fractions

Discover

There are $2\frac{3}{4}$ tonnes of carrots on one trailer and $1\frac{1}{2}$ tonnes on the other.

❶ a) What is the total weight of carrots the farmer has harvested so far?

b) A supermarket orders 5 tonnes of carrots.
The farmer harvests another $\frac{4}{5}$ tonnes of carrots from a different field.
Has the farmer harvested enough carrots to fulfil the order?

152

PUPIL TEXTBOOK 6A PAGE 152

Share

WAYS OF WORKING Whole class teacher led

ASK

- Question ❶ a): *How can you add the parts? What is the common denominator? What do you notice about the answer? Is this a mixed number or an improper fraction?*
- Question ❶ a): *Which method is more efficient?*

IN FOCUS Question ❶ a) introduces two methods for adding mixed numbers where the total of the fractional part is greater than 1. Explore both methods using the fractional strips, emphasising that in Dexter's method, $\frac{5}{4}$ is an improper fraction and needs to be converted to a mixed number before the final step of combining the different parts. With Flo's method, encourage children to work through the calculations converting the mixed numbers to improper fractions. Discuss adding the fractions by finding a common denominator, ensuring that children understand the type of fraction this results in and that they clearly recognise and can explain the difference between mixed numbers and improper fractions.

Question ❶ b) requires children to add a mixed number and a fraction together before comparing their answer with a whole number. Encourage children to work through the calculations of adding the fractions and converting to a mixed number before adding on 4. Explore with children if they could have used the other method, i.e. converting $4\frac{1}{4}$ to an improper fraction before adding $\frac{4}{5}$. Discuss which method is more efficient, ensuring children recognise that the answer would be the same for both methods.

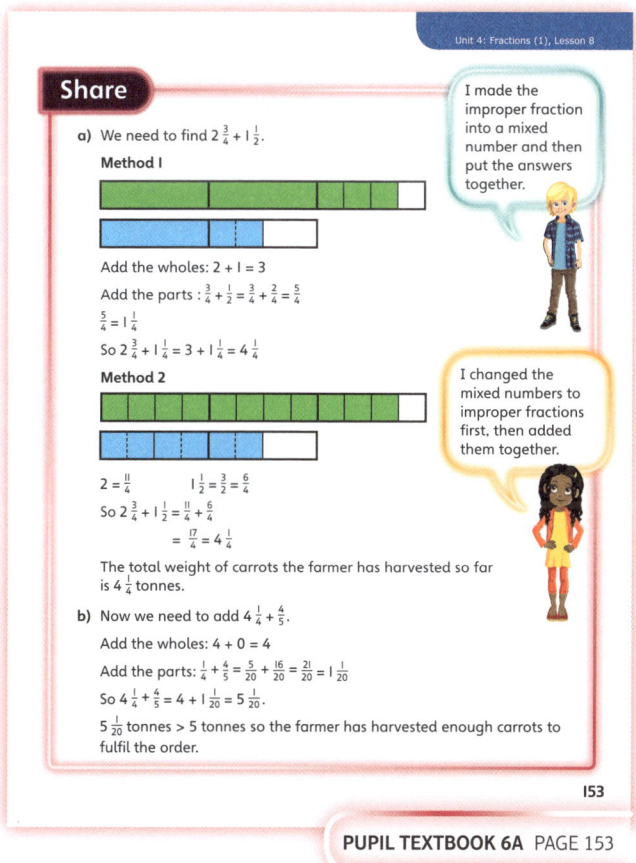

Share

a) We need to find $2\frac{3}{4} + 1\frac{1}{2}$.

Method 1

I made the improper fraction into a mixed number and then put the answers together.

Add the wholes: $2 + 1 = 3$

Add the parts : $\frac{3}{4} + \frac{1}{2} = \frac{3}{4} + \frac{2}{4} = \frac{5}{4}$

$\frac{5}{4} = 1\frac{1}{4}$

So $2\frac{3}{4} + 1\frac{1}{2} = 3 + 1\frac{1}{4} = 4\frac{1}{4}$

Method 2

I changed the mixed numbers to improper fractions first, then added them together.

$2 = \frac{11}{4}$ $1\frac{1}{2} = \frac{3}{2} = \frac{6}{4}$

So $2\frac{3}{4} + 1\frac{1}{2} = \frac{11}{4} + \frac{6}{4}$

$= \frac{17}{4} = 4\frac{1}{4}$

The total weight of carrots the farmer has harvested so far is $4\frac{1}{4}$ tonnes.

b) Now we need to add $4\frac{1}{4} + \frac{4}{5}$.

Add the wholes: $4 + 0 = 4$

Add the parts: $\frac{1}{4} + \frac{4}{5} = \frac{5}{20} + \frac{16}{20} = \frac{21}{20} = 1\frac{1}{20}$

So $4\frac{1}{4} + \frac{4}{5} = 4 + 1\frac{1}{20} = 5\frac{1}{20}$.

$5\frac{1}{20}$ tonnes > 5 tonnes so the farmer has harvested enough carrots to fulfil the order.

153

PUPIL TEXTBOOK 6A PAGE 153

Think together

WAYS OF WORKING Whole class teacher led (I do, We do, You do)

ASK

- Question **1** a): *How can you add the wholes and the parts? How can you find the lowest common multiple of the denominators?*
- Question **1** b): *How can you convert the mixed numbers to improper fractions?*
- Question **2**: *How can you write the amounts of each pizza as a mixed number? As an improper fraction?*
- Question **3**: *Which numbers might add to make* $11\frac{13}{24}$? *Could you start by trying to add some of the numbers?*

IN FOCUS Question **2** is a word problem requiring children to identify the amounts before adding to find the total amount. Encourage them to write the amount of each pizza as either a mixed number or an improper fraction before adding.

Question **3** involves problem solving, requiring children to select the correct numbers to make $11\frac{13}{24}$. The numbers have been carefully selected so that the common denominator could be 24 for all fractions and some of the whole numbers add to 11. Watch out for a common mistake where children may think the answer is $7\frac{3}{4}$ and $4\frac{2}{3}$ because $7 + 4 = 11$, i.e. they have not realised that the total of the fractions will cross the whole and make the answer >12.

STRENGTHEN Encourage children to represent the numbers on bar models and fractional strips. Encourage children to work systematically through the calculations following the structure used in question **1**.

DEEPEN Question **3** can be explored further by giving children another total and asking them which two fractions added together make this total. Alternatively, ask children to add three of the cards together using their preferred method.

ASSESSMENT CHECKPOINT Question **1** assesses children's ability to use both methods for adding mixed numbers where the fractional part makes a total greater than 1. If children struggle with this question, intervention is needed to consolidate understanding.

Question **2** provides an opportunity to assess children's ability to solve word problems involving adding mixed numbers. Look for children confidently translating the problem into number sentences and solving the calculations while clearly explaining the method they have used and why; children who can do this will have mastered the topic.

ANSWERS

Question **1** a): $1\frac{2}{3} + 2\frac{1}{2} = 3 + 1\frac{1}{6} = 4\frac{1}{6}$

Question **1** b): $\frac{5}{3} + \frac{5}{2} = \frac{10}{6} + \frac{15}{6} = \frac{25}{6}$

Change to a mixed number: $\frac{25}{6} = 4\frac{1}{6}$

Question **2**: $3\frac{7}{8} + 4\frac{5}{12} = 8\frac{7}{24}$ pizzas were eaten altogether.

Question **3** a): Isla chose $4\frac{2}{3}$ and $6\frac{7}{8}$.

Question **3** b): $7\frac{3}{4} + 27\frac{17}{24} = 35\frac{11}{24}$; use the method of adding the wholes and adding the parts, as converting the mixed numbers to improper fractions is inefficient because of the large denominator involved.

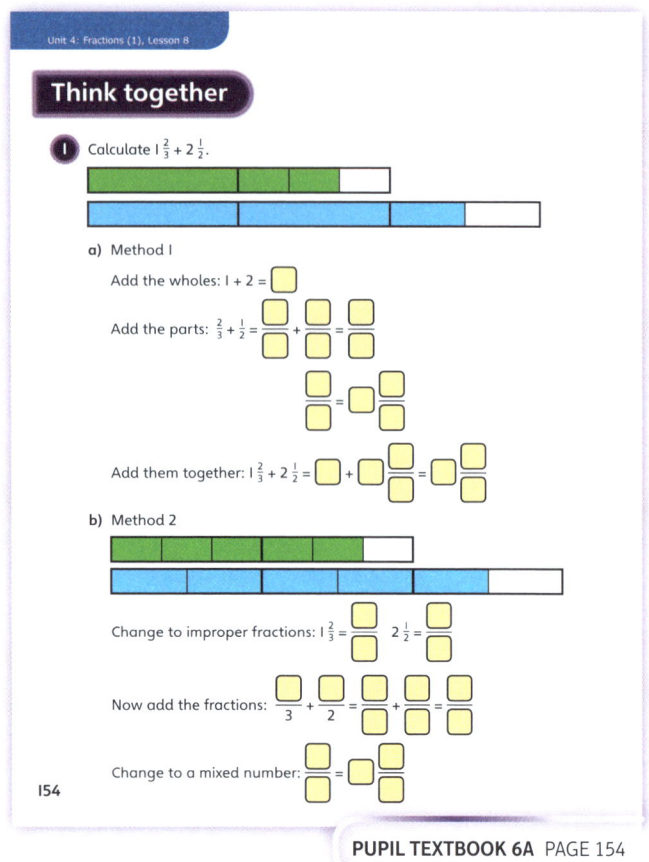

PUPIL TEXTBOOK 6A PAGE 154

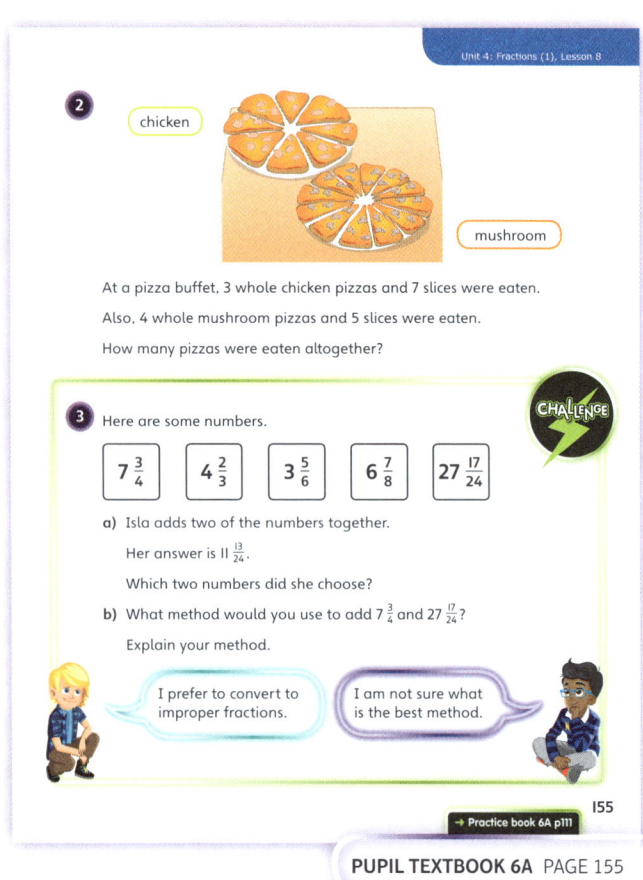

PUPIL TEXTBOOK 6A PAGE 155

186

Practice

Independent thinking

IN FOCUS Question ❶ aims to consolidate children's understanding of adding mixed numbers with pictorial representation as support. Encourage children to use their preferred method and to write down their calculations.

Question ❷ requires children to add mixed numbers with no pictorial support. Encourage children to look carefully at the numbers and to consider if they can simplify any fractions.

Question ❻ incorporates problem solving, with children required to carry out more than one addition and then decide how many packs of fencing are needed. Encourage children to find the distance around the edge of the garden before working out how many packs of fencing are required. Explore with children whether they need to round up or down so that there is enough fencing.

STRENGTHEN Encourage children to use pictorial supports such as bar models to support them with the calculations.

ASSESSMENT CHECKPOINT Questions ❶ and ❷ assess children's ability to add mixed numbers where the fractional part makes a total > 1. Question ❶ is supported with prompting and pictorial representations, while question ❷ has no support. Look for children breaking down question ❷ into the steps shown in question ❶ and systematically adding the parts, using common denominators.

Questions ❸, ❹, ❺ and ❻ assess children's ability to solve word problems which involve adding mixed numbers where the fraction part crosses the whole. Look for children confidently translating the problem into number sentences and solving the calculations, while clearly explaining the methods they have used and why.

ANSWERS Answers for the **Practice** part of the lesson appear in the separate **Practice and Reflect answer guide**.

Reflect

Pair work

IN FOCUS This **Reflect** activity will evaluate children's understanding of adding two mixed numbers. Encourage children to use their preferred method and to explain why they have chosen that method as well as calculating the answer.

ASSESSMENT CHECKPOINT This gives an opportunity to assess children's ability to add two mixed numbers where the fractional part is more than 1. Look for children who can confidently explain how to add two mixed numbers, showing understanding of improper fractions and mixed numbers and explaining why they have chosen the method they have used.

ANSWERS Answers for the **Reflect** part of the lesson appear in the separate **Practice and Reflect answer guide**.

After the lesson

- Can children add mixed numbers using both methods?
- Can they convert between mixed numbers and improper fractions?
- Can they explain which method is more efficient for different calculations?

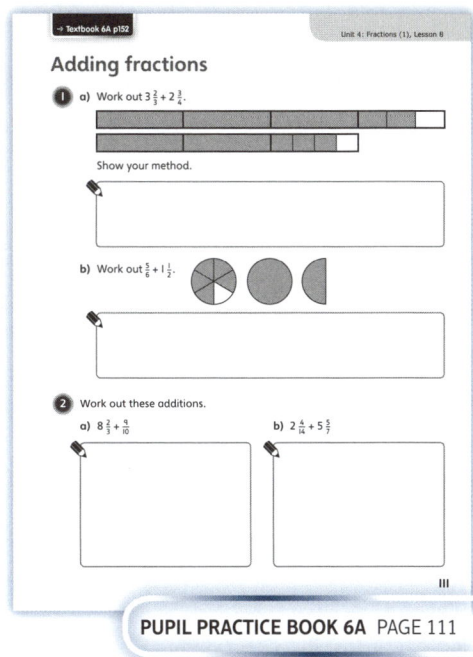

PUPIL PRACTICE BOOK 6A PAGE 111

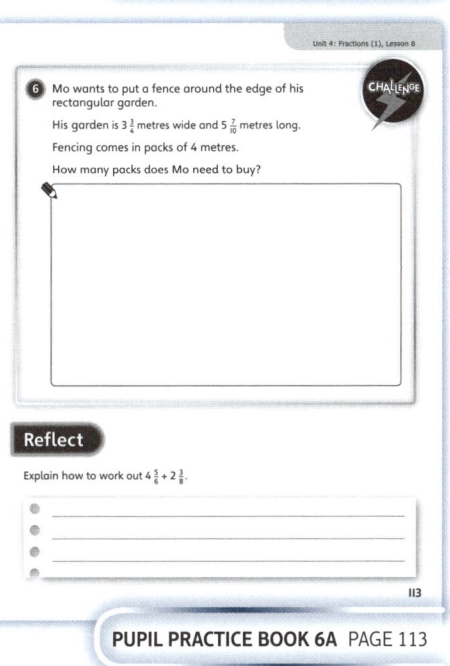

PUPIL PRACTICE BOOK 6A PAGE 112

PUPIL PRACTICE BOOK 6A PAGE 113

Subtracting fractions

Learning focus

In this lesson, children extend their understanding of subtracting mixed numbers and fractions to calculations where the fractional answer crosses the whole and they cannot simply subtract the wholes and subtract the parts.

Small steps

→ Previous step: Adding fractions
→ **This step: Subtracting fractions**
→ Next step: Problem solving – adding and subtracting fractions (1)

NATIONAL CURRICULUM LINKS

Year 6 Number – Fractions

Add and subtract fractions with different denominators and mixed numbers, using the concept of equivalent fractions.

ASSESSING MASTERY

Children can reliably subtract mixed numbers by rewriting the calculation and subtracting the wholes and subtracting the fractions, or by converting to improper fractions. Children can confidently use a number line to subtract mixed numbers.

COMMON MISCONCEPTIONS

Children may subtract the wholes and then the fractions separately but incorrectly change the order of the fractions so that the smaller fraction is always subtracted from the larger fraction. For example, when calculating $5\frac{1}{3} - 2\frac{3}{4}$ children find the difference between 5 and 2 (= 3) and then find the difference between $\frac{1}{3}$ and $\frac{3}{4}$ ($= \frac{5}{12}$), incorrectly writing their answer as 3. Encourage children to realise that the order of the numbers does matter when subtracting and that if the second fraction is greater than the first one they cannot just subtract the smaller from the larger. Use bar models to aid understanding and address this misconception. Ask:

- *Does the order in which you subtract matter?*
- *Can you change the order when subtracting?*
- *Can you use a bar model to help you visualise the calculation?*

STRENGTHENING UNDERSTANDING

When rewriting a mixed number to make it easier to subtract, strengthen understanding using bar models so that children can see why the change is needed. It may also be beneficial to encourage children to use a number line to count on to carry out a subtraction.

GOING DEEPER

Deepen learning with missing number problems involving subtraction, for example $\square\frac{\square}{4} + 2\frac{4}{5} = 6\frac{11}{20}$. Give children some verbal 'think of a number' problems involving subtracting mixed numbers, for example: *I think of a number and add $5\frac{2}{3}$ to it. My answer is $8\frac{13}{24}$. What was my number?*

KEY LANGUAGE

In lesson: wholes, parts, mixed number, improper fraction, convert, subtract, lowest common multiple, common denominator

Other language to be used by the teacher: add on

STRUCTURES AND REPRESENTATIONS

fraction strips, number lines, part-whole models

 In the eTextbook of this lesson, you will find interactive links to a selection of teaching tools.

Before you teach

- Can children subtract fractions with different denominators?
- Can children convert between mixed numbers and improper fractions?
- Can children break down mixed numbers into alternative combinations of wholes and parts, for example, $3\frac{1}{2} = 2 + 1\frac{1}{2} = 2 + \frac{3}{2}$?

188

Discover

WAYS OF WORKING Pair work

ASK

• Question ① a): *Why does Max think you can't do the calculation? What can you do to work out the subtraction?*
• Question ① b): *How can you convert the mixed numbers to improper fractions? What calculation is needed?*

IN FOCUS Question ① introduces children to subtraction of mixed numbers where the first fractional part is smaller than the second fractional part. In these cases, the fraction must be changed – either to an alternative mixed number or to an improper fraction – to complete the subtraction. It is important for children to understand why this change is needed and to recognise that they cannot simply change the order of the fractions.

PRACTICAL TIPS This concept could be introduced in a practical cooking session in the classroom – for example, making fruit smoothies using cups as measures. It will be important with this concept to use visual representations such as bar models and number lines to strengthen understanding that the order of fractions matters when subtracting.

ANSWERS

Question ① a): Max is not correct. $1\frac{5}{6}$ more cups of cherries are needed.

Question ① b): $\frac{10}{3} - \frac{3}{2} = \frac{20}{6} - \frac{9}{6} = \frac{11}{6} = 1\frac{5}{6}$

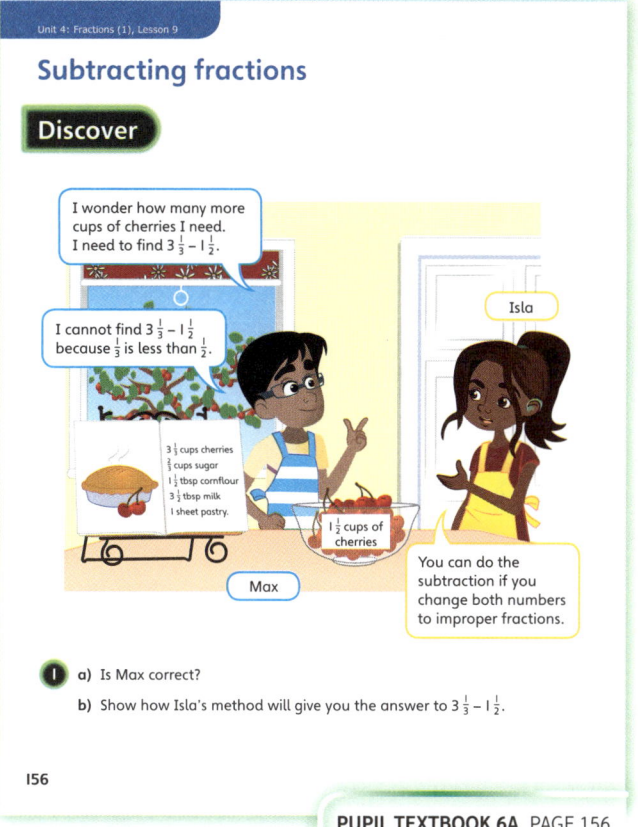

PUPIL TEXTBOOK 6A PAGE 156

Share

WAYS OF WORKING Whole class teacher led

ASK

• Question ① a): *How could you rewrite $3\frac{1}{3}$?*
• Question ① a): *What do you need to subtract first? What next?*
• Question ① b): *How can you convert the mixed numbers to improper fractions? How can you now subtract? What sort of fraction is the answer?*

IN FOCUS This question develops children's understanding of how to subtract mixed numbers and fractions that cross the whole. Question ① a) introduces the concept that when subtracting mixed numbers where the first fractional part is smaller than the second fractional part, it is necessary to change the fractions to carry out the subtraction. Discuss with children the first method of subtracting the wholes and subtracting the parts, as per Unit 4, Lesson 8. It will be important to highlight the fraction strips, emphasising that $\frac{1}{2}$ is greater than $\frac{1}{3}$, which is why Max thinks the subtraction is impossible. Explore how to rewrite $3\frac{1}{3}$ to make the fraction part bigger, so that the subtraction is easier.

Question ① b) focuses on the alternative method of converting mixed numbers to improper fractions and subtracting the fractions. Again, highlight the fraction strips showing the numbers as improper fractions and discuss the calculations needed to convert the mixed numbers to improper fractions.

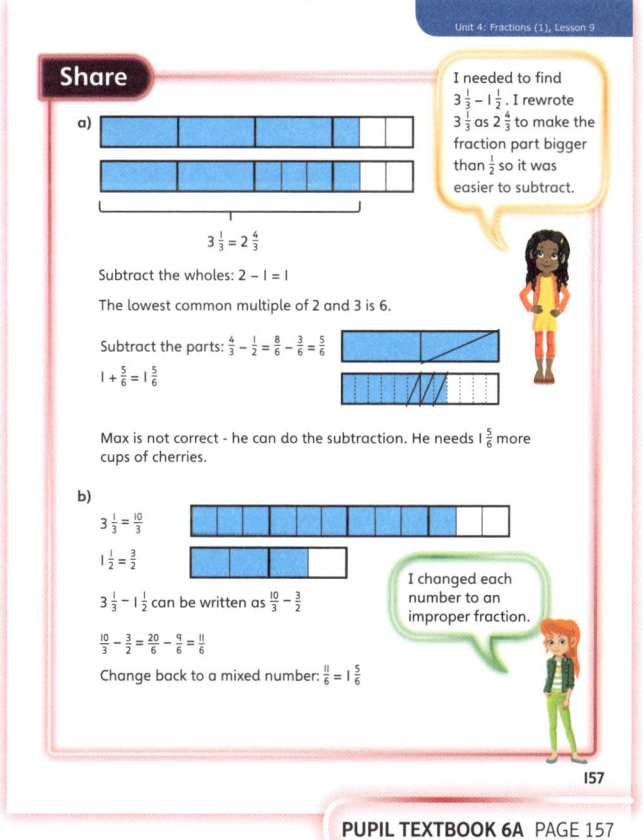

PUPIL TEXTBOOK 6A PAGE 157

Think together

WAYS OF WORKING Whole class teacher led (I do We do, You do)

ASK

- Question **1**: *How can you find the lowest common multiple of each denominator? How can you convert the mixed numbers to improper fractions?*
- Question **2**: *What calculation do you need to do to find $\frac{2}{3}$ less? What method will you use?*
- Question **3**: *What do you add to $2\frac{5}{6}$ to get to 3? What do you add to 3 to get to 5? What do you add to 5 to get to $5\frac{3}{10}$? What do you need to do with all your answers?*

IN FOCUS Question **2** is a word problem where children identify the required calculation with no structure or support. Encourage them to write the fraction in numbers and to show their method. Some children may choose to convert the hours into minutes and calculate the answer this way. This could be a good way for children to check their answer but encourage them to use fractions to find the answer.

Question **3** introduces a third method, using a number line to subtract fractions. Encourage children to explain how this method works, then ask them to use all three methods to work out the answers to part b). Encourage them to explain which method is most efficient.

STRENGTHEN To strengthen learning, encourage children to write down their calculations, breaking the methods into steps and using pictorial support.

DEEPEN Question **2** can be explored further by giving children another time and asking them to find the difference; for example, Alex completes the puzzle in $1\frac{5}{6}$ hours. How much quicker was Alex than Andy? How much quicker was Alex than Jamilla?

ASSESSMENT CHECKPOINT Question **2** provides an opportunity to assess children's ability to subtract mixed numbers where the fractional parts cross the whole. Look for children using either method, confidently working through the steps.

Question **3** provides an opportunity to assess children's ability to subtract mixed numbers. Look for children who can clearly explain the steps and show fluency with all methods and an understanding of why it is necessary to convert the numbers before completing the calculations.

ANSWERS

Question **1** a): $4\frac{1}{3} - 2\frac{3}{4} = 1 + \frac{7}{12} = 1\frac{7}{12}$

Question **1** b): $3\frac{1}{5} = \frac{16}{5}$, $1\frac{1}{2} = \frac{3}{2}$; so $3\frac{1}{5} - 1\frac{1}{2}$ can be written as
$\frac{16}{5} - \frac{3}{2}$
Find a common denominator:
$\frac{16}{5} - \frac{3}{2} = \frac{32}{10} - \frac{15}{10} = \frac{17}{10}$
Change back to a mixed number: $\frac{17}{10} = 1\frac{7}{10}$

Question **2**: It takes Andy $1\frac{7}{12}$ hours to complete the puzzle.

Question **3** a): $5\frac{3}{10} - 2\frac{5}{6} = 2\frac{7}{15}$

Question **3** b): $3\frac{1}{2} - 1\frac{7}{10} = 1\frac{4}{5}$; $26\frac{1}{2} - 18\frac{4}{5} = 7\frac{7}{10}$

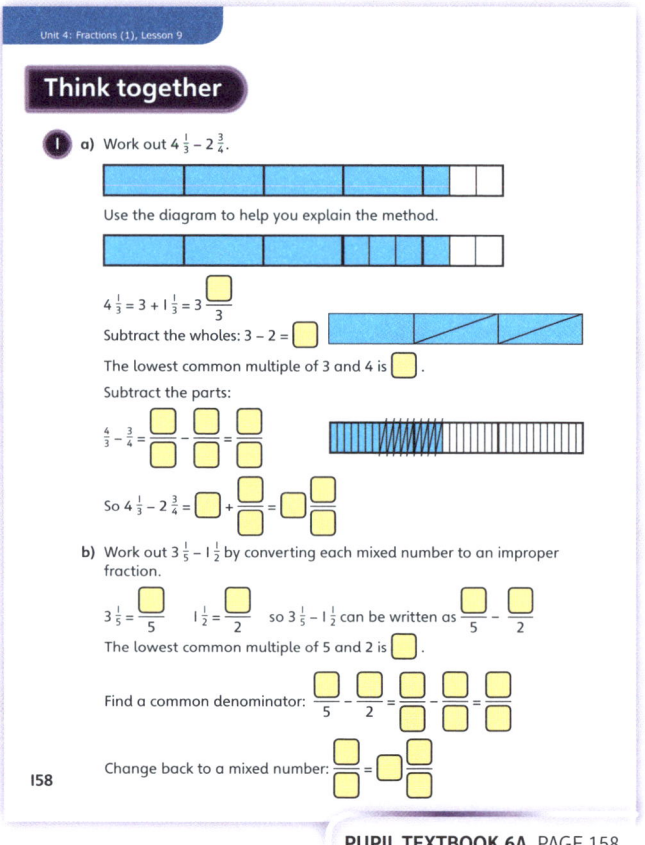

PUPIL TEXTBOOK 6A PAGE 158

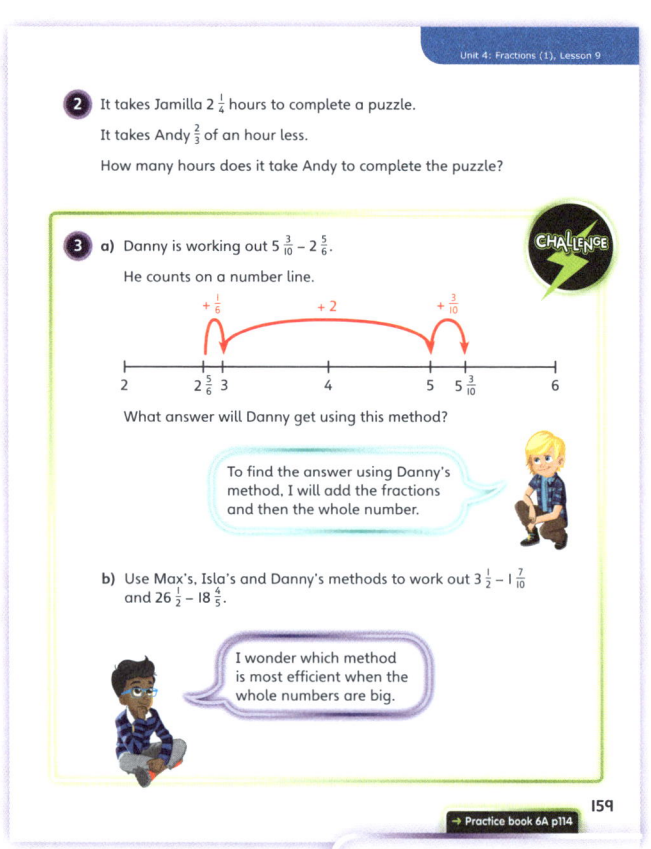

PUPIL TEXTBOOK 6A PAGE 159

190

Practice

WAYS OF WORKING Independent thinking

IN FOCUS Question ❶ aims to consolidate children's understanding of subtracting mixed numbers and fractions. Encourage children to use their preferred method and to write down their calculations.

Question ❷ develops subtraction involving mixed numbers and whole numbers. Encourage children to think carefully about how they could work out the calculations, considering which method is most efficient.

Question ❸ uses part-whole models in subtraction problems, requiring children to find the missing parts. Again, it will be useful to encourage children to write down the calculation required and to show their method.

Question ❹ introduces a context problem involving heights of giraffes, requiring children to find the height of the baby giraffe. This question adds variation by looking at vertical difference rather than horizontal.

Question ❼ develops problem solving in an abstract form, with children required to complete more than one subtraction to work out the value of the symbols.

STRENGTHEN To strengthen learning, encourage children to use bar models or number lines for support. Question ❼ could be adapted, by giving children the first calculation as $5\frac{3}{4} - 3\frac{5}{6} = ★$ so they can see the calculation required.

DEEPEN Question ❼ can be used to deepen learning and develop abstract and algebraic thinking, by giving children another calculation using the value of ♥ to find another symbol: *If $9\frac{3}{8} - ● = ♥$, what is the value of the circle?*

THINK DIFFERENTLY Question ❻ is a problem-solving question. The fractions are represented as shapes, so children need to work out the numbers before they can complete the subtraction calculation.

ASSESSMENT CHECKPOINT Question ❼ provides an opportunity to assess children's ability to subtract mixed numbers in abstract form, demonstrating algebraic reasoning. Look for children being able to comprehend what steps they need to complete and work backwards while clearly explaining their reasoning. Children should be confident in subtracting mixed numbers and fractions using their preferred method.

ANSWERS Answers for the **Practice** part of the lesson appear in the separate **Practice and Reflect answer guide**.

Reflect

WAYS OF WORKING Pair work

IN FOCUS This **Reflect** activity gives children an opportunity to review their learning and describe their preferred method for subtracting mixed numbers, before discussing with a friend why they have chosen that method.

ASSESSMENT CHECKPOINT This activity assesses a child's understanding of subtracting two mixed numbers. Look for children confidently using their preferred method and explaining their reason.

ANSWERS Answers for the **Reflect** part of the lesson appear in the separate **Practice and Reflect answer guide**.

After the lesson ⏸

- Can children subtract mixed numbers and fractions that involve crossing the whole?
- Can children explain the steps in their preferred method for subtracting mixed numbers and say why they prefer that method?

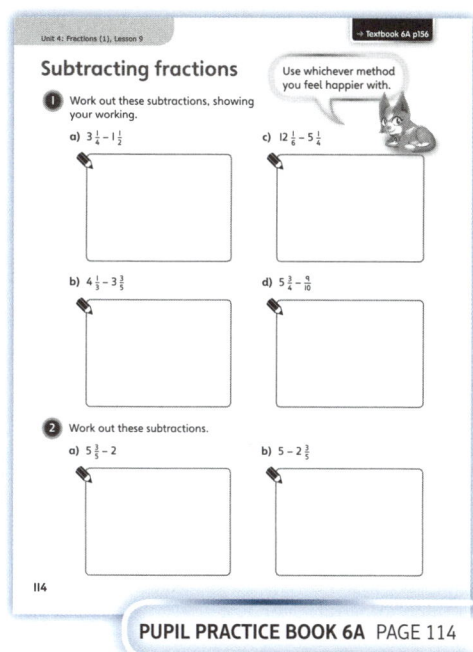

PUPIL PRACTICE BOOK 6A PAGE 114

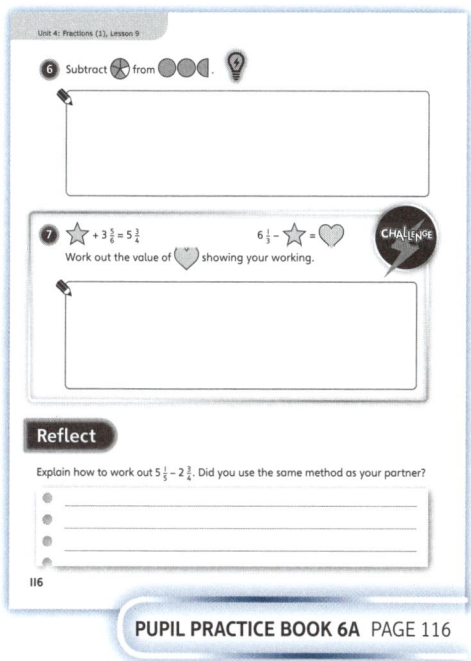

PUPIL PRACTICE BOOK 6A PAGE 115

PUPIL PRACTICE BOOK 6A PAGE 116

Problem solving – adding and subtracting fractions ①

Learning focus

In this lesson, children extend their knowledge of adding and subtracting mixed numbers to solve problems which involve adding and subtracting more than two mixed numbers.

Small steps

→ Previous step: Subtracting fractions

→ **This step: Problem solving – adding and subtracting fractions (1)**

→ Next step: Problem solving – adding and subtracting fractions (2)

NATIONAL CURRICULUM LINKS

Year 6 Number – Fractions

Add and subtract fractions with different denominators and mixed numbers, using the concept of equivalent fractions.

ASSESSING MASTERY

Children can solve problems that incorporate adding and subtracting mixed numbers and fractions, fluently converting between mixed numbers and improper fractions and using equivalent fractions as needed. They can use a variety of methods to solve word and abstract problems, while confidently explaining the steps in their own words.

COMMON MISCONCEPTIONS

Children may struggle to comprehend what calculation a problem is asking them to do; in this case, they will often just add the numbers given. Encourage children to break down the terms and vocabulary used and to use bar models or part-whole diagrams to give a pictorial representation of the problem. Ask:

• *What is the problem asking you to do?*
• *What words are used? What do they mean?*
• *Would a representation help?*
• *Can you write down the number sentences?*

STRENGTHENING UNDERSTANDING

Strengthen understanding by encouraging children to translate problems into number sentences and to use number lines and models to support their understanding.

GOING DEEPER

Children can be encouraged to deepen learning with problems based on other areas of mathematics involving addition and subtraction of fractions and mixed numbers. For example, children could work out the perimeter of a field given the length and width, or they could work out the length of a field given the width and the perimeter.

KEY LANGUAGE

In lesson: wholes, parts, equivalent fraction, compare, denominator

Other language to be used by the teacher: numerator, number sentence, perimeter, multi-step

STRUCTURES AND REPRESENTATIONS

shapes, fraction strips, part-whole models

 In the eTextbook of this lesson, you will find interactive links to a selection of teaching tools.

Before you teach

• Can children add and subtract fractions with different denominators?
• Can children add and subtract mixed numbers?
• Can children convert between mixed numbers and improper fractions?

Discover

WAYS OF WORKING Pair work

WAYS OF WORKING Pair work

ASK

• Question ❶ a): *How can they make purple paint? What are you going to need to do with the numbers?*
• Question ❶ b): *How can you work out if there is enough paint?*

IN FOCUS This question introduces children to adding three numbers together, involving mixed numbers and fractions and using two different methods. Children then compare the fractions using equivalent fractions.

PRACTICAL TIPS This lesson uses the context of mixing paint to introduce the addition of three fractions. This could be introduced in a practical session in the classroom – for example, mixing paint to make purple or making mixed fruit juice cups. Encourage children to use pictorial representations as support.

ANSWERS

Question ❶ a): The children will make $4\frac{17}{20}$ litres of purple paint.

Question ❶ b): $4\frac{17}{20} > 4\frac{14}{20}$ so there will be enough purple paint to paint the roofs and poles.

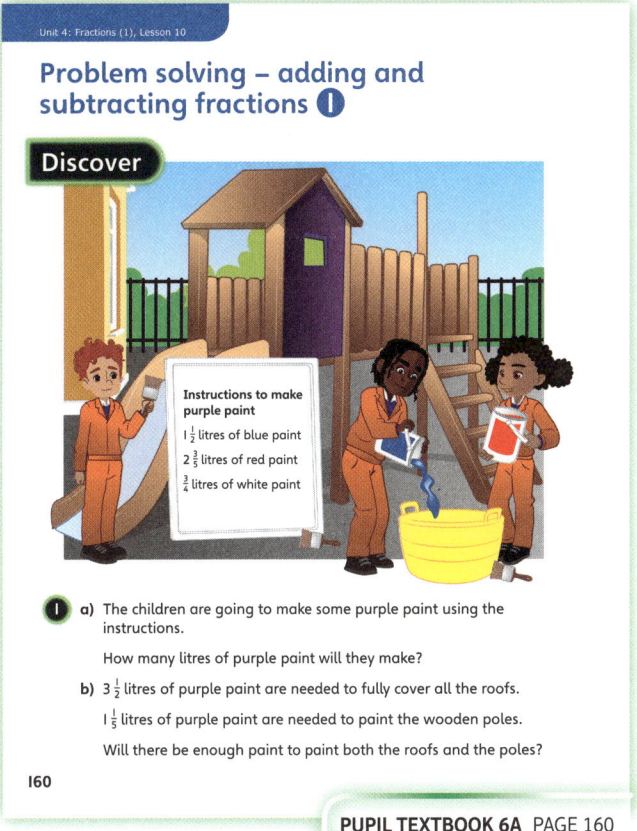

PUPIL TEXTBOOK 6A PAGE 160

Share

WAYS OF WORKING Whole class teacher led

ASK

• Question ❶ a): *What calculation do you need to do? Could you start with just two of the numbers?*
• Question ❶ a): *What do you need to convert to add your subtotal to the other fraction? What units do you need in the answer?*
• Question ❶ a): *Is there a different way of working this out?*
• Question ❶ b): *How can you work out how much paint is needed to cover the roofs and the poles? What do you need to compare this with? How can you compare two fractions?*
• Question ❶ b): *What do you need to do to the denominators? Can you think of another method to work this out?*

IN FOCUS This question introduces children to adding more than two mixed numbers and fractions that cross the whole, using two different methods. In question ❶ a) it will be useful to build on the previous lesson, by discussing with children the first method of working out the total of red and blue paint – by adding the wholes and adding the parts. Highlight that because $\frac{11}{10}$ is an improper fraction, they need to convert it to a mixed number before adding together the red and blue paint and adding on the white paint. Next focus on the alternative method of adding all the numbers together at once; again, highlight the need to convert the improper fraction to a mixed number.

In question ❶ b) encourage children to find the amount of paint needed for the roofs and poles together, before comparing with $4\frac{17}{20}$. Discuss any alternative methods that children come up with. For example, children may subtract $3\frac{1}{2}$ and $1\frac{1}{5}$ from $4\frac{17}{20}$ to work out if there is enough paint.

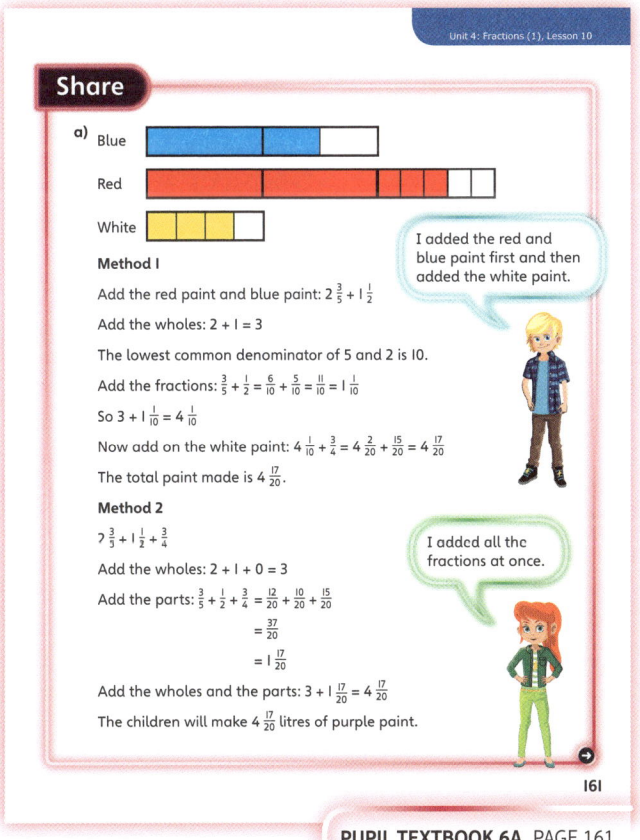

PUPIL TEXTBOOK 6A PAGE 161

Think together

WAYS OF WORKING Whole class teacher led (I do, We do, You do)

ASK

- Question **1**: *How could you work out the height of the cake in two different steps? How about in one step?*
- Question **2**: *What calculations do you need to do to find the missing numbers? How can you check your answer?*
- Question **3**: *Is the blue area smaller or bigger? Are you going to need to do an addition or a subtraction?*

IN FOCUS Question **1** revisits addition of three mixed numbers, using both the methods introduced in the Share section. Encourage children to use the correct units of inches in their answer.

Question **3** involves further problem solving and addresses a common misconception. Children may think that, because the star has been put on top of the square, they must add the numbers together. Encourage children to reason whether the blue area will be bigger or smaller than the rectangle to help them realise that a subtraction calculation is necessary. Encourage children to use the correct units of cm^2 in their answer.

STRENGTHEN In question **2**, encourage children to represent the calculations on a part-whole model to help them identify the calculation needed to solve the problem. Ask: *Can you represent the problem with a part-whole model? What calculation do you need to do to find the answer?*

DEEPEN Question **1** can be explored further by asking children to find the height difference between the tiers of cakes.

ASSESSMENT CHECKPOINT Question **1** gives an opportunity to assess children's ability to add three mixed numbers together, using both methods. Look for children confidently converting between mixed numbers and improper fractions and fluently using common denominators while clearly explaining the steps in the calculations.

Question **3** will assess children's ability to problem solve by adding or subtracting mixed numbers. Look for children being able to translate the problem into a number sentence and deduce what they need to do, before competently using whichever method and operation they prefer to find the answer. Children should be able to explain in their own words their steps and reasoning.

ANSWERS

Question **1**: The total height of the cake is $4\frac{19}{24}$ inches.

Question **2** a): $4\frac{1}{6} - 2\frac{1}{3} = 1\frac{5}{6}$

Question **2** b): $2\frac{1}{3} - \frac{1}{2} = 1\frac{5}{6}$

Question **3**: The area of the blue is $17\frac{1}{6} - 15\frac{4}{9} = 1\frac{13}{18}$ cm^2.

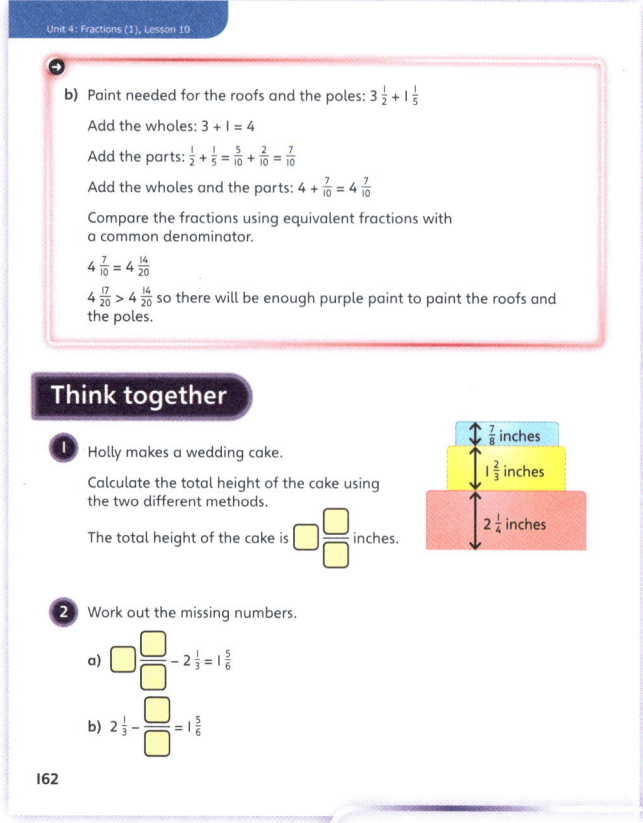

PUPIL TEXTBOOK 6A PAGE 162

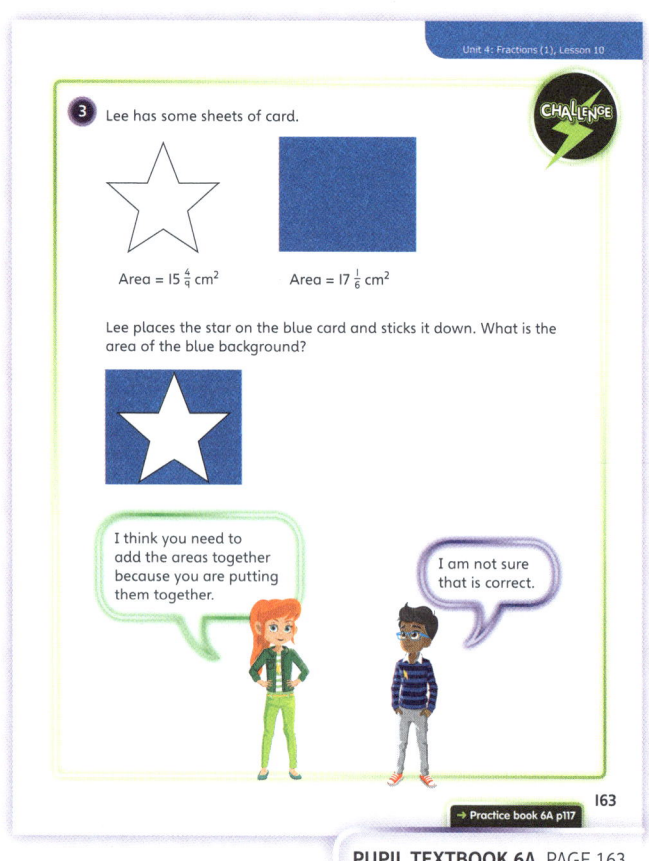

PUPIL TEXTBOOK 6A PAGE 163

Practice

WAYS OF WORKING Independent thinking

IN FOCUS Question ❻ is a multi-step problem that can be solved in more than one way. Ensure children understand what the question is asking and what they need to do to answer it – a common mistake is to only partially answer a problem when it requires multiple steps. Ensure that having found Anna and Georgia's weights they then calculate the difference.

STRENGTHEN In question ❻, strengthen learning by encouraging children to draw three different bar models, one to represent the weight of each baby. This will help them to visualise the calculations they need to do.

DEEPEN Question ❻ can be explored further by giving children a fourth baby and asking them to calculate its mass; for example, ask: *Sophia weighs $\frac{5}{6}$ lb more than Anna. How much does Sophia weigh?*

THINK DIFFERENTLY Question ❹ is a missing number problem in a part-whole model. Explore with children the best place to start, and encourage them to write down the number sentences involved and to explain their reasoning.

ASSESSMENT CHECKPOINT Question ❺ assesses children's ability to add mixed numbers. Look for children confidently using their preferred method and being able to explain their steps in their own words. Note any children who rely on pictorial representation, who will need support to develop their understanding before they move on to abstract calculations.

Question ❻ gives the opportunity to assess children's ability to solve problems that involve adding and subtracting mixed numbers and fractions. Look for children able to translate the problems into number sentences and work systematically through the steps to reach a final answer, while explaining their reasoning in their own words – these children are likely to have mastered the topic.

ANSWERS Answers for the **Practice** part of the lesson appear in the separate **Practice and Reflect answer guide**.

Reflect

WAYS OF WORKING Pair work

IN FOCUS This **Reflect** activity requires children to work backwards and create a problem with a specific answer. Encourage them to use mixed numbers in their problem and to use both addition and subtraction if possible. Ensure children are confident that their answer is $2\frac{1}{3}$ and ask them to give their problem to their partner to check.

ASSESSMENT CHECKPOINT This activity provides an opportunity to assess children's ability to add and subtract mixed numbers. Look for children fluently working backwards to create a problem in abstract format, using their understanding of inverse operations while confidently checking their own work and their friend's.

ANSWERS Answers for the **Reflect** part of the lesson appear in the separate **Practice and Reflect answer guide**.

After the lesson ⏸

- Can children add and subtract mixed numbers confidently?
- Can they solve problems that involve adding and subtracting mixed numbers and fractions?
- Can they break down multi-step problems into the steps and number sentences they need to solve the problem?

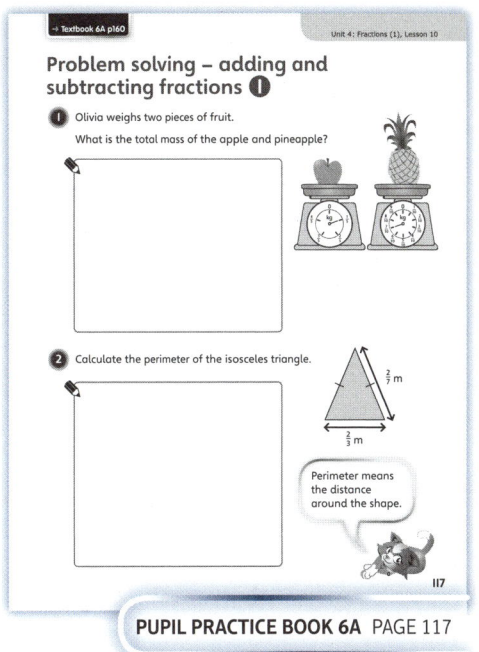

PUPIL PRACTICE BOOK 6A PAGE 117

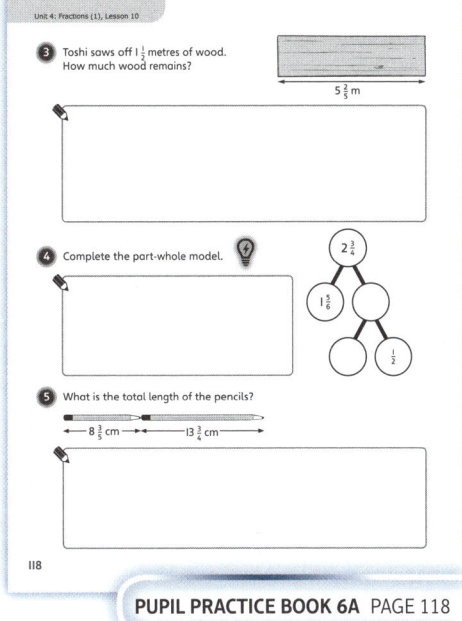

PUPIL PRACTICE BOOK 6A PAGE 118

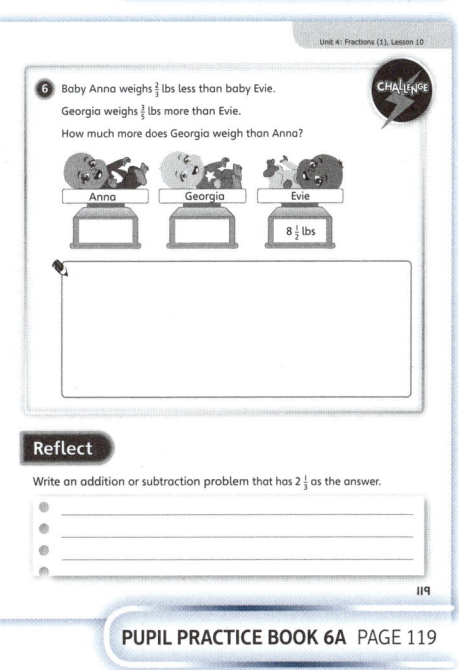

PUPIL PRACTICE BOOK 6A PAGE 119

Problem solving – adding and subtracting fractions ②

Learning focus

In this lesson, children understand how to solve more complex problems that involve adding and subtracting mixed numbers and fractions with more than one step.

Small steps

→ Previous step: Problem solving – adding and subtracting fractions (1)

→ **This step: Problem solving – adding and subtracting fractions (2)**

→ Next step: Multiplying a fraction by a whole number

NATIONAL CURRICULUM LINKS

Year 6 Number – Fractions

Add and subtract fractions with different denominators and mixed numbers, using the concept of equivalent fractions.

ASSESSING MASTERY

Children can solve problems that incorporate adding and subtracting mixed numbers and fractions with more than one calculation, while fluently converting between mixed numbers and improper fractions. Children can clearly explain their steps and methods in their own words.

COMMON MISCONCEPTIONS

A common mistake is that children may not understand what calculation is required, particularly when the information is represented in a problematic way; they will often rush into the problem, perhaps adding all the numbers given. Encourage children to draw diagrams or use number lines to represent the problem. This will help them to understand which calculation is required and to identify which numbers they are going to use before they carry out any calculations. Ask:

- *Would a representation help?*
- *What is the question asking you to do?*
- *What operations do you need to use?*
- *What information do you need to find?*
- *Which numbers are you going to use?*

STRENGTHENING UNDERSTANDING

Children who are struggling with multi-step problems should focus first on problems that involve either addition or subtraction, but not both. Encourage children to use a bar model or a number line to aid their understanding.

GOING DEEPER

Deepen learning with more complex missing number problems that require children to carry out more than one calculation involving addition and subtraction of fractions and mixed numbers, such as $\frac{5}{8} + \frac{\square}{\square} = 1 - \frac{1}{4}$ or $2\frac{1}{5} - \frac{2}{3} = \frac{40}{15} - \frac{\square}{\square}$ or $\square\frac{\square}{\square} + \frac{5}{6} = 4\frac{1}{4} - \frac{2}{3}$ or $\frac{7}{8} - \frac{\square}{\square} = \frac{2}{3} + \frac{\square}{\square}$.

KEY LANGUAGE

In lesson: fraction, mass, perimeter

Other language to be used by the teacher: multi-step, calculation, operation, mixed number, improper fraction, common denominator, numerator, denominator

STRUCTURES AND REPRESENTATIONS

bar models, number lines

 In the eTextbook of this lesson, you will find interactive links to a selection of teaching tools.

Before you teach

- Can children add and subtract mixed numbers?
- Can children use bar models to solve 'missing number' problems?
- Are children confident in solving multi-step problems?

Discover

WAYS OF WORKING Pair work

ASK

• Question ❶ a): *What's the same and what's different about the bowling balls? How can you work out the weight of one yellow bowling ball? Would a bar model help?*
• Question ❶ b): *How could you use your answer from part a) to help with this?*

IN FOCUS Question ❶ introduces children to solving a multi-step problem involving adding and subtracting with mixed numbers; this requires logical thinking and problem solving. Question ❶ a) introduces the first step in the problem, while question ❶ b) requires them to complete the problem using their answer from part a). Encourage children to break down the problem into the steps needed, decoding the language of the question to identify the individual calculations.

PRACTICAL TIPS There are many different everyday contexts for solving multi-step problems involving fractions. This topic uses mass and bowling balls but other contexts could be distance, time or capacity. Whatever the context, it will be important to encourage children to use bar models to represent problems and break them down into the steps needed to find the answers.

ANSWERS

Question ❶ a): The weight of one yellow ball is $4\frac{2}{3}$ kg.

Question ❶ b): The weight of one red striped ball is $1\frac{3}{4}$ kg.

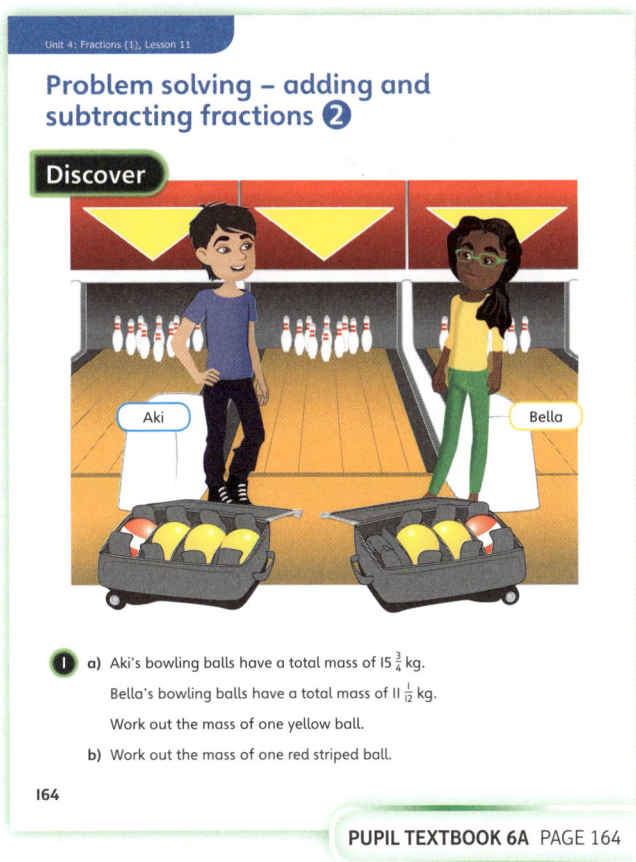

Problem solving – adding and subtracting fractions ❷

Discover

❶ a) Aki's bowling balls have a total mass of $15\frac{3}{4}$ kg.
Bella's bowling balls have a total mass of $11\frac{1}{12}$ kg.
Work out the mass of one yellow ball.

b) Work out the mass of one red striped ball.

164

PUPIL TEXTBOOK 6A PAGE 164

Share

WAYS OF WORKING Whole class teacher led

ASK

• Question ❶ a): *What do the bar models show?*
• Question ❶ a): *What calculation do you need to do to find the mass of one yellow bowling ball? What do you need to find when subtracting mixed numbers?*
• Question ❶ b): *How does knowing the mass of the yellow bowling balls help you to find the mass of the red striped ball? How can you check your answer using the bar model in part a)?*

IN FOCUS Encourage children to use the bar models; discuss how these models have been formed and highlight what is the same and what is different. With question ❶ a), develop prior knowledge by reminding children that they need to find a common denominator to subtract fractions.

In question ❶ b), it will be important for children to realise that they can use their answer from part a) to find the solution. Explore with children how they can calculate the mass of two yellow bowling balls by adding, and then find the mass of the red bowling ball by subtracting; rewriting $11\frac{1}{12}$ will make it easier to subtract. Encourage children to consider the alternative method of converting to improper fractions and subtracting.

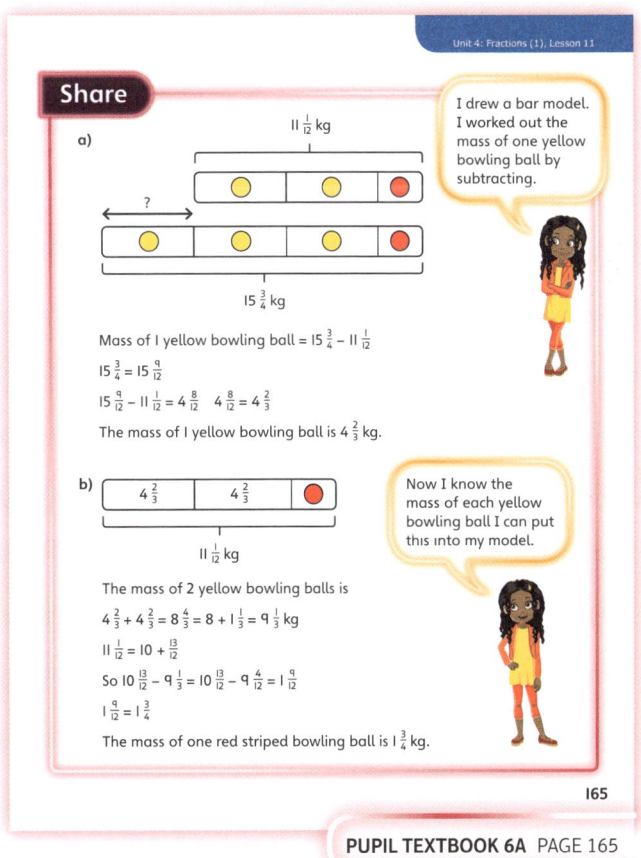

Share

a)
$11\frac{1}{12}$ kg

$15\frac{3}{4}$ kg

I drew a bar model. I worked out the mass of one yellow bowling ball by subtracting.

Mass of 1 yellow bowling ball = $15\frac{3}{4} - 11\frac{1}{12}$

$15\frac{3}{4} = 15\frac{9}{12}$

$15\frac{9}{12} - 11\frac{1}{12} = 4\frac{8}{12}$ $4\frac{8}{12} = 4\frac{2}{3}$

The mass of 1 yellow bowling ball is $4\frac{2}{3}$ kg.

b)
$4\frac{2}{3}$ $4\frac{2}{3}$

$11\frac{1}{12}$ kg

Now I know the mass of each yellow bowling ball I can put this into my model.

The mass of 2 yellow bowling balls is
$4\frac{2}{3} + 4\frac{2}{3} = 8\frac{4}{3} = 8 + 1\frac{1}{3} = 9\frac{1}{3}$ kg

$11\frac{1}{12} = 10\frac{13}{12}$

So $10\frac{13}{12} - 9\frac{1}{3} = 10\frac{13}{12} - 9\frac{4}{12} = 1\frac{9}{12}$

$1\frac{9}{12} = 1\frac{3}{4}$

The mass of one red striped bowling ball is $1\frac{3}{4}$ kg.

165

PUPIL TEXTBOOK 6A PAGE 165

Think together

WAYS OF WORKING Whole class teacher led (I do, We do, You do)

ASK

- Question **1**: *Can you identify what fractions are shaded (or not shaded)? What steps do you need to take to find the answer?*
- Question **3**: *What calculations do you need to do to find the missing numbers?*
- Question **4**: *What do you know about the sides of a square? How can you find the perimeter of a square? Of a rectangle? What calculation are you going to need to do to work out the perimeter?*

IN FOCUS Questions **1** and **2** revisit mixed number problems, similar to those in the Share section. Question **1** uses the context of shapes, requiring children to find the amount not shaded. Encourage children to explain their method; for example, they could add the amounts together to find what is not shaded, or they could find the amount that is shaded and subtract from 1 whole. Look for children who erroneously work out the area of the shape that is shaded, rather than the area not shaded.

Question **3** requires children to identify the calculation needed to work out the missing numbers. Encourage them to find the value of C first and to write down the number sentence before working anything out.

Question **4** involves further contextual problem solving, finding the perimeter of two rooms, using children's prior knowledge of shape. Children may need reminding of the definition of perimeter and the properties of squares.

STRENGTHEN Strengthen learning with question **3** by encouraging children to use the number line to count up to help them find the difference.

DEEPEN Question **4** can be explored further by encouraging children to find other ways of calculating the perimeter of the square room, rather than just adding. Children can also be encouraged to find the perimeters of other rooms on the floor plan, using other mixed number questions or even their own mixed numbers.

ASSESSMENT CHECKPOINT Questions **3** and **4** assess children's ability to solve multi-step number problems involving adding and subtracting. Look for children confidently translating the problems into the number sentences and identifying the steps needed to find the answer, while explaining these steps. Children should be able to convert fluently between mixed numbers and improper fractions and use both methods.

ANSWERS

Question **1**: $\frac{7}{12}$ of the shape is not shaded.

Question **2**: $1\frac{13}{15}$ km

Question **3**: C is $3\frac{2}{3}$ and is $\frac{11}{12}$ greater than A.

Question **4**: The square room has a perimeter of 13 metres. The rectangular room has a perimeter of $12\frac{1}{5}$ metres. The square room has a perimeter that is $\frac{4}{5}$ metres greater.

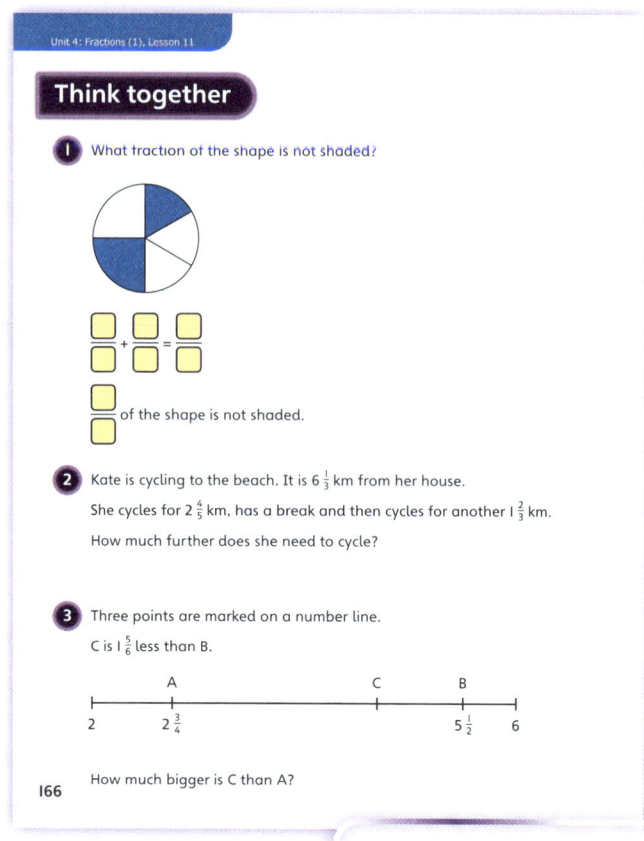

PUPIL TEXTBOOK 6A PAGE 166

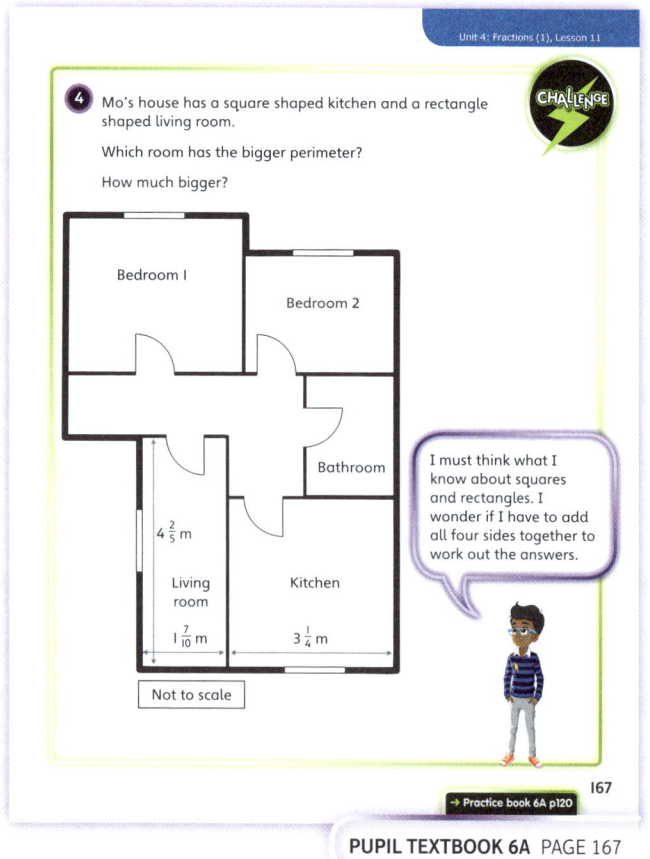

PUPIL TEXTBOOK 6A PAGE 167

Practice

WAYS OF WORKING Independent thinking

IN FOCUS Question ③ introduces a new unit of million, requiring children to find the combined total. Watch out for children who do not complete the problem and only work out the number of downloads on Saturday.

Question ④ requires children to solve a multi-step problem that incorporates both addition and subtraction. Encourage children to solve the problem one step at a time. Watch out for children who do not work out the distance from the top of the drainpipe, but instead just calculate how many metres the spider has climbed up.

Question ⑤ introduces the new notation of AB and BC to identify distances between points on a number line. Explain that this means the distance between A and B and the distance between B and C. This problem requires children to carry out several calculations and it will be beneficial to encourage them to solve the problem one stage at a time.

STRENGTHEN Strengthen learning in question ③ by encouraging children to draw two bar models, to show the number of downloads on Friday and on Saturday; they should label the total of both bars with a question mark.

In question ④, encourage children to draw a diagram to show what is happening to the spider and to help them see what calculations they need to do.

DEEPEN Question ⑤ can be explored further by telling children that B has now been moved so it is exactly half-way between A and C; ask children to calculate the difference between the old value of B and the new value of B.

ASSESSMENT CHECKPOINT Questions ④ and ⑤ require children to use multiple steps and calculations to systematically solve a problem. Children should be able to convert fluently between mixed numbers and improper fractions and use whichever method for adding and subtracting is most suitable for the calculation.

ANSWERS Answers for the **Practice** part of the lesson appear in the separate **Practice and Reflect answer guide**.

Reflect

WAYS OF WORKING Pair work

IN FOCUS This **Reflect** activity encourages children to identify which questions they found difficult. This will provide an opportunity to discuss the difficulties of problem solving – for example, it is easy to make mistakes or miss a step; reassure them that they should not be put off by this. Encourage children to discuss and explain any stumbling blocks or mistakes they have made and to look for ways to correct them.

ASSESSMENT CHECKPOINT Children are likely to identify many different issues, such as not understanding the question, or not completing the question. Look for children who can clearly identify a mistake or stumbling block and suggest strategies for correcting it.

ANSWERS Answers for the **Reflect** part of the lesson appear in the separate **Practice and Reflect answer guide**.

After the lesson ⏸

- Can children confidently translate multi-step problems into the steps and calculations needed?
- Can they solve multi-step problems that incorporate adding and subtracting mixed numbers and fractions?

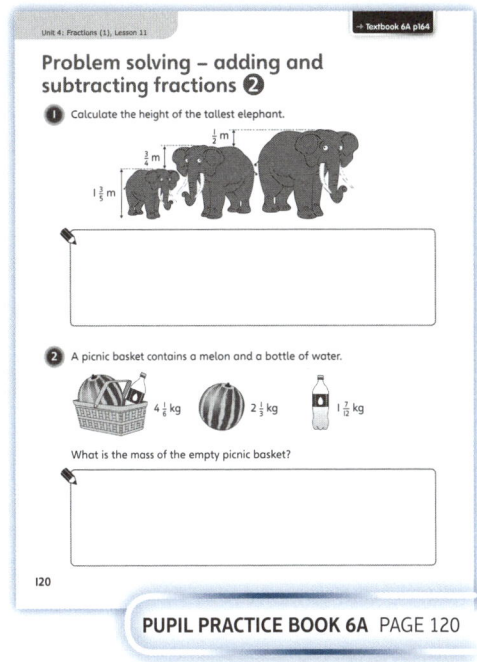

PUPIL PRACTICE BOOK 6A PAGE 120

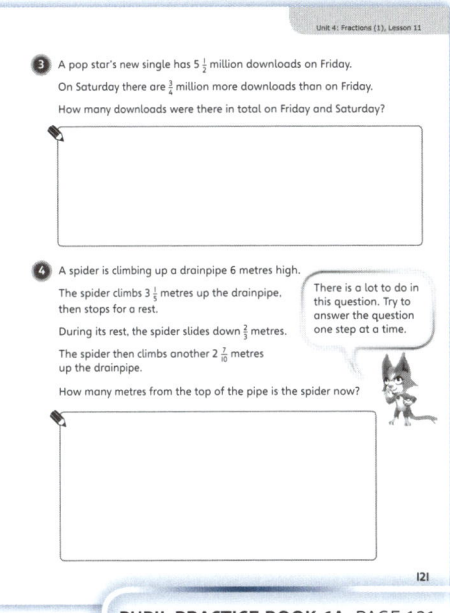

PUPIL PRACTICE BOOK 6A PAGE 121

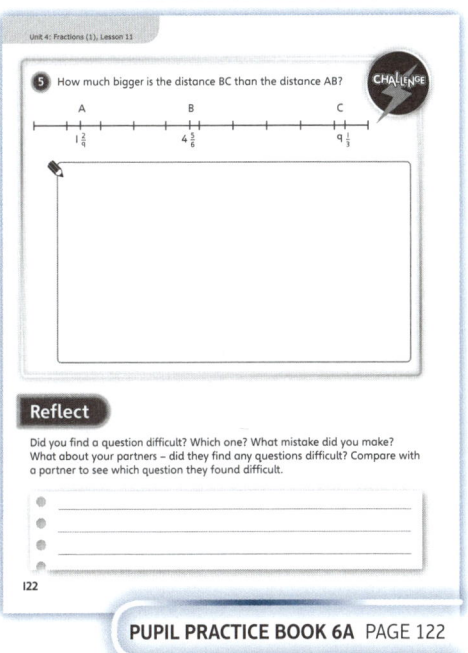

PUPIL PRACTICE BOOK 6A PAGE 122

End of unit check

Don't forget the *Power Maths* unit assessment grid on p26.

WAYS OF WORKING Group work adult led and independent working

IN FOCUS

- Questions **1**, **2** and **4** assess children's ability to add or subtract two proper fractions where the answer exceeds 1. These questions test children's understanding of converting an improper fraction to a mixed number.
- Question **5** requires children to find a missing number; children should realise that they only need to focus on the fractions and look for common denominators by making equivalent fractions.
- Question **7** requires children to problem solve by identifying a fraction from a diagram and choosing the correct calculation. Encourage children to use a bar model to help them visualise what they need to do.

ANSWERS AND COMMENTARY

Children who have mastered this unit will be able to add and subtract fractions and mixed numbers confidently using several formal written methods. They will be able to solve multi-step problems and explain which method is most efficient.

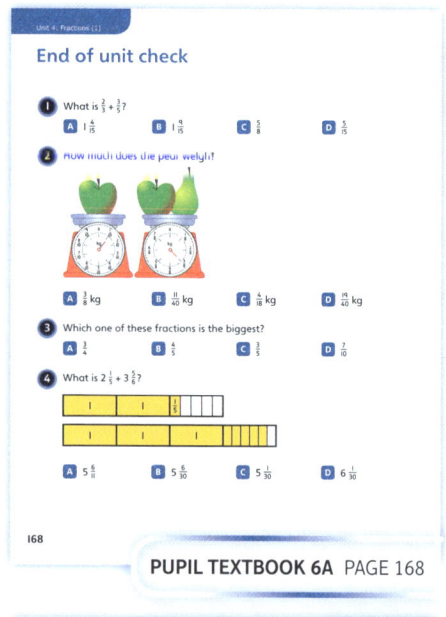

PUPIL TEXTBOOK 6A PAGE 168

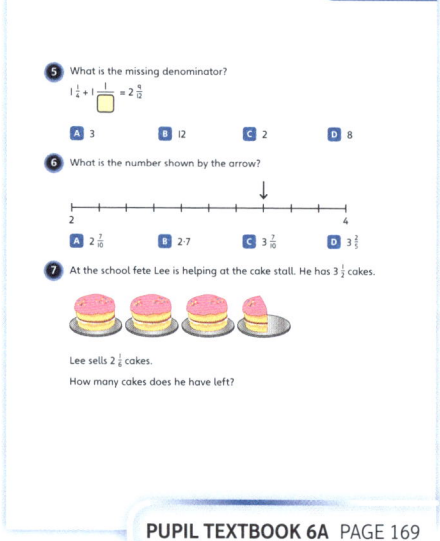

PUPIL TEXTBOOK 6A PAGE 169

Q	A	WRONG ANSWERS AND MISCONCEPTIONS	STRENGTHENING UNDERSTANDING
1	A	B suggests interpreting $\frac{10}{15}$ as 1 whole. C suggests adding of fraction parts. D suggests use of a common denominator without converting numerators.	Support children with finding a common denominator by encouraging them to list the multiples and/or use a bar model.
2	B	A suggests just reading the scale the pear is on. C suggests adding fraction parts. D suggests adding not subtracting.	Encourage children to draw a comparison bar model for both groupings.
3	B	D suggests selecting the fraction with the largest numbers.	Reiterate the need to compare numerators when denominators are equal.
4	D	A suggests adding the wholes, then the numerators and the denominators. B suggests omitting to convert the numerators. C suggests omitting to convert the improper fraction.	Give children some mixed number conversions and ask them to identify which are correct – and to correct the others.
5	C	A suggests dividing 12 by 4. D suggests subtracting 4 from 12.	Encourage children to work with twelfths.
6	D	A suggests children have misread the small gaps; B suggests they have written their answer as a decimal. C suggests they have identified the arrow as > 3 but still counted 7 tenths.	Encourage children to focus on the part of the line between 2 and 3 and identify how many equal parts the line is divided into.
7	$1\frac{1}{3}$	Children may give an unsimplified answer ($1\frac{2}{6}$). They may add instead of subtracting or may ignore the wholes.	Encourage children to subtract the wholes and fractions separately.

My journal

WAYS OF WORKING Independent thinking

ANSWERS AND COMMENTARY

Question ① assesses children's understanding of the steps needed to add and subtract fractions. Look for children fluently using equivalent fractions and converting their answer to a mixed number, while explaining their method. Observe whether children rely on bar models or struggle to find a common denominator.

A) $\frac{2}{3} + \frac{4}{5} = 1\frac{7}{15}$ C) $2\frac{3}{8} - 1\frac{1}{3} = 1\frac{1}{24}$ E) $3\frac{1}{2} + 1\frac{8}{9} = 5\frac{7}{18}$

B) $\frac{7}{10} + \frac{1}{4} = \frac{19}{20}$ D) $3\frac{2}{5} + 4\frac{3}{4} = 8\frac{3}{20}$ F) $\frac{1}{3} + \frac{1}{2} + \frac{1}{4} = 1\frac{1}{12}$

Question ② assesses children's understanding of both methods for subtracting mixed numbers involving conversion of the fractional parts. Look for children clearly explaining both methods and confidently using common denominators and equivalent fractions.

Jamie's method: $5\frac{1}{4} - 3\frac{2}{5} = 4\frac{5}{4} - 3\frac{2}{5} = 4\frac{25}{20} - 3\frac{8}{20} = 1\frac{17}{20}$.

Danny's method: $5\frac{1}{4} - 3\frac{2}{5} = \frac{21}{4} - \frac{17}{5} = \frac{105}{20} - \frac{68}{20} = \frac{37}{20} = 1\frac{17}{20}$.

Power check

WAYS OF WORKING Independent thinking

ASK

• Can you add and subtract fractions using written methods?
• How confident do you feel about converting an improper fraction to a mixed number? Simplifying a fraction? Finding a common denominator?

Power puzzle

WAYS OF WORKING Independent thinking

IN FOCUS Use this to assess whether children can solve missing number fractional calculations, showing understanding of the steps needed to add or subtract unrelated fractions using common denominators.

ANSWERS AND COMMENTARY

① a) The denominators must be 5 and 7, the only factors of 35. The digits needed to make 9 (the wholes) must make 8 or 9 (the fractional parts could be greater than 1); the only options are 3 and 6 or 2 and 6. With 3 and 6, this leaves 2 and 4 for the numerators, giving a fractional answer of $\frac{34}{35}$ (too big) or greater than 1. So the wholes must be 2 and 6 and the fractions must be $\frac{3}{7}$ and $\frac{4}{5}$. Encourage experimentation and answer checking.

① b) Find the total of the top row to give the total of each row or column (7). Then use this to find the missing numbers, starting with rows or columns where only one value is missing.
Second row: $4\frac{3}{4}$, $\frac{2}{3}$, $1\frac{7}{12}$ Third row: $\frac{19}{20}$, $3\frac{5}{6}$, $2\frac{13}{60}$

After the unit ⏸

• Can children confidently compare, order and convert fractions, mixed numbers and improper fractions?
• Can they add and subtract fractions and mixed numbers including solving problems?

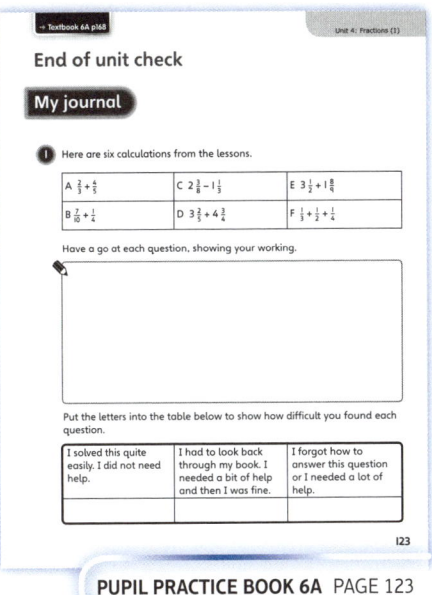

PUPIL PRACTICE BOOK 6A PAGE 123

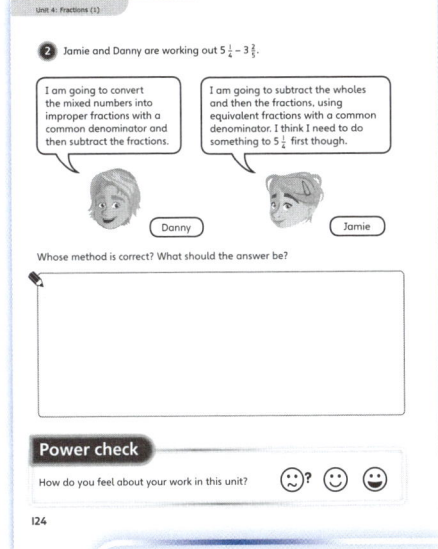

PUPIL PRACTICE BOOK 6A PAGE 124

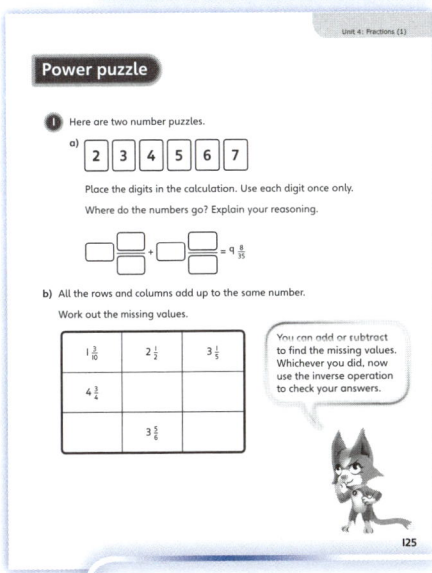

PUPIL PRACTICE BOOK 6A PAGE 125

Strengthen and **Deepen** activities for this unit can be found in the *Power Maths* online subscription.

Unit 5
Fractions ②

Mastery Expert tip! "I found pictorial methods a great way to help children understand what they were doing and the meaning of the question. Too often we just tell children to follow a rule and it is important for them to understand the why behind the methods that they are using. Eventually we hope children will develop the rule for themselves."

Don't forget to watch the Unit 5 video!

WHY THIS UNIT IS IMPORTANT

This unit completes children's primary work on fractions, focusing on multiplying fractions and dividing fractions by a whole number and developing their ability to perform any of the four operations with fractions. Pictorial representations are used throughout to ensure children develop a firm understanding rather than just learning rules. Finally, the unit builds on previous work on fractions of amounts, encouraging children to look for the most efficient methods when solving problems.

WHERE THIS UNIT FITS

→ Unit 4: Fractions (1)
→ **Unit 5: Fractions (2)**
→ Unit 6: Geometry – position and direction

This unit builds on children's fraction work in previous units, including multiplying a proper fraction by a whole number and finding a fraction of an amount. It aims to bring together the four operations with fractions and give children confidence in problem solving with fractions.

Before they start this unit, it is expected that children:
- know how to multiply a proper fraction by a whole number and how to find a fraction of an amount
- have seen a fraction strip above a number line to help add and subtract fractions
- are confident with drawing bar models to represent simple problems
- know the rules for the order of operations
- know how to convert between a mixed number and an improper fraction and vice versa.

ASSESSING MASTERY

Children can multiply any fraction by a whole number and by any other fraction, and divide a fraction by a whole number. They can solve simple and multi-step fraction problems, including problems on fractions of an amount where they are given the fraction and need to find the whole.

COMMON MISCONCEPTIONS	STRENGTHENING UNDERSTANDING	GOING DEEPER
When multiplying a fraction by a whole number, children may multiply both the denominator and the numerator, rather than just the numerator.	Children need to understand that the denominator is unaffected. Use bar models so that they understand the reasons behind the method.	Ask children to compile a set of 'Helpful tips' on multiplying and dividing fractions. Encourage them think about what mistakes others may make.
When presented with questions such as '$\frac{2}{3}$ of a number is 30. What is the number?' children may think that 30 is the whole, not the part, and so find $\frac{2}{3}$ of 30.	Children need to realise that they are finding the whole amount and so need to divide by 2 first and then multiply by 3. Draw a clearly labelled bar model to help them see when they are finding a fraction of an amount and when the whole.	Ask children to explain the most efficient method when solving a problem involving fractions of an amount. For example, when solving 'Which is greater: $\frac{1}{3}$ of 24 or $\frac{1}{4}$ of 24?' children can apply their knowledge of $\frac{1}{3} > \frac{1}{4}$ without working out the values.

WAYS OF WORKING

Use these pages to introduce the unit focus to children. You can use the characters to explore different ways of working.

STRUCTURES AND REPRESENTATIONS

Number line and fraction strip: This model helps children convert between improper fractions and mixed numbers.

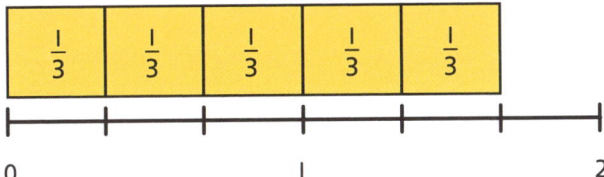

Fraction grid: This model will help children to understand how to multiply fractions, developing the understanding that not only the numerators but also the denominators are multiplied together.

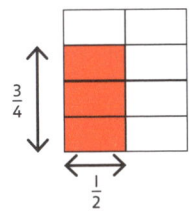

Bar model: This model will help children solve problems involving fractions of an amount, where they are either given the whole or given the fraction of the amount.

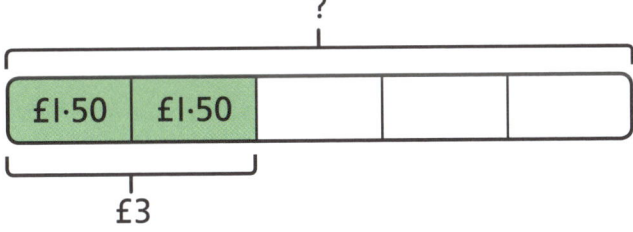

KEY LANGUAGE

There is some key language that children will need to know as part of the learning in this unit:

→ numerator, denominator

→ multiply, divide

→ proper fraction, improper fraction, mixed number, whole number

→ whole, part

→ order of operations

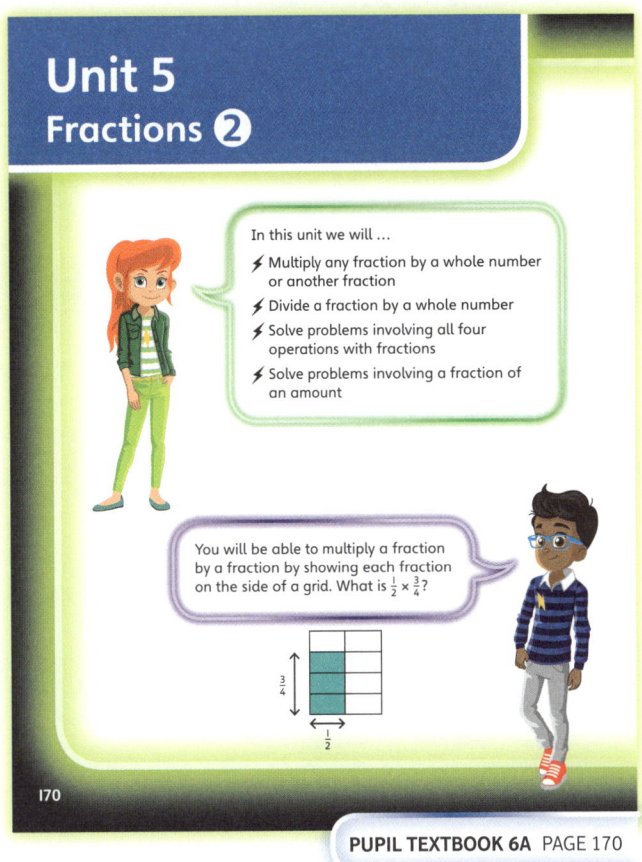

PUPIL TEXTBOOK 6A PAGE 170

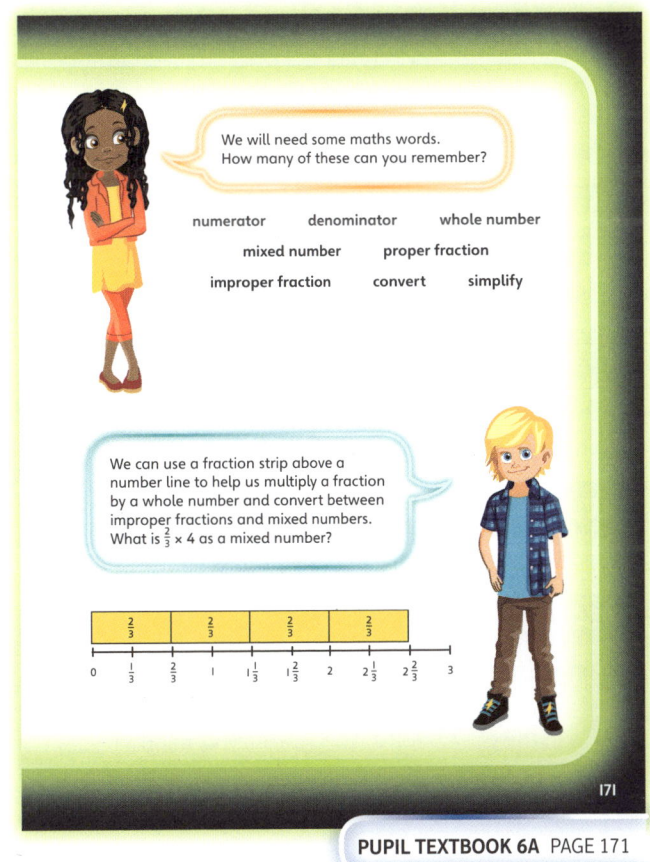

PUPIL TEXTBOOK 6A PAGE 171

Multiplying a fraction by a whole number

Learning focus

In this lesson, children will learn to multiply proper and improper fractions and mixed numbers by a whole number.

Small steps

→ Previous step: Problem solving – adding and subtracting fractions (2)
→ **This step: Multiplying a fraction by a whole number**
→ Next step: Multiplying a fraction by a fraction (1)

NATIONAL CURRICULUM LINKS

Year 5 Number – Fractions

Multiply proper fractions and mixed numbers by whole numbers, supported by materials and diagrams.

ASSESSING MASTERY

Children can multiply any fraction by a whole number, including improper fractions and mixed numbers.

COMMON MISCONCEPTIONS

Children may multiply both the numerator and denominator by the whole number, for example $\frac{2}{3} \times 4 = \frac{8}{12}$. Ask:
- *Can you show me $\frac{2}{3}$ using fraction strips? How can you show $\frac{2}{3} \times 4$ using fraction strips?*
- *The fraction wall shows that $\frac{2}{3}$ and $\frac{8}{12}$ are equivalent. If you multiply a number by 4, will be answer be larger, smaller or the same as the original number? Can $\frac{2}{3} \times 4$ be $\frac{8}{12}$?*

STRENGTHENING UNDERSTANDING

Use fraction strips above number lines and diagrams of fractions to help secure understanding. Discuss with children each step in finding the answer using a pictorial method. Link the representation clearly to the abstract calculation. For example, to show $\frac{2}{3} \times 4$ draw a fraction strip split in thirds, above a number line. Shade in $\frac{2}{3}$ four times. Ask what fraction of the strip is shaded. Use the number line to convert this to a mixed number.

GOING DEEPER

Ask children missing number problems such as $\frac{3}{8} \times \square = \frac{9}{8}$, $\frac{\square}{6} \times 3 = \frac{33}{6}$, $\frac{3}{4} \times \square = 3\frac{3}{4}$. Extend to problems where the answer has been simplified, for example $\frac{5}{6} \times \square = \frac{10}{3}$, $\frac{3}{10} \times \square = 3\frac{3}{5}$.

KEY LANGUAGE

In lesson: whole number, fraction, multiply, mixed number, improper fraction, numerator

Other language to be used by the teacher: denominator, proper fraction

STRUCTURES AND REPRESENTATIONS

bar model, number line

RESOURCES

Optional: fraction strips, fraction circles

 In the eTextbook of this lesson, you will find interactive links to a selection of teaching tools.

Before you teach ⏸

- Do children know their times-tables?
- Do children remember how to multiply a fraction by a whole number from Year 4?
- Can children represent a fraction using a fraction strip?

Discover

ASK

• Question ❶ a): *What information do you need? What do you have to do to get the answer?*
• Question ❶ b): *What is the same about this question? What is different? How will you work out the total time?*

IN FOCUS Question ❶ a) requires a fraction to be multiplied by a whole number. Some information is given in the question text which needs to be combined with relevant information from the image. Discuss the approaches that children may take to answer this question, for example multiplication or addition. Encourage children to draw diagrams or use fraction shapes to explain their methods. Question ❶ b) requires a grids number to be multiplied by a whole number, using all the information given in the image. Again, discuss the different approaches that children may take to answer the question.

PRACTICAL TIPS Have fraction strips or fraction circles available for children to fit together.

ANSWERS

Question ❶ a): $1\frac{2}{3}$ tanks of fuel are used in a day.

Question ❶ b): The total duration of the boat trips is $6\frac{1}{4}$ hours.

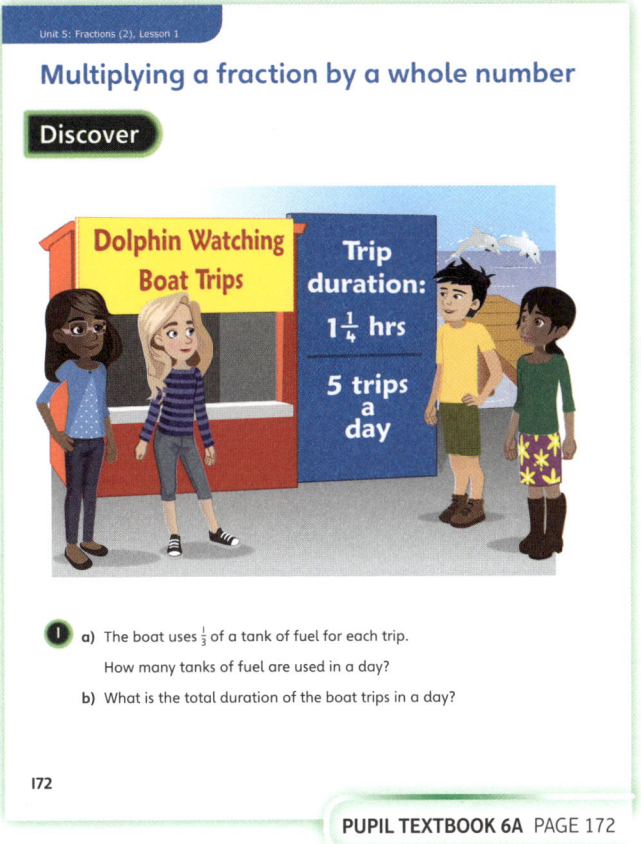

Multiplying a fraction by a whole number

Discover

❶ a) The boat uses $\frac{1}{3}$ of a tank of fuel for each trip. How many tanks of fuel are used in a day?

b) What is the total duration of the boat trips in a day?

172

PUPIL TEXTBOOK 6A PAGE 172

Share

ASK

• Question ❶ a): *What are the two methods shown? Why do they both give the same answer?*
• Question ❶ a): *How does the diagram show the answer to the question? Did anyone show it with their equipment in a different way?*
• Question ❶ a): *Why is the answer given as $1\frac{2}{3}$, not $\frac{5}{3}$?*
• Question ❶ b): *Can you draw a diagram to explain method 1? What about method 2? Why do they give the same answer? Which method do you prefer?*

IN FOCUS Question ❶ a) shows two methods: addition and multiplication. Discuss how the diagram shows both methods, establishing that repeated addition is the same as multiplication. Explain that it makes more sense to give the answer as a mixed number than as an improper fraction.

Question ❶ b) also shows two methods: multiplying the whole and the fraction separately and then adding them; or converting the mixed number to an improper fraction before multiplying by the number. Both methods require the conversion of an improper fraction into a mixed number. Also use the diagrams to show the total of $\frac{25}{4}$. Explain that it makes sense to change the answer into a mixed number as we would say 6 and a quarter hours rather than 25 quarter hours.

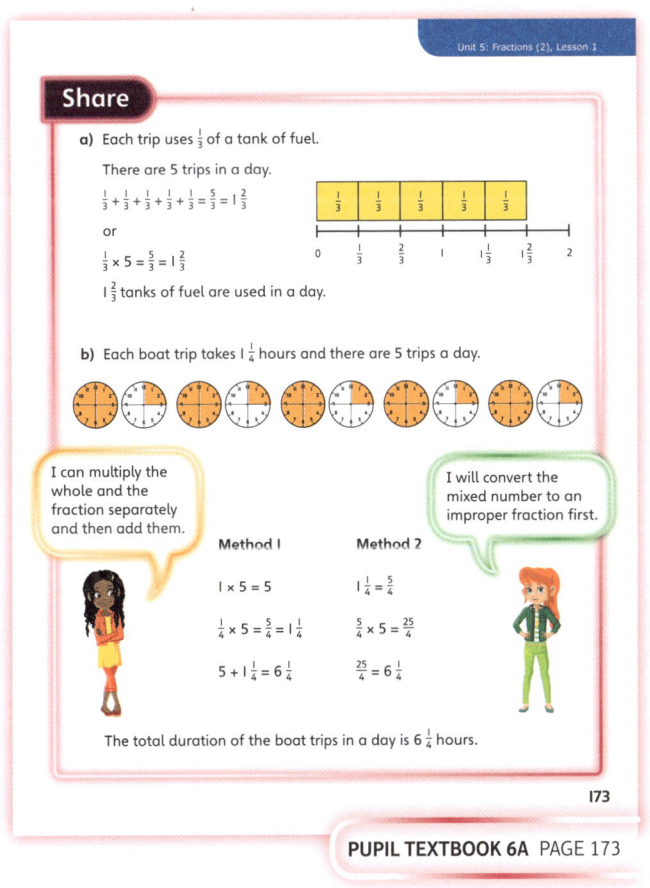

Share

a) Each trip uses $\frac{1}{3}$ of a tank of fuel.

There are 5 trips in a day.

$\frac{1}{3} + \frac{1}{3} + \frac{1}{3} + \frac{1}{3} + \frac{1}{3} = \frac{5}{3} = 1\frac{2}{3}$

or

$\frac{1}{3} \times 5 = \frac{5}{3} = 1\frac{2}{3}$

$1\frac{2}{3}$ tanks of fuel are used in a day.

b) Each boat trip takes $1\frac{1}{4}$ hours and there are 5 trips a day.

I can multiply the whole and the fraction separately and then add them.

I will convert the mixed number to an improper fraction first.

Method 1

$1 \times 5 = 5$

$\frac{1}{4} \times 5 = \frac{5}{4} = 1\frac{1}{4}$

$5 + 1\frac{1}{4} = 6\frac{1}{4}$

Method 2

$1\frac{1}{4} = \frac{5}{4}$

$\frac{5}{4} \times 5 = \frac{25}{4}$

$\frac{25}{4} = 6\frac{1}{4}$

The total duration of the boat trips in a day is $6\frac{1}{4}$ hours.

173

PUPIL TEXTBOOK 6A PAGE 173

Think together

WAYS OF WORKING Whole class teacher led (I do, We do, You do)

ASK

- Question **1**: *How does this question differ from question **1** a) in **Discover**? How many more thirds do you need to add to the fraction strip? How should the answer be presented?*
- Question **2**: *What do you need to multiply? What do you do in the first method? What about the second method? Which method do you prefer? Why?*
- Question **2**: *What do you notice about the numerator of the fraction and the numerator of the final answer?*
- Question **3** a): *What connection can you see between the numerator of the fraction, the whole number and the numerator of the answer?*
- Question **3** b): *How can you use what you discovered in part a) to help you? Can you find more than one answer for each number?*

IN FOCUS Question **2** requires children to use both methods of multiplying a mixed number by a whole number that were described in **Share**. Some children may notice that they multiply the numerator by the whole number when they are working through these methods. Question **3** a) is designed to develop this understanding by asking children to spot patterns. They should see that the numerator of the answer is the same as the numerator of the fraction multiplied by the whole number. They should also see that the denominator does not change. Explore these findings using fraction strips. Children then use this understanding to answer question **3** b).

STRENGTHEN In question **2**, give children fraction circles to model each method. Discuss how the way that they are manipulating the circles relates to the steps in the method.

DEEPEN In question **3** b), ask children to find more than one answer for each number. Challenge them to find a number that can be made in four ways by multiplying a fraction and a whole number. They can give their numbers to a friend to work out.

ASSESSMENT CHECKPOINT Use questions **1** and **2** to assess whether children can multiply a fraction and a mixed number by a whole number. They should know different methods for getting to the answer.

ANSWERS

Question **1**: $\frac{1}{3} \times 7 = \frac{7}{3} = 2\frac{1}{3}$. $2\frac{1}{3}$ tanks of fuel are used.

Question **2**: Method 1: $1 \times 4 = 4, \frac{2}{5} \times 4 = \frac{8}{5} = 1\frac{3}{5}, 4 + 1\frac{3}{5} = 5\frac{3}{5}$

Method 2: $1\frac{2}{5} = \frac{7}{5}, \frac{7}{5} \times 4 = \frac{28}{5}, \frac{28}{5} = 5\frac{3}{5}$

The boat travels $5\frac{3}{5}$ km.

Question **3** a): $\frac{5}{4}, \frac{9}{4}; \frac{10}{6}, \frac{25}{6}, \frac{35}{6}$

Question **3** b): Numerous answers, for example:

$\frac{1}{8} \times 5, \frac{5}{8} \times 1, \frac{1}{16} \times 10, \frac{1}{24} \times 15$

$\frac{1}{9} \times 10, \frac{2}{9} \times 5, \frac{5}{9} \times 2, \frac{10}{9} \times 1$

$\frac{1}{5} \times 6, \frac{2}{5} \times 3, \frac{3}{5} \times 2, \frac{6}{5} \times 1$

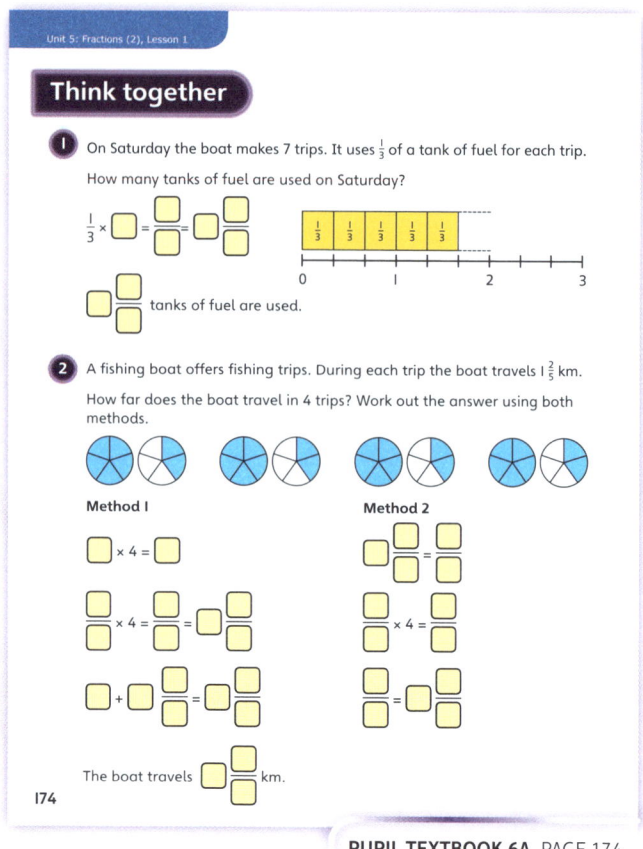

PUPIL TEXTBOOK 6A PAGE 174

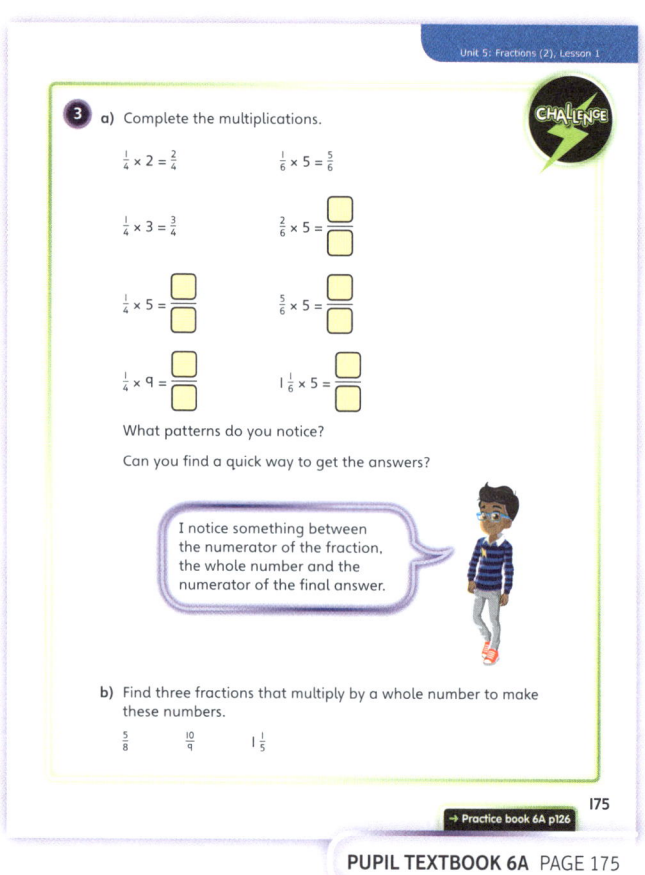

PUPIL TEXTBOOK 6A PAGE 175

Practice

→ Textbook 6A p172

WAYS OF WORKING Independent thinking

IN FOCUS All the fractions in questions **1** and **2** are less than 1. Question **1** includes images and representations to help children, whereas question **2** requires them to begin to recognise and use the fact that the numerator of the answer is the same as the numerator multiplied by the whole number.

Question **3** requires children to use both methods to multiply a mixed number by a whole number. In question **4**, they need to select a method for themselves. Discuss with children which method they have chosen and why. This could extend into a discussion of times when one method is more appropriate.

Question **5** has been designed to highlight a common misconception, where children multiply the denominator as well as the numerator by the whole number.

STRENGTHEN Provide children with fraction strips and fraction circles to model questions **1** and **2**. Providing a number line with the fraction strips will help them convert between the improper fractions and mixed numbers. Use the fraction strips to establish why the denominator stays the same, but the numerator changes.

DEEPEN Ask children questions that elicit a deeper understanding. For example, ask which method they would use to work out $125\frac{2}{7} \times 8$. Ask them to explain whether changing the fraction to an improper fraction would be an efficient method.

THINK DIFFERENTLY In question **6**, children need to realise that the answer is not $\frac{11}{5}$ or $2\frac{1}{5}$ as the question asks how many bags the owner needs to buy. The owner therefore needs 3 bags.

ASSESSMENT CHECKPOINT Use question **2** to assess whether children can multiply a proper fraction by a whole number. Use question **4** to assess whether they can multiply improper fractions and mixed numbers by a whole number. Check whether they are using efficient methods.

ANSWERS Answers for the **Practice** part of the lesson appear in the separate **Practice and Reflect answer guide**.

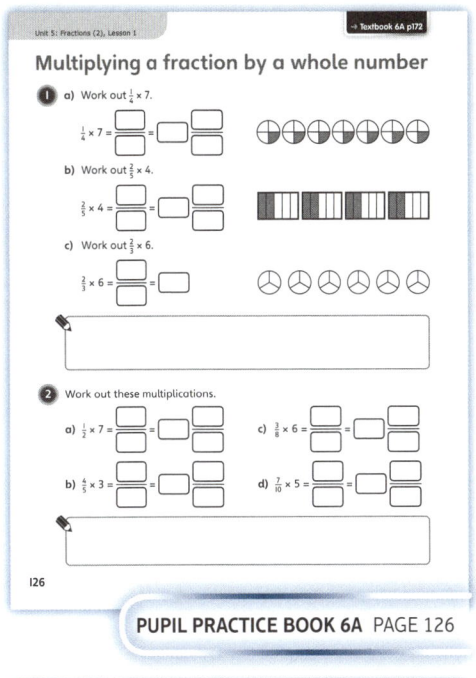

PUPIL PRACTICE BOOK 6A PAGE 126

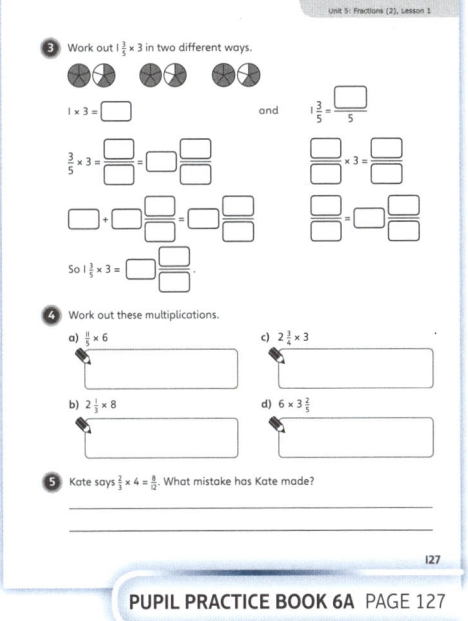

PUPIL PRACTICE BOOK 6A PAGE 127

Reflect

WAYS OF WORKING Independent thinking

IN FOCUS Children can use manipulatives, diagrams and words to help explain why $1\frac{2}{3} \times 4$ is equal to $6\frac{2}{3}$. They can use either method of multiplying a mixed number by a whole number.

ASSESSMENT CHECKPOINT Children should be able to show that they can multiply a mixed number by a whole number. They must provide some reasoning and not just show the answer. Look for explanations that discuss the rule of multiplying the numerator by the whole number to get the numerator of the final answer.

ANSWERS Answers for the **Reflect** part of the lesson appear in the separate **Practice and Reflect answer guide**.

After the lesson ⏸

- Can children use diagrams to explain why a proper fraction multiplied by a whole number gives a particular answer?
- Can they multiply a mixed number by a whole number?

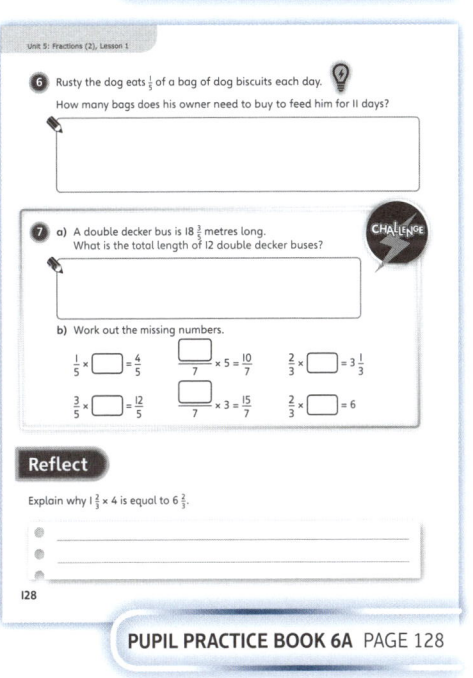

PUPIL PRACTICE BOOK 6A PAGE 128

Multiplying a fraction by a fraction ❶

Learning focus

In this lesson, children will learn to multiply a fraction by a fraction. They will use visual aids such as divided squares to support their understanding.

Small steps

→ Previous step: Multiplying a fraction by a whole number

→ **This step: Multiplying a fraction by a fraction (1)**

→ Next step: Multiplying a fraction by a fraction (2)

NATIONAL CURRICULUM LINKS

Year 6 Number – Fractions

Multiply simple pairs of proper fractions, writing the answer in its simplest form [for example, $\frac{1}{4} \times \frac{1}{2} = \frac{1}{8}$].

ASSESSING MASTERY

Children can multiply a fraction by a fraction by drawing diagrams, and express their answers in their simplest form. They understand that when a proper fraction is multiplied by a proper fraction the answer will be smaller.

COMMON MISCONCEPTIONS

Children may think that they have to make the denominators the same (as in addition and subtraction). Although this will lead to a correct answer, it is not efficient. Ask:

• *How can you divide a square to show $\frac{2}{3} \times \frac{1}{4}$? What do you need to divide the side and bottom of the square into?*

STRENGTHENING UNDERSTANDING

To strengthen understanding, give children a simple calculation where both numerators are 1 (for example, $\frac{1}{3} \times \frac{1}{2}$). Draw two squares, one divided into three bars and one into two columns. Give children a blank square. In pairs, ask them to discuss how they need to divide each side of the square to show the calculation. Establish that the square shows that the answer will be smaller than either of the fractions.

GOING DEEPER

Give children a calculation (for example, $\frac{5}{6} \times \frac{3}{4}$). Ask them to find the answer by drawing two different diagrams (for example, a 4 by 6 square and a 6 by 4 square) and to explain why this shows that the order of the fractions does not matter. Encourage them to simplify their answer when possible.

KEY LANGUAGE

In lesson: whole, proper fraction, numerator, denominator, simplest form

Other language to be used by the teacher: fraction of a fraction

RESOURCES

Optional: paper plates, square templates

 In the eTextbook of this lesson, you will find interactive links to a selection of teaching tools.

Before you teach ⏸

• Do children understand what a fraction is?
• Can children draw a diagram to represent a fraction?
• Do children understand the concept of multiplication?

Discover

WAYS OF WORKING Pair work

ASK

• Question ❶ a): *If Bella and Amal use half of the oats in this bag, will there be any oats left in the bag? Will this be more or less than when they started? How do you know?*
• Question ❶ b): *What diagram could you draw to help you find out how much butter Bella and Amal need?*
• Question ❶ b): *In part a), you were asked for a half of a half, Do you think the answer is going to be bigger or smaller than the answer to part a)? How do you know?*

IN FOCUS Question ❶ a) has been designed to draw out the understanding that a fraction of a fraction is smaller than either of the fractions. The children can see this from the bag of oats because there will only be a quarter of the bag left.

PRACTICAL TIPS Ask children to look at baking ingredients and different recipes. Bring in the half bag of oats and other ingredients to provide a visual aid. Encourage children to draw on diagrams or draw a pictorial representation of the oats.

ANSWERS

Question ❶ a): Bella and Amal need to use $\frac{1}{4}$ of a bag of oats.

Question ❶ b): Bella and Amal need to use $\frac{3}{8}$ of a block of butter.

PUPIL TEXTBOOK 6A PAGE 176

Share

WAYS OF WORKING Whole class teacher led

ASK

• Question ❶ a): *What does the whole square represent? Why has Dexter divided the square into 4? How many small squares represent the oats in the bag before Bella and Amal start cooking? What does the diagram show?*
• Question ❶ a): *Why does Flo say that $\frac{1}{2}$ of $\frac{1}{2}$ means the same as $\frac{1}{2} \times \frac{1}{2}$?*
• Question ❶ b): *Is there more than one way to divide the diagram up to show the answer?*

IN FOCUS

Question ❶ a) makes the link between a 'fraction of a fraction' and a 'fraction multiplied by a fraction'. The fractions in question ❶ b) have been chosen to be harder to visualise than those in part a), so that children understand the advantage of using a diagram. Discuss whether it makes any difference whether the diagram is divided into 2 columns and 4 rows or 4 columns and 2 rows.

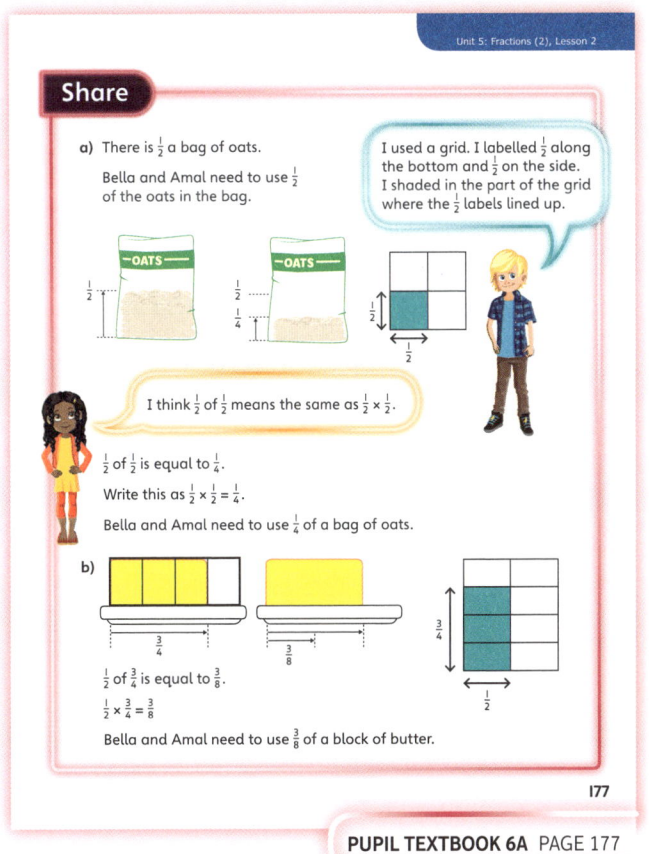

PUPIL TEXTBOOK 6A PAGE 177

Think together

WAYS OF WORKING Whole class teacher led (I do, We do, You do)

ASK

- Question ①: *What does the whole square represent? Why has it been divided into 9?*
- Question ②: *How many blocks have you shaded? How many blocks are there in total? What fraction of an hour do the shaded blocks represent? Can you simplify your answer?*
- Question ③ a): *What is the same and what is different about the two diagrams?*
- Question ③ b): *How else can you write $\frac{2}{5}$ of $\frac{1}{4}$?*

IN FOCUS Question ① requires children to multiply two unit fractions. Encourage children to draw a diagram for themselves, rather than just using the one in the **Textbook**. Question ② uses two non-unit fractions, so children need to shade an array of blocks in their diagram. It also changes the context to time, giving children a different way to visualise the concept by looking at a circular diagram. The fractions have been chosen so that the answer can be simplified from $\frac{6}{12}$ to $\frac{1}{2}$. Question ③ gives children the opportunity to begin to think about multiplying fractions without a context. The two diagrams both show the same multiplication, to emphasise that multiplication is commutative. All the answers in question ③ can be simplified.

STRENGTHEN To support understanding in question ②, provide children with paper plates that they can draw on or cut up to model the representation. Ask them how they could show the time on the clock divided into quarters and how they could use this to find $\frac{3}{4}$. In question ③, make the link from $\frac{1}{4} \times \frac{2}{5}$ in part a) to $\frac{2}{5}$ of $\frac{1}{4}$ in part b).

DEEPEN Ask children to explore using a circular diagram (similar to the clock shown in question ②) to build conceptual flexibility. Encourage them to think about which diagram they prefer and which diagram is more efficient for different questions. Challenge them to write a question that will be difficult to represent on a circular diagram.

ASSESSMENT CHECKPOINT Use questions ① and ② to assess whether children can use a diagram to multiply two fractions. Check whether they can explain why the answer is smaller than the starting fractions and whether they recognise the link between 'a fraction of a fraction' and 'a fraction multiplied by a fraction'. Use question ③ to assess whether they are confident working on abstract calculations.

ANSWERS

Question ①: $\frac{1}{3} \times \frac{1}{3} = \frac{1}{9}$
Bella will use $\frac{1}{9}$ of the bag of sugar.

Question ②: $\frac{3}{4} \times \frac{2}{3} = \frac{6}{12} (= \frac{1}{2})$
$\frac{6}{12}$ (or $\frac{1}{2}$) of the hour will have passed.

Question ③ a): Both of the diagrams are correct, because they both show 2 parts shaded out of 20 equal parts.

Question ③ b): $\frac{3}{12}$ (or $\frac{1}{4}$) $\frac{2}{20}$ (or $\frac{1}{10}$)

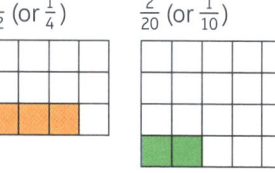

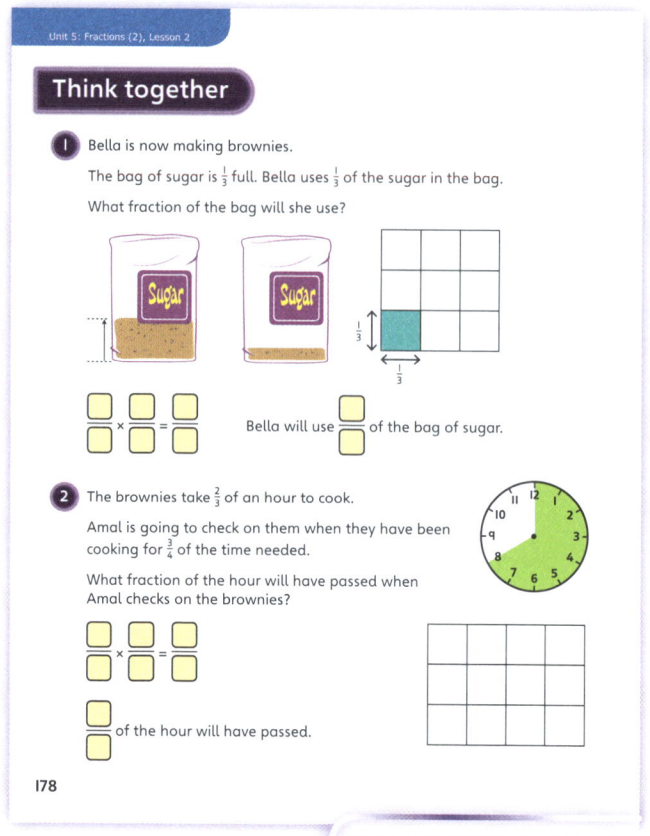

PUPIL TEXTBOOK 6A PAGE 178

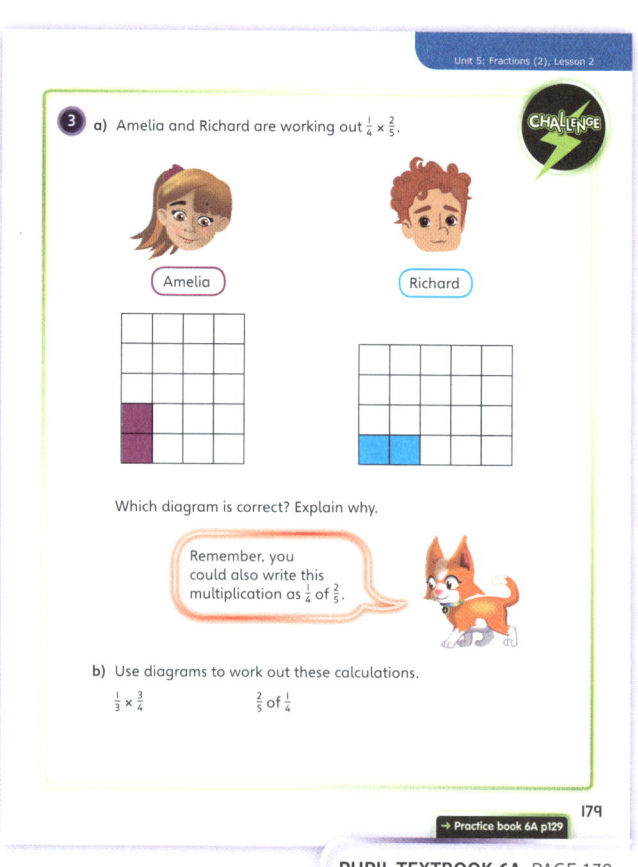

PUPIL TEXTBOOK 6A PAGE 179

Practice

WAYS OF WORKING Independent thinking

IN FOCUS Question ❶ aims to consolidate children's understanding of calculating a fraction of a fraction and links back to the context of ingredients. Question ❷ is abstract, gradually reducing the amount of scaffolding provided. It also emphasises the link between '×' and 'of'. The fractions in question ❸ have been chosen such that the answers can be simplified. Children may need reminding of the meaning of 'simplest form'. Question ❺ incorporates a great deal of reasoning to provide an explanation why this is true. Encourage children to explore with various examples, trying to find examples where this is false.

STRENGTHEN Provide children with templates of squares for them to draw on, especially for the challenge question. Encourage them to consolidate their understanding by drawing arrays in all the questions, rather than trying to find short cuts or follow processes they may have discovered.

DEEPEN Encourage children to spot patterns between the fractions in a question and the fraction in the answer. Can they recognise for themselves that the numerators have been multiplied and the denominators have been multiplied? If they see this pattern, ask them to use the diagram to explain this finding.

THINK DIFFERENTLY Question ❹ provides the diagram and wants to know what the question *could* be. Encourage children to reflect on whether their answer is the only possibility and, if not, to work out a different answer. This may require them to rearrange the shaded blocks to show different fractions.

ASSESSMENT CHECKPOINT Use questions ❶ and ❷ to assess whether children can use a diagram to multiply two fractions. Use questions ❷ and ❸ to assess whether they can draw the diagrams for themselves without scaffolding. Throughout, children should be able to refer back to a practical context such as a bag of sugar to ensure understanding. Use question ❺ to assess whether children have a conceptual understanding.

ANSWERS Answers for the **Practice** part of the lesson appear in the separate **Practice and Reflect answer guide**.

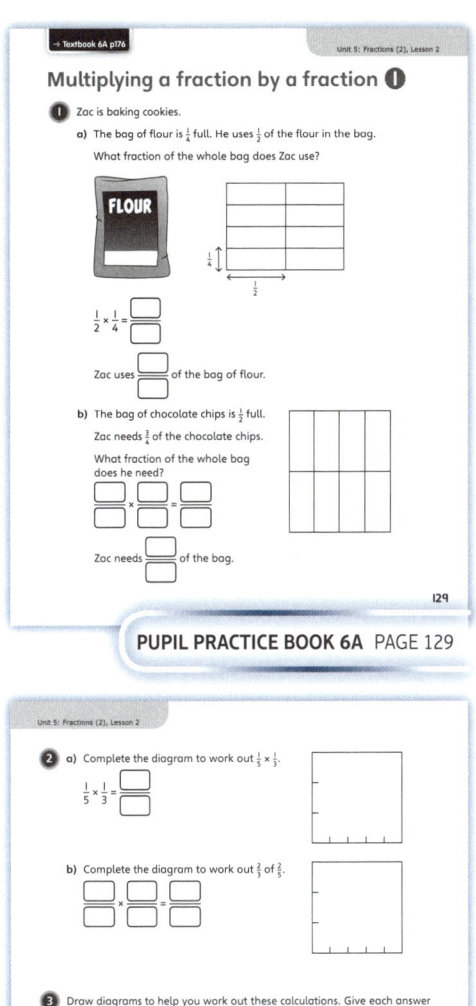

PUPIL PRACTICE BOOK 6A PAGE 129

PUPIL PRACTICE BOOK 6A PAGE 130

Reflect

WAYS OF WORKING Independent thinking

IN FOCUS This activity checks that children understand how to multiply two proper fractions and that they have not completed the practice through just following a process. If they cannot draw a suitable diagram then their understanding is unlikely to be deep enough to move on.

ASSESSMENT CHECKPOINT Check that children can draw a diagram and use it to explain why the answer is $\frac{3}{10}$. They may also be able to explain which parts of the fraction have been multiplied and where that can be seen in the diagram.

ANSWERS Answers for the **Reflect** part of the lesson appear in the separate **Practice and Reflect answer guide**.

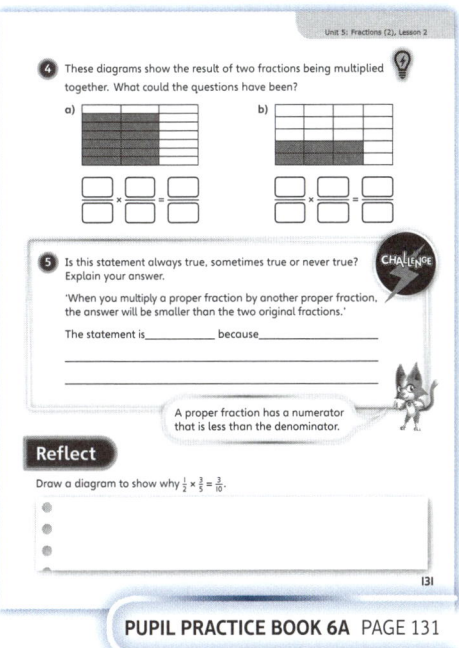

PUPIL PRACTICE BOOK 6A PAGE 131

After the lesson

- Can children find a fraction of a fraction by drawing a diagram?
- Can children multiply a fraction by a fraction by drawing a diagram?
- Can children explain, using a diagram, why when they multiply two fractions together the answer will be smaller than either of the starting fractions?

Multiplying a fraction by a fraction ❷

Learning focus

In this lesson, children will learn to multiply a fraction by a fraction by multiplying the numerators and multiplying the denominators.

Small steps

➜ Previous step: Multiplying a fraction by a fraction (1)
➜ **This step: Multiplying a fraction by a fraction (2)**
➜ Next step: Dividing a fraction by a whole number (1)

NATIONAL CURRICULUM LINKS

Year 6 Number – Fractions

Multiply simple pairs of proper fractions, writing the answer in its simplest form [for example, $\frac{1}{4} \times \frac{1}{2} = \frac{1}{8}$].

ASSESSING MASTERY

Children can multiply together two or more fractions by multiplying the numerators and multiplying the denominators. They can explain their understanding by drawing a diagram.

COMMON MISCONCEPTIONS

Children may think that they have to make the denominators the same (as in addition and subtraction). Although this will lead to a correct answer, it is not efficient. Some children may mix up which process they need to use for multiplying fractions if they do not have a conceptual understanding. As a result, they may try to use the process used for addition, subtraction or division. Ask:

• *How can you divide a square to show $\frac{3}{4} \times \frac{1}{2}$? What will the denominator of the answer be? What will the numerator of the answer be?*

STRENGTHENING UNDERSTANDING

To strengthen understanding, model a calculation such as $\frac{3}{5} \times \frac{3}{4}$ using a square divided into a 5×4 array. Establish that the whole is made up of 5 × 4 parts and link this to multiplying the denominators. Similarly, establish that the number of shaded parts is 3 × 3, which is the two numerators multiplied together.

GOING DEEPER

Challenge children to create a pictorial representation for multiplying together three fractions. They may need a hint that the representation should be 3D. Ask them to explain how their diagram shows that it does not matter in which order the fractions are multiplied together.

KEY LANGUAGE

In lesson: numerator, denominator, simplest form, simplify

Other language to be used by the teacher: whole, fraction of a fraction, proper fraction, equivalent fractions, area

RESOURCES

Optional: square templates, counters, fraction walls

 In the eTextbook of this lesson, you will find interactive links to a selection of teaching tools.

Before you teach ⏸

• Can children draw a diagram to represent a fraction?
• Do children understand the concept of multiplication?
• Do children understand equivalent fractions?
• Can children multiply two fractions together using a diagram?

Discover

WAYS OF WORKING Pair work

ASK

- Question ❶ a): *What could you draw to help you? What would the diagram show?*
- Question ❶ b): *How could you prove that your answers are correct? What operation have you used?*
- Question ❶ b): *What do you notice about the numerators in the question and the numerator in the answer? What do you notice about the denominators?*

IN FOCUS Question ❶ b) encourages children to look at their answers and the questions, and try to spot links. This gives all children the opportunity to make the link for themselves. By using a diagram, children will be able to see what part the denominator represents and what part the numerators represent. By linking to area, children will be able to see why the operation is multiplication. Discussing the two methods will allow you to talk about efficiency and in what circumstances they would want to draw a diagram and why it is important to be able to do so.

PRACTICAL TIPS Use templates of squares and counters to illustrate different multiplications. Refer back to the context of baking ingredients from the previous lesson.

ANSWERS

Question ❶ a): $\frac{1}{6}$, $\frac{3}{20}$, $\frac{2}{10}$ $(= \frac{1}{5})$

Question ❶ b): The numerators can be multiplied together and the denominators can be multiplied together. The third answer can be simplified.

Share

WAYS OF WORKING Whole class teacher led

ASK

- Question ❶ a): *What does the total number of blocks represent? What do the shaded blocks represent?*
- Question ❶ b): *Is there another way of explaining why the answer is $\frac{1}{5}$? ($\frac{1}{2}$ of $\frac{2}{5}$ is $\frac{1}{5}$) If I said $\frac{1}{2}$ of $\frac{2}{5}$, what diagram could you draw to represent this? How can the answer be $\frac{2}{10}$ AND $\frac{1}{5}$?*

IN FOCUS By modelling the multiplication using a square the link is made between the denominator of the answer being the total number of blocks and the numerator being the number of shaded blocks. This develops the understanding that the numerators are multiplied together and similarly the denominators. Links can be made to finding the area of a rectangle. The last calculation has been chosen to show that some answers can be simplified.

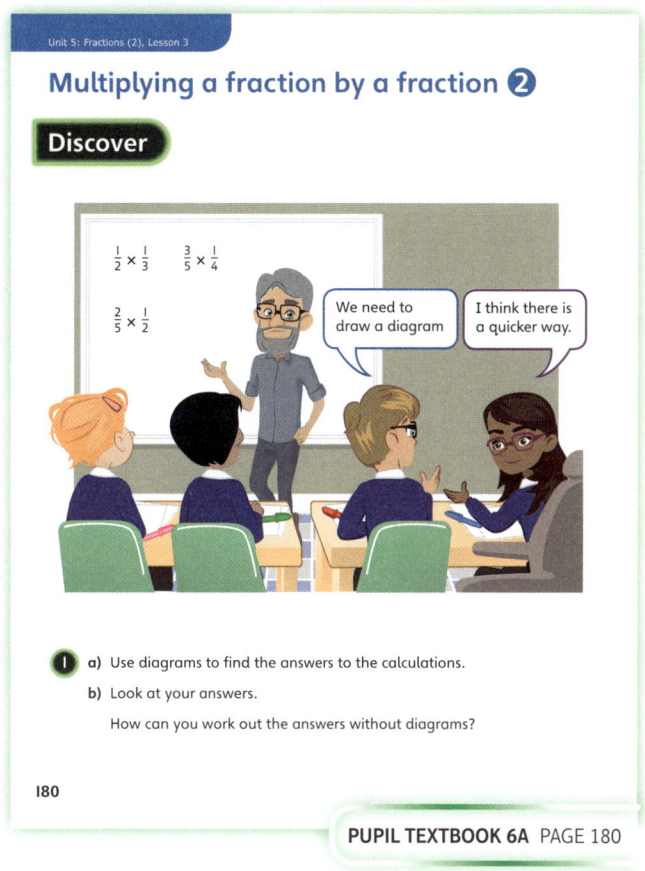

PUPIL TEXTBOOK 6A PAGE 180

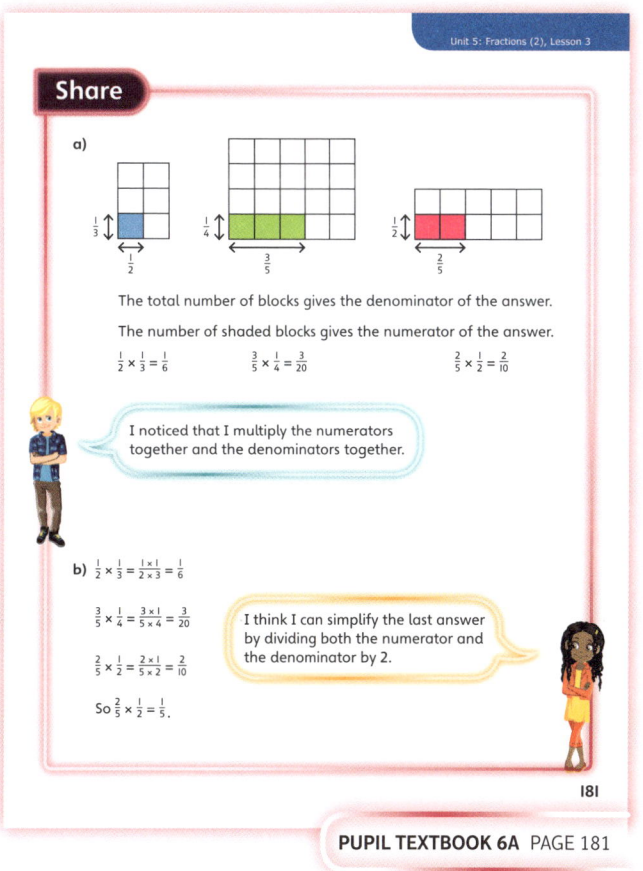

PUPIL TEXTBOOK 6A PAGE 181

Think together

Whole class teacher led (I do, We do, You do)

ASK

- Question **1**: *What is the same and what is different about the two methods?*
- Question **2** a): *How can you simplify your answer?*
- Question **2** b): *How would you find the answer using a diagram? How does this compare with multiplying the numerators and multiplying the denominators?*
- Question **2** c): *How can you use what you know to multiply three fractions? Can you represent this using a diagram?*
- Question **3** a): *Do the numerators always need to multiply together to make 5? Why?*

IN FOCUS Question **2** starts to bring in larger denominators and calculations with more than two fractions, showing that a diagram is not the most efficient method. Question **3** gives children the opportunity to begin to think about multiplying fractions without a context in a more abstract way: it starts with the answer and requires children to use the process of multiplying the numerators and denominators to find possible pairs of fractions. It encourages them to think about equivalent fractions in order to provide more than one possible answer.

STRENGTHEN In question **1**, encourage children to reflect on their diagram by asking them what is representing the denominator of the answer and what is representing the numerator. Encourage them to think about the operation they are using. Allow children to draw diagrams for question **2**, because it will help them to understand that it is not an efficient method for $\frac{9}{10} \times \frac{2}{17}$. In question **3**, some children may need to explore equivalent fractions in relation to $\frac{1}{6}$ and $\frac{1}{4}$, for example using a fraction wall.

DEEPEN Ask children what they notice about their answers for question **3** b). They should recognise that the missing fraction will always simplify to $\frac{1}{4}$. Change the known fraction to $\frac{2}{5}$ and ask them to find a possible fraction (for example, $\frac{5}{12}$) and to explain their reasoning.

ASSESSMENT CHECKPOINT Use question **2** to assess whether children can multiply two fractions by multiplying the numerators and the denominators. Check that they understand why this gives the correct answer. Use question **3** to assess whether children can work flexibly by recognising equivalent fractions.

ANSWERS

Question **1**: $\frac{8}{15}$

Question **2** a): $\frac{3 \times 5}{7 \times 6} = \frac{15}{42} = \frac{5}{14}$

Question **2** b): $\frac{9 \times 2}{10 \times 17} = \frac{18}{170} = \frac{9}{85}$

Question **2** c): $\frac{1}{12}$

Drawing a diagram for part b) would not be efficient because the denominators are large, meaning the diagram would have 170 blocks.

Question **3** a): Possible answers include: $\frac{1}{1} \times \frac{5}{9}$, $\frac{2}{3} \times \frac{5}{6}$, $\frac{4}{6} \times \frac{5}{6}$, $\frac{2}{3} \times \frac{10}{12}$

Question **3** b): Possible answers include $\frac{1}{4}, \frac{2}{8}, \ldots$ (all fractions equivalent to $\frac{1}{4}$)

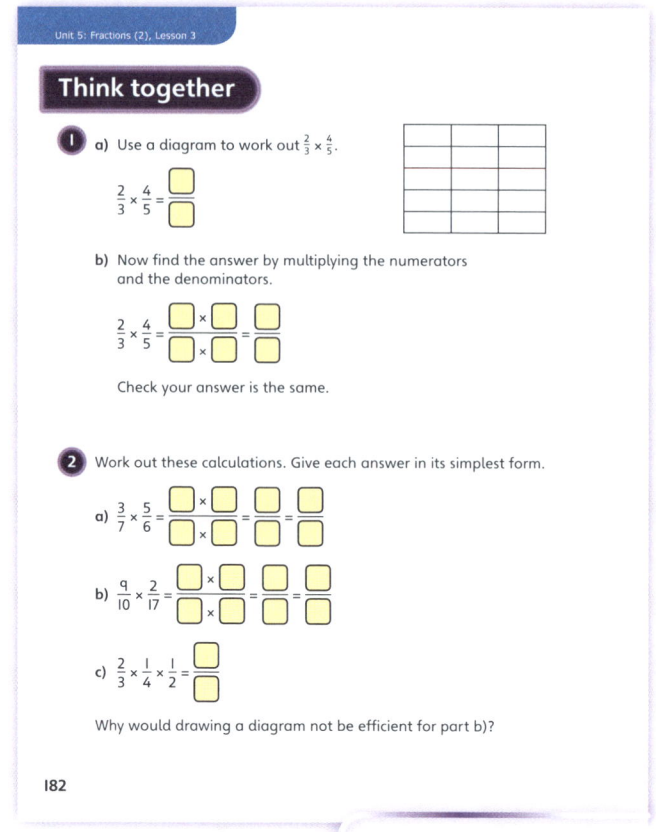

1 a) Use a diagram to work out $\frac{2}{3} \times \frac{4}{5}$.

$\frac{2}{3} \times \frac{4}{5} = \dfrac{\square}{\square}$

b) Now find the answer by multiplying the numerators and the denominators.

$\frac{2}{3} \times \frac{4}{5} = \dfrac{\square \times \square}{\square \times \square} = \dfrac{\square}{\square}$

Check your answer is the same.

2 Work out these calculations. Give each answer in its simplest form.

a) $\frac{3}{7} \times \frac{5}{6} = \dfrac{\square \times \square}{\square \times \square} = \dfrac{\square}{\square} = \dfrac{\square}{\square}$

b) $\frac{9}{10} \times \frac{2}{17} = \dfrac{\square \times \square}{\square \times \square} = \dfrac{\square}{\square} = \dfrac{\square}{\square}$

c) $\frac{2}{3} \times \frac{1}{4} \times \frac{1}{2} = \dfrac{\square}{\square}$

Why would drawing a diagram not be efficient for part b)?

182

PUPIL TEXTBOOK 6A PAGE 182

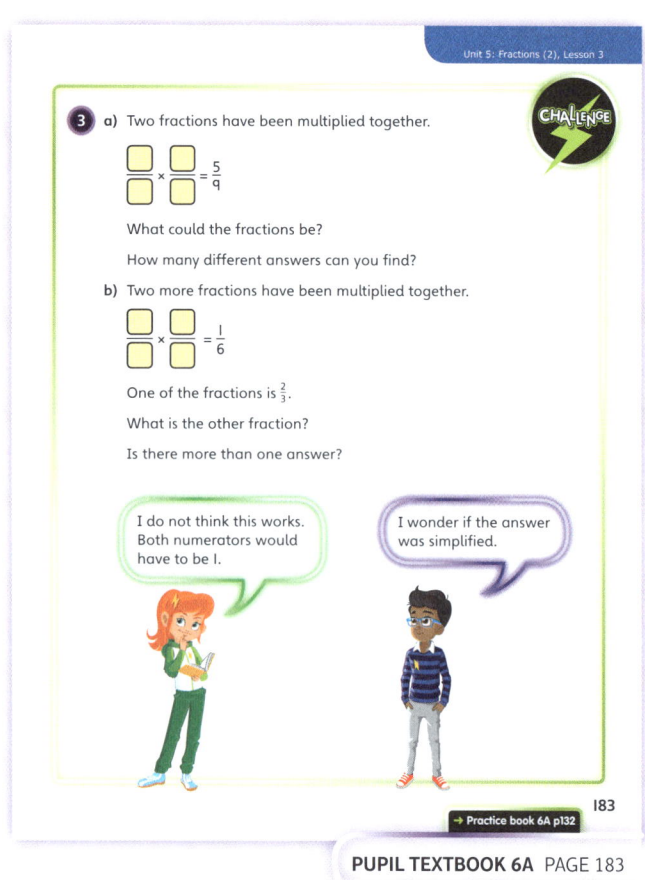

3 a) Two fractions have been multiplied together.

$\dfrac{\square}{\square} \times \dfrac{\square}{\square} = \dfrac{5}{9}$

What could the fractions be?

How many different answers can you find?

b) Two more fractions have been multiplied together.

$\dfrac{\square}{\square} \times \dfrac{\square}{\square} = \dfrac{1}{6}$

One of the fractions is $\frac{2}{3}$.

What is the other fraction?

Is there more than one answer?

> I do not think this works. Both numerators would have to be 1.

> I wonder if the answer was simplified.

183

→ Practice book 6A p132

PUPIL TEXTBOOK 6A PAGE 183

Practice

WAYS OF WORKING Independent thinking

IN FOCUS These questions are designed to incorporate variation to ensure children are constantly thinking. Children have the option to refer back to a diagram for support, although question **3** f) will present a problem because the denominators are large. Question **4** relies on children understanding how the numerator and denominator are worked out when multiplying fractions, in order to find missing numbers. Question **5** is designed to address a misconception, where children add the numerators instead of multiplying them. Encourage children to provide their explanation in a full sentence. Question **6** aims to allow all children to reason and problem solve, by requiring more than one possible answer.

STRENGTHEN To strengthen understanding, encourage children to use diagrams. However, in order for them to be ready to use a more abstract approach, discuss their diagrams and the link to abstract calculation. In question **6**, provide support by giving one of the fractions and asking children to find the other fraction.

DEEPEN In question **6**, ask children to find more than two possible answers and whether it is possible to find them all. This may prompt a discussion about equivalent fractions.

ASSESSMENT CHECKPOINT Use questions **2** and **3** to assess whether children can multiply two fractions by multiplying the numerators and the denominators. Check that they can explain why they need to multiply the numerators and denominators. Use questions **4** and **6** to assess whether children can apply their knowledge of how to multiply fractions and of equivalent fractions to find fractions that multiply to give a specific answer.

ANSWERS Answers for the **Practice** part of the lesson appear in the separate **Practice and Reflect answer guide**.

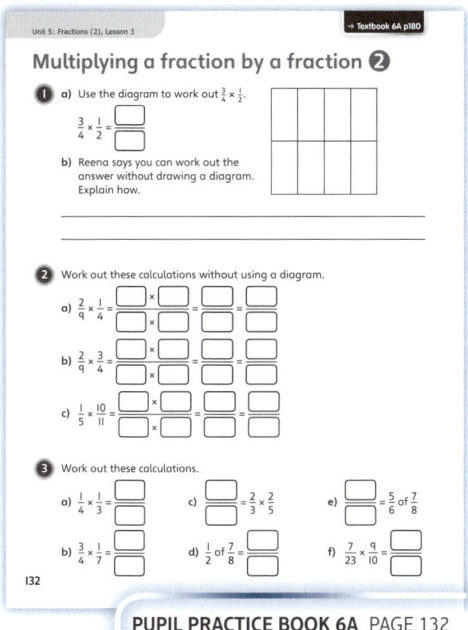

PUPIL PRACTICE BOOK 6A PAGE 132

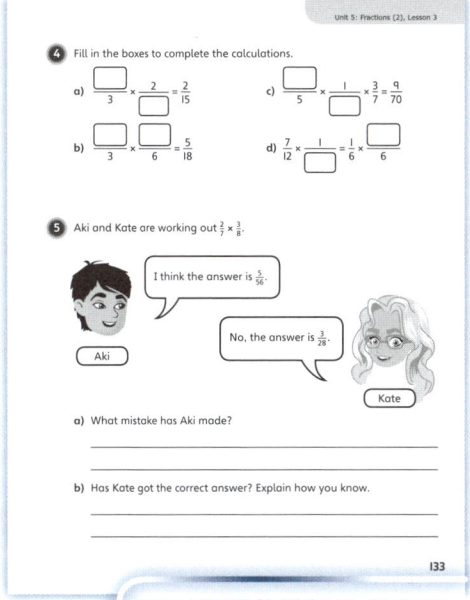

PUPIL PRACTICE BOOK 6A PAGE 133

Reflect

WAYS OF WORKING Pair work

IN FOCUS This activity requires children to either use a diagram or multiply the numerators together and the denominators together. Comparing their method with a partner may show them an alternative method.

ASSESSMENT CHECKPOINT Check that children can fully explain, in full sentences, what their method is and why it works.

ANSWERS Answers for the **Reflect** part of the lesson appear in the separate **Practice and Reflect answer guide**.

After the lesson ⏸

- Can children multiply two fractions by drawing a diagram?
- Can they multiply two fractions without using a diagram?
- Can they explain a more abstract process for multiplying two fractions?
- Can they work fluently with equivalent fractions?

PUPIL PRACTICE BOOK 6A PAGE 134

215

Dividing a fraction by a whole number ❶

Learning focus

In this lesson, children will learn to divide unit fractions by a whole number. They will practise recording the original fractions in a diagram and then dividing one of the sections. They will be exposed to a pattern between denominators and the number they are dividing by.

Small steps

→ Previous step: Multiplying a fraction by a fraction (2)
→ **This step: Dividing a fraction by a whole number (1)**
→ Next step: Dividing a fraction by a whole number (2)

NATIONAL CURRICULUM LINKS

Year 6 Number – Fractions

Divide proper fractions by whole numbers [for example, $\frac{1}{3} \div 2 = \frac{1}{6}$].

ASSESSING MASTERY

Children can divide a unit fraction by a whole number, first using circular and rectangular diagrams and then multiplying the denominator by the whole number. Children should be able to answer questions in real-life contexts and from given diagrams.

COMMON MISCONCEPTIONS

Some children may divide one section of the shape by the whole number but not the remaining sections so they have unequal parts (for example, divide a third of a circle into 2, but not the other two thirds, giving $\frac{1}{4}$, not $\frac{1}{6}$). Ask:
• *Are all your parts the same size as the shaded part? What is the shaded part as a fraction of the whole?*

Some children may divide the denominator by the whole number instead of multiplying it. Ask:
• *What happens to a number when you divide it by a whole number – does it get smaller or larger?*
• *Does a smaller unit fraction have a larger or smaller denominator?*

STRENGTHENING UNDERSTANDING

Give children paper circles and strips to physically fold and cut. Use a paper strip to model dividing $\frac{1}{4}$ by 3, emphasising the need to create equal parts that are the same size as a third of one of the quarters.

GOING DEEPER

Encourage children to find a variety of ways to create a unit fraction by using the pattern between denominators and the whole number. Ask them to explain how they can be sure they have found all the combinations.

KEY LANGUAGE

In lesson: denominator, whole number, divide, part, whole

Other language to be used by the teacher: numerator, unit fraction, multiply

RESOURCES

Mandatory: paper strips, paper plates

Optional: paper circles, squares and strips

 In the eTextbook of this lesson, you will find interactive links to a selection of teaching tools.

Before you teach

• Can children divide shapes equally?
• Can they show fractions using shapes?
• Do they understand the importance of equal parts?
• Do they know what happens to a number when it is divided by a whole number?

Discover

Pair or small group work

ASK

- Question **1** a): *What is each part of the circle worth? How do you know?*
- Question **1** b): *What is the same, compared with part a)? What is different?*
- Question **1** b): *What is each part of the strip worth? How do you know?*
- Question **1** b): *How can you show the section to be glued?*

IN FOCUS Questions **1** a) highlights whether children understand that the two parts of one third must be equal. Question **1** b) is not more difficult than part a), just set out differently.

PRACTICAL TIPS Take a group of six children and divide them into three equal groups of two children, ensuring that children understand that each group represents $\frac{1}{3}$ of the whole. Divide one of the groups into two single children. Discuss what fraction of the whole each child represents. Establish that the other groups need to be divided in the same way to make equal parts.

ANSWERS

Question **1** a): $\frac{1}{6}$ of the penguin's body is white.

Question **1** b): $\frac{1}{8}$ of the strip of paper is covered in glue.

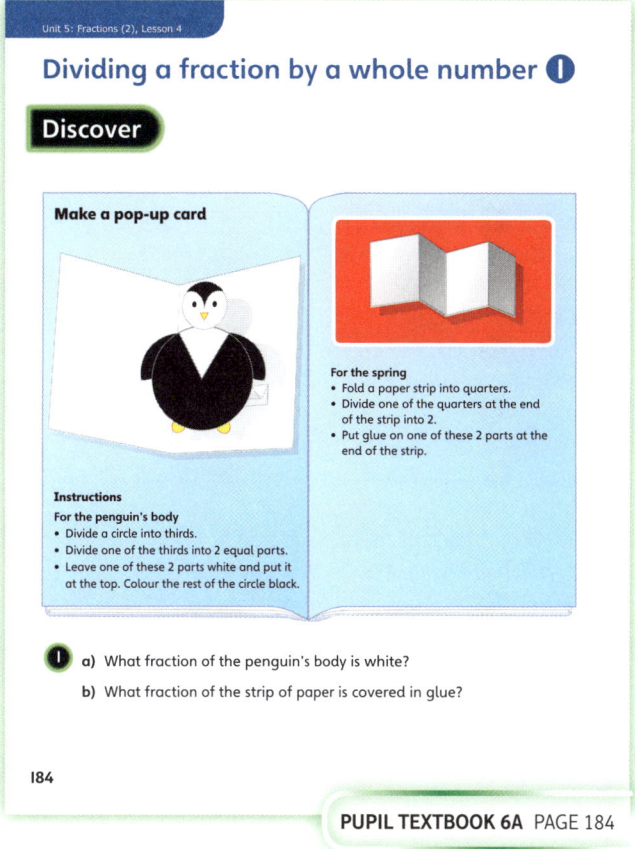

PUPIL TEXTBOOK 6A PAGE 184

Share

Whole class teacher led

ASK

- Question **1** a): *The second diagram shows one shaded part and three unshaded – why is the answer not $\frac{1}{4}$? Can you leave only one section with two equal parts? What do you need to do to the other sections? How much is each section worth now? What calculation represents what you have done?*
- Question **1** b): *What is each part of the strip worth now? How many parts would be covered in glue?*

IN FOCUS Both parts a) and b) draw attention to the fact dividing into two equal parts is the same as dividing by 2. Emphasis should be placed on the need to divide every part by the same number to give equal parts after the division. Children may spot the pattern that when a fraction is divided by 2 the number of original parts doubles, and so the denominator doubles.

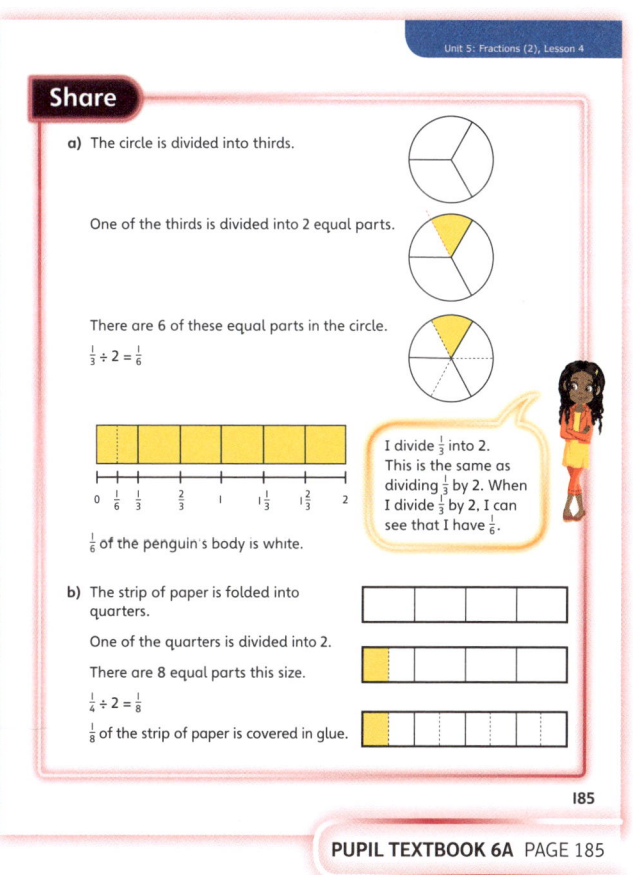

PUPIL TEXTBOOK 6A PAGE 185

Think together

Unit 5: Fractions (2), Lesson 4

Think together

WAYS OF WORKING Whole class teacher led (I do, We do, You do)

ASK

• Question **1**: *What has the quarter been divided into? What calculation is this representing? What do you need to do to the other quarters? How many parts does this mean the whole is now divided into?*

• Question **1**: *Has the original number of parts been doubled like they were in the* ***Discover*** *section? Why not?*

• Question **2**: *How is this similar to question* **1**? *How is this different from question* **1**? *Looking at your answer to Question* **1**, *can you estimate what this answer will be? What has the whole been divided into after you have divided by 3? What is each of these parts worth? Can you spot a pattern when dividing by 3?*

• Question **3** b): *What patterns can you see? What do you think will happen when you divide a fraction by 5? By 7? Why?*

IN FOCUS Question **3** draws on all the divisions that children have done in **Share** and in **Think together**. Use questions **1** and **2** to encourage them to identify the pattern between the denominators and the number they are dividing by. This then encourages them to use the pattern they have found to move towards a more abstract/mental way of working.

STRENGTHEN Give children paper circles and strips to annotate, fold and/or cut up to help them see the fractions more clearly.

DEEPEN Ask children to investigate whether their 'pattern' works when dividing by any whole number. Ask them to use diagrams to explain why the pattern works.

ASSESSMENT CHECKPOINT Use questions **1** and **2** to assess whether children can use diagrams to divide a unit fraction by a whole number. Check whether they can explain what is happening. Use question **3** to assess whether children can divide a unit fraction by a whole number without using a diagram.

ANSWERS

Question **1**: $\frac{1}{4} \div 3 = \frac{1}{12}$
$\frac{1}{12}$ of the circle is shaded.

Question **2**: $\frac{1}{3} \div 3 = \frac{1}{9}$
$\frac{1}{9}$ of the paper is shaded.

Question **3** a): All the fractions are unit fractions. All the fractions have been divided by a whole number.
The answers are all smaller than the original fraction.
The first two calculations divide by 2. The second two calculations divide by 3.

Question **3** b): The denominator of the answer is the denominator of the original fraction multiplied by the number the fraction is being divided by.
$\frac{1}{6} \div 2 = \frac{1}{12}, \frac{1}{4} \div 4 = \frac{1}{16}, \frac{1}{5} \div 3 = \frac{1}{15}$

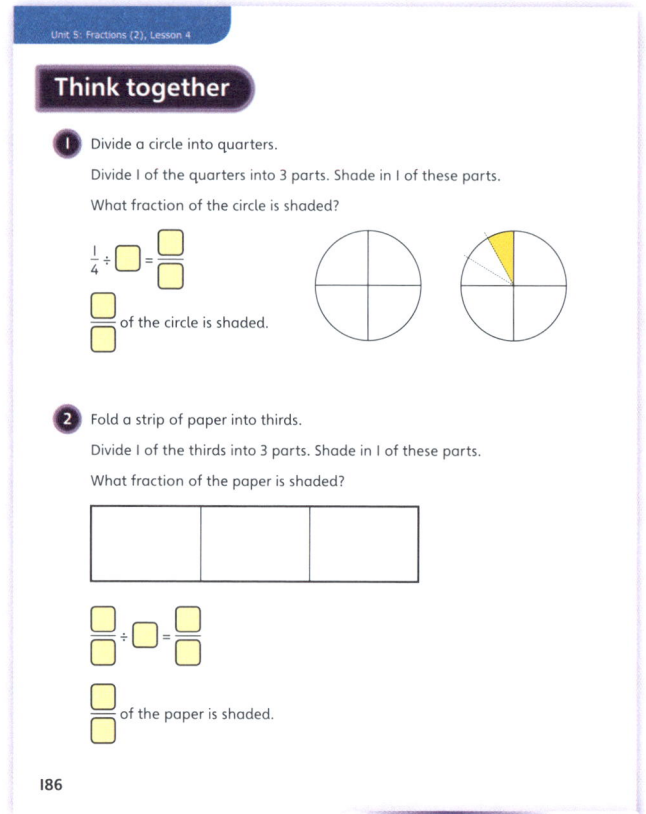

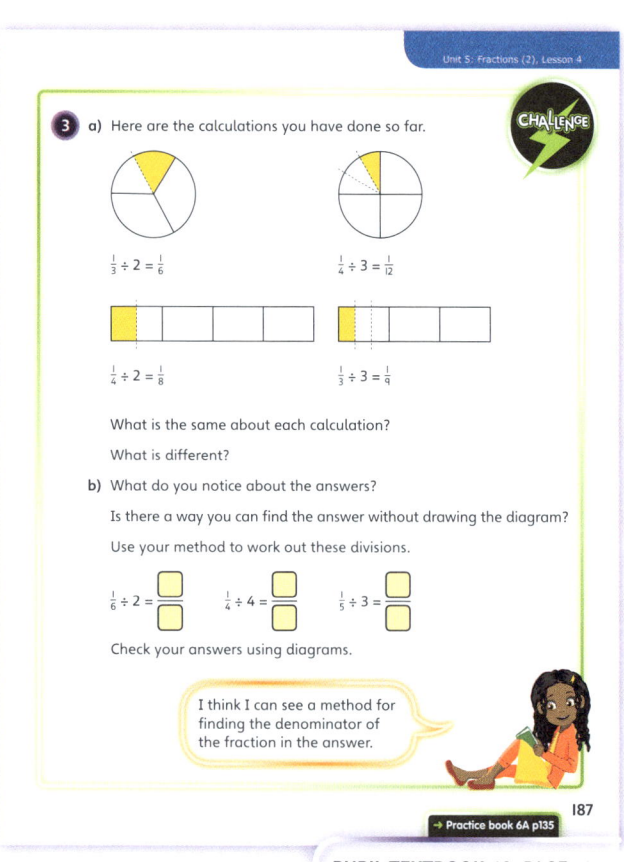

218

Practice

WAYS OF WORKING Independent thinking

IN FOCUS Questions **1**, **2** and **3** provide more opportunities to practise methods that have been worked on in previous examples. Scaffolding is provided through diagrams that are already divided into fractions, enabling children to focus on the dividing element of the question. In question **4**, children need to interpret the diagram to work out the division that is shown, requiring them to demonstrate the understanding they have developed through drawing diagrams. In question **5**, the first six parts are standard division of unit fractions by a whole number. The last three parts require children to use the pattern they have identified to find the missing numbers.

STRENGTHEN For question **5**, give children paper circles and strips to annotate, fold and/or cut up. In part g), suggest that they cut the same circle/strip into four and eight parts and compare the size of $\frac{1}{4}$ and $\frac{1}{8}$ to work out the missing number.

DEEPEN Ask those children who spotted the pattern quickly and understand why it works to estimate answers before checking them using the given diagrams in questions **1** to **3**. In question **3**, ask children to investigate whether the way the shape is divided affects the answer or whether the section they divide affects the answer. In question **5**, ask children how they can be sure that they have found all the possible solutions.

ASSESSMENT CHECKPOINT Use questions **1**, **2** and **3** to assess whether children can use diagrams to divide a unit fraction by a whole number. Check that children are aware what section they are dividing and how many parts are in the whole after the division. Use question **5** to assess whether children can children divide a unit fraction by a whole number without using a diagram.

ANSWERS Answers for the **Practice** part of the lesson appear in the separate **Practice and Reflect answer guide**.

Reflect

WAYS OF WORKING Independent thinking

IN FOCUS This question explores a common misconception, where children divide the denominator by the whole number instead of multiplying it. Children can practise their reasoning skills to prove why the statement is false.

ASSESSMENT CHECKPOINT Children should be able to answer true or false by working out the answer for themselves, but their reasoning will show their understanding of the concept.

ANSWERS Answers for the **Reflect** part of the lesson appear in the separate **Practice and Reflect answer guide**.

After the lesson

- Can children correctly divide sections?
- Do children understand what the whole has now been divided into?
- Can children see the relationship between the denominators and the whole number?
- Can children use diagrams to solve calculations and/or prove answers?

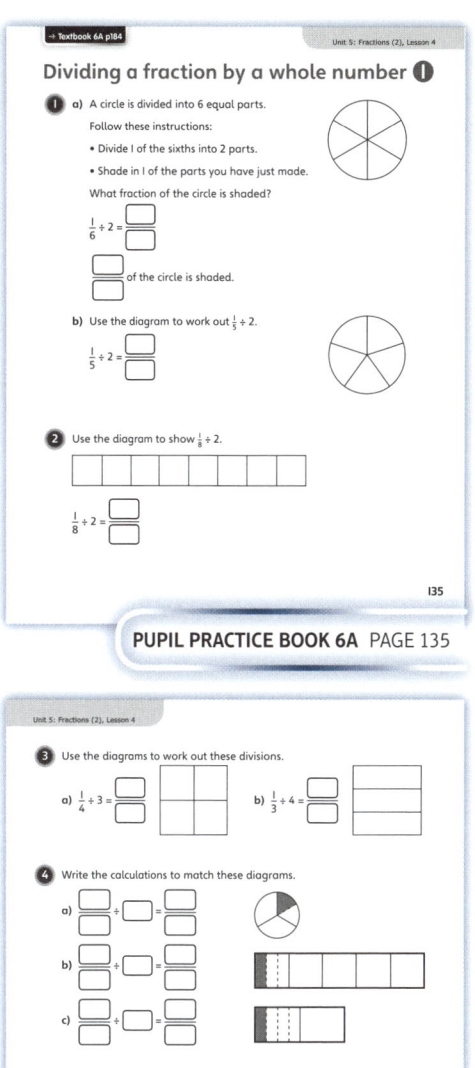

PUPIL PRACTICE BOOK 6A PAGE 135

PUPIL PRACTICE BOOK 6A PAGE 136

PUPIL PRACTICE BOOK 6A PAGE 137

Dividing a fraction by a whole number 2

Learning focus

In this lesson, children will learn how to divide a non-unit fraction by a whole number when the numerator is a multiple of the whole number. They will build on their work using diagrams in the previous lesson. They will start to identify the pattern between numerators and the number they are dividing by.

Small steps

→ Previous step: Dividing a fraction by a whole number (1)

→ **This step: Dividing a fraction by a whole number (2)**

→ Next step: Dividing a fraction by a whole number (3)

NATIONAL CURRICULUM LINKS

Year 6 Number – Fractions

Divide proper fractions by whole numbers [for example, $\frac{1}{3} \div 2 = \frac{1}{6}$].

ASSESSING MASTERY

Children can divide a non-unit fraction by a whole number when the numerator is a multiple of the whole number. First they will use diagrams and then move on to dividing the numerator by the whole number.

COMMON MISCONCEPTIONS

Some children may divide both the numerator and the denominator by the whole number. Ask:
• *What happens if you divide the numerator and denominator of a fraction by the same number? What do we call these fractions? What can you tell me about these fractions?*
• *What happens to a number when you divide it by a whole number – does it get smaller or larger?*

STRENGTHENING UNDERSTANDING

Model $\frac{4}{5} \div 2$ by placing four counters in the segments of a circle divided into five equal parts. Establish that this shows $\frac{4}{5}$. Ask children how they would divide the number of counters by 2. Remove half of the counters and ask how many counters are left. Establish that the counters now represent $\frac{2}{5}$. Repeat for other calculations.

GOING DEEPER

Ask children to produce word problems that involve dividing a non-unit fraction by a whole number for a partner to solve. Ensure that they understand that the numerator of the non-unit fraction must be a multiple of the whole number.

KEY LANGUAGE

In lesson: fraction, whole number, numerator, share

Other language to be used by the teacher: non-unit fraction, denominator, part

RESOURCES

Optional: counters, paper circles, paper strips, whiteboard pens

 In the eTextbook of this lesson, you will find interactive links to a selection of teaching tools.

Before you teach

• Do children know how to divide a shape into equal parts?
• Do they understand the importance of equal parts?
• Can they divide a unit fraction by a whole number using diagrams?
• Can they explain the relationship between the denominators and whole number when dividing a unit fraction?

Discover

WAYS OF WORKING Pair work

ASK

- Question **1** a): *How many fifths of juice are there? How many are you sharing it between? How are you going to share it? How can you show the sharing?*
- Question **1** b): *How is this question different from part a)? How much baby food is there to start with? How much will be used?*

IN FOCUS These questions are designed show whether children understand how to share equally. Their responses will highlight whether they understand that the total they are sharing is $\frac{4}{5}$ and not 1, and that each part they have shared is $\frac{1}{5}$ and not 1. Question **1** b) draws out this point further by showing a full jar in the picture, but stating that only $\frac{9}{10}$ of the jar is to be used.

PRACTICAL TIPS Give children the opportunity to model question **1** a) by supplying a bottle partially filled with liquid and two cups to help them investigate and solve the problem. Similarly, for question **1** b) provide a partially filled jar and three empty bowls.

ANSWERS

Question **1** a): $\frac{2}{5}$ of the original jug is in each cup.

Question **1** b): $\frac{3}{10}$ of the jar of baby food should be put into each bowl.

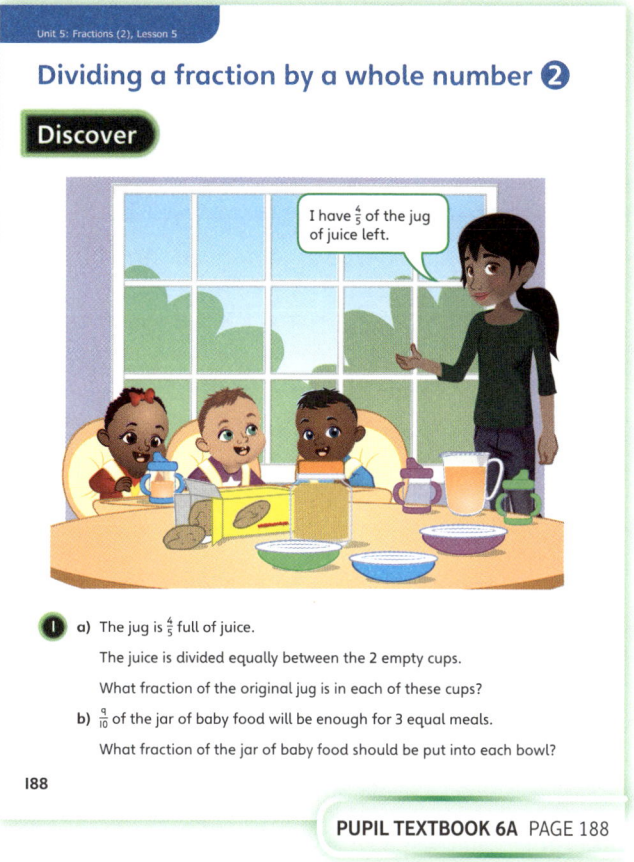

PUPIL TEXTBOOK 6A PAGE 188

Share

WAYS OF WORKING Whole class teacher led

ASK

- Question **1** a): *How does the first diagram represent the jug of juice? Why do the two diagrams for the cups of juice each have two shaded blocks? Can there be more juice in one cup than the other? How can you use the diagrams to work out how much juice goes in each cup? How can you write this as a calculation?*
- Question **1** b): *How are these diagrams different from those in part a)? Do the diagrams show the same method? Do you find the bars or the circles easier for dividing a fraction?*

IN FOCUS The diagrams in question **1** a) emphasise that it is the fraction that is being divided into equal parts, not the whole. Children should see that the parts of the whole can be shared equally without needing to divide each part as they did in the previous lesson. There could be a discussion as to whether the method of dividing each part would work – if so, is it an effective method? Question **1** b) shows the same method but using circles instead of bars.

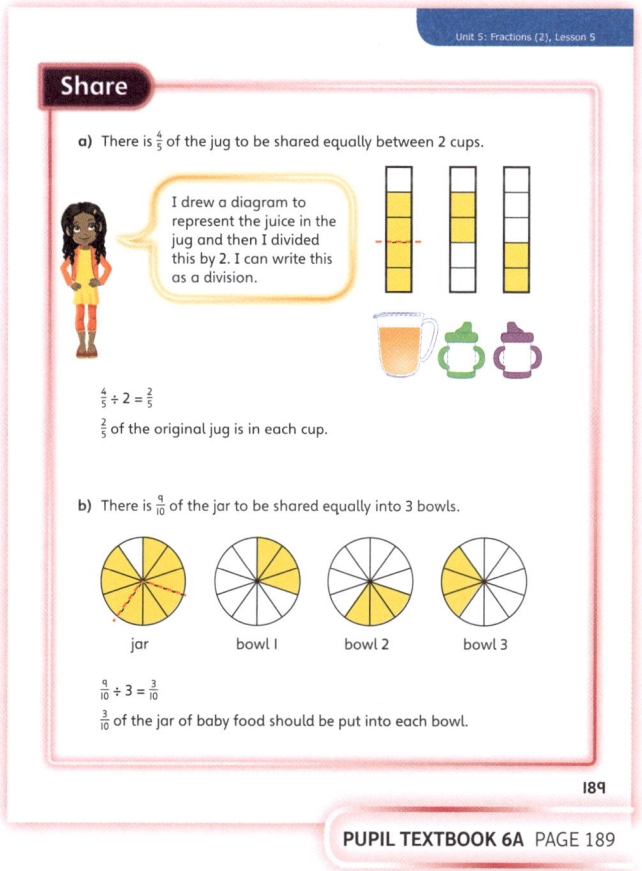

PUPIL TEXTBOOK 6A PAGE 189

Think together

Think together

WAYS OF WORKING Whole class teacher led (I do, We do, You do)

ASK

- Question ❶: *What is each part worth?*
- Question ❶: *How many parts do you have? What are you dividing/sharing those parts into? Do you need to divide each part by 3 or is there a more efficient way to share?*
- Question ❷: *How many parts has your shape been divided into? How many of these parts do you have? How many groups have the parts been shared into? How much are the parts worth?*
- Question ❸ a): *Do you agree with Ash's statement? What could the link be?*
- Question ❸ b): *How can you work out the fraction for the last question? What operation will you need to use?*

IN FOCUS Question ❶ allows children to practise sharing the parts using the most efficient method. Watch to see whether children are identifying how many parts would be in each group straight away or sharing each part one at a time. Encourage children to think about division facts for whole numbers. Question ❷ is designed to deepen understanding by requiring children to work out the original fraction, how many groups it was shared into and how many parts are in each group. Question ❸ a) requires children to complete four more divisions and then encourages them to spot a pattern between the numerator and the whole number. This enables them to move on to working more abstractly in question ❸ b), where they are prompted by Dexter to use pictorial representations to check and prove answers.

STRENGTHEN Provide children with paper circles and strips to draw on, fold, cut up and share into groups. Cutting each part up into the number being divided by will help the children to see that sharing each part gives the same answer as sharing the total number of parts.

DEEPEN Challenge children to investigate whether the pattern between the numerator and whole number works with any numerator, denominator or whole number. Ask them to give examples of when it works and when it does not work. They should be able to explain that it only works when the numerator is a multiple of the whole number.

ASSESSMENT CHECKPOINT Questions ❶ and ❸ a) assess whether children can divide a non-unit fraction by a whole number. Check whether they can explain the most efficient way to share the fraction. Question ❸ assesses whether children can explain the link between the numerator and the whole number and work out divisions using this method.

ANSWERS

Question ❶: $\frac{6}{7} \div 3 = \frac{2}{7}$

Each baby gets $\frac{2}{7}$ of the packet.

Question ❷ a): $\frac{9}{10} \div 3 = \frac{3}{10}$

Question ❷ b): $\frac{4}{6} \div 4 = \frac{1}{6}$

Question ❸ a): $\frac{3}{5} \div 3 = \frac{1}{5}, \frac{5}{8} \div 5 = \frac{1}{8}, \frac{8}{10} \div 4 = \frac{2}{10} = \frac{1}{5}, \frac{10}{11} \div 5 = \frac{2}{11}$

The numerator divided by the whole number gives the numerator of the answer. The denominator stays the same.

Question ❸ b): $\frac{3}{4} \div 3 = \frac{1}{4}, \frac{8}{9} \div 2 = \frac{4}{9}, \frac{12}{25} \div 3 = \frac{4}{25}, \frac{8}{9} \div 4 = \frac{2}{9}$

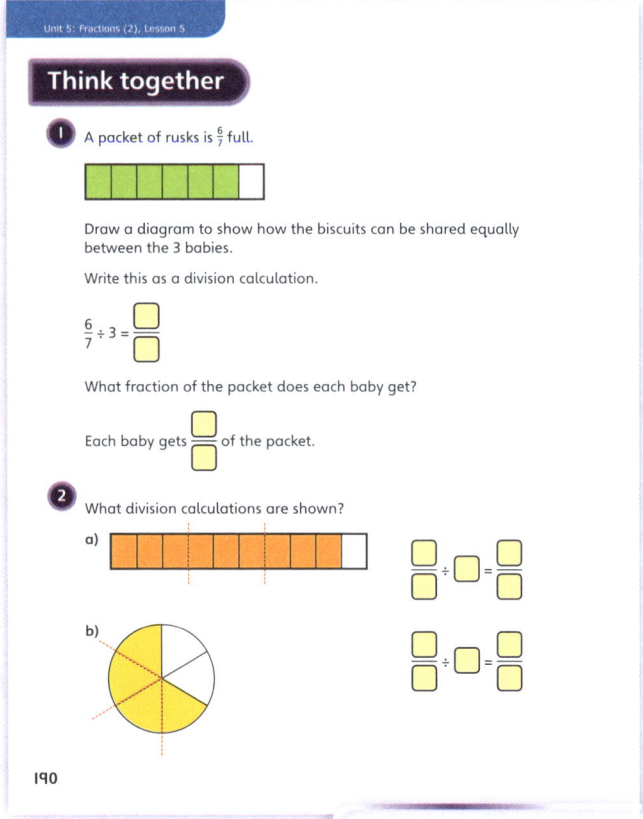

PUPIL TEXTBOOK 6A PAGE 190

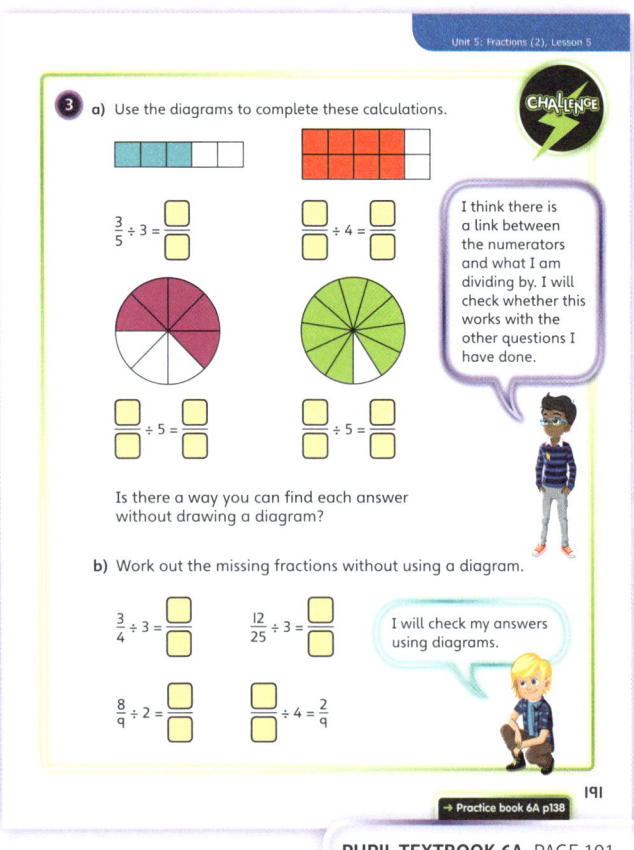

PUPIL TEXTBOOK 6A PAGE 191

Practice

Independent thinking

IN FOCUS Questions **1**, **2** and **3** give children the opportunity to practise the method using a variety of 2D shapes, gradually reducing the amount of scaffolding. Question **1** is worded to help the children understand what is happening at each point and where each number in the calculation comes from. Question **4** is designed to highlight children's understanding by asking them to write the whole calculation represented by an image. This enables them to demonstrate that they understand what each part of the diagram is showing and the order information is presented in. Questions **5** and **7** encourage children to work more mentally. In question **6**, children start to work out different combinations of numerators and whole numbers that give a required answer. This could lead into an investigation into how many ways a particular fraction can be divided by a whole number while keeping the denominator the same.

STRENGTHEN Provide children with counters that can be shared into groups. They could use whiteboard pens to write down the worth of each part on the counter to help them remember that the amount in each group is a fraction. When they have shared the counters into groups, they can then add up the number of parts in each group.

DEEPEN When children have completed question **6**, ask them to find other numbers they could use for the numerators and whole number in part d). Ask them to explain how they know they have found them all. Challenge them to produce similar sets of number sentences, where there are exactly 2, 3 or 4 possible number sentences.

ASSESSMENT CHECKPOINT Use questions **2** and **3** to assess whether children can use diagrams to divide non-unit fractions by a whole number. Use questions **5** and **7** to assess whether they can divide non-unit fractions by a whole number without the support of a diagram.

ANSWERS Answers for the **Practice** part of the lesson appear in the separate **Practice and Reflect answer guide**.

Reflect

Independent thinking

IN FOCUS This question uses a misconception to provide children with an opportunity to demonstrate they understand the method. Children need to realise that Danny has divided both the numerator and the denominator by 5, rather than just the numerator.

ASSESSMENT CHECKPOINT Check whether children work out the correct answer by drawing a diagram or by dividing the numerator. If they choose to divide the numerator, check that they can explain why this gives the correct answer.

ANSWERS Answers for the **Reflect** part of the lesson appear in the separate **Practice and Reflect answer guide**.

After the lesson

- Can children divide non-unit fractions by whole numbers where the numerator is a multiple of the number? Can they estimate answers and check their answers using diagrams?
- Do children understand the relationship between the numerator and the whole number? Do they recognise that only multiples of the numerator work using the method of dividing the numerator?

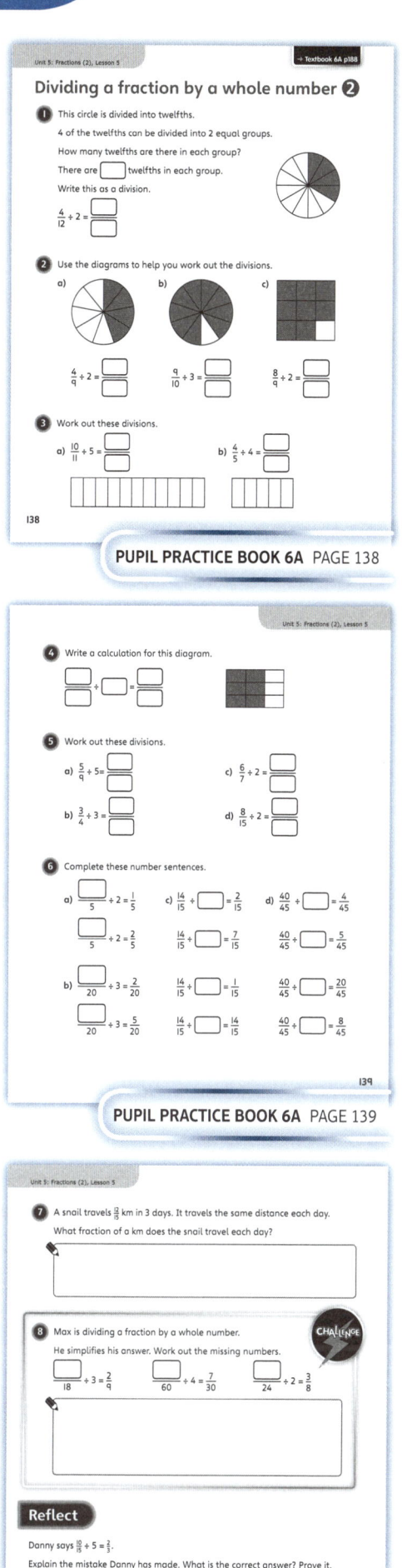

PUPIL PRACTICE BOOK 6A PAGE 138

PUPIL PRACTICE BOOK 6A PAGE 139

PUPIL PRACTICE BOOK 6A PAGE 140

Dividing a fraction by a whole number ③

Learning focus

In this lesson, children will build on the previous two lessons and learn to divide any fraction by a whole number. They will continue to use both diagrams and abstract methods to solve calculations and show their working out.

Small steps

→ Previous step: Dividing a fraction by a whole number (2)

→ **This step: Dividing a fraction by a whole number (3)**

→ Next step: Four rules with fractions

NATIONAL CURRICULUM LINKS

Year 6 Number – Fractions

Divide proper fractions by whole numbers [for example, $\frac{1}{3} \div 2 = \frac{1}{6}$].

ASSESSING MASTERY

Children can divide any fraction by a whole number. They understand what is happening when they are sharing a fraction and can use diagrams to explain their thinking.

COMMON MISCONCEPTIONS

Some children may try to divide the denominator by the whole number. Ask:

• *What happens to a fraction when you divide it by a whole number – does it get smaller or larger? If the denominator of a fraction gets smaller, does the fraction get larger or smaller?*

STRENGTHENING UNDERSTANDING

To strengthen children's understanding, provide them with paper circles and strips to annotate, fold, cut up and physically share.

GOING DEEPER

Ask children to produce a poster that explains how to divide any fraction by a whole number, drawing on this lesson and the previous two lessons.

KEY LANGUAGE

In lesson: fraction, whole number, numerator, equally, share

Other language to be used by the teacher: denominator, multiple

RESOURCES

Optional: concrete materials to represent length of bamboo (for example, sticks, wool, ribbon, card, cubes), pictures of pandas, paper circles, paper squares, paper strips, fraction strips, counters, whiteboard pens

 In the eTextbook of this lesson, you will find interactive links to a selection of teaching tools.

Before you teach

• Can children represent fractions using diagrams?
• Are they confident dividing unit fractions by whole numbers and non-unit fractions by multiples of the numerator?
• Can they record working out using diagrams?

Discover

Unit 5: Fractions (2), Lesson 6

WAYS OF WORKING Pair work

ASK

- Question ❶ a): *What fraction do you have to begin with? Can you draw a diagram to show this fraction?*
- Question ❶ a): *What are you dividing the fraction into? Can the parts of the fraction be shared equally? Is the numerator a multiple of the whole number? Can you show the division on your diagram?*
- Question ❶ a): *Which parts of the fraction need splitting up?*
- Question ❶ b): *How does this new information change the calculation in part a)? How does the diagram need changing? What is the size of each part now?*

IN FOCUS These questions will show whether children recognise that the parts of the fraction cannot be shared equally as in the previous lesson. Children should then realise that each part of the fraction needs to be divided in a similar way to dividing unit fractions in Lesson 4, and that this changes the number of parts. When children are sharing they need to remember that everything they share is a part/fraction, not a whole number.

PRACTICAL TIPS Give children something concrete to represent the bamboo: sticks, wool, ribbon, card, cubes etc. They could also be given pictures of pandas, sorting hoops or similar to clearly share the fraction between.

ANSWERS

Question ❶ a): Each panda will get $\frac{2}{9}$ m of bamboo shoot.

Question ❶ b): Each panda will get $\frac{1}{6}$ m of bamboo shoot.

Share

WAYS OF WORKING Whole class teacher led

ASK

- Question ❶ a): *What is represented by the diagram? Can the parts of the fraction be shared equally? Explain.*
- Question ❶ a): *How many parts do you need to divide each part into? What is each part worth now you have divided it further? What is $\frac{2}{3}$ equivalent to? Can the new parts be shared equally? How many parts are in each group? How much does this equal?*
- Question ❶ b): *Does the diagram need to change? How many parts are in each group now? What is each part worth? What is the total in each group?*
- Question ❶ b): *What did Ash do that was different? Why did his method work?*

IN FOCUS Question ❶ a) shows how each part in the diagram can be divided by the whole number and these smaller parts shared equally between the pandas. This links back to Lesson 4, as children need to understand that the size of each smaller part is the unit fraction divided by the whole number, and to Lesson 5, as they divide the numerator of the equivalent fraction by the whole number. In question ❶ b), Ash shows an alternative way of dividing the diagram when the numerator of the fraction and the whole number share a factor.

Dividing a fraction by a whole number ❸

Discover

❶ a) The bamboo shoots are $\frac{2}{3}$ m long.

   If the pandas share one bamboo shoot equally, how much will each panda get?

 b) Another panda comes along to share the bamboo shoot.

   How much will each panda get now?

192

PUPIL TEXTBOOK 6A PAGE 192

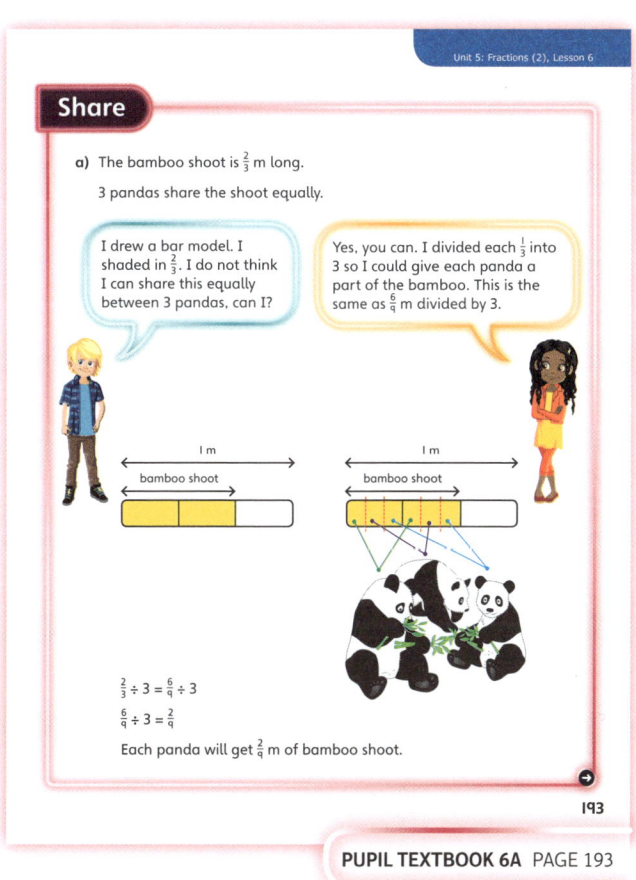

Share

a) The bamboo shoot is $\frac{2}{3}$ m long.

 3 pandas share the shoot equally.

 I drew a bar model. I shaded in $\frac{2}{3}$. I do not think I can share this equally between 3 pandas, can I?

 Yes, you can. I divided each $\frac{1}{3}$ into 3 so I could give each panda a part of the bamboo. This is the same as $\frac{6}{9}$ m divided by 3.

 1 m

 bamboo shoot

 1 m

 bamboo shoot

 $\frac{2}{3} \div 3 = \frac{6}{9} \div 3$

 $\frac{6}{9} \div 3 = \frac{2}{9}$

 Each panda will get $\frac{2}{9}$ m of bamboo shoot.

193

PUPIL TEXTBOOK 6A PAGE 193

Think together

WAYS OF WORKING Whole class teacher led (I do, We do, You do)

ASK

- Question ❶: *What fraction do you have? How many groups are you sharing into? Can the parts of this fraction be shared equally? Why not? How many are you going to have to divide each part into? How much is each part worth now? How many parts are in each group? What is the total amount in each group?*
- Question ❷: *How can you annotate the bar model to find the answer? What is each part worth? How many groups are you sharing into? What is the total fraction in each group? Are the totals equal?*
- Question ❸ a): *How are the methods different? Can you predict whether the answers will be the same or different? Why do you think that? How will you represent Max's method using a diagram? How will you represent Ambika's method using a diagram? What do you notice about the answers?*

IN FOCUS In question ❶, the fraction and groups to share into are given so the children can concentrate on each step within the method. Question ❷ requires children to do a little more for themselves using the information given. Once they have identified the number of groups to share into, children will again be able to practise the method and work through each step. Question ❸ introduces a variation of the method, giving children the chance to investigate a different way to find the answer. Their answers to question ❸ a) will indicate their understanding of equivalent/simplifying fractions.

STRENGTHEN To support children's understanding, provide them with an appropriate fraction strip for each question. Shading and cutting up the fraction strip gives the children a clear visual representation of the parts getting smaller and the denominator changing. Physically sharing the parts into piles enables children to check for equal amounts and to see clearly how many parts are in each pile.

DEEPEN In question ❸, challenge children to find other calculations that both Max and Ambika's methods work for. Children may look for and spot patterns linking the numerator, the denominator and the whole number and investigate any patterns or links they find.

ASSESSMENT CHECKPOINT Use question ❷ to assess whether children can divide a fraction by a whole number. Check that they can identify and represent the fraction they are starting with and how many groups they are dividing it into.

ANSWERS

Question ❶: $\frac{5}{6} \div 3 = \frac{15}{18} \div 3 = \frac{5}{18}$
Each panda will get $\frac{5}{18}$ m of bamboo shoot.

Question ❷: $\frac{1}{10}$

Question ❸ a): Yes, the children get the same answer. Ambika's answer ($\frac{1}{9}$) is Max's answer ($\frac{2}{18}$) simplified.

Question ❸ b): Max's method: $\frac{3}{24}$, Ambika's method: $\frac{1}{8}$

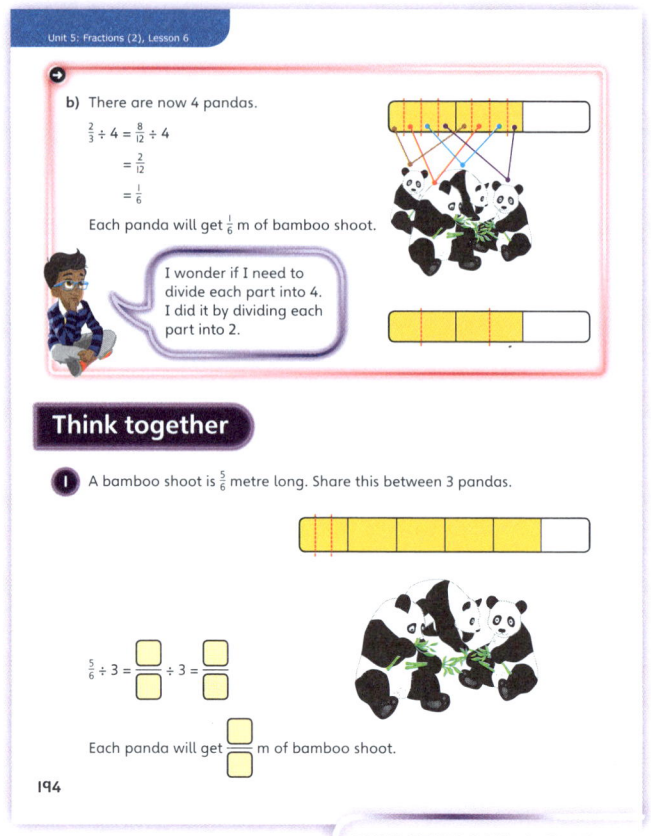

PUPIL TEXTBOOK 6A PAGE 194

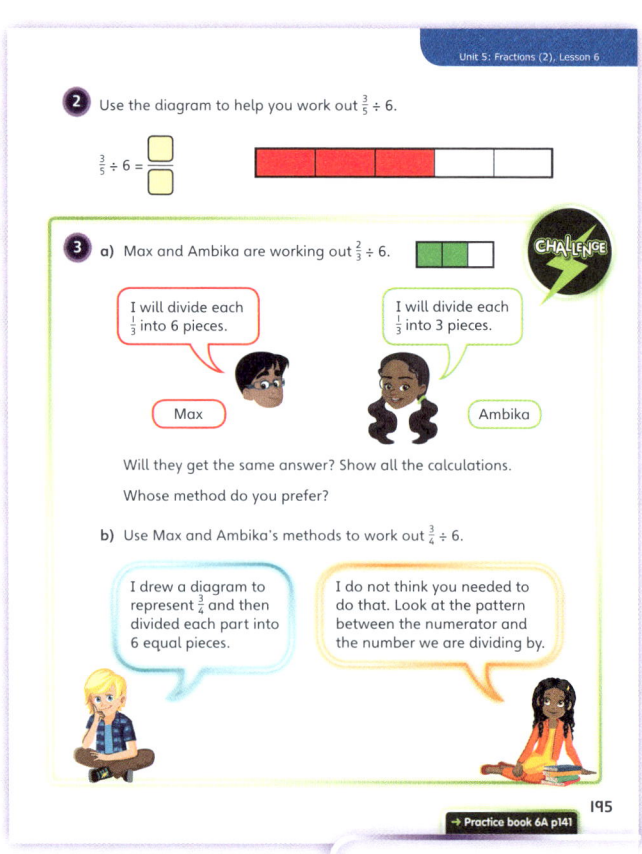

PUPIL TEXTBOOK 6A PAGE 195

Practice

WAYS OF WORKING Independent thinking

IN FOCUS Questions ❶ and ❷ provide diagrams to support children's reasoning. Question ❸ is more abstract, requiring children to understand how to use equivalent fractions. Encourage them to draw diagrams if they lack confidence. The divisions in question ❹ have been chosen to allow more confident children to spot patterns when the numerator and whole number share a common factor.

STRENGTHEN Model a division using fraction strips and counters. For example, when working out $\frac{4}{5} \div 3$, place 3 counters on 4 of the 5 parts on a fifths fraction strip. Use a whiteboard pen to write the fraction represented by each counter ($\frac{1}{5}$) on the counters. Then share the counters into 3 equal piles and establish that each pile represents $\frac{4}{15}$.

DEEPEN In question ❹, ask children whether they can solve these divisions using a diagram in a different way. If necessary, refer them back to Max and Ambika's methods in **Think together** question ❸. Ask them if they can see a pattern between the numerators, denominators and whole numbers. Give them a division such as $\frac{4}{7} \div 8$ and ask them to predict the denominator of the answer.

ASSESSMENT CHECKPOINT Use questions ❶ and ❷ to assess whether children can use a diagram to divide a fraction by a whole number. Use questions ❸ and ❹ to assess whether children can use equivalent fractions to divide a fraction by a whole number without using a diagram. Check whether they can explain how the method works.

ANSWERS Answers for the **Practice** part of the lesson appear in the separate **Practice and Reflect answer guide**.

Reflect

WAYS OF WORKING Independent thinking

IN FOCUS This question highlights whether children understand how to use the information given by the fractions/numbers in a division calculation. Some children may choose to draw a diagram to help them, while others will go straight to equivalent fractions.

ASSESSMENT CHECKPOINT Assess whether children can confidently divide a fraction by a whole number. Check whether they give the answer in its simplest form. Look for children who realise that they can use the relationship between 2 and 4 to find the answer.

ANSWERS Answers for the **Reflect** part of the lesson appear in the separate **Practice and Reflect answer guide**.

After the lesson ⏸

- Can children divide fractions by whole numbers, explaining and modelling using diagrams how each method works?
- Are children aware what each fraction/number represents within a calculation?
- Do they know when parts of a fraction can be shared equally and when they need dividing further?

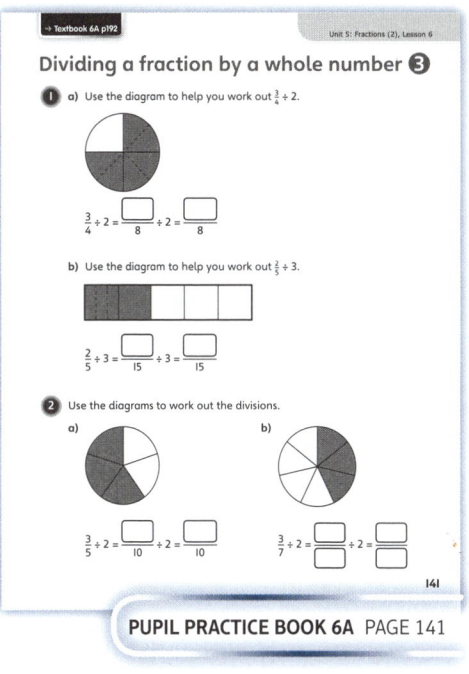

PUPIL PRACTICE BOOK 6A PAGE 141

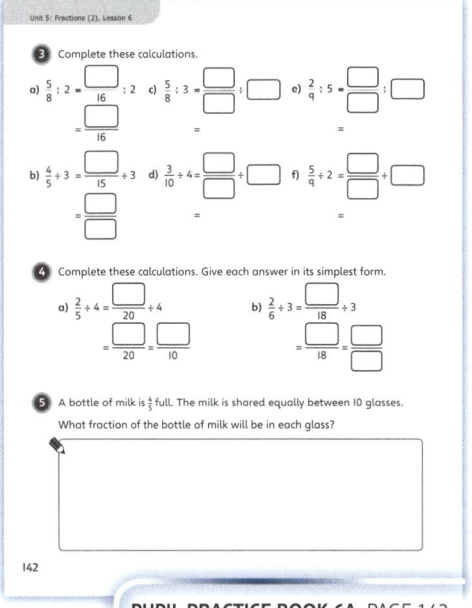

PUPIL PRACTICE BOOK 6A PAGE 142

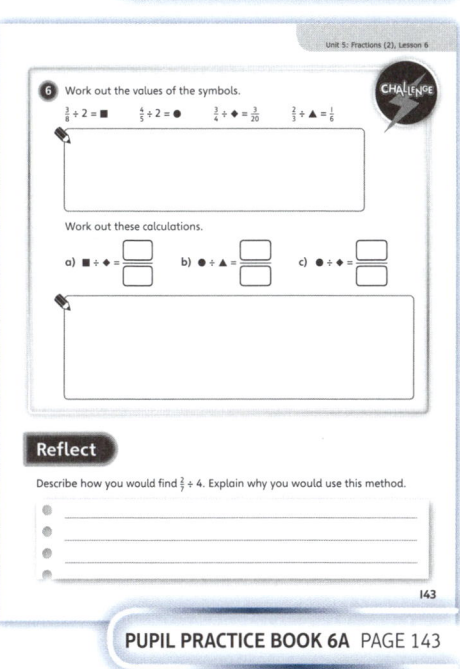

PUPIL PRACTICE BOOK 6A PAGE 143

Four rules with fractions

Learning focus

In this lesson, children will solve fraction problems involving addition, subtraction, multiplication and division. They will use the order of operations and visual aids such as bar models to support their understanding.

Small steps

→ Previous step: Dividing a fraction by a whole number (3)
→ **This step: Four rules with fractions**
→ Next step: Calculating fractions of amounts

NATIONAL CURRICULUM LINKS

Year 6 Number – Fractions
- Add and subtract fractions with different denominators and mixed numbers, using the concept of equivalent fractions.
- Multiply simple pairs of proper fractions, writing the answer in its simplest form [for example, $\frac{1}{4} \times \frac{1}{2} = \frac{1}{8}$].

Year 5 Number – Addition, Subtraction, Multiplication and Division
- Use their knowledge of the order of operations to carry out calculations involving the four operations.

ASSESSING MASTERY

Children can multiply, divide, add and subtract fractions when applied to various contexts, including perimeter and areas of 2D shapes. They can follow the correct order of operations when calculating with fractions. They can appreciate that there are multiple ways to solve a problem: for example, multiplying a fraction by 3 is the same as adding three of these fractions together.

COMMON MISCONCEPTIONS

Some children may not apply the order of operations when calculating with fractions. Ask:
- In $4 + 2 \times 3$, which operation do you do first? In $\frac{2}{3} + \frac{4}{5} \times \frac{1}{2}$, which operation do you do first? In $\frac{2}{3} + \frac{4}{5} \div 2$, which operation do you do first?

STRENGTHENING UNDERSTANDING

Use a bar model to help children to recognise the link between multiplication and repeated addition. Emphasise that there may be more than one effective method for solving a problem. If children use a different method from their partner, encourage them to discuss the methods to confirm that both are valid.

GOING DEEPER

Encourage children to always give their answer as a mixed number where possible and to simplify answers fully. When appropriate, challenge children to show two different ways of solving the problem and to explain why both methods work.

KEY LANGUAGE

In lesson: add, multiply, area, perimeter, square, triangle, isosceles triangle

Other language to be used by the teacher: order of operations, subtract, divide, mixed number, simplify

STRUCTURES AND REPRESENTATIONS

bar model, number line

 In the eTextbook of this lesson, you will find interactive links to a selection of teaching tools.

Before you teach

- Can children use all four operations with fractions?
- Can children find the areas of rectangles, squares and triangles?
- Can children find the perimeters of polygons?
- Do children understand and follow the order of operations with whole numbers?

Discover

WAYS OF WORKING Pair work

ASK

- Question **1** a): *How far did Luis walk each day? How can you work out how far Luis walked from Monday to Friday? What operation do you need to do to work out the total? Can you use a diagram to show how to work it out?*
- Question **1** b): *How can you work out how far Luis walked in total in the week? How do you know from your total whether Luis achieved his target?*

IN FOCUS Question **1** a) can be solved using either addition or multiplication: adding $\frac{2}{3}$ repeatedly 5 times or multiplying $\frac{2}{3}$ by 5. Encourage children to discuss how they could draw a diagram to show these two methods. Question **1** b) builds on the learning taking place in part a) as it is also a problem that can be solved in more than way, using either addition or multiplication followed by addition. Question **1** b) also requires a comparison of a mixed number with a whole number.

PRACTICAL TIPS Use the context to explore further questions, such as: *What if his target was double?* Children could take part in a similar activity, to see how far they could walk in one week, which would link to other areas of the curriculum.

ANSWERS

Question **1** a): Luis walked $3\frac{1}{3}$ km from Monday to Friday.

Question **1** b): $4\frac{8}{9} < 5$, so Luis did not meet his target.

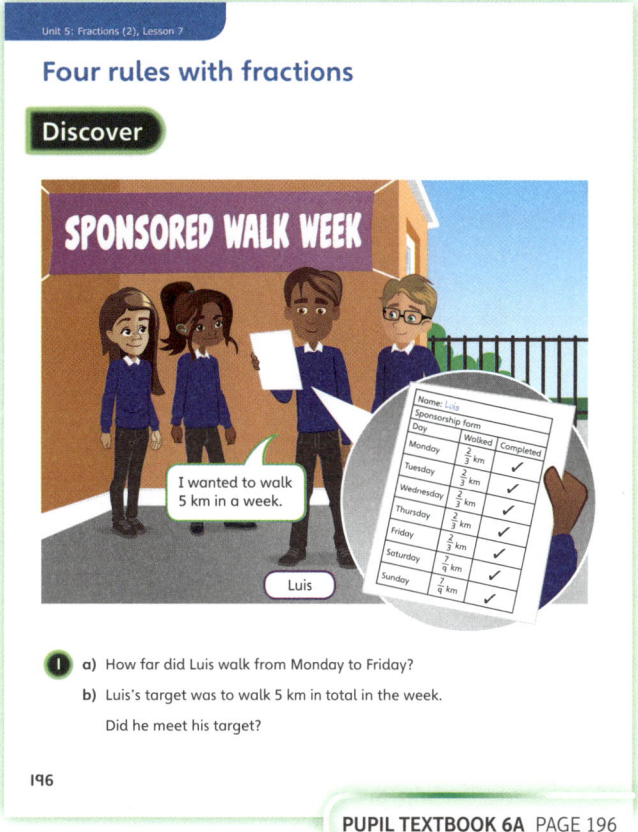

PUPIL TEXTBOOK 6A PAGE 196

Share

WAYS OF WORKING Whole class teacher led

ASK

- Question **1** a): *How does the bar model show how far Luis walked Monday to Friday? What would the bar model look like if you wanted to work out how far he walked Monday to Wednesday? Why can you work out the answer using addition **or** multiplication?*
- Question **1** b): *What does the diagram show? What operation(s) do you need to do to find the total? Can you write a number sentence using just addition to show how to find the answer? Can you write a number sentence using multiplication and addition? Do you need to use brackets?*
- Question **1** b): *Why does Flo use a common denominator to add the fractions?*
- Question **1** b): *How do you know that Luis has not met his target?*
- Question **1** b): *Can you use subtraction or division? Why not?*

IN FOCUS Question **1** a) shows two ways of finding the answer, addition or multiplication. In contrast, question **1** b) only shows one method: multiplication followed by addition. Discuss whether this could also be solved using just addition. Model the two methods, showing the calculations and matching bar models to emphasise why both methods are valid. Check that children understand that the final answer is 'yes' or 'no', not just the distance that Luis walked in the week.

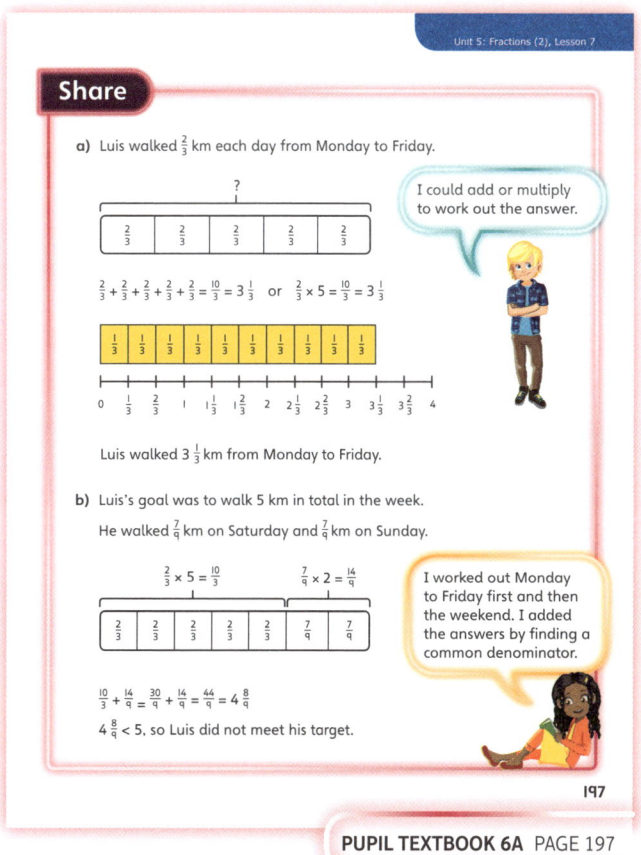

PUPIL TEXTBOOK 6A PAGE 197

Think together

Whole class teacher led (I do, We do, You do)

ASK

- Question **1**: *Is your answer in its simplest form?*
- Question **2**: *What is the same as question* **1**? *What is different? What do you do to multiply three fractions?*
- Question **3** a): *Can you tell me the correct order of operations? Which part of the calculation do you need to do first?*
- Question **3** b): *What do you think Alex did first?*

IN FOCUS Questions **1** and **2** both show how this skill is relevant to other areas of the curriculum, whilst still practising the calculations the children have been doing in **Discover**. Question **3** asks children to think about the misconception of completing the addition before the multiplication, which would result in an incorrect answer. Encourage children to explain their reasoning in full sentences.

STRENGTHEN In question **1**, suggest that children use a bar model to support their understanding. Discuss whether there are any alternative methods (repeated addition) and encourage them to reflect on which method they would have chosen. Model both methods using bar models to demonstrate that they give the same answer. In question **3**, encourage children to draw diagrams to help them with the addition and multiplication.

DEEPEN Ask children to think about multiple ways of answering questions **1** and **2**, and to explain which method they think is more efficient. In question **2**, encourage children to explore why multiplying by $\frac{1}{2}$ has the same result as dividing by 2.

ASSESSMENT CHECKPOINT Use questions **1** and **2** to assess whether children can solve problems using fractions. Check that they can decide what operation is needed, and recognise that the operation could be multiplication or addition when finding the perimeter. Check that they can convert from improper fractions to mixed numbers and give the answers in their simplest form. Use question **3** to assess whether children can use the correct order of operations to calculate with fractions.

ANSWERS

Question **1**: $\frac{3}{4} \times 2 = \frac{6}{4}$, $\frac{1}{6} \times 2 = \frac{2}{6}$
$\frac{6}{4} + \frac{2}{6} = \frac{18}{12} + \frac{4}{12} = \frac{22}{12} = 1\frac{10}{12} = 1\frac{5}{6}$
The perimeter of the rectangle is $1\frac{5}{6}$ m.
Some children may simplify fractions earlier.

Question **2**: $\frac{1}{2} \times \frac{1}{4} \times \frac{1}{2} = \frac{1}{16}$
The area of the triangle is $\frac{1}{16}$ m².

Question **3** a): Jamilla's answer is correct.

Question **3** b): Jamilla answered the question by completing the multiplication first and then adding $\frac{1}{5}$. Alex added $\frac{1}{5}$ and $\frac{3}{5}$ first and then multiplied. This is incorrect because the multiplication needs to be done first.

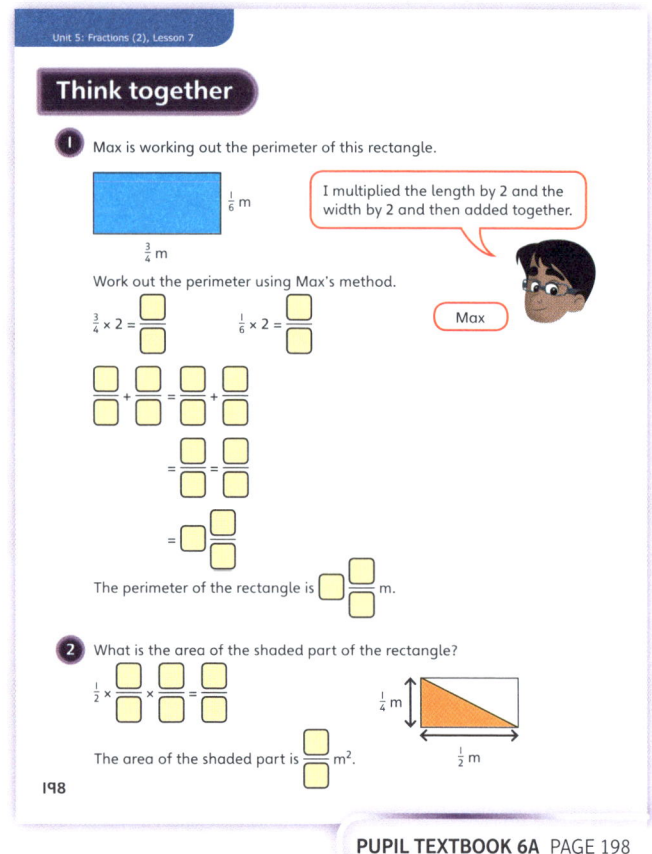

PUPIL TEXTBOOK 6A PAGE 198

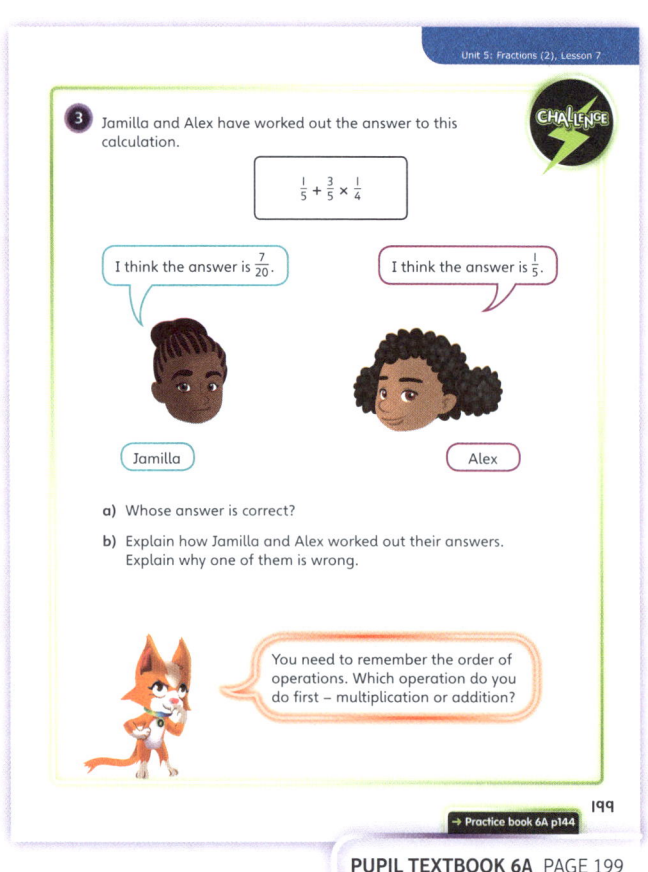

PUPIL TEXTBOOK 6A PAGE 199

Practice

WAYS OF WORKING Independent thinking

IN FOCUS Questions ❶ and ❷ aim to consolidate what children have covered in the **Textbook**, with question ❶ covering multiplication of a fraction by a whole number and question ❷ multiplication and addition of fractions. Questions ❷ onwards require children to decide which method to use. Question ❻ incorporates both addition and subtraction, challenging children to think about how they can represent a whole 1 as a fraction.

STRENGTHEN In question ❶, where the scaffolding encourages children to multiply one side by the number of sides, rather than use addition, use a bar model to show that the answer will be the same. Encourage children to use diagrams in all questions to help them to calculate with fractions, as well as helping them to see what operation is needed. Display a poster reminding them of the methods that they have used to calculate with fractions in this and the previous unit.

DEEPEN In question ❹, ask children to explain what common mistakes might be made in relation to the order of operations. Ask children to create their own calculations for a partner to solve. Suggest that they include brackets in some of their calculations.

THINK DIFFERENTLY Question ❺ introduces subtraction and division for the first time in the lesson, as children are given the perimeter and asked to find a missing side. They need to use their knowledge of isosceles triangles and squares to work out what numbers they need to multiply and divide by. Suggest that they write any information that they know on the diagram.

ASSESSMENT CHECKPOINT Use questions ❷ to ❻ to assess whether children can decide what operation is needed and can complete a calculation involving fractions with more than one operation.

ANSWERS Answers for the **Practice** part of the lesson appear in the separate **Practice and Reflect answer guide**.

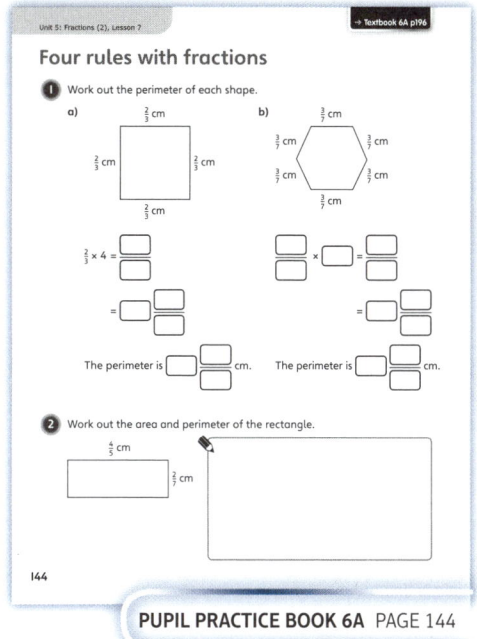

PUPIL PRACTICE BOOK 6A PAGE 144

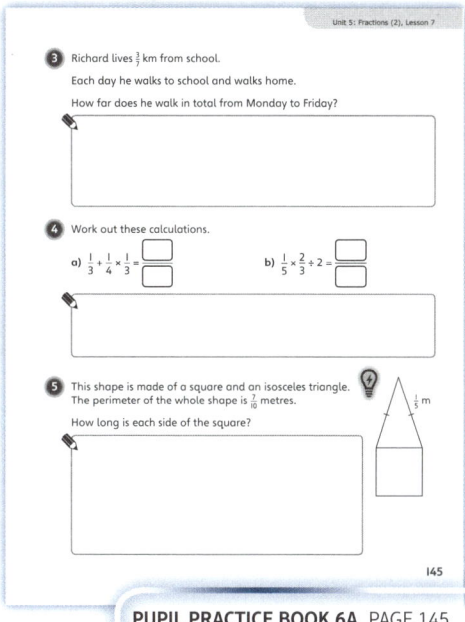

PUPIL PRACTICE BOOK 6A PAGE 145

Reflect

WAYS OF WORKING Independent thinking

IN FOCUS This activity checks whether children can successfully follow the order of operations when calculating with fractions. They need to understand that when multiplication and addition are in the same calculation, the multiplication is completed first.

ASSESSMENT CHECKPOINT Assess whether children can explain, in full sentences, the mistake that has been made. Check that they can answer the question correctly.

ANSWERS Answers for the **Reflect** part of the lesson appear in the separate **Practice and Reflect answer guide**.

After the lesson

- Can children decide independently what operation is required?
- Do children recognise the correct order of operations?
- Can children complete a calculation that has more than one operation?
- Can children add, subtract and multiply fractions, and divide fractions by whole numbers?

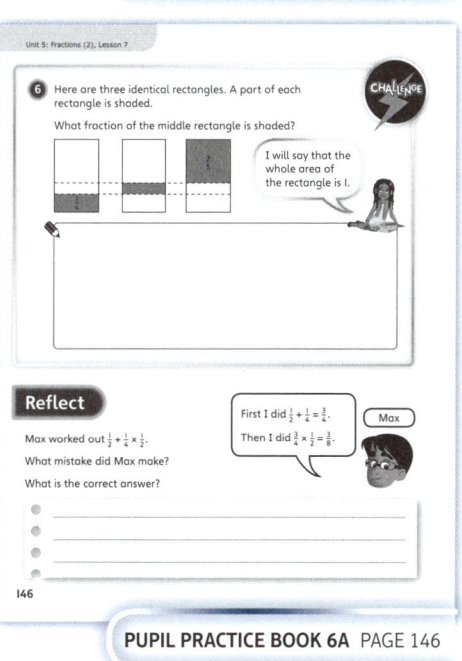

PUPIL PRACTICE BOOK 6A PAGE 146

231

Calculating fractions of amounts

Learning focus

In this lesson, children will learn to find fractions of amounts in various contexts. They will use visual aids such as bar models to solve problems and support their understanding.

Small steps

→ Previous step: Four rules with fractions
→ **This step: Calculating fractions of amounts**
→ Next step: Problem solving – fractions of amounts

NATIONAL CURRICULUM LINKS

Year 6 Number – Fractions

Use written division methods in cases where the answer has up to two decimal places.

ASSESSING MASTERY

Children can find fractions of amounts involving unit and non-unit fractions by using a bar model. They can use the bar model to explain their understanding. Children can apply their skills of finding a fraction of an amount to unfamiliar problems and multi-step problems.

COMMON MISCONCEPTIONS

Some children may misinterpret what the question is asking. For example, if a question says, 'There are 30 students in a class. $\frac{2}{5}$ of the class are going swimming. How many children are not going swimming?', children may calculate $\frac{2}{5}$ instead of $\frac{3}{5}$. Say:

• *Tell me what the question is asking for. Can you draw a bar model to show the number of children? How can you divide your bar into those going swimming and those not going swimming? Which part represents the children not going swimming?*

STRENGTHENING UNDERSTANDING

Give children strips of paper and counters to help them find fractions of an amount. Tell children that the strip of paper represents the whole and can be divided into whatever denominator the question asks for. Children can then share the counters, which represent the items in the question, equally between the parts. Encourage the children to think about what one-quarter, one-third or one-half might be before asking for non-unit fractions.

GOING DEEPER

Add additional reasoning questions to the problems. For example, if the question was, 'There are 36 children in a swimming class. $\frac{1}{3}$ of the class are boys. How many of the class are girls?', ask additional questions such as, 'How many more girls than boys are in the swimming class?' to stretch children further. Where appropriate, ask children to show more than one method for finding the solution.

KEY LANGUAGE

In lesson: fraction of, equal parts, share

Other language to be used by the teacher: numerator, denominator, division, multiply

STRUCTURES AND REPRESENTATIONS

bar model

RESOURCES

Optional: paper strips, counters, baskets, apples

 In the eTextbook of this lesson, you will find interactive links to a selection of teaching tools.

Before you teach

• Can children draw a diagram to represent a fraction?
• Can they divide an amount into equal parts?
• Can they multiply two integers?

Discover

Pair work

ASK

• Question **1** a): *How many baskets does each year have? What is different about Year 6? What diagram could you draw to help you?*

• Question **1** b): *How is this question the same as part a)? How is it different? What diagram could you draw for this question? How many parts will it have? Did the children eat more apples in the morning or in the afternoon?*

IN FOCUS Questions **1** a) and **1** b) both illustrate the importance of reading the question carefully: part a) requires children to understand that the total needs to be divided into 5 equal parts, not 4; and part b) requires them to realise that they are being asked for $\frac{7}{10}$ of the total, not $\frac{3}{10}$. Question **1** a) tests children's ability to divide by asking them to share the apples equally into 5 baskets. The question does not talk about fractions, just equal parts. A link could be made between 5 equal parts and fifths, and the fraction that Year 6 represents. Children may misinterpret the question and draw 4 boxes in their bar model, because there are 4 year groups. Emphasise the importance of labelling diagrams clearly to avoid misinterpreting the question.

PRACTICAL TIPS Carry out a similar question practically, by bringing in 5 baskets and a number of apples – a multiple of 5. Demonstrate the apples being shared equally among the baskets and then count the apples in 2 baskets. Ask children to draw a diagram to represent the baskets and the apples.

ANSWERS

Question **1** a): The Year 6 children will get 80 apples.

Question **1** b): The Year 6 children eat 56 apples in the afternoon.

Share

WAYS OF WORKING Whole class teacher led

ASK

• Question **1** a): *Why does the bar model have 5 parts? How many parts does Year 6 have? Does there need to be an equal number of apples in each part? How many apples are there in 1 part? What fraction does each part represent? What fraction of the total apples does Year 6 have? How many apples does Year 3 get? What fraction is this?*

• Question **1** b): *How many apples are in each part? What operations do you need to use for this question? Why does Flo say that she only needs to find $\frac{7}{10}$? Can you explain whether her method is more efficient?*

IN FOCUS Question **1** b) draws attention to two possible approaches: find $\frac{7}{10}$ of 80, or find $\frac{3}{10}$ of 80 and subtract it from 80. Show how the bar model illustrates both these methods and discuss which is more efficient. Use Dexter's and Flo's statements to discuss what common mistakes may be made with these questions.

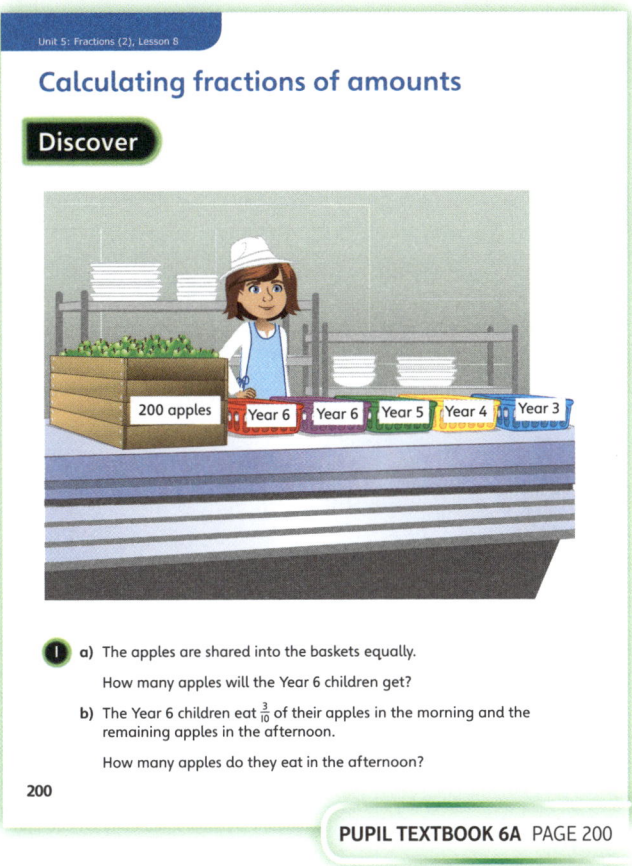

PUPIL TEXTBOOK 6A PAGE 200

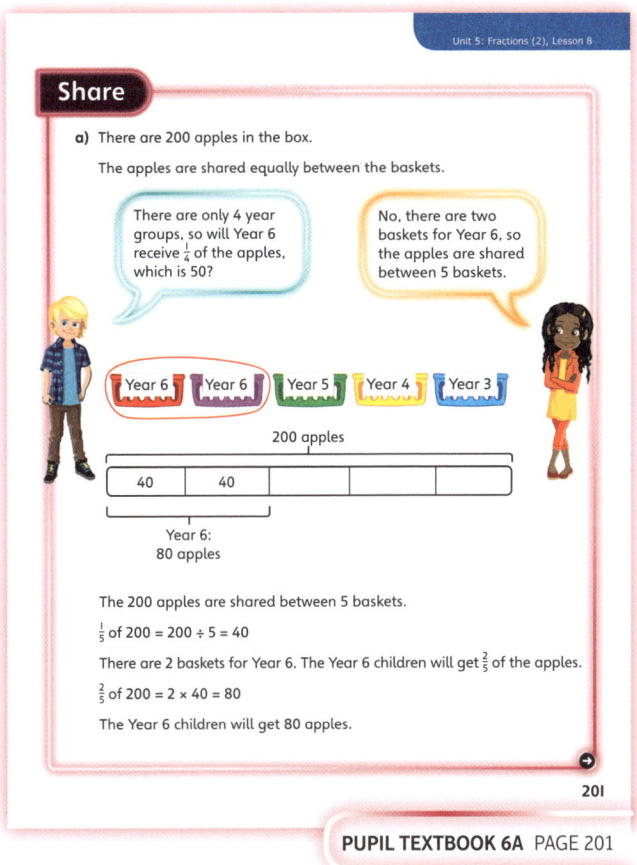

PUPIL TEXTBOOK 6A PAGE 201

Think together

WAYS OF WORKING Whole class teacher led (I do, We do, You do)

ASK

- Question ❶: *How many parts do you need to divide the bar into? How do you know?*
- Question ❷: *Can you explain what you need to find? What diagram could you draw to help you? What fraction of the children are not going on the trip?*
- Question ❸: *What has Richard found? How can you explain to Mo and Richard the mistakes they have made?*

IN FOCUS Question ❶ requires children to find a fraction of a quantity in its simplest form, where the fraction is a non-unit fraction. The scaffolding encourages children to find $\frac{1}{6}$ of the amount and then use that to find $\frac{5}{6}$. Question ❷ requires children to read the question carefully to avoid misinterpreting what they are being asked for. The fraction in the question text is $\frac{5}{7}$, but the amount being asked for is the remaining amount, $\frac{2}{7}$. The question challenges children to think about how they could complete the questions without subtraction and therefore prompts them for two different methods. A bar model will help children to see that $\frac{2}{7}$ of the amount remains. Question ❸ is a two-step problem: children need to find a fraction of an amount and then a fraction of that number. A bar model will help them to understand what the question is asking and to identify the common mistakes illustrated by Mo and Richard.

STRENGTHEN To strengthen understanding, use the scaffolding provided in question ❶ as an example of how children could approach questions ❷ and ❸ through smaller steps. Encourage children to use bar models and to label their diagrams clearly and to shade what the question is asking for. This is particularly important for questions ❷ and ❸, which are more complex.

DEEPEN In question ❷, ask children whether they could have completed the question without subtraction. Discuss similarities and differences in the calculations for both methods. Ask children to create a question similar to question ❸ for a partner to solve.

ASSESSMENT CHECKPOINT Use questions ❶ and ❷ to assess whether children can use a diagram to find a fraction of an amount, labelling the whole, the parts and what the question is asking for. Use question ❷ to assess whether children can find a fraction of an amount when the fraction is not given directly.

ANSWERS

Question ❶: $\frac{1}{6}$ of 300 g is 300 ÷ 6 = 50 g

$\frac{5}{6}$ of 300 g is 5 × 50 = 250 g

250 g of flour is needed.

Question ❷: 8 children are not going on the trip.

Question ❸: Mo found $\frac{1}{2}$ of 36, without first finding out how many of the class are boys. Richard found $\frac{1}{3}$ of 36, which is the number of boys in the class, but has not continued the question to find out how many of the boys wear goggles. 6 boys wear goggles.

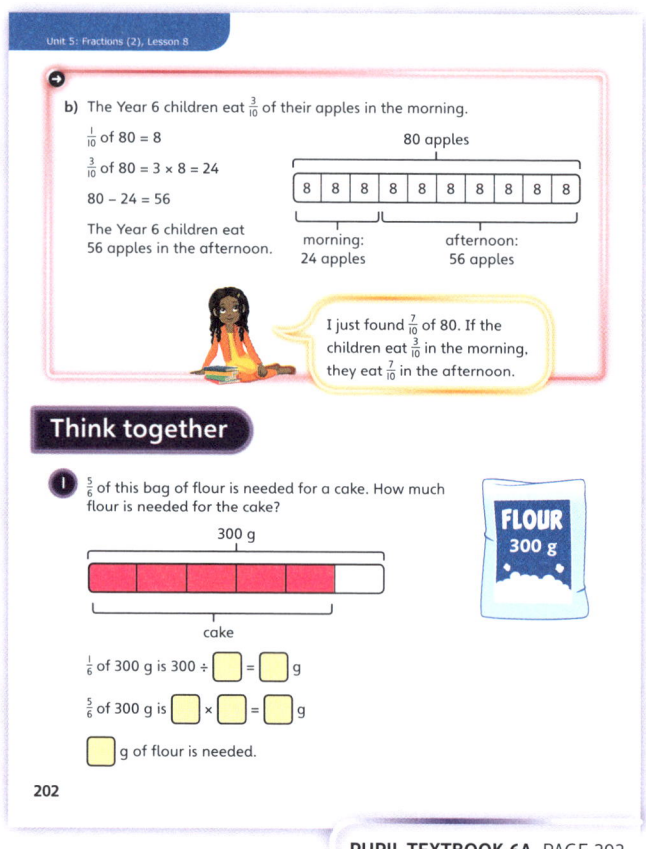

PUPIL TEXTBOOK 6A PAGE 202

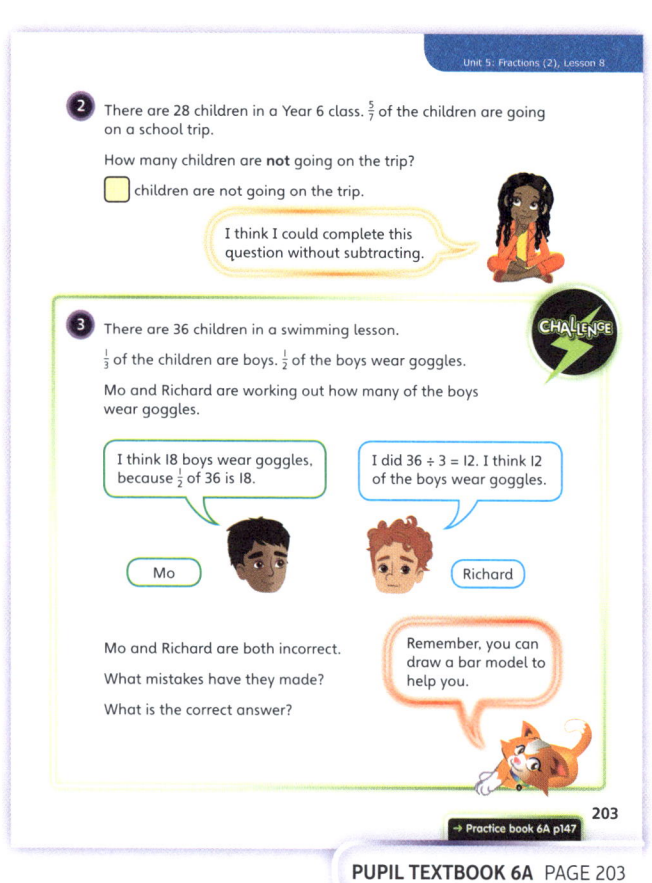

PUPIL TEXTBOOK 6A PAGE 203

Practice

WAYS OF WORKING Independent thinking

IN FOCUS Most of the questions are not scaffolded, reflecting that there are multiple ways of solving these problems. The emphasis is for children to find the most efficient method to solve each problem. In questions **1** and **2**, children can either first find the fraction given in the question and subtract it from the total or first work out the fraction represented by the answer. It might be useful to stop the class after a few problems and discuss the methods used. In question **6**, the fractions in each comparison are of the same total, giving children the opportunity to realise that they can just look at the two fractions without finding fractions of an amount.

STRENGTHEN Read through a question together. Draw a simple bar model on the board, explaining each feature of the model. Emphasise that the bar is the total and they need to look at the denominator of the fraction to work out how many parts to divide the bar into. Agree that the numerator shows how many parts are represented. Read the question again, drawing attention to what is being asked for, and shade these parts on the diagram.

For questions that involve comparisons (for example, question **3**), suggest to children that they draw one bar model above the other using bars of the same length.

DEEPEN When children have completed question **7**, ask them to work out what fraction of the total Amelia gave to her mum. Ask them to explore the relationship between the fractions in the question and this fraction. They could extend this exploration to the problem in **Think together** question **3**, which is a similar problem, and explain their findings using a bar model.

THINK DIFFERENTLY In question **6**, children have to put < or > between two calculations. In part a) they may see straight away that $\frac{3}{7}$ of 70 is less than $\frac{5}{7}$ of 70 and will not need to draw a diagram to help them. Part b) will give them the opportunity to practise drawing fraction grids.

ASSESSMENT CHECKPOINT Use questions **1** to **6** to assess whether children can solve problems involving finding fractions of amounts. Look for children using the most efficient methods.

ANSWERS Answers for the **Practice** part of the lesson appear in the separate **Practice and Reflect answer guide**.

Reflect

WAYS OF WORKING Whole class

IN FOCUS Children need to think about the questions they have answered and say which question they found the most challenging and why. Use this as a class discussion, so children can see that maths can be challenging and that everyone struggles sometimes. For those children who say it was all straightforward, discuss whether they think they used the most efficient method.

ASSESSMENT CHECKPOINT Assess whether children can describe how to find a fraction of an amount, looking for the most efficient method.

ANSWERS Answers for the **Reflect** part of the lesson appear in the separate **Practice and Reflect answer guide**.

After the lesson

- Can children find a fraction of an amount?
- Can they represent a problem using a bar model and use this to find the most efficient strategy?

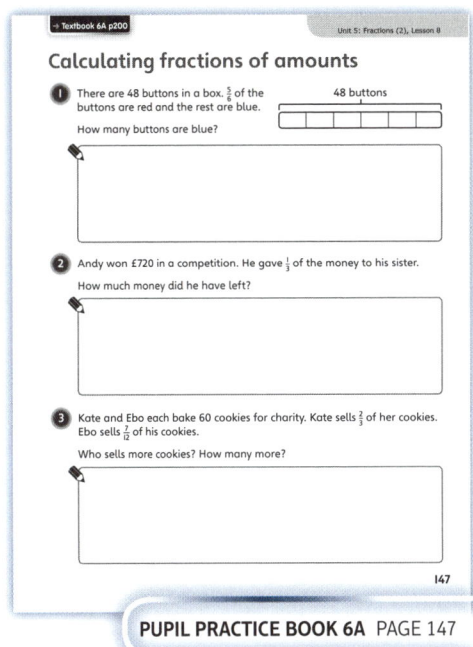

PUPIL PRACTICE BOOK 6A PAGE 147

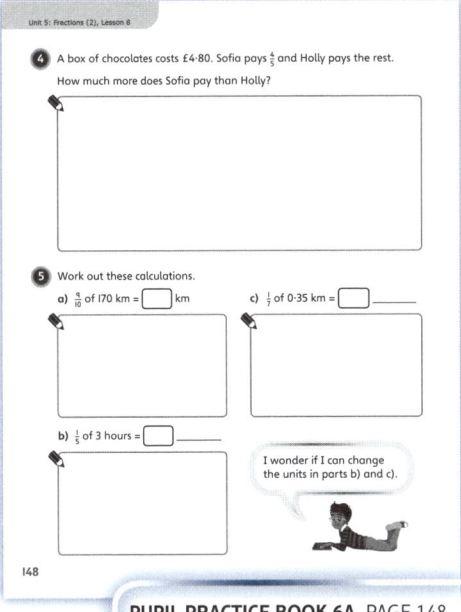

PUPIL PRACTICE BOOK 6A PAGE 148

PUPIL PRACTICE BOOK 6A PAGE 149

Problem solving – fractions of amounts

Learning focus

In this lesson, children will solve problems involving finding fractions of amounts, including problems where children have to find the whole given information about a part.

Small steps

→ Previous step: Calculating fractions of amounts
→ **This step: Problem solving – fractions of amounts**
→ Next step: Plotting coordinates in the first quadrant

NATIONAL CURRICULUM LINKS

Year 6 Number – Fractions

Use written division methods in cases where the answer has up to two decimal places.

COMMON MISCONCEPTIONS

Some children may always just find a fraction of an amount regardless of the question. For example, for the question, '$\frac{2}{5}$ of a number is 10. What is the number?' they will work out $\frac{2}{5}$ of 10 = 4. Ask:

• *Is 10 the whole number or a fraction of it? Is $\frac{2}{5}$ of a number smaller or larger than the number? Will the answer be more or less than 10?*

STRENGTHENING UNDERSTANDING

In order to support children with understanding whether they are finding a fraction of an amount or the whole, ask children to underline the key information in the question. Draw a single bar and ask them if they know the whole bar or just part. Give them concrete objects such as counters to help. Divide the model into the number of parts. Write the information they know on the model (whole amount with a brace or the parts). The picture should then help them work out what steps they need to take.

GOING DEEPER

Give children more complex problems to solve, such as '$\frac{1}{3}$ of 60 = $\frac{2}{5}$ of □.' This requires them to both find a part and work out the whole from a part in the same question. Move on to questions like, 'Chris's height has increased by $\frac{1}{5}$. If his new height is 1·8 metres, what was his height before?'

KEY LANGUAGE

In lesson: fraction of

Other language to be used by the teacher: numerator, denominator, whole, part, multiply, divide

STRUCTURES AND REPRESENTATIONS

bar model

RESOURCES

Optional: play coins, counters, multilink cubes

 In the eTextbook of this lesson, you will find interactive links to a selection of teaching tools.

Before you teach

• Can children find a simple fraction of an amount?
• Can they use bar models to represent fraction of amount questions?
• Do they know their times-tables off by heart to support them with the calculations?

Discover

WAYS OF WORKING Pair work

ASK

- Question **1** a): *How does this question differ from the ones you were doing in the previous lesson? What do you know about £1·60? Can you represent the situation with a bar model? How can you work out how much pocket money Lee had?*
- Question **1** b): *What information are you given? How many sweets are in $\frac{1}{5}$ of the jar? How did you work this out? How many sweets are in the whole jar? What other way could you get the same answer?*

IN FOCUS This lesson builds on the previous lesson, progressing to working out the whole amount given a fraction of the amount, using the same models. Question **1** a) uses a unit fraction, so the whole can be found by multiplying by the denominator. Look for children who just find a quarter of the amount instead of multiplying up. Encourage children to draw a bar model to represent the situation. Question **1** b) gives a non-unit fraction, so children can explore ways of working out the whole. For example, they may double and add on half or they could halve and then multiply by 5.

PRACTICAL TIPS Act out the scenario using toy coins and counters or cubes to represent the sweets.

ANSWERS

Question **1** a): Lee had £6·40 to begin with.

Question **1** b): There were 75 sweets in the jar when it was full.

Share

WAYS OF WORKING Whole class teacher led

ASK

- Question **1** a): *What information were you given? How does the bar model show the situation? What mistake did Dexter make? What could you do to help you not make that mistake?*
- Question **1** a): *Why do you multiply by 4? What method do you know of multiplying by 4?*
- Question **1** b): *How is this question different from part a)? How is it the same? Why do you not just multiply by 5? What do you have to do first?*
- Question **1** b): *Can you explain where Astrid has got her numbers from? What has she done?*

IN FOCUS In part a), focus on the difference between this type of question and the ones from the last lesson. By explaining what Dexter has done wrong, children can develop an understanding of how to use the bar model correctly. In part b), some children may want to multiply by 5, forgetting that $\frac{2}{5}$ is not a unit fraction. Discuss that this time they have to work out what $\frac{1}{5}$ of the jar is first, before they can multiply. The bar model helps explain why they first need to divide by 2. Use Astrid's method as a discussion point to explain other ways you could approach the question.

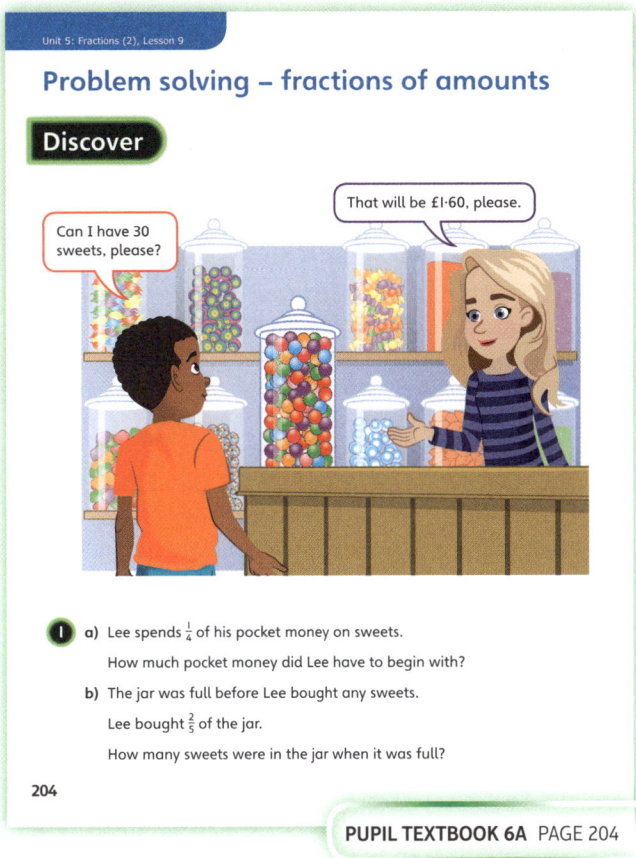

PUPIL TEXTBOOK 6A PAGE 204

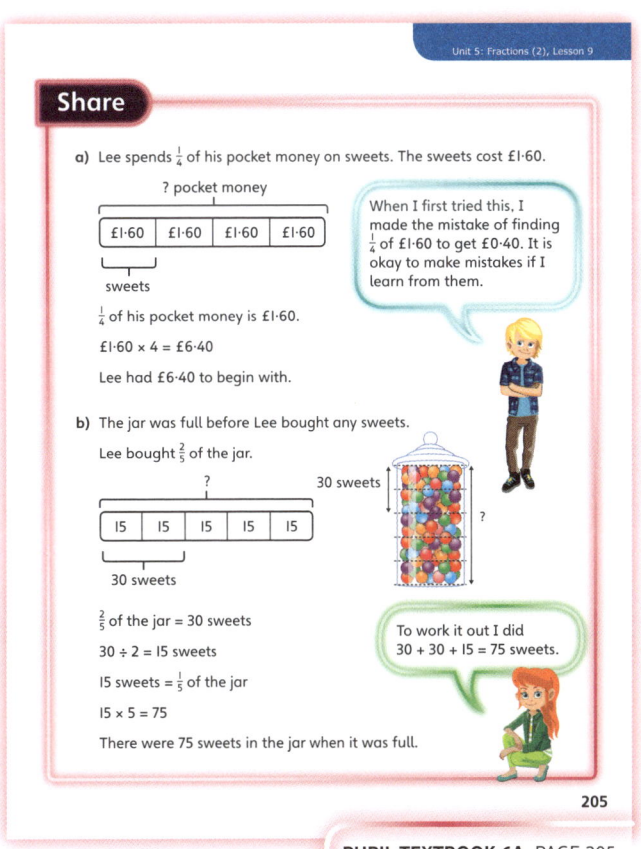

PUPIL TEXTBOOK 6A PAGE 205

Think together

Whole class teacher led (I do, We do, You do)

ASK

- Question **1**: *Do you know the whole? What should you do first? Then what? How can you work out the total number of darts in the red and yellow sections? What about the blue section? How can you work out the number of points once you know this?*
- Question **2**: *Do you know the whole rope length? How can you use a bar model to help you? How can you find the length of rope A? What about rope B?*
- Question **3**: *What information are you given? Are you given the whole? How could you work out the original number (the whole)? Once you have worked this out, what can you do then? How do you think Ash worked it out?*

IN FOCUS These problems are multi-step and require children to determine first whether they know the whole and have to find a fraction of the amount or whether they have to find the whole. Encourage children to see this as the first step.

Question **1** gives children the whole from which they need to find the number of darts in each section. They then need to use their knowledge of the scores for each section to work out the total score. In question **2**, children need to realise that they have to work out both wholes in order to compare the lengths of rope. In this question they are working with decimals, so some children might need additional support.

Question **3** provides children with the opportunity to use their problem solving skills to find more than one way of working out the answer. For example, they may find the whole (45) and then work out $\frac{8}{9}$ of the whole. Ask what facts they know about the number; for example, can they find $\frac{1}{3}$, $\frac{1}{6}$, $\frac{1}{9}$ of the number? Some children may use their knowledge of equivalent fractions and realise that $\frac{2}{3} = \frac{6}{9}$ and so $\frac{8}{9}$ is $\frac{2}{9}$ more. Explore different ways of approaching the question.

STRENGTHEN Sit with children and ask them to highlight the key information in the question. Ask if they know the whole or if they need to work out the whole. Provide them with concrete materials to help. Guide them through drawing a bar model to illustrate the question. Once children have a picture of the situation this should help them work out what steps they need to take.

DEEPEN Question **3** asks children to make up their own problems for their friends to solve. Encourage children to create more complex problems than just '$\frac{3}{8}$ of my number is 15. What is my number?' Ask children to solve their partner's problems in more than one way, explaining why each method works.

ASSESSMENT CHECKPOINT Use questions **1**, **2** and **3** to assess whether children can solve a variety of multi-step problems involving fractions of an amount.

ANSWERS

Question **1**: Kate scores 138 points.

Question **2**: Rope A is longer by 0·2 m. (Rope A: 5·4 m, rope B: 5·2 m.)

Question **3**: $\frac{8}{9}$ of Amelia's number is 40.

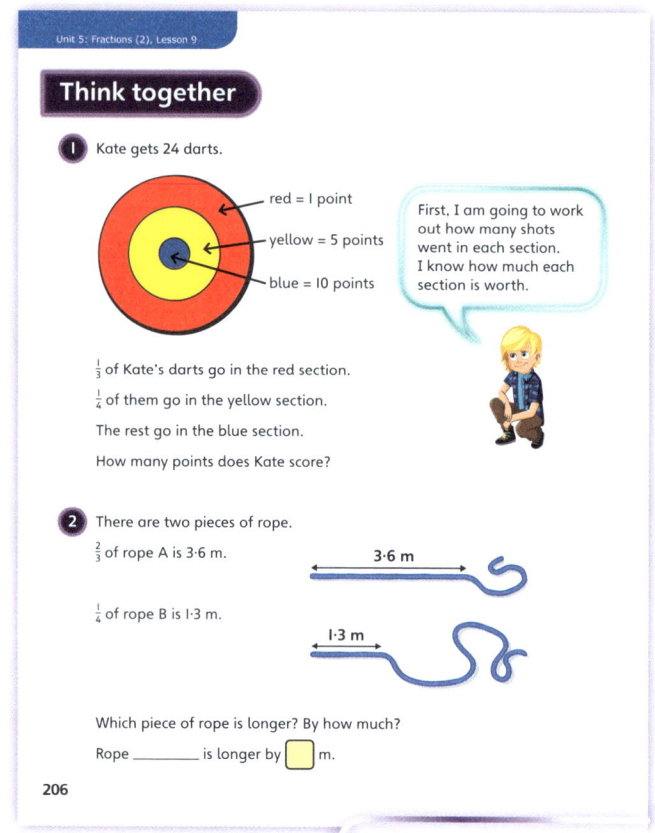

PUPIL TEXTBOOK 6A PAGE 206

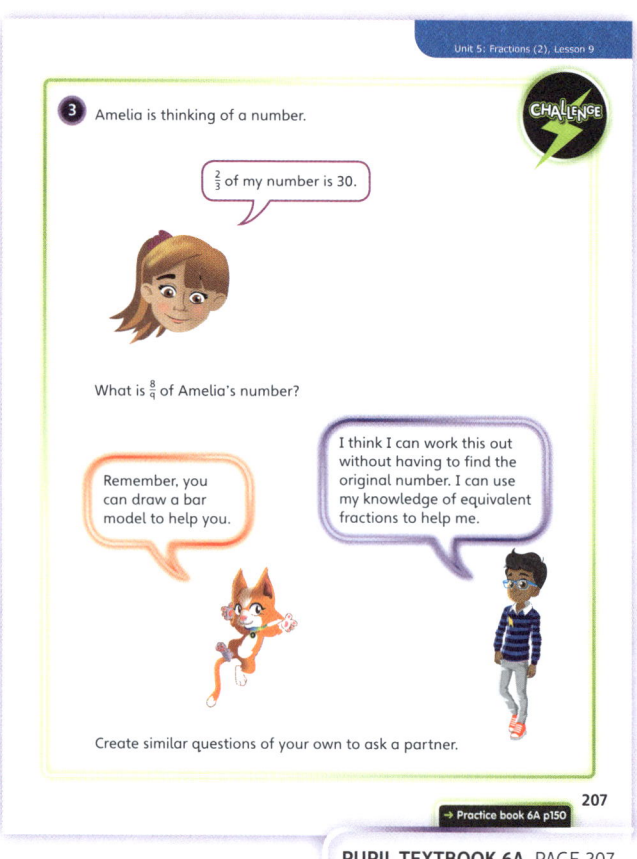

PUPIL TEXTBOOK 6A PAGE 207

Practice

WAYS OF WORKING Independent thinking

IN FOCUS Questions ❶, ❷ and ❸ are simple problems where children have to find the whole, given a fraction of an amount. Question ❶ uses a unit fraction, so children simply have to multiply. In questions ❷ and ❸, children need to divide first. For all questions, encourage children to draw a bar model. Question ❺ provides more abstract practice. Each question steps up in difficulty, with part (d) requiring children to realise that there are several steps involved in finding the answer.

STRENGTHEN For questions that are more abstract (for example, questions ❷ and ❺), children might find it easier if they create their own context to fit the numbers. Provide them with counters or cubes to support their scenario.

DEEPEN Ask children to solve problems such as $\frac{1}{3}$ of $\frac{1}{4}$ of □ = 10. Ask them to represent this using a bar model.

ASSESSMENT CHECKPOINT Use questions ❷ to ❼ to assess whether children can solve multi-step questions on fractions of an amount, including ones where they have to find the whole.

ANSWERS Answers for the **Practice** part of the lesson appear in the separate **Practice and Reflect answer guide**.

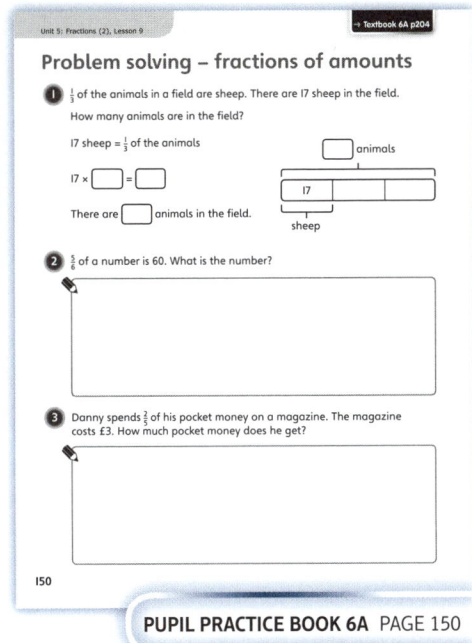

PUPIL PRACTICE BOOK 6A PAGE 150

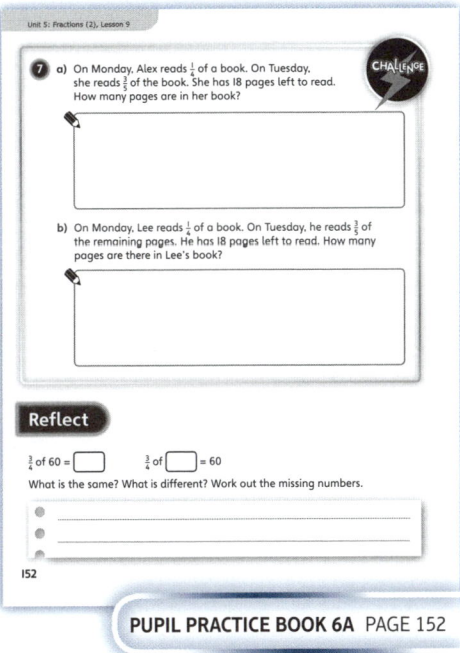

PUPIL PRACTICE BOOK 6A PAGE 151

Reflect

WAYS OF WORKING Independent thinking

IN FOCUS Children compare questions where they are given the whole and where they need to find the whole, requiring them to focus on the difference between the methods. Although the questions look similar, they have a different underlying structure.

ASSESSMENT CHECKPOINT Children should know when they are finding a fraction of an amount and when they are working out the whole. They should use a bar model or other pictorial representation to explain the key difference.

ANSWERS Answers for the **Reflect** part of the lesson appear in the separate **Practice and Reflect answer guide**.

After the lesson

- Can children find the whole, given a fraction of an amount?
- Can they interpret questions to work out whether they are finding a part or the whole?
- Can they solve more complicated multi-step problems involving fractions of an amount?

PUPIL PRACTICE BOOK 6A PAGE 152

End of unit check

> **Don't forget the *Power Maths* unit assessment grid on p26.**

WAYS OF WORKING Group work adult led

IN FOCUS

- Questions **1** to **7** are fluency-based questions to check children's understanding of methods for multiplying and dividing fractions. Many of the incorrect answers are likely to indicate that children have confused the two methods.
- Questions **8** and **9** are SATs-style questions. In question **8**, children need to multiply the two fractions and then use their knowledge of simplifying fractions to ensure that the answer is given in its simplest form. Question **9** requires children to read the question carefully to identify what they need to find.

ANSWERS AND COMMENTARY

Children should be able to multiply any fraction by a whole number and by any other fraction, and divide a fraction by a whole number. They should be able to solve simple and multi-step fraction problems, including problems on fractions of an amount where they are given the fraction and need to find the whole.

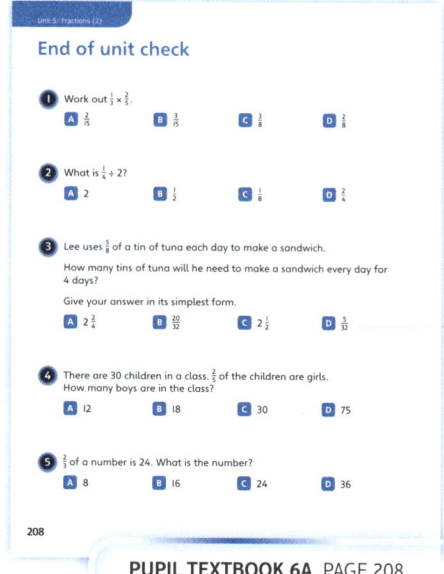

PUPIL TEXTBOOK 6A PAGE 208

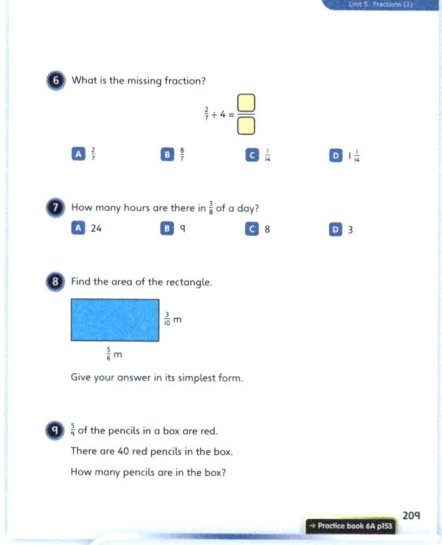

PUPIL TEXTBOOK 6A PAGE 209

Q	A	WRONG ANSWERS AND MISCONCEPTIONS	STRENGTHENING UNDERSTANDING
1	A	B suggests numerators added instead of multiplied; C both numerators and denominators added; D denominators added.	To support children with multiplying fractions, divide a grid to show the fractions on adjacent sides. Shade the relevant squares and ask children what fraction this is. How do they know?
2	C	A suggests dividing 4 by 2; B denominator divided by 2; D numerator multiplied by 2.	
3	C	A is not fully simplified; B suggests both numerator and denominator multiplied by 4; D indicates denominator multiplied by 4.	To support children with dividing a fraction by a whole number, shade in the fraction on a fraction strip. Divide the shaded parts by the whole number. Discuss the size of of the parts and how many of them make up the answer.
4	B	A suggests calculating the number of girls. D suggests having found the whole if $\frac{2}{5}$ is 30.	
5	D	A suggests found $\frac{1}{3}$ of 24, and B $\frac{2}{3}$ of 24.	
6	C	A suggests that the child has not divided by 4; B suggests they have multiplied by 4 instead of divided; and D suggests having found the answer correctly, but put 1 in front of it.	For problems on fractions of an amount, clearly label a bar model with the known information, using a question mark to show the quantity that is being asked for.
7	C	B suggests a calculation error; C suggests $\frac{1}{3}$ found; D suggests $\frac{1}{8}$ found.	
8	$\frac{1}{4}$ m²	Some children may find the perimeter instead of the area.	
9	72	Some children may find $\frac{4}{9}$ of 40.	

My journal

WAYS OF WORKING Independent thinking

ANSWERS AND COMMENTARY This journal activity brings together all the work that children have done on fractions in the last two units. Children often struggle to remember the different methods, and this journal attempts to try to recap all of them. Use the journal as a diagnostic tool to check:

- which children understand which questions
- the most common questions children are struggling with
- whether children are getting methods confused
- the methods that children use to get the answers
- whether children are giving their answers in the simplest form.

You might want to ask children to show their understanding of a method by drawing a diagram. You might also want to pair children up to teach each other and explain their method.

For the last two questions look for children following the correct order of operations.

$\frac{1}{5} \times 3 = \frac{3}{5}$ $\frac{1}{3} \div 4 = \frac{1}{12}$ $\frac{2}{3} \div 4 = \frac{1}{6}$ $\frac{7}{10} + \frac{2}{5} \times \frac{1}{2} = \frac{9}{10}$

$\frac{2}{3} \times \frac{3}{8} = \frac{1}{4}$ $\frac{4}{5} \div 2 = \frac{2}{5}$ $\frac{7}{10} + \frac{2}{5} = \frac{11}{10} = 1\frac{1}{10}$ $\frac{7}{10} \times \frac{2}{5} + \frac{1}{2} = \frac{39}{50}$

Power check

WAYS OF WORKING Independent thinking

ASK

- *Are you able to multiply a fraction by a fraction?*
- *Can you divide a fraction by a whole number?*
- *Can you find a fraction of an amount?*
- *Given $\frac{2}{5}$ of a number is 8, can you find the number?*

Power puzzle

WAYS OF WORKING Independent working

IN FOCUS This tests children's knowledge of all the work in this unit using problems in which each answer leads to the next. Emphasise the importance of double checking answers, as an error early on will lead to many wrong answers. Encourage children to start thinking about whether an answer is likely or not: complicated calculations or long decimal answers are likely to imply that they have made a mistake somewhere.

ANSWERS AND COMMENTARY

Question **1**:

A	B	C	D	E	F	G	H
36	18	27	15	4·5 or $4\frac{1}{2}$	8	2	$\frac{1}{20}$

Question **2**: Children first need to work out the distance between points A and B to find the total that they are finding $\frac{2}{3}$ of C = 150.

After the unit

- How useful did children find the diagrams in calculating with fractions?
- Which diagrams did they find most useful?
- Can they use diagrams to explain why a method works?

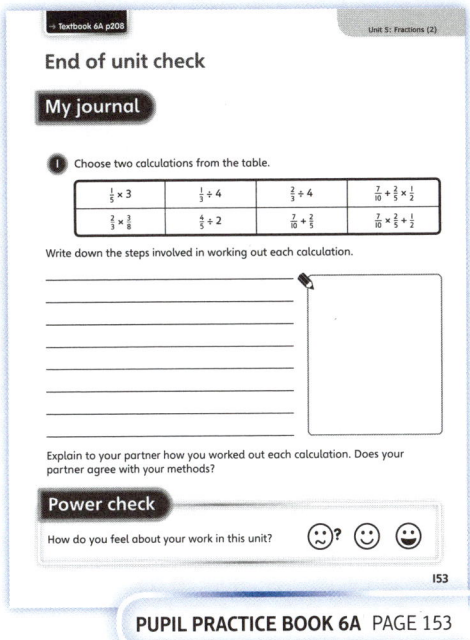

PUPIL PRACTICE BOOK 6A PAGE 153

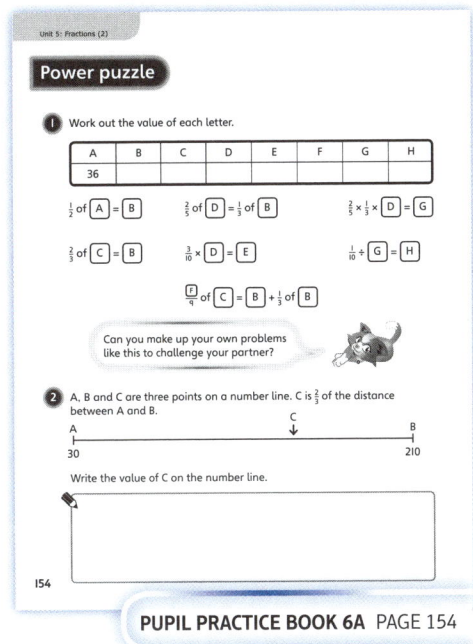

PUPIL PRACTICE BOOK 6A PAGE 154

Strengthen and **Deepen** activities for this unit can be found in the *Power Maths* online subscription.

Unit 6
Geometry – position and direction

Don't forget to watch the Unit 6 video!

WHY THIS UNIT IS IMPORTANT

This unit exposes children to coordinates in all four quadrants for the first time. Children are encouraged to combine their knowledge of properties of shapes with their coordinate knowledge and to reason and solve problems involving shapes on a coordinate grid. This provides a great opportunity to develop their problem solving and reasoning skills and allows them to make connections between areas of mathematics. Finally, this unit is important as it exposes children to reflections and translations on a coordinate grid for the first time.

WHERE THIS UNIT FITS

→ Unit 5 – Fractions (2)
→ **Unit 6 – Geometry – position and direction**
→ Unit 7 – Decimals

This unit builds on work in Year 5, where children were introduced to coordinates being used to describe the positions of points on grids, and they developed the skill of plotting coordinates in the first quadrant. It also builds upon work on properties of shape, and it encourages children to make connections between properties of shape and coordinates to solve increasingly complex problems involving shapes in all four quadrants.

Before they start this unit, it is expected that children:

• know that a pair of coordinates describes the position of a point within a grid
• can plot coordinates in the first quadrant
• can read coordinates in the first quadrant
• understand properties of a range of shapes, for example, the number of vertices, the relationship between side lengths, and the number of sides of a range of common regular and irregular polygons.

ASSESSING MASTERY

Children who have mastered this unit can plot and read coordinates in all four quadrants. They can identify coordinates that form the vertices of a range of common shapes, and they can solve increasingly complex problems involving shapes in all four quadrants. They can reflect points and shapes on a coordinate grid in the x- and y-axes as well as in simple diagonal lines, and they can carry out multi-step translations and reflections. Finally, children are able to extend these skills to problems where they are given just the coordinates and no coordinate grid.

COMMON MISCONCEPTIONS	STRENGTHENING UNDERSTANDING	GOING DEEPER
Children may start with the y-axis, leading to transposed coordinates, for example, (⁻3,4) becomes (4,⁻3).	Encourage children to use a mnemonic such as 'y's up that x is across' or 'go along the corridor then up the stairs'.	Increase the complexity of the problems one stage at a time. For example, children could consider what the different possibilities are to complete a shape based on two given pairs of coordinates. They should also be encouraged to begin to create their own problems involving shapes and reflections or translations for others to solve.
Children may incorrectly identify properties of shapes and therefore make incorrect connections and assumptions when seeing them on a coordinate grid.	Encourage children to draw or manipulate the shapes separately and identify their properties, then support children in applying these properties to the problem they are facing involving a coordinate grid.	

WAYS OF WORKING

Whole class: Go through the unit starter pages of the Pupil Book. Talk through the key learning points and the key vocabulary.

STRUCTURES AND REPRESENTATIONS

Coordinate grid with one quadrant: Children are reintroduced to coordinate grids with just one quadrant which they have learnt about previously. They will use these to plot coordinates in the first quadrant.

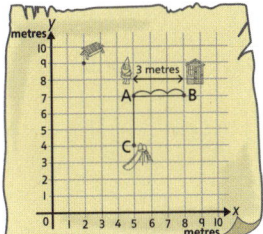

Coordinate grid with four quadrants: Children are then introduced to coordinate grids which show all four quadrants. They will use these to plot coordinates in all four quadrants, work out missing coordinates in shapes and reason about shapes using their coordinates.

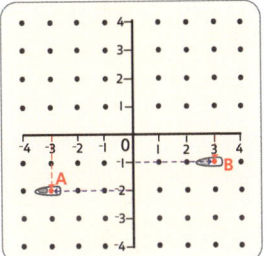

Zero-centred number line: Children may also benefit from using a 0-centred number line and thinking about how it relates to the x- and y-axes of a coordinate grid, so that they are able to correctly identify where to plot coordinates with positive or negative values.

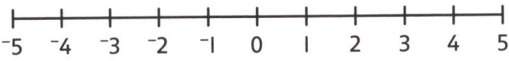

KEY LANGUAGE

There is some key language that children will need to know as part of the learning in this unit:

→ plotting, coordinates, quadrant, point, axis, x-axis, y-axis, grid, x-coordinate, y-coordinate

→ vertices, vertex, square, side, rectangle, triangle, equilateral, oblong, shape, irregular, hexagon, identical, similar, parallelogram

→ perimeter, metre (m), distance, length, long

→ horizontal, vertical

→ halfway, line, properties, value, reason

→ negative, positive

→ translation, reflection, original, left, down, up, right, mirror, away, diagonal

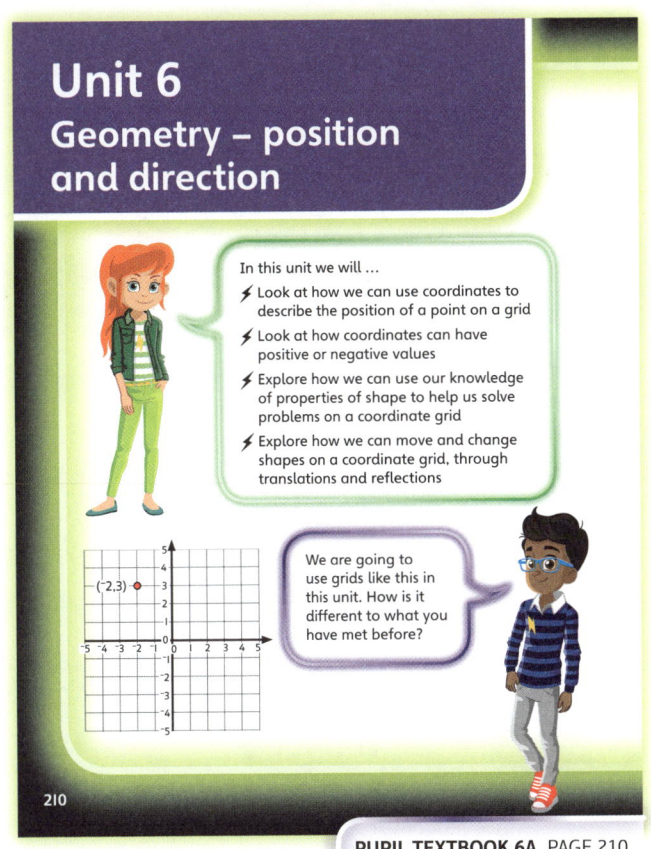

PUPIL TEXTBOOK 6A PAGE 210

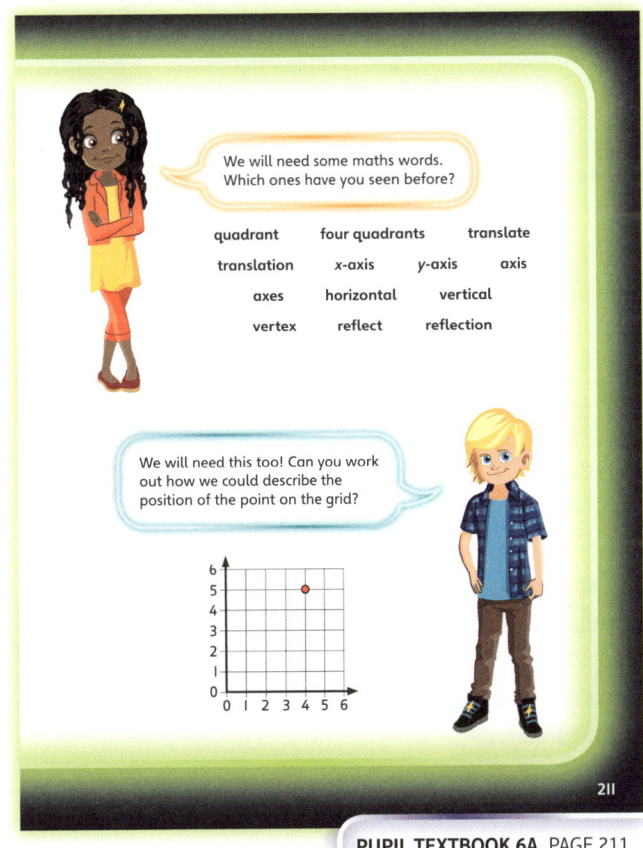

PUPIL TEXTBOOK 6A PAGE 211

Plotting coordinates in the first quadrant

Learning focus

In this lesson children will re-visit how to plot coordinates in the first quadrant. They will solve problems that involve reasoning using properties of shape and their coordinate knowledge.

Small steps

→ Previous step: Problem solving – fractions of amounts
→ **This step: Plotting coordinates in the first quadrant**
→ Next step: Plotting coordinates

NATIONAL CURRICULUM LINKS

Year 6 Geometry – Position and Direction

Describe positions on the full coordinate grid (all four quadrants).

ASSESSING MASTERY

Children can accurately plot coordinates in the first quadrant. Children can solve problems that involve completing missing vertices of shapes, applying their knowledge of shape properties.

COMMON MISCONCEPTIONS

Children may misremember the names of each axis, calling the *x*-axis the *y*-axis and vice versa. Ask:

· *What do you call the horizontal axis? Is there a saying you could use to help you remember that this is the horizontal axis, such as 'You go along the corridor (x) before you go up the stairs (y)'?*

Children may think the coordinates are written with the *y*-axis value first, rather than the *x*-axis value. Ask:

· *Which axis value do you write first? How can you help yourself to remember this?*

STRENGTHENING UNDERSTANDING

Encourage children to physically plot coordinates in the first quadrant and to use matchsticks and other items to create shapes. They can then use these to reason about the properties of shapes and investigate the answers to problems presented in this lesson. For example, they can use matchsticks to represent two side lengths of a square and investigate what this means about the values of the other vertices.

GOING DEEPER

Encourage children to create their own problems involving shapes on a coordinate grid for others to solve. Ask children if they can work out the minimum amount of information they must give for the problem to be solvable.

KEY LANGUAGE

In lesson: plotting, coordinates, quadrant, point, vertices, vertex, horizontal, vertical, axis, *x*-axis, *y*-axis, grid, identical, symmetrical

Other language to be used by the teacher: properties, oblong

STRUCTURES AND REPRESENTATIONS

coordinate grids with the first quadrant

RESOURCES

Mandatory: coordinate grids with the first quadrant

Optional: matchsticks, counters

 In the eTextbook of this lesson, you will find interactive links to a selection of teaching tools.

Before you teach

· Are children secure with their knowledge of properties of shapes that was covered in Year 5?
· Are children able to plot and identify coordinates in the first quadrant?

Discover

WAYS OF WORKING Pair work

ASK

- Question **1** a): *What information can you use from the diagram to help you solve this problem?*
- Question **1** a): *What do you know about the properties of squares? How can you use this information to help you solve the problem?*
- Question **1** a): *How can you work out the coordinates for the missing vertex?*
- Question **1** b): *What does 'perimeter' mean? How can you use the information from the diagram to help you calculate it?*

IN FOCUS Question **1** a) encourages students to visually reason about properties of a square and asks them to complete the square, based on two given vertices. Children are expected to identify that the sides of a square are all the same length, and therefore they are able to work out the coordinates of the missing vertex.

PRACTICAL TIPS Provide children with practical experiences of plotting coordinates and completing shapes, including using counters or other objects to mark coordinates, and matchsticks to create polygons on the coordinate grid and explore their properties.

ANSWERS

Question **1** a): The coordinates (8,4) take you to the treasure at D.

Question **1** b): The perimeter of the square is 12 metres.

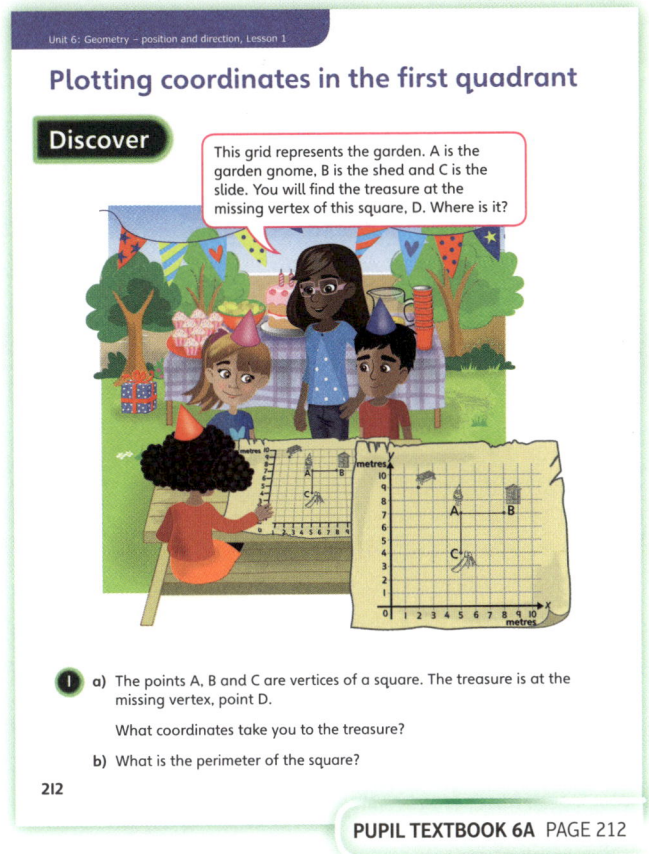

PUPIL TEXTBOOK 6A PAGE 212

Share

WAYS OF WORKING Whole class teacher led

ASK

- Question **1** a): *How can you use your knowledge of the properties of squares to help you?*
- Question **1** a): *Could you work out the coordinates of the missing vertex in a different way?*
- Question **1** b): *What information could you use to help you work out the perimeter of the square?*

IN FOCUS In question **1** a), children are encouraged to work out the length of a given side of the square, and therefore reason that the missing points must be the same distance down from (8,7) or left from (5,4). It is important that children understand that they could use either (8,7) or (5,4) and their knowledge of the side length in order to calculate the coordinates of the missing vertex.

In question **1** b), children are encouraged to use their knowledge that the perimeter of a square is 4 times the length of one side, and to use this and the given side length to calculate the perimeter of the square.

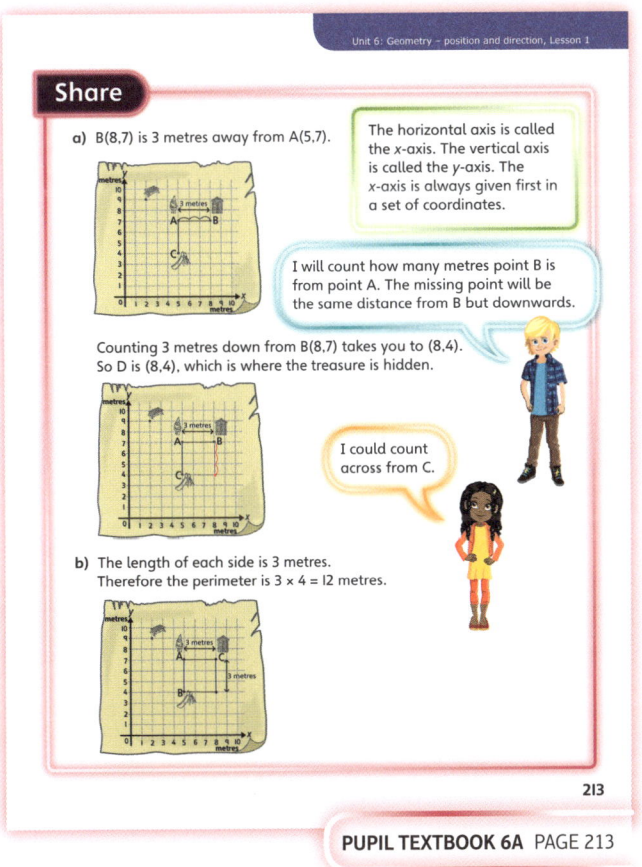

PUPIL TEXTBOOK 6A PAGE 213

Think together

Whole class teacher led (I do, We do, You do)

ASK

- Question **1** a) and b): *What do you know about the properties of rectangles and squares that could help you answer this question?*
- Question **1** a): *Can you point to where the missing vertex will be? What coordinates is this point at?*
- Question **1** b): *How long is each side of this square? Can you use this to help you work out the missing vertices?*
- Question **1** b): *Is there more than one possible set of answers for this question? Why is this?*
- Question **2**: *How can you work out the missing coordinates without a coordinate grid? Can you use the properties of a rectangle to help you?*

IN FOCUS Question **2** introduces children to working out the missing coordinates of shapes that are not presented on a grid. This requires them to apply their knowledge and strategies developed in **Discover**, **Share** and question **1** of this section, and to complete the missing coordinates by using the properties of shapes to calculate them. For example, they should understand that point C is vertically above point B, and therefore will share the same *x*-axis value as point B.

STRENGTHEN To support children in working out the missing vertices in question **2** when a grid is not provided, initially create a similar style question on a coordinate grid. Ask children what they notice about the *x* and *y* values of each vertex, and then encourage them to apply the same reasoning to question **2**.

DEEPEN Children should be encouraged to reason about and use the properties of a wider range of shapes in order to solve coordinate problems in the first quadrant. Question **3** provides some initial exposure to this. Children could be encouraged to create their own problems for partners that use a wider range of shapes.

ASSESSMENT CHECKPOINT Use questions **1** a) and **1** b) to assess whether children can apply their knowledge of properties of shapes to solve problems presented on a coordinate grid. Do they understand why Dexter says he could count up or across?

ANSWERS

Question **1** a): (4,3)

Question **1** b): (9,6) and (9,2) or (1,6) and (1,2)

Question **2**: B(8,4), C(8,7) and D(3,7)

Question **3**: A(13,7) and B(4,13)

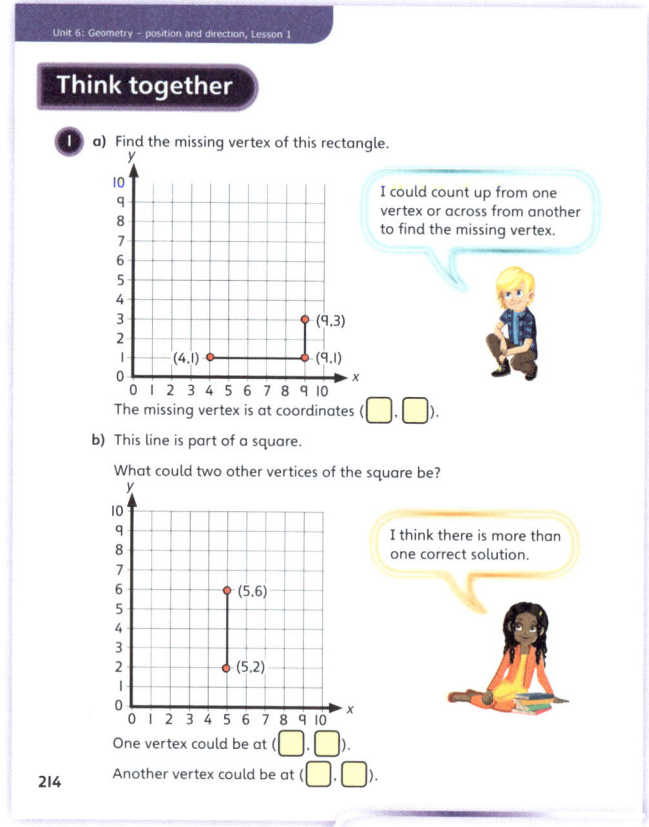

PUPIL TEXTBOOK 6A PAGE 214

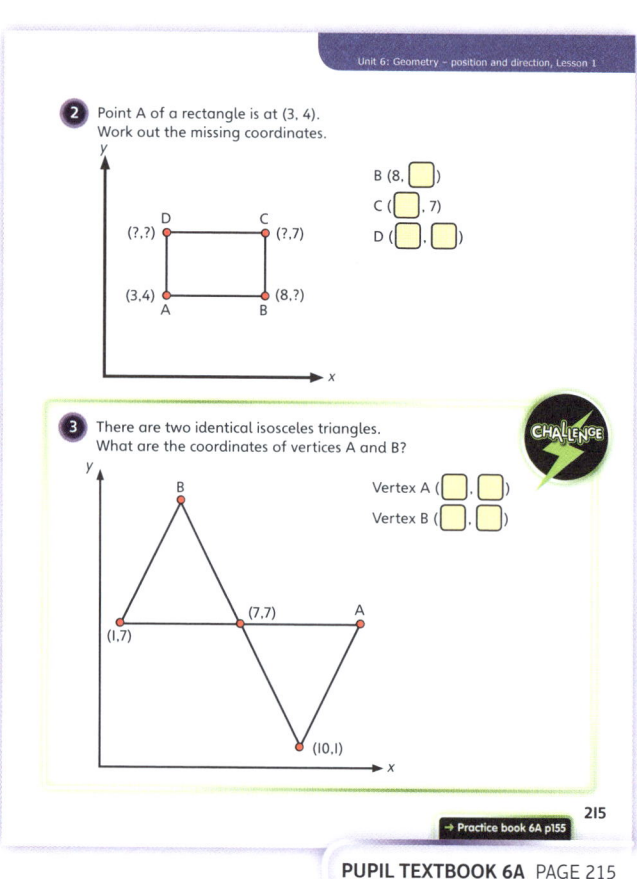

PUPIL TEXTBOOK 6A PAGE 215

Practice

WAYS OF WORKING Independent thinking

IN FOCUS Question **4** and the first grid of question **2** introduce children to shapes that are presented on a diagonal on a coordinate grid. Children should be encouraged to calculate the length of each side based on the *x* and *y* values of the two given vertices, and to check their complete shape looks correct visually (it looks like a square).

STRENGTHEN Question **3** a) presents a problem that could have two different solutions. To help children find both sets of coordinates, encourage them to physically represent the problem on a coordinate grid, using matchsticks or other items to represent the given line. Encourage them to manipulate the item(s) used to represent the side, exploring where the coordinates would be if the square extended either side of the given line.

DEEPEN Children should be encouraged to extend their knowledge to multiple shapes on the same grid. Question **5** provides some opportunity to explore this. Ask: *If shapes are identical, what does this mean? How can you use this to help you solve more complex problems involving coordinates and properties of shape?*

ASSESSMENT CHECKPOINT Use question **2** to assess whether children can calculate the missing vertices on different shapes. Check that they remember which coordinate to write first.

ANSWERS Answers for the **Practice** part of the lesson appear in the separate **Practice and Reflect answer guide**.

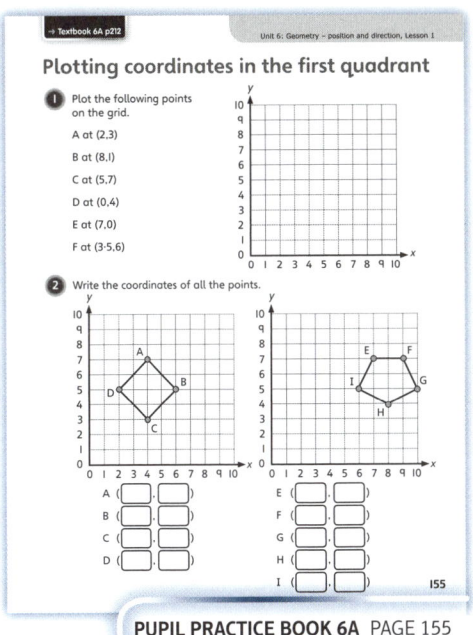

PUPIL PRACTICE BOOK 6A PAGE 155

PUPIL PRACTICE BOOK 6A PAGE 156

Reflect

WAYS OF WORKING Independent thinking

IN FOCUS This **Reflect** activity is designed to draw out children's understanding of points placed on the axes: that if a point is on the *x*-axis then the *y*-coordinate is equal to 0, and vice versa.

ASSESSMENT CHECKPOINT Use this question to assess whether children understand how the axes relate to each other, and that a point placed on an axis will have a 0 coordinate.

ANSWERS Answers for the **Reflect** part of the lesson appear in the separate **Practice and Reflect answer guide**.

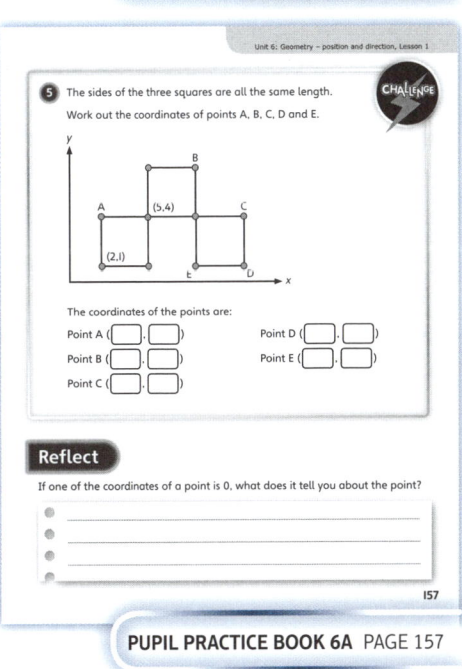

After the lesson

- Are all children secure at plotting and identifying points in the first quadrant? How will you address any misconceptions about this through same-day intervention before children are exposed to coordinates in four quadrants in Lesson 2?
- Can you make cross-curricular links between coordinates in one quadrant and other subjects, for example GPS coordinates in Geography or PE?

PUPIL PRACTICE BOOK 6A PAGE 157

Plotting coordinates

Learning focus

In this lesson children are introduced to plotting coordinates in all four quadrants.

Small steps

→ Previous step: Plotting coordinates in the first quadrant
→ **This step: Plotting coordinates**
→ Next step: Plotting translations and reflections

NATIONAL CURRICULUM LINKS

Year 6 Geometry – Position and Direction

Describe positions on the full coordinate grid (all four quadrants).

ASSESSING MASTERY

Children can read the coordinates of points that are plotted in all four quadrants and can plot these. Children can begin to develop their problem solving and reasoning abilities, including simple reasoning about shapes and identifying common errors.

COMMON MISCONCEPTIONS

When plotting coordinates, children may plot the first coordinate against the *y*-axis rather than the *x*-axis. Ask:
• *Which axis do you plot against first? What do you call this axis? Is there a saying you could use to help you remember that you plot against the horizontal axis first?*

Children may plot negative values as positive values. For example, they may plot (⁻3,⁻4) at (3,4). Ask:
• *What is the value of the coordinate? Is it before or after 0? Where is this on the axis? What are the coordinates where both axes cross each other?*

STRENGTHENING UNDERSTANDING

Encourage children to physically plot coordinates in all four quadrants. If you have a gridded area on your school playground, this would be an ideal opportunity to children to stand at given coordinates.

When plotting or reading values, it can be helpful to make the link between the axis and a 0-centred number line. Children can also record a '+' (positive) and '–' (negative) sign at the appropriate ends of each axis, to help them remember which coordinates are positive and which are negative. Drawing the axes with two different colours (one for the positive and one for the negative section of each axis) can also help children when plotting coordinates that have a negative value.

GOING DEEPER

Encourage children to begin to solve problems involving properties of shapes in all four quadrants. For example, children could try to find the missing vertex of a square or rectangle.

KEY LANGUAGE

In lesson: plotting, coordinates, point, grid, negative, *x*-axis, *y*-axis, **quadrants**, positive, value

Other language to be used by the teacher: vertices, vertex, horizontal, vertical

STRUCTURES AND REPRESENTATIONS

coordinate grids with all four quadrants, 0-centred number line

RESOURCES

Mandatory: coordinate grids with all four quadrants

Optional: matchsticks

 In the eTextbook of this lesson, you will find interactive links to a selection of teaching tools.

Before you teach

• Are children secure with plotting and reading coordinates in the first quadrant from Lesson 1 in this unit?
• Are there any common misconceptions around plotting coordinates from Lesson 1? If so, how will you adapt your teaching so that these can be addressed in this lesson?

Discover

WAYS OF WORKING Pair work

ASK

- Question **1** a): *What do you notice about the coordinate grid in this problem?*
- Question **1** a): *How can you use the information on the diagram to help you?*
- Question **1** a): *How do you read coordinates? Does this work for plotting points on a four-quadrant grid?*
- Question **1** b): *Why do you think this coordinate has a negative value? Where would this point be represented along the x-axis?*

IN FOCUS This activity introduces children to a coordinate grid that has more than one quadrant, and to coordinates that have a negative value. Children should be encouraged to make connections between plotting coordinates in the first quadrant and plotting coordinates in all four quadrants.

PRACTICAL TIPS Provide children with practical experiences of plotting coordinates in all four quadrants, alongside the experience of reading and pointing to coordinates that they will gain through the **Discover**, **Share** and **Think together** sections.

ANSWERS

Question **1** a): A($^-$3,$^-$2) and B(3,$^-$1)

Question **1** b):

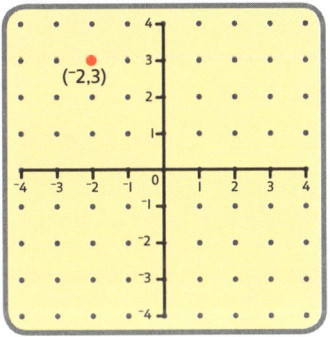

Share

WAYS OF WORKING Whole class teacher led

ASK

- Question **1** a): *How can you work out the coordinates of ships A and B?*
- Question **1** a): *How can you use your knowledge of plotting coordinates from our last maths lesson to help you?*
- Question **1** a): *How can you work out the x and y values for each point? What happens when a value is below 0?*
- Question **1** b): *How can you make sure you are plotting this point against the correct places on the x- and y-axes?*

IN FOCUS Children are exposed to a range of coordinates in all four quadrants, including those that have two negative values, for example ($^-$3,$^-$2), as well as those that have a positive and negative value, for example (3,$^-$1). Some children may not understand why there are four quadrants and how these relate to negative and positive values.

Plotting coordinates

Discover

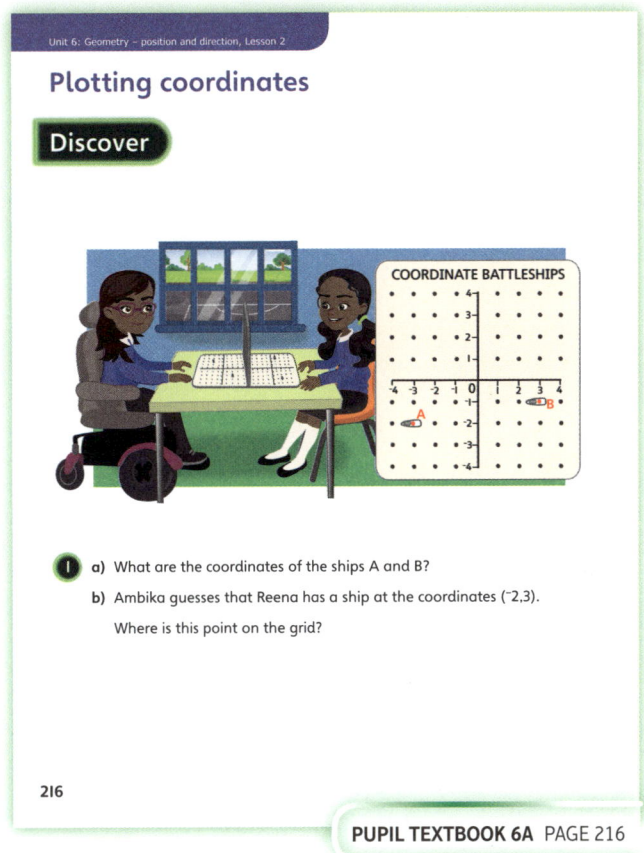

1 a) What are the coordinates of the ships A and B?

b) Ambika guesses that Reena has a ship at the coordinates ($^-$2,3).
Where is this point on the grid?

216

PUPIL TEXTBOOK 6A PAGE 216

Share

a)

I think coordinates can also have negative values.

I remember that I should read the x-axis coordinate first, and then the y-axis coordinate.

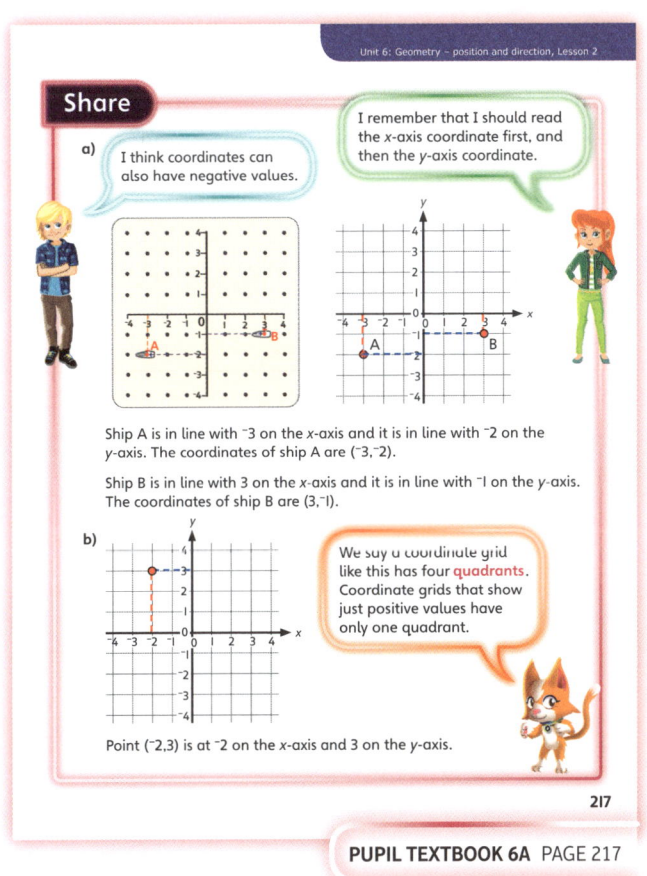

Ship A is in line with $^-$3 on the x-axis and it is in line with $^-$2 on the y-axis. The coordinates of ship A are ($^-$3,$^-$2).

Ship B is in line with 3 on the x-axis and it is in line with $^-$1 on the y-axis. The coordinates of ship B are (3,$^-$1).

b)

We say a coordinate grid like this has four quadrants. Coordinate grids that show just positive values have only one quadrant.

Point ($^-$2,3) is at $^-$2 on the x-axis and 3 on the y-axis.

217

PUPIL TEXTBOOK 6A PAGE 217

Think together

Whole class teacher led (I do, We do, You do)

ASK

- Question **1** a): *Do you record the x- or y-axis value first?*
- Question **1** a): *How do the lines drawn on the grid for ship A help you?*
- Question **1** b): *If the value is negative, which part of the x- or y-axis is it on?*
- Question **2** : *Has Mark plotted all the coordinates in the right order? Has he plotted all the negative values correctly?*

IN FOCUS Question **2** explores some common misconceptions and errors that occur when plotting in all four quadrants and how they can be avoided.

STRENGTHEN Children may be unsure where to plot positive and negative values. Encourage them to write a plus or minus sign at the appropriate ends of each axis in their books, as a reminder.

DEEPEN Give children missing vertex problems and challenge them to reason using properties of shapes in grids with four quadrants.

ASSESSMENT CHECKPOINT Use all questions to assess whether children can read and plot coordinates in all four quadrants. Point out the dashed lines in **1** a), which are there to help. Do children understand how positive and negative values relate to the quadrants?

ANSWERS

Question **1** a): A(5,⁻2), B(⁻4,⁻2), C(⁻3,3), D(3,2)

Question **1** b):

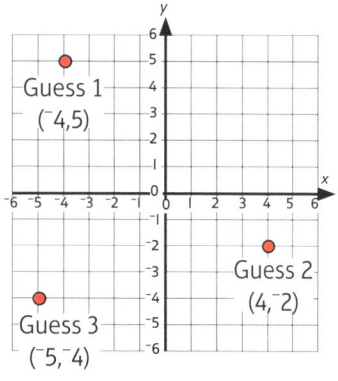

Question **2** : Points A, C and D are incorrect. Mark gave the y-axis coordinate before the x-axis coordinate for A and D; he mistook negative for positive for C.

Question **3** : (2,1), (2,⁻1), (1,⁻1) and (1,1)

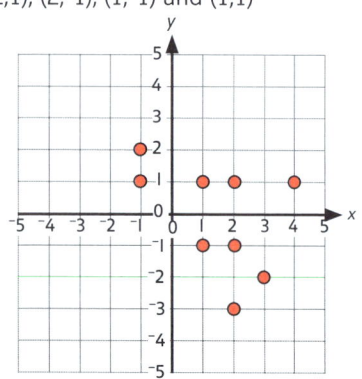

Think together

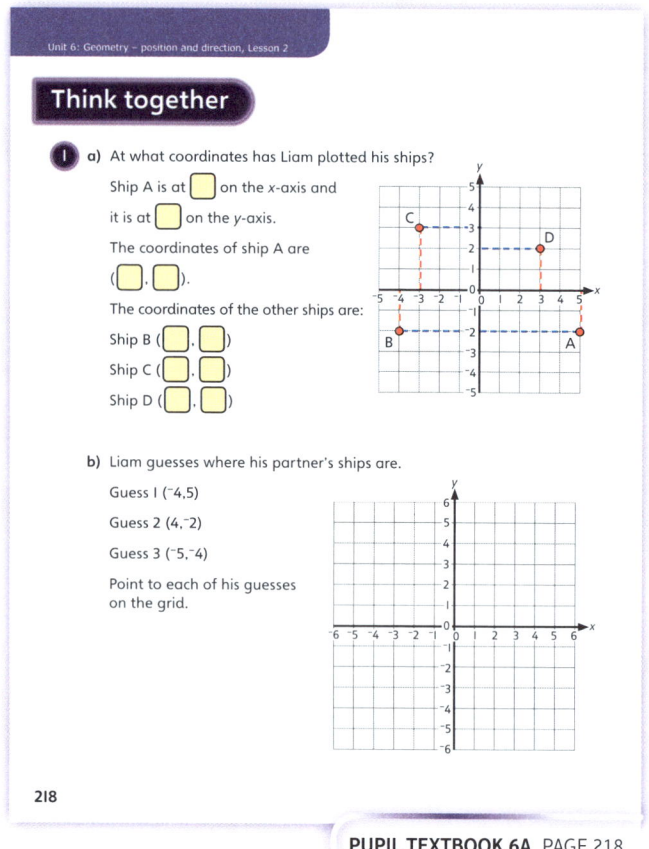

1 a) At what coordinates has Liam plotted his ships?

Ship A is at ☐ on the x-axis and it is at ☐ on the y-axis.

The coordinates of ship A are (☐ , ☐).

The coordinates of the other ships are:

Ship B (☐ , ☐)

Ship C (☐ , ☐)

Ship D (☐ , ☐)

b) Liam guesses where his partner's ships are.

Guess 1 (⁻4,5)

Guess 2 (4,⁻2)

Guess 3 (⁻5,⁻4)

Point to each of his guesses on the grid.

218

PUPIL TEXTBOOK 6A PAGE 218

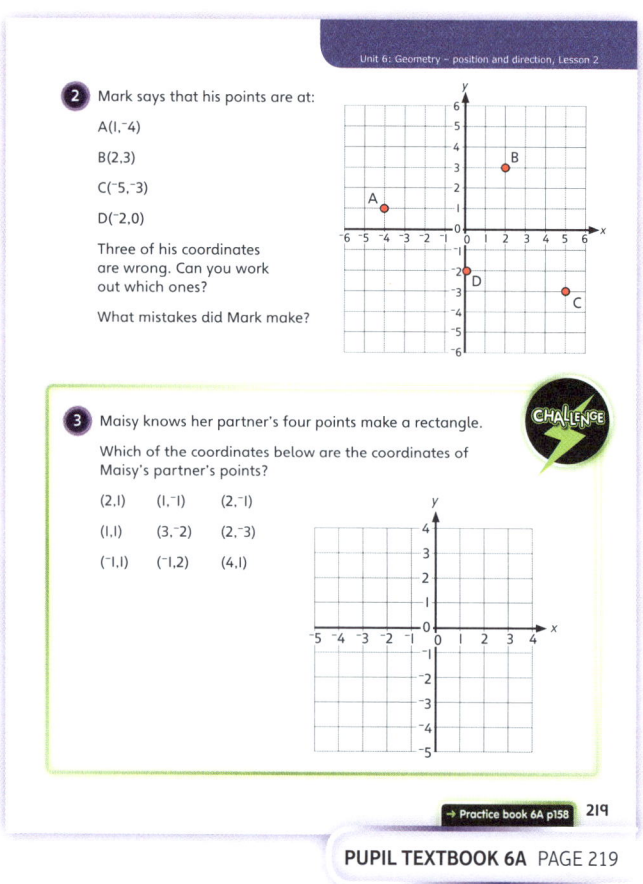

2 Mark says that his points are at:

A(1,⁻4)

B(2,3)

C(⁻5,⁻3)

D(⁻2,0)

Three of his coordinates are wrong. Can you work out which ones?

What mistakes did Mark make?

3 Maisy knows her partner's four points make a rectangle. **CHALLENGE**

Which of the coordinates below are the coordinates of Maisy's partner's points?

(2,1)   (1,⁻1)   (2,⁻1)

(1,1)   (3,⁻2)   (2,⁻3)

(⁻1,1)   (⁻1,2)   (4,1)

→ Practice book 6A p158   219

PUPIL TEXTBOOK 6A PAGE 219

Practice

WAYS OF WORKING Independent thinking

IN FOCUS Question ② introduces children to basic reasoning about shapes based on coordinates. They are expected to plot the coordinates on a coordinate grid and then to identify which shape the coordinates could be the vertices of.

STRENGTHEN To help children identify the shapes formed by the coordinates in question ②, encourage them to use matchsticks or other objects to join the vertices so that they are able to see the outline of the shape.

DEEPEN Children should be encouraged to begin to solve missing vertices and coordinates problems, reasoning about the properties of shapes to help them. For example, ask: *What coordinates would I need to plot to complete this square?*

THINK DIFFERENTLY Question ③ asks children to explain a common misconception when plotting coordinates. They will need to understand that you get to a different point if you do not write or plot coordinates in the correct order.

ASSESSMENT CHECKPOINT Use question ② to assess if children can accurately plot coordinates in all four quadrants. They should be able to see by the resulting shapes if they have gone wrong and why.

ANSWERS Answers for the **Practice** part of the lesson appear in the separate **Practice and Reflect answer guide**.

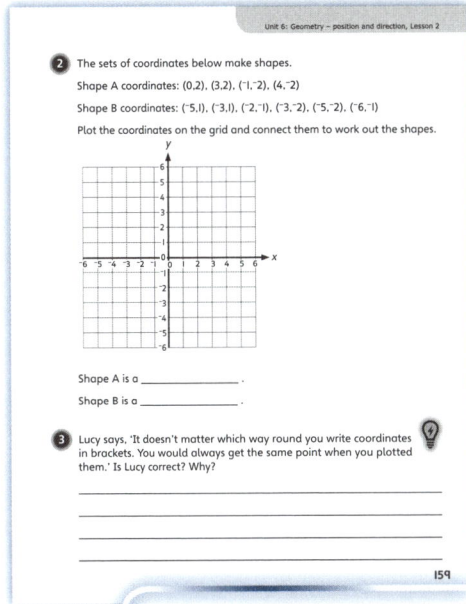

PUPIL PRACTICE BOOK 6A PAGE 158

PUPIL PRACTICE BOOK 6A PAGE 159

Reflect

WAYS OF WORKING Independent thinking

IN FOCUS This **Reflect** section activity encourages children to reflect on the similarities and differences between plotting coordinates in one quadrant and plotting them in four quadrants.

ASSESSMENT CHECKPOINT Use this question to assess whether children can connect their knowledge of plotting coordinates in one quadrant with the knowledge gained in this lesson about plotting coordinates in all four quadrants.

ANSWERS Answers for the **Reflect** part of the lesson appear in the separate **Practice and Reflect answer guide**.

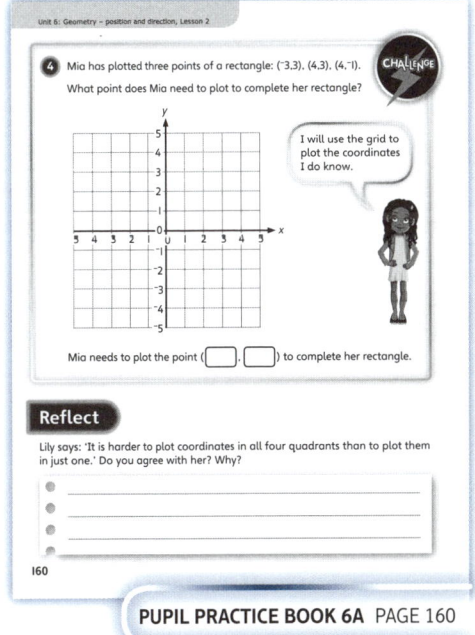

After the lesson ⏸

- Are all children secure in plotting coordinates in all four quadrants?
- How could you address any insecurities through same-day intervention, before children are expected to use these skills as part of Lesson 3 in this unit?

PUPIL PRACTICE BOOK 6A PAGE 160

251

Plotting translations and reflections

Learning focus

In this lesson, children are introduced to reflecting and translating shapes on a coordinate grid.

Small steps

→ Previous step: Plotting coordinates
→ **This step: Plotting translations and reflections**
→ Next step: Reasoning about shapes with coordinates

NATIONAL CURRICULUM LINKS

Year 6 Geometry – Position and Direction

Draw and translate simple shapes on the coordinate plane, and reflect them in the axes.

ASSESSING MASTERY

Children can translate and reflect shapes that are presented on a coordinate grid. They can explain translations and reflections in terms of a coordinate grid, for example 'the shape has been translated 2 to the right and 3 down' or 'the shape has been reflected in the *y*-axis'.

COMMON MISCONCEPTIONS

Children may identify a reflection as a translation. Draw children's attention to the fact that a shape after a translation is identical to the original shape, whereas as a shape after a reflection is a mirror image. Ask:

• *What do you notice about shapes after a translation? How about after a reflection? Let's translate this shape by [for example] 4 right. Does this look the same as the reflected shape?*

Children may struggle to identify the axis or line that a shape has been reflected in. Encourage children to use a mirror to investigate the impact that reflecting the shape in each axis/line would have on the original shape, and which image matches the reflection they are trying to describe. Ask:

• *How could you check if this shape is reflected in [for example] the x-axis?*

STRENGTHENING UNDERSTANDING

Encourage children to use mirrors to help them identify and describe the effect of reflections. For example, when reflecting in the *x*-axis, invite children to place a mirror along the *x*-axis, and note the effect this reflection has on the shape, and the position that the reflected shape is in. Then invite them to draw this reflection on the coordinate grid, before drawing their attention to the effect the reflection has had on the coordinates of each vertex of the shape.

GOING DEEPER

Encourage children to explore problems that combine both a translation and a reflection.

KEY LANGUAGE

In lesson: plotting, translation, reflection, *x*-axis, *y*-axis, mirror, coordinates, vertices, irregular, grid, identical

Other language to be used by the teacher: quadrants

STRUCTURES AND REPRESENTATIONS

coordinate grids with all four quadrants

RESOURCES

Mandatory: coordinate grids with all four quadrants

Optional: mirror

 In the eTextbook of this lesson, you will find interactive links to a selection of teaching tools.

Before you teach

• Are children secure with plotting and reading coordinates in all four quadrants from Lesson 2 in this unit? How will you address any common misconceptions at the start of this lesson?
• Are children secure in their knowledge of reflections from Year 5, where they experienced reflections that were not on a coordinate grid?

Discover

WAYS OF WORKING Pair work

ASK

- Question ① a): *What information can you use from the diagram to help you solve this problem?*
- Question ① a): *What does it mean to reflect the shape? What do you think it means to reflect the shape in the x-axis?*
- Question ① a): *Is there anything you could use to explore the effect of reflecting this shape?*
- Question ① b): *What do you think it means when it asks us to translate a shape?*

IN FOCUS This activity encourages children to begin to explore both the reflection and the translation of shapes in all four quadrants. They are introduced to the vocabulary of reflecting shapes across an axis for the first time.

PRACTICAL TIPS Provide children with practical experiences of reflecting and translating shapes, including on their own coordinate grids. Ensure children have had the experience of using a mirror to reflect shapes and explore the effect of this reflection and the position of the reflected image.

ANSWERS

Question ① a): Olivia would draw these vertices to make the shape: E(1,⁻3), F(3,⁻3), G(4,⁻5) and H(2,⁻6).

Question ① b): Olivia would draw these vertices to make the shape: E(⁻3,⁻2), F(⁻1,⁻2), G(0,0), H(⁻2,1).

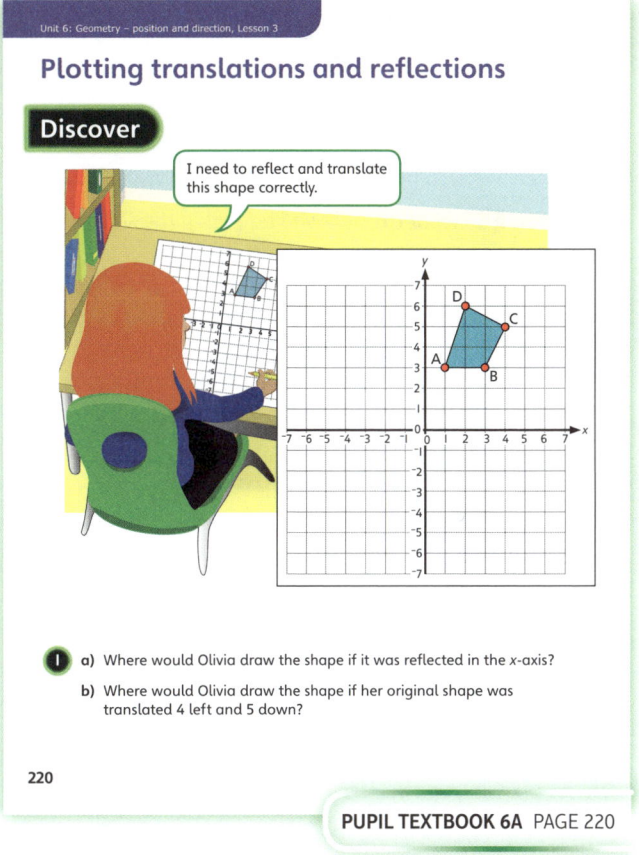

PUPIL TEXTBOOK 6A PAGE 220

Share

WAYS OF WORKING Whole class teacher led

ASK

- Question ① a): *How can you work out the coordinates of the reflected shape? Can anyone see a different way?*
- Question ① a): *Can you point to the axis that you are reflecting the shape in?*
- Question ① a): *Can you place a mirror on the x-axis? What do you notice about the position of the reflected shape in relation to the x-axis?*
- Question ① a): *Does the reflected shape look identical to the original shape?*
- Question ① b): *If you translate a shape, what does it mean you are doing to it?*
- Question ① b): *How can you work out where you would draw the translated shape?*
- Question ① b): *Can anyone work out the position of the translated shape in another way?*
- Question ① b): *Does the translated shape look identical to the original shape?*

IN FOCUS Children are encouraged to explore the effect of reflecting and translating shapes. They should be encouraged to identify that when they reflect shapes, the reflected shape is the same distance away from the axis of reflection as the original shape.

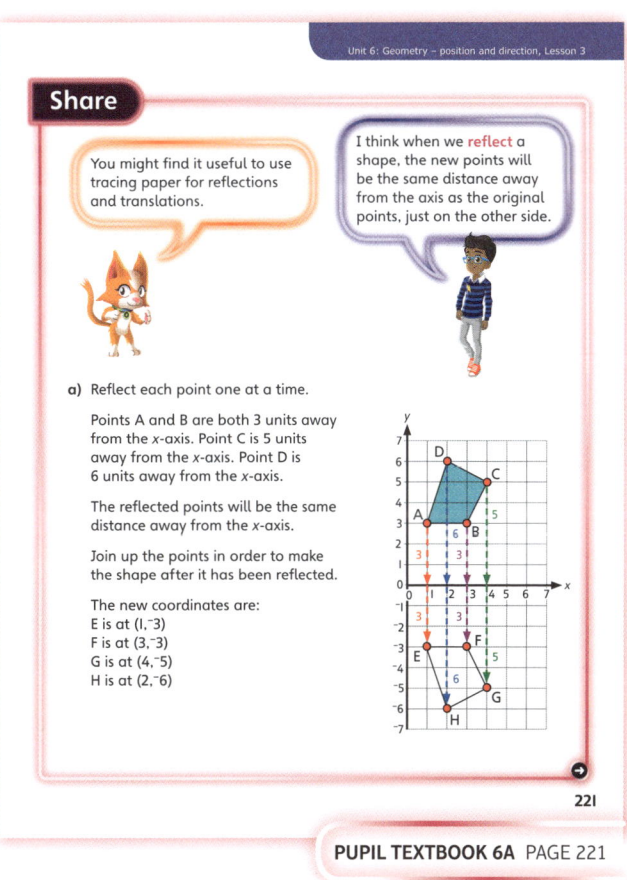

PUPIL TEXTBOOK 6A PAGE 221

Think together

Whole class teacher led (I do, We do, You do)

ASK

- Question **1** : *What axes are you reflecting the shapes in? Can you point to them?*
- Question **1** b): *Does reflecting the shape in the y-axis produce the same result as reflecting it in the x-axis? Why?*
- Question **2** a): *Which point are you going to move first?*
- Question **2** b): *How can you identify the translation?*

IN FOCUS Question **2** b) invites children to identify a translation based on two identical shapes (rather than complete a translation themselves). They should be encouraged to explore how one of the vertices of the shape has been moved to make the corresponding vertex of the second shape, and then check that this translation and relationship is the same for the other vertices of the shapes.

STRENGTHEN To support children when translating shapes, it can be useful to physically represent the problem with the shape made out of paper or other material. This shape can then be physically moved along the course of the translation, which helps to reinforce that the shapes are identical after the translation.

DEEPEN Children should be encouraged to begin to reflect shapes in lines other than the axes, for example, reflecting shapes in a diagonal line. Question **3** introduces children to this concept, and they should be encouraged to initially explore the effect of reflecting in a diagonal line by using a mirror on the diagonal line. Once children are secure with this, they can then be encouraged to explore the impact that the angle and position of the line of reflection have on the reflected shape.

ASSESSMENT CHECKPOINT Use questions **1** a) and **1** b) to assess whether children can accurately reflect shapes in both axes.

ANSWERS

Question **1** a): The reflected image of shape A has coordinates (2,⁻2), (4,⁻4) and (6,⁻1).

Question **1** b): The reflected image of shape A has coordinates (⁻2,2), (⁻4,4) and (⁻6,1).

Question **2** a):

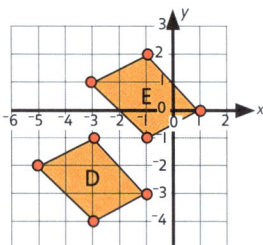

Question **2** b): Shape F has been translated 2 right and 5 down to become shape G.

Question **3** a): The diagonal line has not been used for the reflection. Shape B should have coordinates (1,2), (1,0), (3,0), (3,2).

Question **3** b): The translation in the x-direction is ⁻4, not 4. Shape D should have coordinates (1,1), (2,2), (4,1), (5,⁻1), (3,⁻1).

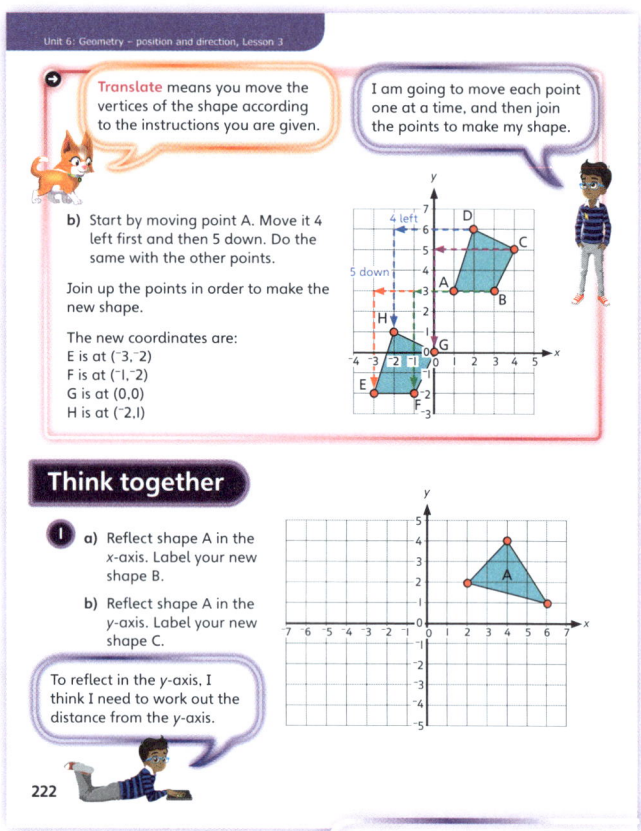

PUPIL TEXTBOOK 6A PAGE 222

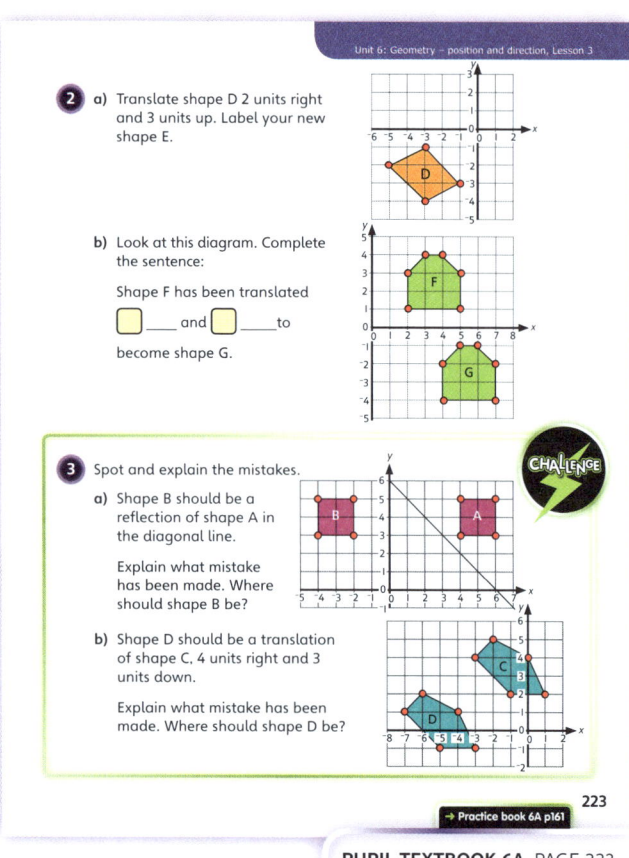

PUPIL TEXTBOOK 6A PAGE 223

Practice

WAYS OF WORKING Independent thinking

IN FOCUS Question **5** exposes children to reflecting shapes on a coordinate grid, where the gridlines and axis numbering have been removed. Children will need to draw on their learning from earlier questions in order to develop their approach for working out the coordinates.

STRENGTHEN In question **4**, to help children identify the effect of a reflection in a diagonal line, ensure that children explore this initially using a mirror, and are encouraged to draw the position of the reflected shape based on what they have seen in the mirror. Having two copies of the problem and working in pairs (so one can hold the mirror on one copy of the problem, and the other can draw the answer on the second copy, and then switch over) can be helpful.

DEEPEN Children should be encouraged to explore reflections that are followed by translations and vice versa. They should notice that carrying out a reflection followed by a translation does not give the same result as carrying out the same translation followed by the same reflection.

THINK DIFFERENTLY Question **4** exposes children to reflecting a shape in a diagonal line. Children should be encouraged to use mirrors to check that they are plotting the reflection in the diagonal line correctly.

ASSESSMENT CHECKPOINT Use question **3** to assess whether children can identify reflections and translations.

ANSWERS Answers for the **Practice** part of the lesson appear in the separate **Practice and Reflect answer guide**.

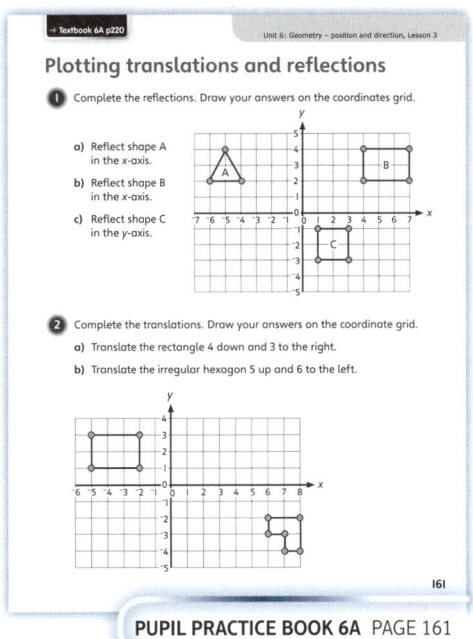

PUPIL PRACTICE BOOK 6A PAGE 161

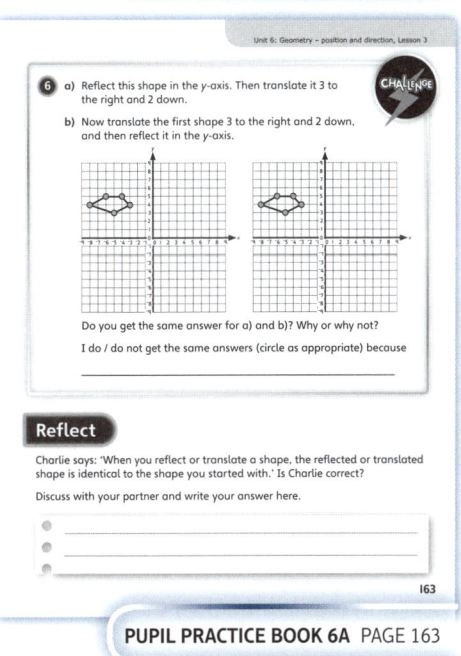

PUPIL PRACTICE BOOK 6A PAGE 162

Reflect

WAYS OF WORKING Independent thinking

IN FOCUS This **Reflect** activity encourages children to discuss whether a reflected or translated image is exactly the same as the original shape. You might like to discuss how having the same side lengths and angles makes two shapes the same, even if they are facing opposite ways.

ASSESSMENT CHECKPOINT Use this question to assess whether children can explain what effect reflecting or translating has on the resulting shapes.

ANSWERS Answers for the **Reflect** part of the lesson appear in the separate **Practice and Reflect answer guide**.

After the lesson

- Are all children secure in translating and reflecting shapes? How will you address any individual misconceptions through same-day intervention before children use these skills as part of Lesson 4?
- Are there any common misconceptions that you need to incorporate into your teaching of the next lesson when children will use and apply their knowledge of translating shapes to a wider range of problems?

PUPIL PRACTICE BOOK 6A PAGE 163

Reasoning about shapes with coordinates

Learning focus

In this lesson, children extend their ability to reason about shapes based on their properties, to solve problems that involve coordinates in all four quadrants.

Small steps

→ Previous step: Plotting translations and reflections
→ **This step: Reasoning about shapes with coordinates**
→ Next step: Multiplying by 10, 100 and 1,000

NATIONAL CURRICULUM LINKS

Year 6 Geometry – Position and Direction

Draw and translate simple shapes on the coordinate plane, and reflect them in the axes.

ASSESSING MASTERY

Children can solve problems that involve reasoning about shapes in all four quadrants. They can find the coordinates of missing vertices of shapes, including where a grid is not provided. Children can also reason about identical and similar shapes and use their reasoning to solve problems and find missing coordinates.

COMMON MISCONCEPTIONS

Children may misremember the names of each axis, calling the x-axis the y-axis and vice versa, which will mean errors when plotting and reading coordinates. Ask:

• *What do you call the vertical axis? Is there a saying you could use to help you remember that this is the vertical axis? Which axis do you record first in your coordinates?*

Children may not realise that identical shapes share all the same properties. They may therefore not be able to apply the properties between identical shapes in order to help them solve missing coordinate problems. Ask:

• *What does it mean when shapes are identical? What is the same about them? How can you use this to help you solve problems?*

STRENGTHENING UNDERSTANDING

Encourage children to physically plot coordinates in all four quadrants, and to use matchsticks or other items to create shapes, so that they are able to physically manipulate the shapes and begin to reason about their properties. They can also use these resources to create identical shapes, and see the connections between these shapes, including seeing that they have all the same properties, such as side lengths.

GOING DEEPER

Encourage children to create their own problems involving shapes in all four quadrants for others to solve. Ask children what makes the problem they have created hard or easy to solve and why?

KEY LANGUAGE

In lesson: coordinates, properties, x-axis, y-axis, vertex, x-coordinate, y-coordinate, translation, grid, vertices

Other language to be used by the teacher: quadrant, horizontal, vertical

STRUCTURES AND REPRESENTATIONS

coordinate grids with all four quadrants, 0-centred number line

RESOURCES

Mandatory: coordinate grids with all four quadrants

Optional: matchsticks, counters

 In the eTextbook of this lesson, you will find interactive links to a selection of teaching tools.

Before you teach

• Are children secure with plotting and reading coordinates in all four quadrants from Lesson 2 in this unit?
• Were there any common misconceptions about properties of shapes and using these on a coordinate grid in the first quadrant from Lesson 1? How will you address these during this lesson?

Discover

ASK

- Question ① a): *What information can you use from the diagram to help you solve this problem?*
- Question ① a): *What do you know about the properties of squares?*
- Question ① a): *How could you work out the length of the side of the shape when you don't have a grid to help you?*
- Question ① b): *What do you think it means when the two squares are identical? Can you use what you know about the first square to help you find the coordinates for points F, G and H? Can you work out how the first square has been translated to form the second square?*

IN FOCUS This activity encourages children to visually reason about properties of two identical squares and asks them to find the coordinates of different vertices of two squares. Children are exposed to the vocabulary of 'identical shapes' for the first time in Year 6 and need to understand that identical shapes share all the same properties.

PRACTICAL TIPS Provide children with practical experiences of plotting coordinates and completing shapes in all four quadrants, including using counters or other objects to mark coordinates and matchsticks or other objects to create polygons on the coordinate grid.

ANSWERS

Question ① a): C(4,5), D(1,5)

Question ① b): F(6,⁻5), G(6,⁻2), H(3,⁻2)

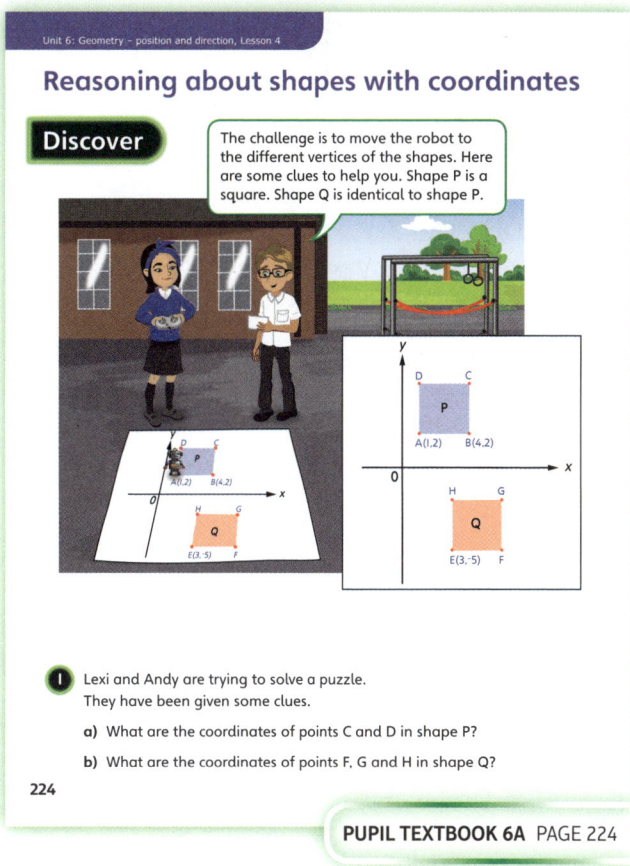

PUPIL TEXTBOOK 6A PAGE 224

Share

ASK

- Question ① a): *How can you work out the coordinates of the missing vertices?*
- Question ① a): *How can you use your knowledge of the properties of squares to help you?*
- Question ① a): *How can you work out the length of the sides? How can you use these to help you work out the coordinates of points C and D?*
- Question ① b): *What do you know about the second square?*
- Question ① b): *How can you work out the coordinates of points F, G and H? Can you use your knowledge of translating shapes to help you?*

IN FOCUS In question ① a), children are encouraged first of all to work out the length of a given side of the square that is in the first quadrant. A grid is not provided, meaning children need to reason about the length of the sides of the shapes based on the *x* and *y* values of the coordinates of given vertices. In question ① b), they are then expected to use the knowledge that the two squares are identical, along with the properties of squares (all side lengths the same) to work out the translation that would move the first square to the second, and therefore work out missing coordinates of the vertices in the second square.

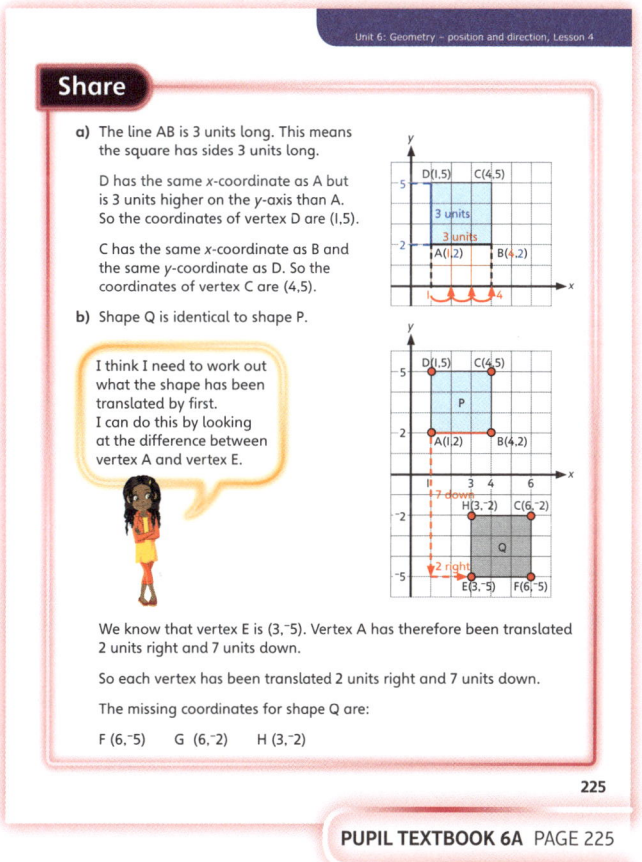

PUPIL TEXTBOOK 6A PAGE 225

Think together

Whole class teacher led (I do, We do, You do)

ASK

- Question **1** : *What do you know about the properties of a square that could help you answer this question?*
- Question **1** a): *How can you work out the length of a side of this square? What do you know about the sides?*
- Question **1** a): *Can you use the length of the sides of the square to help you work out points B and C?*
- Question **1** b): *What does it mean when you say the squares are 'identical'?*
- Question **1** b): *Can you work out how the first square has been translated to make the second square?*

IN FOCUS Question **2** asks children to work out the value of missing coordinates of shapes that are not presented on a grid. This requires them to apply their knowledge and strategies developed in **Discover**, **Share** and question **1** of this section, and to complete the missing coordinates by using the properties of an isoceles triangle (i.e. that it is symmetrical).

STRENGTHEN Children may become confused when calculating a transition that goes into a different quadrant and may not realise that this may create a negative value for the coordinates, for example (‾1,3) becomes (1,3). Encourage children to use a 0-centred number line to help them calculate the impact of the translations on the coordinates. For example, when translating a *y*-axis value of 4 down by 6 units, use a number line to support the calculation of $4 - 6 = ‾2$.

DEEPEN Children should be encouraged to reason about and use the properties of a wider range of shapes in order to solve coordinate problems in all four quadrants. This should include identical shapes and shapes with points lying on an axis where one coordinate is 0, for example, (1,0). Question **3** provides some initial exposure to this. Children could be encouraged to create their own problems for partners that use a wider range of shapes.

ASSESSMENT CHECKPOINT Use questions **1** a) and **1** b) to assess whether children can apply their knowledge of properties of shapes to solve problems in all four quadrants.

ANSWERS

Question **1** a): B(4,1), C(4,4)

Question **1** b): E(‾6,‾5), F(‾3,‾5), G(‾3,‾2)

Question **2** a): The coordinates of point C are (3,2).

Question **2** b): (‾2,7) and (‾1,4)

Question **3** : C(1,0), D(3,0), E(1,5)

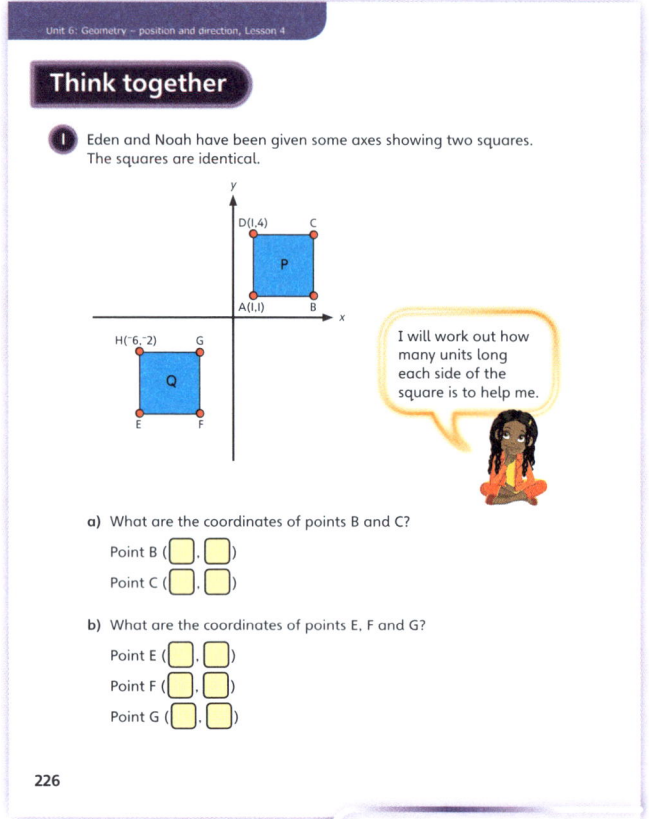

PUPIL TEXTBOOK 6A PAGE 226

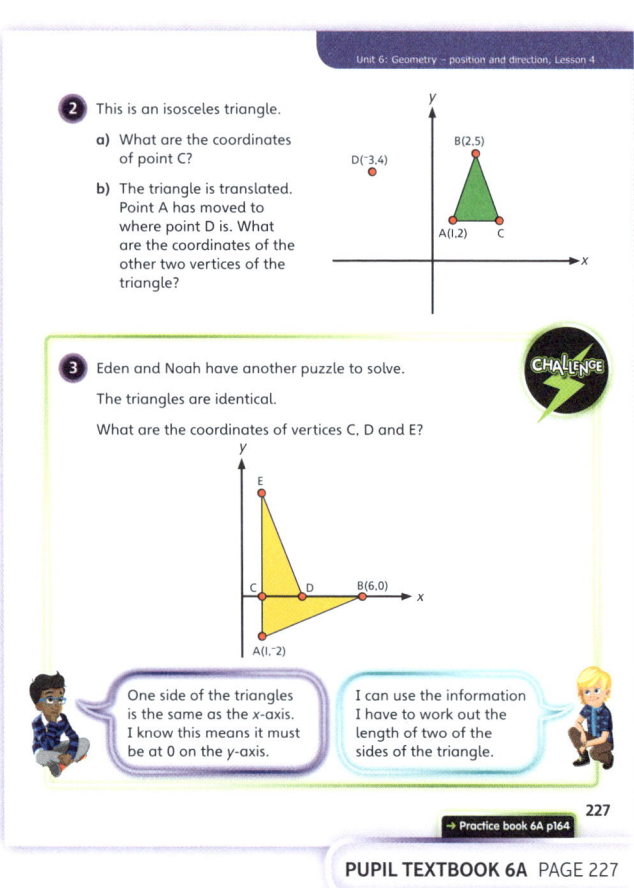

PUPIL TEXTBOOK 6A PAGE 227

Practice

WAYS OF WORKING Independent thinking

IN FOCUS Question **3** exposes children to coordinates of a point located on one of the axes. Children should be able to identify that this represents a '0' value for the 'opposite' coordinate to the axis on which it lies: i.e. if a point lies on the y-axis, then this means its x-axis value is 0.

Question **4** introduces children to shapes that are related to each other by a scale factor (in this case a scale factor of 2, or double). They need to make the connection between this and understand the effect this has on the length of the sides, i.e. that the length of the sides on the larger shape will be twice the length of the sides on the smaller shape.

STRENGTHEN To help children identify the effect that a given translation has on a coordinate (i.e. that the translation increases or decreases the value of a coordinate) encourage them to mark a '+' (positive) and '-' (negative) on the relevant ends of both axes. Ask: *Is the shape being moved towards the positive or negative end of the axis? Does this mean you need to add or subtract to work out the translated coordinates?*

DEEPEN Children should be encouraged to extend their knowledge to a wider range of shapes, and reason about more than two shapes on the same axes. Question **5** provides some opportunity to explore this. Ask: *If shapes are identical, what does this mean? What do you know about the properties of parallelograms?* Finally, invite children to begin to create their own similar problems for others to solve.

ASSESSMENT CHECKPOINT Use question **3** to assess whether children can extend their knowledge to a range of different shapes by answering this question about isosceles (symmetrical) triangles. They should understand what the half-way coordinate is on the x-axis.

ANSWERS Answers for the **Practice** part of the lesson appear in the separate **Practice and Reflect answer guide**.

Reflect

WAYS OF WORKING Independent thinking

IN FOCUS This **Reflect** section activity encourages children to reflect on what they found challenging in the lesson. It may be they found it difficult to identify and draw reflections and translations without the help of a grid or numbers on the two axes. They can solve problems without a grid and how they will go about this.

ASSESSMENT CHECKPOINT Use this question to assess whether children have got to grips with the most challenging parts of the lesson. Encourage them to say whether they found a problem too challenging to solve.

ANSWERS Answers for the **Reflect** part of the lesson appear in the separate **Practice and Reflect answer guide**.

After the lesson ⏸

- Are all children secure at solving problems involving properties of shapes in all four quadrants? How will you address any misconceptions through same-day interventions?
- This completes children's work on coordinates in Year 6. How will you incorporate work on coordinates into other curriculum areas in order to help children practise and apply their knowledge in different contexts?

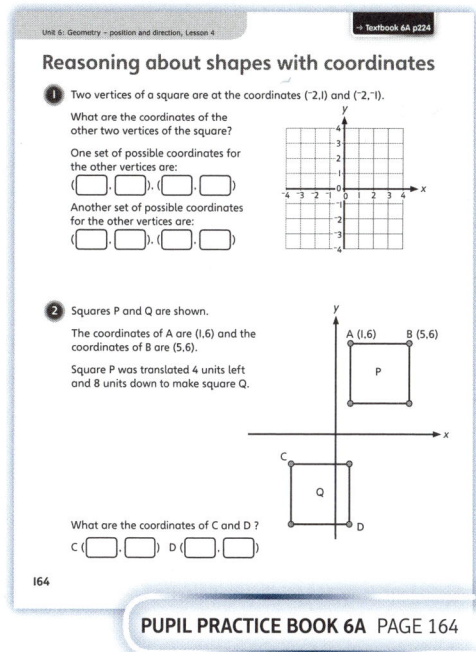

PUPIL PRACTICE BOOK 6A PAGE 164

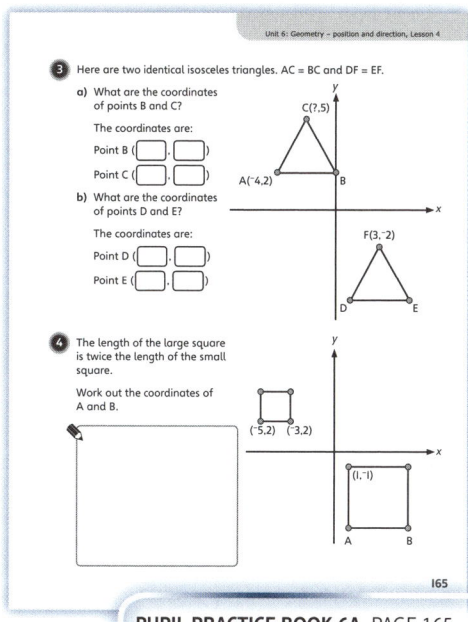

PUPIL PRACTICE BOOK 6A PAGE 165

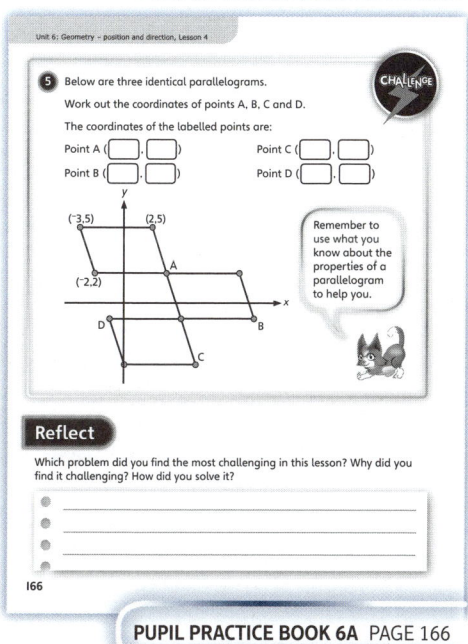

PUPIL PRACTICE BOOK 6A PAGE 166

End of unit check

Don't forget the *Power Maths* unit assessment grid on p26.

WAYS OF WORKING Group work adult led

IN FOCUS

- This end of unit check will allow you to focus on children's understanding of geometry: the position and direction of coordinates plotted in coordinate grids and relating this to their understanding of the properties of shapes.
- Question **5** is presented in the style of a SATs question.

ANSWERS AND COMMENTARY

Children who have mastered this unit will be able to:
- plot and read coordinates in all four quadrants
- identify coordinates that form the vertices of a range of common shapes
- solve increasingly complex problems involving shapes in all four quadrants, including working out the positions of vertices and the lengths of the sides of shapes based on the coordinate information given
- reflect points and shapes on a coordinate grid in the *x*- and *y*-axes as well as on simple diagonal lines, and carry out multi-step translations
- extend these skills to problems where they are given the values of coordinates only, but not a coordinate grid.

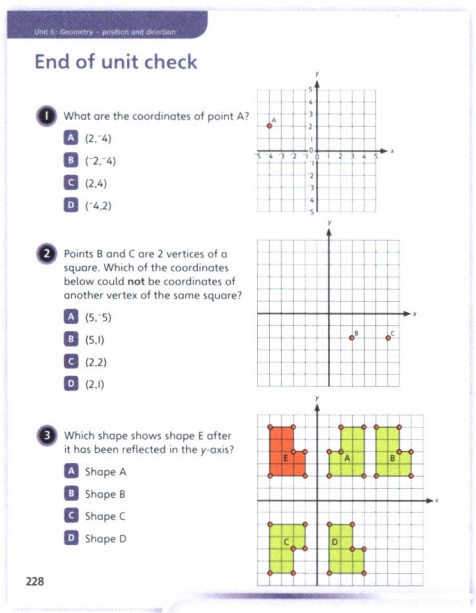

PUPIL TEXTBOOK 6A PAGE 228

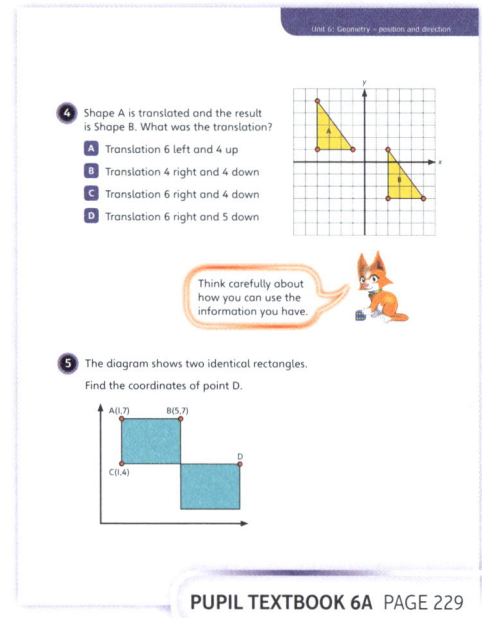

PUPIL TEXTBOOK 6A PAGE 229

Q	A	WRONG ANSWERS AND MISCONCEPTIONS	STRENGTHENING UNDERSTANDING
1	D	A suggests children have transposed the coordinates and have recorded the *y*-axis value first. B suggests that children are not confident using negative coordinates.	Encourage children to create the known side of the shape out of matchsticks, and to then manipulate this side from the given vertices in order to check the possible locations of the other two vertices of the shape.
2	C	D suggests that children have not considered that the square could be completed above the given vertices. A or B suggest that children have miscalculated the side of the square.	
3	A	C suggests children have confused the *x*- and *y*-axes. D suggests that children have confused translation with reflection.	Encourage children to check the results of the reflection using a mirror.
4	C (9,4)	Wrong answers suggest incorrect counting or counting in the wrong direction.	
5		As there are no coordinates for the rectangle containing D, or a marked grid, children might think they cannot find D.	

My journal

WAYS OF WORKING Independent thinking

ANSWERS AND COMMENTARY

Question **1** : Children should identify that Kate's statement is incorrect. Children should first identify that, given the two coordinates that are provided, they can work out the coordinates of the missing vertices are (1,1) and (3,⁻1). They should then be able to make the connection to the x-coordinates being the same distance away from the y-axis on both sides of the reflection, to give the values of the vertices of the reflected points as (⁻1,⁻1), (⁻3,⁻1), (⁻3,1), (⁻1,1).

Question **2** : Children should identify that there are eight different possible rectangles that could be drawn which meet the given criteria. They should be encouraged to work systematically, drawing one side length in one orientation (for example 5 units to the right of point A forms one possible side) and then complete the other sides of the rectangle based on this. They should be encouraged to consider rectangles that are both vertically and horizontally aligned (i.e. 3 units from point A to the left and right would both form acceptable sides of the rectangle).

Power check

WAYS OF WORKING Independent thinking

ASK

- *What do you know now that you did not know at the start of this unit?*
- *How confident do you feel about solving problems involving coordinates in all four quadrants?*

Power play

WAYS OF WORKING Pair work

IN FOCUS Use this **Power play** to identify whether children can apply their knowledge of properties of squares (that all side lengths are the same) and their skill in plotting coordinates in all four quadrants to a game-based situation.

ANSWERS AND COMMENTARY Exact answers for this **Power play** depend on the squares that have been drawn by the child and their partner. Assess if children are making connections between given coordinates and possible side lengths in order to then make more informed choices regarding the possible locations of other vertices of their partner's squares.

After the unit

- How can you continue to expose children to coordinates in other areas of the curriculum and school day so that this learning is used in different contexts, such as Geography, Science, Computing or PE?
- Will you create a classroom display, so children can reflect back on this unit when necessary?

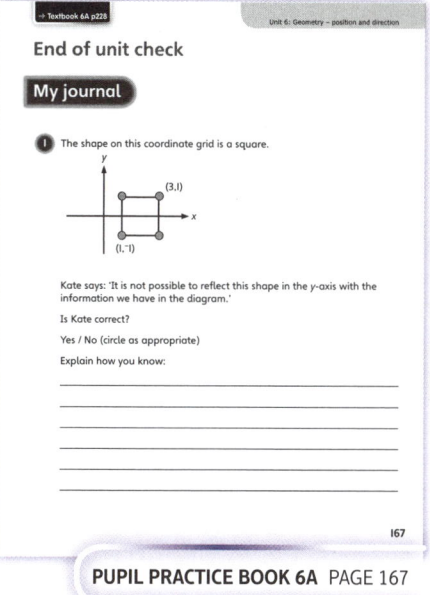

PUPIL PRACTICE BOOK 6A PAGE 167

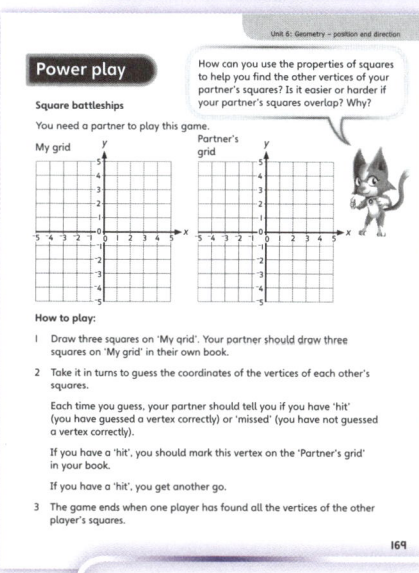

PUPIL PRACTICE BOOK 6A PAGE 168

PUPIL PRACTICE BOOK 6A PAGE 169

Strengthen and **Deepen** activities for this unit can be found in the *Power Maths* online subscription.

Published by Pearson Education Limited, 80 Strand, London, WC2R 0RL.

www.pearsonschools.co.uk

Text © Pearson Education Limited 2018
Edited by Pearson, Little Grey Cells Publishing Services and Haremi Ltd
Designed and typeset by Kamae Design
Original illustrations © Pearson Education Limited 2018
Illustrated by Diago Diaz and Nadene Naude at Beehive Illustration; Emily Skinner at Graham-Cameron Illustration;
and Kamae.
Cover design by Pearson Education Ltd
Back cover illustration © Diago Diaz and Nadene Naude at Beehive Illustration.

Series Editor: Tony Staneff
Consultants: Professor Liu Jian and Professor Zhang Dan

The rights of Tony Staneff, Liu Jian, Josh Lury, Zhou Da, Zhang Dan, Zhu Dejiang, Tim Handley, Kate Henshall, Wei Huinv, Hou Huiying, Zhang Jing, Stephanie Kirk, Huang Lihua, Yin Lili, Liu Qimeng, Timothy Weal and Zhu Yuhong to be identified as authors of this work have been asserted by them in accordance with the Copyright, Designs and Patents Act 1988.

First published 2018

22 21 20 19
10 9 8 7 6 5 4 3 2

British Library Cataloguing in Publication Data
A catalogue record for this book is available from the British Library

ISBN 978 0 435 19041 5

Printed in Slovakia by Neografia

www.activelearnprimary.co.uk

Note from the publisher
Pearson has robust editorial processes, including answer and fact checks, to ensure the accuracy of the content in this publication, and every effort is made to ensure this publication is free of errors. We are, however, only human, and occasionally errors do occur. Pearson is not liable for any misunderstandings that arise as a result of errors in this publication, but it is our priority to ensure that the content is accurate. If you spot an error, please do contact us at resourcescorrections@pearson.com so we can make sure it is corrected.